BARRON'S

AP*

STATISTICS

7TH EDITION

Martin Sternstein, Ph.D.
Professor of Mathematics
Ithaca College
Ithaca, New York

AUTHOR'S NOTE

In 1997, 7,667 students took the AP Statistics exam, and as enrollment in AP Statistics classes increased at a higher rate than in any other AP class, 152,699 students took the exam in 2012. The number of students required to take statistics in college has surpassed the number of students required to take calculus. High schools across the country have recognized this trend and are developing and expanding their statistics offerings. The new Common Core mathematics standards feature statistics and probability in a primary role throughout the high school curriculum. This Barron's book is intended both as a topical review during the year and for final review in the weeks before the AP exam. Step-by-step solutions with detailed explanations are provided for the many illustrative examples and practice problems as well as for the six practice tests, including a diagnostic test.

Special thanks are due to Steve Hanson, Dave Bock, Lee Kucera, Ruth Reece, Diann Resnick, and Jane Viau (and her KIPP students!) for their many useful suggestions. Thanks to Linda Turner, senior editor at Barron's, for her guidance. Thanks to my brother, Allan, my sons, Jonathan and Jeremy, my daughters-in-law, Cheryl and Asia, and my grandson, Jaiden, for their heartfelt love and support. Most thanks of all are due to my wife, Faith, whose love, warm encouragement, and always calm and optimistic perspective on life provide a home environment in which deadlines can be met and goals easily achieved.

Ithaca College
Spring 2013

Martin Sternstein

All inquiries should be addressed to:
Barron's Educational Series, Inc.
250 Wireless Boulevard
Hauppauge, New York 11788
www.barronseduc.com

ISBN (book only): 978-1-4380-0202-6
ISBN (with CD-ROM): 978-1-4380-9315-4

ISSN: 2163-4629 (print)
ISSN: 2161-668X (print with CD-ROM)

PRINTED IN THE UNITED STATES OF AMERICA
9 8 7 6 5 4 3 2 1

Contents

As you review the content in this book and work toward earning that **5** on your AP STATISTICS exam, here are five things that you **MUST** know:

1

Graders want to give you credit—help them! Make them understand *what* you are doing, *why* you are doing it, and *how* you are doing it. Don't make the reader guess at what you are doing.

- **Communication** is just as important as statistical knowledge!
- Be sure you understand **exactly what you are being asked to do or find or explain**.
- *Naked* or *bald answers* will receive little or **no** credit! You must show where answers come from.
- On the other hand, don't give more than one solution to the same problem—you will receive credit only for the weaker one.

2

Random sampling and random assignment are different ideas!

- Random sampling is use of chance in selecting a sample from a population.
  - A *simple random sample* (SRS) is when every possible sample of a given size has the same chance of being selected.
  - A *stratified random sample* is when the population is divided into homogeneous units called strata, and random samples are chosen from each strata.
  - A *cluster sample* is when the population is divided into heterogeneous units called clusters, and a random sample of the clusters is chosen.
- Random assignment in experiments is when subjects are randomly assigned to treatments.
  - This randomization evens out effects over which we have no control.
  - *Randomized block design* refers to when the randomization occurs only within groups of similar experimental units called blocks

3

Distributions describe variability! Understand the difference between:

- a *population distribution* (variability in an entire population),
- a *sample distribution* (variability in a particular sample), and
- a *sampling distribution* (variability between samples).
- The larger the sample size, the more the **sample distribution** looks like the population distribution.
- Central Limit Theorem: the larger the sample size, the more the **sampling distribution** (probability distribution of the sample means) looks like a normal distribution.

4

Check assumptions!

- Be sure the assumptions to be checked are stated correctly, but **don't just state them!**
- Verifying assumptions and conditions means more than simply listing them with little check marks—you must show work or give some reason to confirm verification.
- If you refer to a graph, whether it is a histogram, boxplot, stemplot, scatterplot, residuals plot, normal probability plot, or some other kind of graph, you should **roughly draw it.** It is not enough to simply say, "I did a normal probability plot of the residuals on my calculator and it looked linear."

5

Calculating the *P*-value is not the final step of a hypothesis test!

- There must be a *decision* to reject or fail to reject the null hypothesis.
- You must indicate how you interpret the *P*-value, that is, you need *linkage*. So, "Given that $P = .007$, I reject ..." isn't enough. You need something like, "Because $P = .007$ is less than .05, there is sufficient evidence to reject ..."
- Finally, you need a conclusion *in context* of the problem.

Introduction

The contents of this book cover the topics recommended by the AP Statistics Development Committee. A review of each of the 15 topics is followed by multiple-choice and free-response questions on that topic. Detailed explanations are provided for all answers. It should be noted that some of the topic questions are not typical AP exam questions but rather are intended to help review the topic. Finally, there is a diagnostic exam, and there are five full-length practice exams, totaling 276 questions, all with instructive, complete answers. An optional disk contains two new, full-length exams with 92 more questions.

Several points with regard to particular answers should be noted. First, step-by-step calculations using the given tables sometimes give minor differences from calculator answers due to round-off error. Second, there are examples where the case may be made for using *t*-scores or *z*-scores, again with minor differences in resulting answers. Third, calculator packages easily handle degrees of freedom that are not whole numbers, also resulting in minor answer differences. In all of the above cases, multiple-choice answers in this book have only one reasonable correct answer, and written explanations are necessary when answering free-response questions.

Students taking the AP Statistics Examination will be furnished with a list of formulas (from descriptive statistics, probability, and inferential statistics) and tables (including standard normal probabilities, *t*-distribution critical values, χ^2 critical values, and random digits). While students will be expected to bring a graphing calculator with statistics capabilities to the examination, answers should not be in terms of calculator syntax. Furthermore, many students have commented that calculator usage was less than they had anticipated. However, even though the calculator is simply a tool, to be used sparingly, as needed, students should be proficient with this technology.

The examination will consist of two parts: a 90-minute section with 40 multiple-choice problems and a 90-minute free-response section with five open-ended questions and an investigative task to complete. In grading, the two sections of the exam will be given equal weight. Students have remarked that the first section involves "lots of reading," while the second section involves "lots of writing." The percentage of questions from each content area is approximately 25% data analysis, 15% experimental design, 25% probability, and 35% inference. Questions in both sections may involve reading generic computer output.

Note that in the multiple-choice section the questions are much more conceptual than computational, and thus use of the calculator is minimal.

In the free-response section, students must show all their work, and communication skills go hand in hand with statistical knowledge. Methods must be clearly indicated, as the problems will be graded on the correctness of the methods as well as on the accuracy of the results and explanation. That is, the free-response answers should address *why* a particular test was chosen, not just *how* the test is performed. Even if a statistical test is performed on a calculator such as the TI-84, formulas should still be stated. Choice of test, in inference, must include confirmation of underlying assumptions, and answers must be stated in context, not just as numbers. Work is graded *holistically;* that is, a student's complete response is considered as a whole, with positive scores depending on if the answer is complete, substantial, developing, or minimal.

The score on the multiple-choice section is based on the number of correct answers, with no points deducted for incorrect answers. Blank answers are ignored. Free-response questions are scored on a 0 to 4 scale, with each open-ended question counting 15% of the total free-response score and the investigative task counting 25% of the free-response score. The first open-ended question is typically the most straightforward, and after doing this one to build confidence, students might consider looking at the investigative task since it counts more. Each completed AP examination paper will receive a grade based on a 5-point scale, with 5 the highest score and 1 the lowest score. Most colleges and universities accept a grade of 3 or better for credit or advanced placement or both.

While a review book such as this can be extremely useful in helping prepare students for the AP exam (practice problems, practice more problems, and practice even more problems are the three strongest pieces of advice), nothing can substitute for a good high school teacher and a good textbook. This author personally recommends the following texts from among the many excellent books on the market: *Workshop Statistics* by Rossman, *Activity-Based Statistics* by Scheaffer, *Introduction to Statistics and Data Analysis* by Devore, Olsen, and Peck, *Stats, Modeling the World* by Bock, Velleman, and DeVeaux, *Statistics: The Art and Science of Learning from Data* by Agretsi and Franklin, *Introduction to the Practice of Statistics* by Moore and McCabe, *Statistics in Action: Understanding a World of Data* by Cobb, Scheaffer, and Watkins and *The Practice of Statistics* by Starnes, Yates, and Moore.

Other wonderful sources of information are the College Board's websites: *www.collegeboard.com* for students and parents, and *www.apcentral.collegeboard.com* for teachers. After registering at the AP Central site, teachers should especially note the sections (1) AP Statistics Course Description, (2) The Statistics Exam, and (3) Exam Tips: Statistics.

A good piece of advice is for the student from day one to develop critical practices (like checking assumptions and conditions), to acquire strong technical skills, and to always write clear and thorough, yet to the point, interpretations in context. Final answers to most problems should be not numbers, but rather sentences explaining and analyzing numerical results. To help develop skills and insights to tackle AP free response questions (which often choose contexts students haven't seen before), pick up newspapers and magazines and figure out how to apply what you are learning to better understand articles in print that reference numbers, graphs, and statistical studies.

The student who uses this Barron's review book should study the text and illustrative examples carefully and try to complete the practice problems before referring to the solution keys. Simply reading the detailed explanations to the answers without first striving to work through the problems on one's own is not the best approach. There is an old adage: *Mathematics is not a spectator sport!* Teachers clearly may use this book with a class in many profitable ways. Ideally, each individual topic review, together with practice problems, should be assigned after the topic has been covered in class. The full-length practice exams should be reserved for final review shortly before the AP examination.

Answer Sheet

DIAGNOSTIC EXAMINATION

1. Ⓐ Ⓑ Ⓒ Ⓓ Ⓔ	11. Ⓐ Ⓑ Ⓒ Ⓓ Ⓔ	21. Ⓐ Ⓑ Ⓒ Ⓓ Ⓔ	31. Ⓐ Ⓑ Ⓒ Ⓓ Ⓔ
2. Ⓐ Ⓑ Ⓒ Ⓓ Ⓔ	12. Ⓐ Ⓑ Ⓒ Ⓓ Ⓔ	22. Ⓐ Ⓑ Ⓒ Ⓓ Ⓔ	32. Ⓐ Ⓑ Ⓒ Ⓓ Ⓔ
3. Ⓐ Ⓑ Ⓒ Ⓓ Ⓔ	13. Ⓐ Ⓑ Ⓒ Ⓓ Ⓔ	23. Ⓐ Ⓑ Ⓒ Ⓓ Ⓔ	33. Ⓐ Ⓑ Ⓒ Ⓓ Ⓔ
4. Ⓐ Ⓑ Ⓒ Ⓓ Ⓔ	14. Ⓐ Ⓑ Ⓒ Ⓓ Ⓔ	24. Ⓐ Ⓑ Ⓒ Ⓓ Ⓔ	34. Ⓐ Ⓑ Ⓒ Ⓓ Ⓔ
5. Ⓐ Ⓑ Ⓒ Ⓓ Ⓔ	15. Ⓐ Ⓑ Ⓒ Ⓓ Ⓔ	25. Ⓐ Ⓑ Ⓒ Ⓓ Ⓔ	35. Ⓐ Ⓑ Ⓒ Ⓓ Ⓔ
6. Ⓐ Ⓑ Ⓒ Ⓓ Ⓔ	16. Ⓐ Ⓑ Ⓒ Ⓓ Ⓔ	26. Ⓐ Ⓑ Ⓒ Ⓓ Ⓔ	36. Ⓐ Ⓑ Ⓒ Ⓓ Ⓔ
7. Ⓐ Ⓑ Ⓒ Ⓓ Ⓔ	17. Ⓐ Ⓑ Ⓒ Ⓓ Ⓔ	27. Ⓐ Ⓑ Ⓒ Ⓓ Ⓔ	37. Ⓐ Ⓑ Ⓒ Ⓓ Ⓔ
8. Ⓐ Ⓑ Ⓒ Ⓓ Ⓔ	18. Ⓐ Ⓑ Ⓒ Ⓓ Ⓔ	28. Ⓐ Ⓑ Ⓒ Ⓓ Ⓔ	38. Ⓐ Ⓑ Ⓒ Ⓓ Ⓔ
9. Ⓐ Ⓑ Ⓒ Ⓓ Ⓔ	19. Ⓐ Ⓑ Ⓒ Ⓓ Ⓔ	29. Ⓐ Ⓑ Ⓒ Ⓓ Ⓔ	39. Ⓐ Ⓑ Ⓒ Ⓓ Ⓔ
10. Ⓐ Ⓑ Ⓒ Ⓓ Ⓔ	20. Ⓐ Ⓑ Ⓒ Ⓓ Ⓔ	30. Ⓐ Ⓑ Ⓒ Ⓓ Ⓔ	40. Ⓐ Ⓑ Ⓒ Ⓓ Ⓔ

Diagnostic Examination

SECTION I

Questions 1–40

Spend 90 minutes on this part of the exam.

Directions: The questions or incomplete statements that follow are each followed by five suggested answers or completions. Choose the response that best answers the question or completes the statement.

1. The statistician for a professional basketball team calculates the percentages of points scored through 3-point shots, 2-point shots, and foul shots over two seasons. They are summarized in the following segmented bar chart.

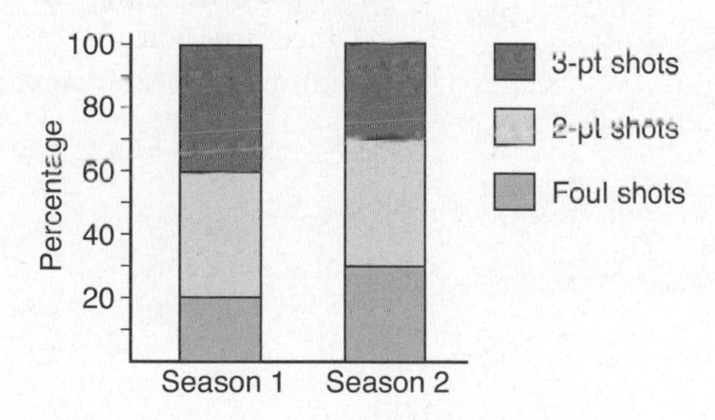

Which of the following is an *incorrect* conclusion?

(A) More points were scored through foul shots in the second season than in the first.
(B) In the first season, twice as many points were scored through 2-point shots than through foul shots.
(C) In the second season, the same number of points were scored through 3-point shots and through foul shots.
(D) In both seasons, the same proportion of total points were scored through 2-point shots.
(E) In the first season, a greater proportion of the points were scored through 3-point shots than in the second season.

GO ON TO THE NEXT PAGE ➤

2. Is there a linear relationship between calories and sodium content in beef hot dogs? A study of 20 beef hot dogs gives the following regression output:

```
Dependent variable is: Sodium

Predictor      Coef   SE Coef       T       P
Constant    -228.33     77.97   -2.93   0.009
Calories     4.0133    0.4922    8.15   0.000

S = 48.5799    R-Sq = 78.7%    R-Sq(adj) = 77.5%
```

Which of the following gives a 99% confidence interval for the slope of the regression line?

(A) $4.0133 \pm 2.861\left(\frac{0.4922}{\sqrt{20}}\right)$
(B) $4.0133 \pm (2.861)(0.4922)$
(C) $4.0133 \pm (2.878)(0.4922)$
(D) $4.0133 \pm 2.861\left(\frac{48.5799}{\sqrt{20}}\right)$
(E) $4.0133 \pm 2.878\left(\frac{48.5799}{\sqrt{20}}\right)$

3. In tossing a fair coin, which of the following sequences is more likely to appear?

(A) HHHHH
(B) HTHTHT
(C) HTHHTTH
(D) TTHTHHTH
(E) All are equally likely.

4. An entomologist hypothesizes that the mean life expectancy of a particular species of insect is 12.5 days. Researchers believing that the true mean is less than 12.5 days plan a hypothesis test at the 5% significance level on a random sample of 50 of these insects. If the alternative hypothesis is correct, for which of the following values of μ will the power of the test be greatest?

(A) 9
(B) 11
(C) 12.5
(D) 14
(E) 17

5. A simple random sample is defined by

(A) the method of selection
(B) how representative the sample is of the population
(C) whether or not a random number generator is used
(D) the assignment of different numbers associated with the outcomes of some chance situation
(E) examination of the outcome

6. Can shoe size be predicted from height. In a random sample of 50 adults, the standard deviation in heights was 8.7 cm, while the standard deviation in shoe size was 2.3. The least squares regression equation was: *Predicted shoe size* = –33.6 + 0.25 (*Height in cm*). What was the correlation?

   (A) $\frac{(0.25)(8.7)}{2.3}$

   (B) $\frac{(0.25)(2.3)}{8.7}$

   (C) $\frac{2.3}{8.7/\sqrt{50}}$

   (D) $\frac{8.7}{2.3/\sqrt{50}}$

   (E) There is not enough information to calculate the correlation.

Questions 7–9 refer to the following situation:

A researcher would like to show that a new oral diabetes medication he developed helps control blood sugar level better than insulin injection. He plans to run a hypothesis test at the 5% significance level.

7. What would be a Type I error?

   (A) The researcher concludes he has evidence his new medication helps more than insulin injection, and his medication really is better than insulin injection.
   (B) The researcher concludes he has evidence his new medication helps more than insulin injection, when in reality his medication is not better than insulin injection.
   (C) The researcher concludes he has no evidence his new medication helps more than insulin injection, and his medication really is not better than insulin injection.
   (D) The researcher concludes he has no evidence his new medication helps more than insulin injection, when in reality his medication is better than insulin injection.
   (E) The researcher concludes he has evidence his new medication controls blood sugar level the same as insulin injection, and in reality there is a difference.

GO ON TO THE NEXT PAGE ➤

8. What would be a Type II error?

   (A) The researcher concludes he has evidence his new medication helps more than insulin injection, and his medication really is better than insulin injection.
   (B) The researcher concludes he has evidence his new medication helps more than insulin injection, when in reality his medication is not better than insulin injection.
   (C) The researcher concludes he has no evidence his new medication helps more than insulin injection, and his medication really is not better than insulin injection.
   (D) The researcher concludes he has no evidence his new medication helps more than insulin injection, when in reality his medication is better than insulin injection.
   (E) The researcher concludes he has evidence his new medication controls blood sugar level the same as insulin injection, and in reality there is a difference.

9. The researcher thinks he can improve his chances by running five such identical hypotheses tests, each using a different group of diabetic volunteers, hoping that at least one of the tests will show that his new oral diabetes medication helps control blood sugar level better than insulin injection. What is the probability of committing at least one Type I error?

   (A) .05
   (B) .204
   (C) .226
   (D) .774
   (E) .95

10. A financial analyst determines the yearly research and development investments for 50 blue chip companies. She notes that the distribution is distinctly not bell-shaped. If the 50 dollar amounts are converted to *z*-scores, what can be said about the standard deviation of the 50 *z*-scores?

    (A) It depends on the distribution of the raw scores.
    (B) It is less than the standard deviation of the raw scores.
    (C) It is greater than the standard deviation of the raw scores.
    (D) It is equal to the standard deviation of the raw scores.
    (E) It equals 1.

11. Tossing a fair die has outcomes {1,2,3,4,5,6} with mean 3.5 and standard deviation 1.708. If a fair die is thrown three times, and the mean of the resulting triplet is calculated, the mean and standard deviation of the set of all possible such triplets is

    (A) $\bar{x} = 3.5$, $\sigma_{\bar{x}} = 0.569$
    (B) $\bar{x} = 3.5$, $\sigma_{\bar{x}} = 0.986$
    (C) $\bar{x} = 6.062$, $\sigma_{\bar{x}} = 1.208$
    (D) $\bar{x} = 10.5$, $\sigma_{\bar{x}} = 0.569$
    (E) $\bar{x} = 10.5$, $\sigma_{\bar{x}} = 0.986$

GO ON TO THE NEXT PAGE ➤

12. A 40-question multiple-choice statistics exam is graded as number correct minus ¼ number incorrect, so scores can range from –10 to +40. Suppose the standard deviation for one class's results is reported to be –3.14. What is the proper conclusion?

 (A) More students received negative scores than positive scores.
 (B) At least half the class received negative scores.
 (C) Some students must have received negative scores.
 (D) Some students must have received positive scores.
 (E) An error was made in calculating the standard deviation.

13. Of the 423 seniors graduating this year from a city high school, 322 plan to go on to college. When the principal asks an AP student to calculate a 95% confidence interval for the proportion of this year's graduates who plan to go to college, the student says that this would be inappropriate. Why?

 (A) The independence assumption may have been violated (students tend to do what their friends do).
 (B) There is no evidence that the data come from a normal or nearly normal population (GPAs help determine college admission and may be skewed).
 (C) Randomization was not used.
 (D) There is a difference between a confidence interval and a hypothesis test with regard to the proportion of graduates planning on college.
 (E) Some other reason.

14. An AP Statistics student in a large high school plans to survey his fellow students with regard to their preference between using a laptop or using an iPad. Which of the following survey methods would result in an unbiased result?

 (A) The student comes to school early and surveys the first 50 students who arrive.
 (B) The student passes a survey card to every student with instructions to fill it out at home and drop the filled out card in a box by the school entrance the next day.
 (C) The student posts the survey on his Facebook page, asking everyone to respond.
 (D) The student goes to all of the high school sports events for a week, hands out the survey, and waits for each student to fill it out and hand it back.
 (E) All the above would lead to biased results.

15. The mean combined SAT score for students in one state is 1758 with a standard deviation of 213, while for a second state the mean is 1725 with a standard deviation of 228. Assuming both distributions are approximately normal, what is the probability that a randomly selected student in the first state scores higher than a randomly selected student in the second state?

 (A) 0.458
 (B) 0.500
 (C) 0.524
 (D) 0.542
 (E) 0.559

GO ON TO THE NEXT PAGE ➤

16. In a random sample of 10 insects of a newly discovered species, an entomologist measures an average life expectancy of 17.3 days with a standard deviation of 2.3 days. Assuming all conditions for inference are met, what is a 95% confidence interval for the mean life expectancy for insects of this species?

(A) $17.3 \pm 1.96\left(\frac{2.3}{\sqrt{9}}\right)$

(B) $17.3 \pm 1.96\left(\frac{2.3}{\sqrt{10}}\right)$

(C) $17.3 \pm 2.228\left(\frac{2.3}{\sqrt{9}}\right)$

(D) $17.3 \pm 2.228\left(\frac{2.3}{\sqrt{10}}\right)$

(E) $17.3 \pm 2.262\left(\frac{2.3}{\sqrt{10}}\right)$

17. A coin is weighted so that heads is twice as likely to occur as tails. The coin is flipped repeatedly until a tail occurs. Let X be the number of flips made. What is the most probable value for X?

(A) 1
(B) 2
(C) 3
(D) 4
(E) 5

18. Suppose we are interested in determining whether or not a student's score on the AP Statistics Exam is a reasonable predictor of the student's GPA in the first year of college. Which of the following is the best statistical test?

(A) Two-sample t-test of population means
(B) Linear regression t-test
(C) Chi-square test of independence
(D) Chi-square test of homogeneity
(E) Chi-square test of goodness of fit

19. Following are the graphs of three normal curves and three cumulative distribution graphs:

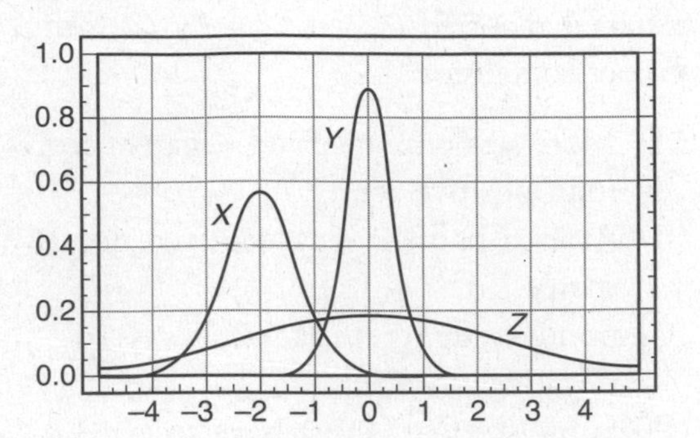

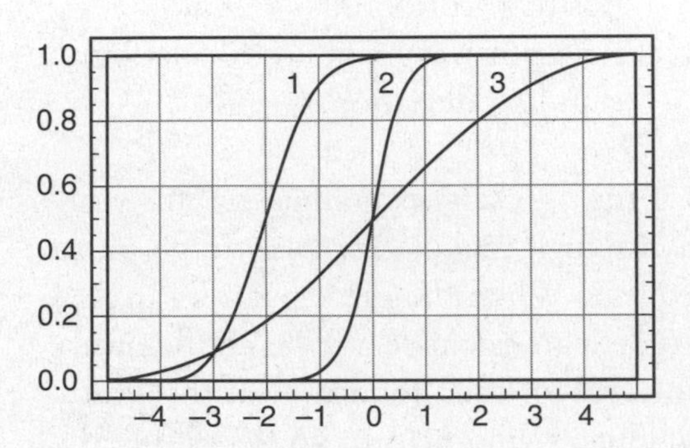

Which normal curve corresponds to which cumulative curve?

(A) X-1, Y-2, Z-3
(B) X-1, Y-3, Z-2
(C) X-2, Y-1, Z-3
(D) X-3, Y-1, Z-2
(E) X-3, Y-2, Z-1

GO ON TO THE NEXT PAGE ➤

20. A campus has 55% male students. Suppose 30% of the male students pick basketball as their favorite sport compared to 20% of the females. If a randomly chosen student picks basketball as the student's favorite sport, what is the probability the student is male?

(A) $\frac{.30}{.30+.20}$

(B) $\frac{.55}{.30+.20}$

(C) $\frac{.30}{(.55)(.30)+(.45)(.20)}$

(D) $\frac{.55}{(.55)(.30)+(.45)(.20)}$

(E) $\frac{(.55)(.30)}{(.55)(.30)+(.45)(.20)}$

21. The kelvin is a unit of measurement for temperature; 0 K is absolute zero, the temperature at which all thermal motion ceases. Conversion from Fahrenheit to Kelvin is given by $K = 5/9 \times (F - 32) + 273$. The average daily temperature in Monrovia, Liberia, is 78.35°F with a standard deviation of 6.3°F. If a scientist converts Monrovia daily temperatures to the Kelvin scale, what will be the new mean and standard deviation?

(A) Mean, 25.75 K; standard deviation, 3.5 K
(B) Mean, 231.75 K; standard deviation, 3.5 K
(C) Mean, 298.75 K; standard deviation, 3.5 K
(D) Mean, 298.75 K; standard deviation, 258.72 K
(E) Mean, 298.75 K; standard deviation, 276.5 K

22. A cattle veterinarian is considering two experimental designs to compare two sources of bovine growth hormone, or BVH, to spur increased milk production in Guernsey cattle. Design 1 involves flipping a coin as each cow enters the stockade, and if *heads*, giving it BVH from bovine cadavers, and if *tails*, giving it BVH from engineered *E. coli*. Design 2 involves flipping a coin as each cow enters the stockade, and if *heads*, giving it BVH from bovine cadavers for a specified period of time and then switching to BVH from engineered *E. coli* for the same period of time, and if *tails*, the order is reversed. With both designs, daily milk production is noted. Which of the following is accurate?

(A) Neither design uses randomization since there is no indication that cows will be randomly picked from the population of all Guernsey cattle.
(B) Design 1 is a *completely randomized design*, while Design 2 is a *block design*.
(C) Both designs use *double-blinding*, but neither uses a *placebo*.
(D) In the second design, BVH from bovine cadavers and BVH from engineered *E. coli* are *confounded*.
(E) One of the two designs is actually an *observational study*, while the other is an *experiment*.

23. The purpose of the linear regression *t*-test is

(A) to determine if there is a linear association between two numerical variables
(B) to find a confidence interval for the slope of a regression line
(C) to find the *y*-intercept of a regression line
(D) to be able to calculate residuals
(E) to be able to determine the consequences of Type I and Type II errors

GO ON TO THE NEXT PAGE ➤

24. A fair die is tossed 12 times, and the number of 3's is noted. This is repeated 200 times. Which of the following distributions is the most likely to occur?

(A)

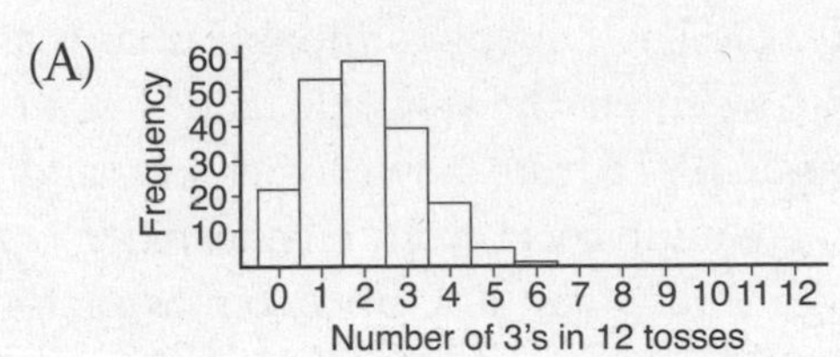

(B)

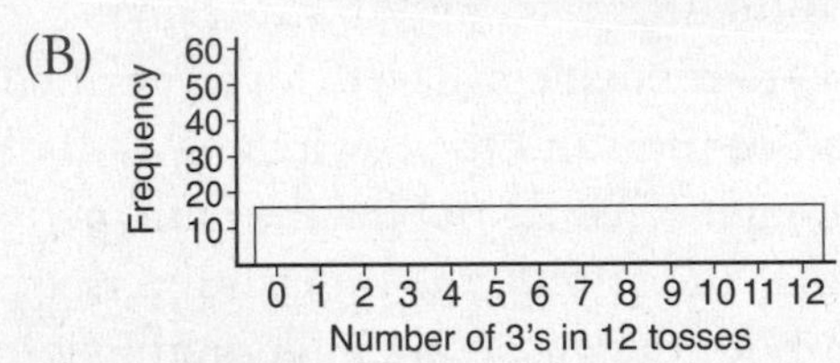

(C)

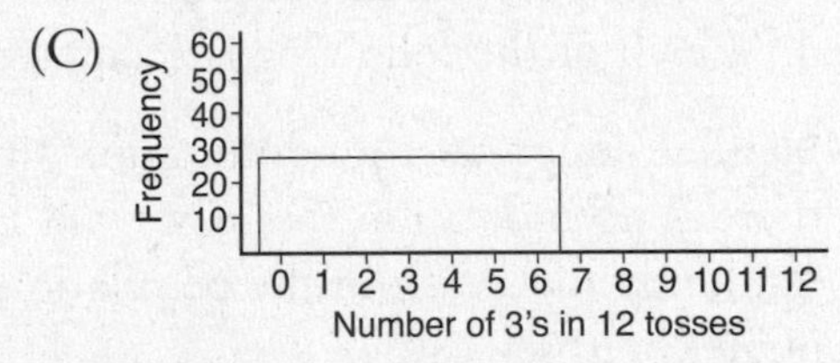

(D)

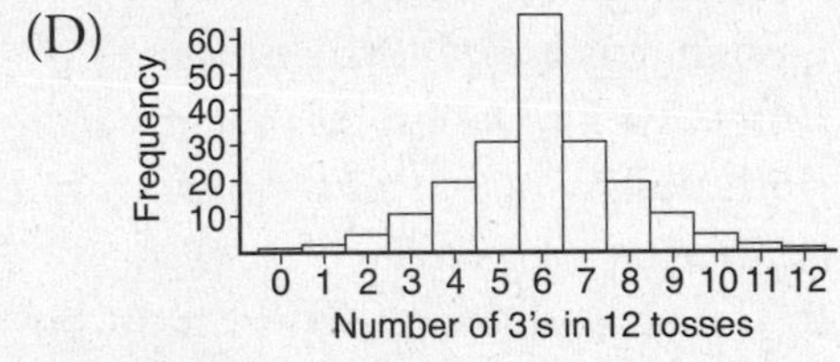

(E)

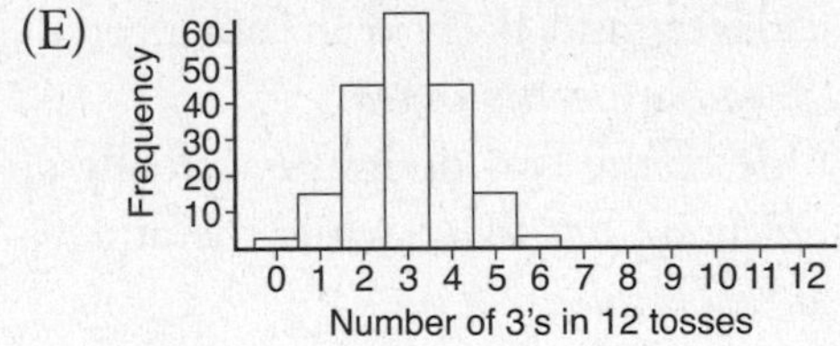

25. Which of the following is a true statement about sampling?

(A) If the sample is random, the size of the sample doesn't matter.
(B) If the sample is random, the size of the population doesn't matter.
(C) A sample of less than 1% of the population is too small for statistical inference.
(D) A sample of more than 10% of the population is too large for statistical inference.
(E) All of the above are true statements.

26. Suppose, in a study of mated pairs of soldier beetles, it is found that the measure of the elytron (hardened forewing) length is always 0.5 millimeters longer in the female. What is the correlation between elytron lengths of mated females and males?

(A) –1
(B) –0.5
(C) 0
(D) 0.5
(E) 1

GO ON TO THE NEXT PAGE ➤

27. A random sample of 100 individuals who were singled out at an international airport security checkpoint is reviewed, and the individuals are classified according to country of origin:

Country of origin	United States	Europe	Arabic	Asia, non-Arabic	Other
Number singled out	41	19	15	13	12

The proportion of travelers in each category who use this airport follows:

Country of origin	United States	Europe	Arabic	Asia, non-Arabic	Other
Proportion	0.64	0.12	0.08	0.09	0.07

We wish to test whether the distribution of people singled out is the same as the distribution of people who use the airport with regard to country of origin. What is the appropriate χ^2 statistic?

(A) $\frac{(41-64)^2}{64}+\frac{(19-12)^2}{12}+\frac{(15-8)^2}{8}+\frac{(13-9)^2}{9}+\frac{(12-7)^2}{7}$

(B) $\frac{(41-64)^2}{41}+\frac{(19-12)^2}{19}+\frac{(15-8)^2}{15}+\frac{(13-9)^2}{13}+\frac{(12-7)^2}{12}$

(C) $\frac{(0.41-0.64)^2}{0.64}+\frac{(0.19-0.12)^2}{0.12}+\frac{(0.15-0.08)^2}{0.08}+\frac{(0.13-0.09)^2}{0.09}+\frac{(0.12-0.07)^2}{0.07}$

(D) $\frac{(0.41-0.64)^2}{0.41}+\frac{(0.19-0.12)^2}{0.19}+\frac{(0.15-0.08)^2}{0.15}+\frac{(0.13-0.09)^2}{0.13}+\frac{(0.12-0.07)^2}{0.12}$

(E) $\frac{(41-64)^2}{20}+\frac{(19-12)^2}{20}+\frac{(15-8)^2}{20}+\frac{(13-9)^2}{20}+\frac{(12-7)^2}{20}$

GO ON TO THE NEXT PAGE ➤

28. The age distribution for a particular debilitating disease has a mean greater than the median. Which of the following graphs most likely illustrates this distribution?

(A)

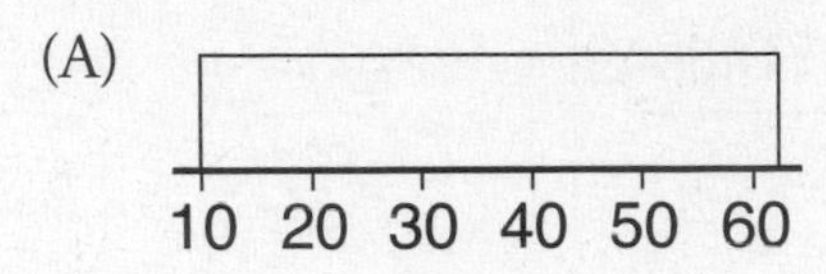

(B)

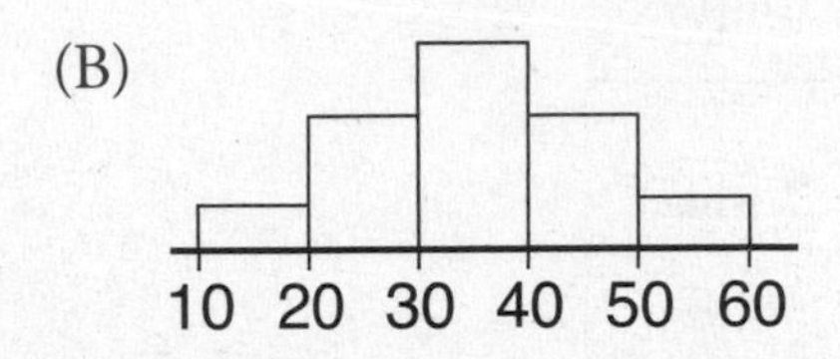

(C)

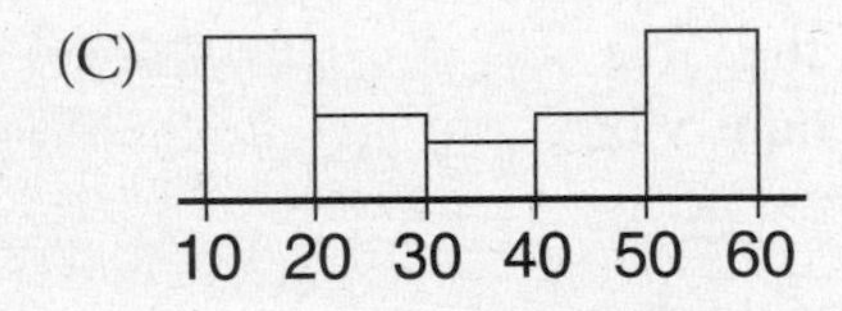

(D)

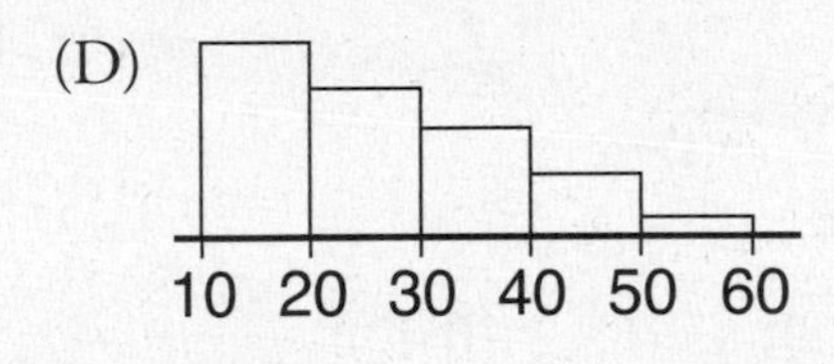

(E)

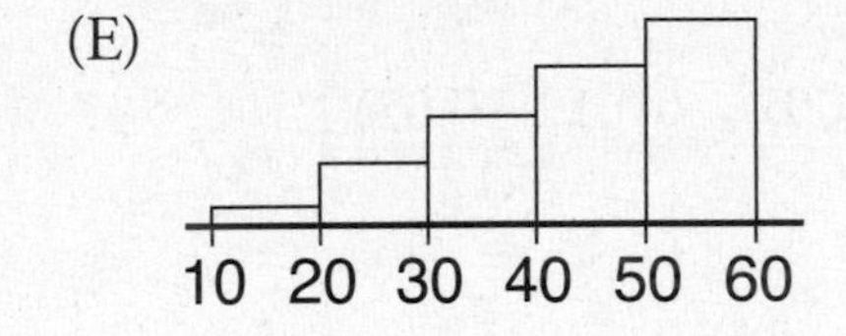

29. Suppose we have a random variable X where the probability associated with the value k is $\binom{10}{k}(.38)^k(.62)^{10-k}$ for $k = 0, \ldots, 10$.

What is the mean of X?

(A) 0.38
(B) 0.62
(C) 3.8
(D) 5.0
(E) 6.2

30. It is hypothesized that high school varsity pitchers throw fastballs at an average of 80 mph. A random sample of varsity pitchers are timed with radar guns resulting in a 95% confidence interval of (74.5, 80.5). Which of the following is a correct statement?

(A) There is a 95% chance that the mean fastball speed of all varsity pitchers is 80 mph.
(B) There is a 95% chance that the mean fastball speed of all varsity pitchers is 77.5 mph.
(C) Most of the interval is below 80, so there is evidence at the 5% significance level that the mean of all varsity pitchers is something different than 80 mph.
(D) The test H_0: $\mu = 80$, H_a: $\mu \neq 80$ is not significant at the 5% significance level, but it would be at the 1% level.
(E) It is likely that the true mean fastball speed of all varsity pitchers is within 3 mph of the sample mean fastball speed.

31. A recent study noted prices and battery lives of 10 top-selling tablet computers. The data follow:

	1	2	3	4	5	6	7	8	9	10
Cost	303	450	260	480	540	390	350	400	600	450
Battery life (hr)	8.5	10	7	11	10	9	8	9.5	11	9.5

The residual plot of the least squares model is

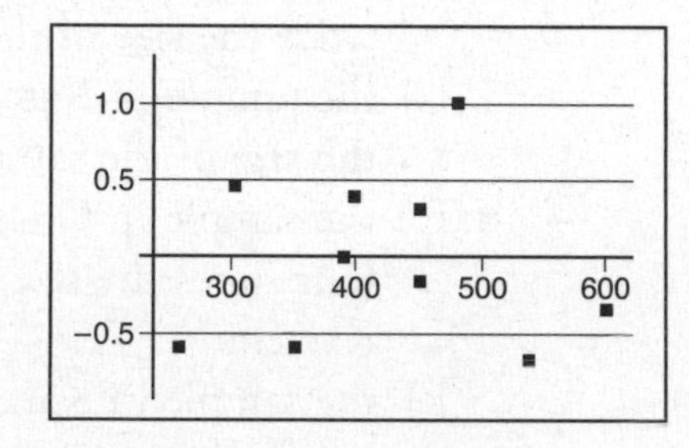

What is the model's predicted battery life for the tablet computer costing $480?

(A) 10 hr
(B) 10.5 hr
(C) 11 hr
(D) 11.5 hr
(E) 12 hr

32. Should college athletes be required to give their coaches their Facebook IDs and passwords? A survey of student-athletes is to be taken. The statistician believes that Division I, II, and III players may differ in their views, so she selects a random sample of athletes from each Division to survey. This is a

(A) simple random sample
(B) stratified sample
(C) cluster sample
(D) systematic sample
(E) convenience sample

33. Which of the following use of a random number table would be appropriate to simulate tossing 3 fair coins and noting the number of heads?

(A) Assign "0,1" to 0 heads, "2,3" to 1 head, "4,5" to 2 heads, "6,7" to 3 heads, and ignore "8,9."
(B) Assign "0,1" to 0 heads, "2,3,4" to 1 head, "5,6,7" to 2 heads, and "8,9" to 3 heads.
(C) Assign "0" to 0 heads, "1,2" to 1 head, "3,4" to 2 heads, "5" to 3 heads, and ignore "6,7,8,9."
(D) Assign "0" to 0 heads, "1,2,3" to 1 head, "4,5,6" to 2 heads, "7" to 3 heads, and ignore "8,9."
(E) Assign "0" to 0 heads, "1,2,3,4" to 1 head, "5,6,7,8" to 2 heads, and "9" to 3 heads.

34. The 2012 population of the Greater Tokyo area is 34,400,000 and of Karachi is 17,200,000. A random sample of citizens is to be taken in each city, and confidence intervals for the mean age in each city will be calculated. Assuming roughly equal sample standard deviations, to obtain the same margin of error for each confidence interval,

(A) the sample sizes should be the same
(B) the sample in Greater Tokyo should be twice the size of the sample in Karachi
(C) the sample in Karachi should be twice the size of the sample in Greater Tokyo
(D) the sample in Greater Tokyo should be four times the size of the sample in Karachi
(E) the sample in Karachi should be four times the size of the sample in Greater Tokyo

35. The *midhinge* is defined to be the average of the first and third quartiles. If the midhinge is 20 and the interquartile range is also 20, what is the first quartile?

(A) 0
(B) 10
(C) 20
(D) 30
(E) Impossible to determine from the given information

GO ON TO THE NEXT PAGE ➤

36. Which of the following is an *incorrect* statement?

    (A) Statistics are random variables with their own probability distributions.
    (B) The standard error does not depend on the size of the population.
    (C) Bias means that, on average, our estimate of a parameter is different from the true value of the parameter.
    (D) There are some statistics for which the sampling distribution is not approximately normal, no matter how large the sample size.
    (E) The larger the sample size, the closer the sample distribution is to a normal distribution.

37. For male Air Force cadets, the recommended fitness level with regard to the number of push-ups is 34. In a test whether or not current classes of recruits can meet this standard, a *t*-test of H_0: $\mu = 34$ against H_a: $\mu < 34$ gives a *P* value of 0.068. Using this data, among the following, which is the largest level of confidence for a two-sided confidence interval that does not contain 34?

    (A) 85%
    (B) 90%
    (C) 92%
    (D) 95%
    (E) 96%

38. A particular car is tested for stopping distance in feet on wet pavement at 30 mph using tires with one tread design and then tires with another tread design. For each set of tires, the test is repeated 30 times, and the following parallel boxplots give a comparison of the resulting five-number summaries.

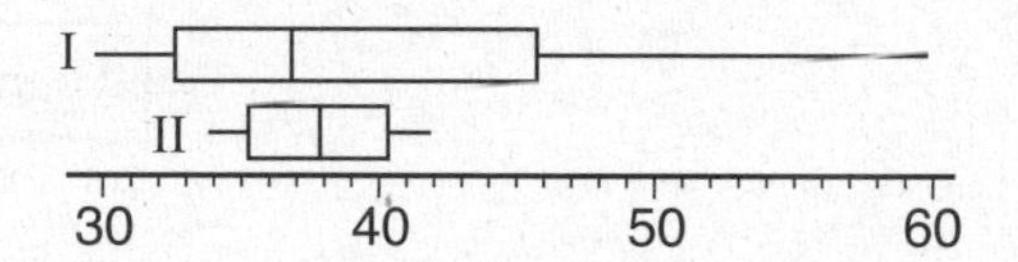

Which of the following is a reasonable conclusion?

    (A) Distribution I is skewed right, while distribution II is bell-shaped.
    (B) Distribution I is skewed left, while distribution II is a normal distribution.
    (C) The mean of distribution I is greater than the mean of distribution II.
    (D) The range of distribution I is approximately 46 – 33 = 13.
    (E) The upper 50% of the values in distribution I are all greater than the lower 50% of the values in distribution II.

39. In American roulette there are 18 red pockets, 18 black pockets, and two green pockets (labeled 0 and 00). The ball is equally likely to land in any of the 38 pockets. What is the probability that a player ends up with a positive outcome, that is, makes money, after 50 equal bets on "red" (that is, for each of 50 spins of the wheel, the player wins or loses the specified identical dollar bet depending on whether or not the ball lands in a red or non-red pocket, respectively)?

    (A) 0.105
    (B) 0.212
    (C) 0.303
    (D) 0.408
    (E) 0.500

GO ON TO THE NEXT PAGE ➤

40. Do middle school and high school students have different views on what makes someone popular? Random samples of 100 middle school and 100 high school students yield the following counts with regard to three choices: lots of money, good at sports, and handsome or pretty:

	Money	Sports	Looks
Middle school	22	48	30
High school	36	24	40

A chi-square test of homogeneity yields which of the following test statistics?

(A) $\frac{(22-29)^2}{22}+\frac{(48-36)^2}{48}+\frac{(30-35)^2}{30}+\frac{(36-29)^2}{36}+\frac{(24-36)^2}{24}+\frac{(40-35)^2}{40}$

(B) $\frac{(22-29)^2}{29}+\frac{(48-36)^2}{36}+\frac{(30-35)^2}{35}+\frac{(36-29)^2}{29}+\frac{(24-36)^2}{36}+\frac{(40-35)^2}{35}$

(C) $\left(\frac{22}{29}\right)^2+\left(\frac{48}{36}\right)^2+\left(\frac{30}{35}\right)^2+\left(\frac{36}{29}\right)^2+\left(\frac{24}{36}\right)^2+\left(\frac{40}{35}\right)^2$

(D) $\frac{(22)(29)}{58}+\frac{(48)(36)}{72}+\frac{(30)(35)}{70}+\frac{(36)(29)}{58}+\frac{(24)(36)}{72}+\frac{(40)(35)}{70}$

(E) $\sqrt{(22-29)^2+(48-36)^2+(30-35)^2+(36-29)^2+(24-36)^2+(40-35)^2}$

STOP

If there is still time remaining, you may review your answers.

SECTION II

Part A

Questions 1–5

Spend about 65 minutes on this part of the exam.
Percentage of Section II grade—75

You must show all work and indicate the methods you use. You will be graded on the correctness of your methods and on the accuracy of your results and explanations.

1. A horticulturist plans a study on the use of compost tea for plant disease management. She obtains 16 identical beds, each containing a random selection of five mini-pink rose plants. She plans to use two different composting times (two and five days), two different compost preparations (aerobic and anaerobic), and two different spraying techniques (with and without adjuvants). Midway into the growing season she will check all plants for rose powdery mildew disease.

   (a) List the complete set of treatments.
   (b) Describe a completely randomized design for the treatments above.
   (c) Explain the advantage of using only mini-pink roses in this experiment.
   (d) Explain a disadvantage of using only mini-pink roses in this experiment.

2. A top-100, 7.0-rated tennis pro wishes to compare a new Wilson N1 racquet against his current model. He strings the new racquet with the same Luxilon strings at 60 pounds tension that he uses on his old racquet. From past testing he knows that the average forehand cross court volley with his old racquet is 82 miles per hour (mph). On an indoor court, using a ball machine set at 70 mph, the same speed he had his old racquet tested against, he takes 47 swings with the new racquet. An associate with a speed gun records an average of 83.5 mph with a standard deviation of 3.4 mph. Assuming that the 47 swings represents a random sample of his swings, is there statistical evidence that his speed with the new racquet is an improvement over the old? Justify your answer.

3. In October 2008, a comprehensive residential college in upstate New York reported undergraduate enrollment by ethnic/racial categories as follows: 2.7% non-Hispanic Black, 3.7% Asian or Pacific Islander, 4.0% Hispanic, 80.0% non-Hispanic White, and the rest other/unknown. While racial/ethnic status is not considered in the admissions process, an admissions counselor is interested in whether or not the makeup of the new freshman class will change, and plans to do a statistical analysis on an appropriately drawn simple random sample.

   (a) What statistical test/procedure should be used?
   (b) State the null and alternative hypotheses. Is the test appropriate for an intended sample size of 200?
   (c) If the admissions counselor performs the indicated test on the following data, is there statistical evidence of a change in ethnic/racial composition? Explain.

	Non-Hispanic Black	Asian or Pacific Islander	Hispanic	Non-Hispanic White	Other/Unknown
Number of Students	3	4	14	150	29

   (d) Suppose the data was obtained by noting the racial/ethnic status of a simple random sample of 200 potential new students visiting the campus during fall 2008. Did the test/procedure target the intended population? Explain.

4. Concrete is made by mixing sand and pebbles with water and cement and then hardening through hydration. Different densities result from different proportions of the aggregates. Assume that concrete densities are normally distributed with mean 2317 kilograms per cubic meter and standard deviation 128 kilograms per cubic meter.

   (a) What is the probability that a given concrete density is over 2400 kg/m^3?
   (b) In a random sample of five independent concrete densities, what is the probability that a majority have densities over 2400 kg/m^3?
   (c) What is the probability that the mean of the five independent concrete densities is over 2400 kg/m^3?

GO ON TO THE NEXT PAGE ➤

5. A small art gallery in Laguna Beach has the choice of stocking either oil paintings or finger paintings for a given tourist season. The oil paintings require a substantial investment, but the potential returns are also greater. The return (profit or loss) depends on whether or not the tourists that season are primarily serious art collectors or more casual buyers. A sales analysis gives the following expectations.

Season return ($1000)	Type of tourists	
Stock decision	Art collectors	Casual buyers
Oil paintings	135	–35
Finger paintings	–5	25

Let p be the probability that the type of tourist is primarily art collectors, so $(1 - p)$ is the probability of primarily casual buyers.

(a) As a function of p, what is the expected return for stocking oil paintings?
(b) As a function of p, what is the expected return for stocking finger paintings?
(c) For what value of p are the two expected returns the same, and what does it mean in context for p to be greater or less than this value?
(d) In a random sample of similar establishments in similar tourist regions, 33 out of 150 reported seasons with tourists who were primarily art collectors. Construct a 95% confidence interval for the proportion of similar establishments with tourists who were primarily art collectors.
(e) Use the above results to justify a decision to stock finger paintings.

SECTION II

Part B

Question 6

Spend about 25 minutes on this part of the exam.
Percentage of Section II grade—25

6. A national retail chain classifies cashiers as "entry level" for the first ten years. The series of boxplots below shows the relationship between yearly wages (in $1000) and years of experience (YOE) for a random sample of these employees. Below the boxplots are computer regression outputs.

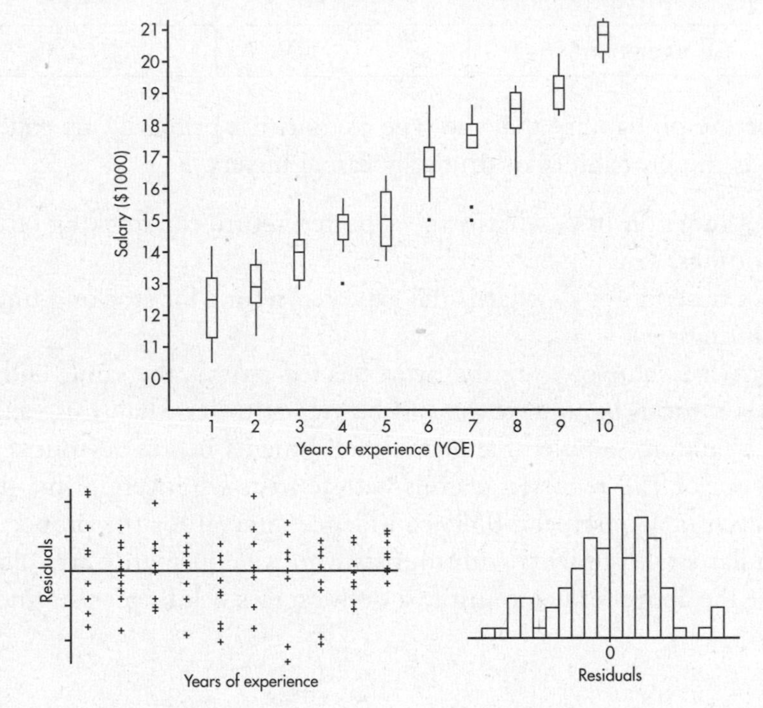


```
Dependent variable is: WAGES
Var        Coef       s.e. Coef   t      p
Constant   11.1113    0.2031      54.7   ≤ 0.0001
YOE        0.910485   0.03273     27.8   ≤ 0.0001
R-sq = 88.8%   R-sq(adj) = 88.6%
s = 0.9402 with 100 – 2 = 98 degrees of freedom
```

(a) Discuss the relationship between salary and experience based on the boxplots.
(b) Discuss how conditions for regression inference are met.
(c) Determine a 95% confidence interval for the regression slope, and interpret in context.
(d) Using only the given information, give a rough estimate of the probability that a salary is at least $1000 over what is predicted by the regression line.

If there is still time remaining, you may review your answers.

Answer Key

Section I

1. **A**	9. **C**	17. **A**	25. **B**	33. **D**
2. **C**	10. **E**	18. **B**	26. **E**	34. **A**
3. **A**	11. **B**	19. **A**	27. **A**	35. **B**
4. **A**	12. **E**	20. **E**	28. **D**	36. **E**
5. **A**	13. **E**	21. **C**	29. **C**	37. **A**
6. **A**	14. **E**	22. **B**	30. **E**	38. **C**
7. **B**	15. **D**	23. **A**	31. **A**	39. **C**
8. **D**	16. **E**	24. **A**	32. **B**	40. **B**

Answers Explained

Section I

1. **(A)** Without knowing the actual number of points scored each season, only proportions, not numbers of points, can be compared between seasons.

2. **(C)** The critical t-values with $df = 20 - 2 = 18$ are $\pm$ 2.878. Thus, we have $b_1 \pm t^* \times SE(b_1) = 4.0133 \pm (2.878)(0.4922)$.

3. **(A)** The shortest sequence has a greater probability than any longer sequence.

4. **(A)** Power, the probability of rejecting a false null hypothesis, will be the greatest for parameter values farthest from the hypothesized value, in the direction of the alternative hypothesis.

5. **(A)** A simple random sample may or may not be representative of the population. It is a method of selection in which every possible sample of the desired size has an equal chance of being selected.

6. **(A)** A formula relating the given statistics is $b = r\frac{s_y}{s_x}$ which in this case gives $0.25 = r\frac{2.3}{8.7}$ and thus $r = \frac{(0.25)(8.7)}{2.3}$.

7. **(B)** The null hypothesis is that the new medication is no better than insulin injection, while the alternative hypothesis is that the new medication is better. A Type I error means a mistaken rejection of a true null hypothesis.

8. **(D)** A Type II error means a mistaken failure to reject a false null hypothesis.

9. **(C)** Running a hypothesis test at the 5% significance level means that the probability of committing a Type I error is .05. Then the probability of not committing a Type I error is .95. Assuming the tests are independent, the probability of not committing a Type I error on any of the five tests is $(.95)^5 = .77378$, and the probability of at least one Type I error is $1 - .77378 = .22622$.

10. **(E)** No matter what the distribution of raw scores, the set of z-scores always has mean 0 and standard deviation 1.

11. **(B)** In the sampling distribution of $\bar{x}$, the mean is equal to the population mean, and the standard deviation is equal to the population standard deviation divided by the square root of the sample size, in this case, $\sigma_{\bar{x}} = \dfrac{1.708}{\sqrt{3}} = 0.986$.

12. **(E)** The standard deviation can never be negative.

13. **(E)** When a complete census is taken (all 423 seniors were in the study), the population proportion is known and a confidence interval has no meaning.

14. **(E)** The method described in (A) is a convenience sample, (B) and (C) are voluntary response surveys, and (D) suffers from undercoverage bias.

15. **(D)** Since each set is normally distributed so is the set of differences, $X_1 - X_2$. We calculate $\mu_{x_1 - x_2} = 1758 - 1725 = 33$ and $\sigma_{x_1 - x_2} = \sqrt{213^2 + 228^2} = 312.0$.

    So $P(X_1 - X_2) = P(X_1 - X_2 > 0) = P\left(z > \dfrac{0 - 33}{312.0}\right) = P(z > -0.1058) = 0.542$.

16. **(E)** With $df = n - 1 = 10 - 1 = 9$ and 95% confidence, the critical t-values are ± 2.262. Also $SE(\bar{x}) = \dfrac{s}{\sqrt{n}} = \dfrac{2.3}{\sqrt{10}}$.

17. **(A)** $P(X = 1) = \dfrac{1}{3}$, $P(X = 2) = \left(\dfrac{2}{3}\right)\left(\dfrac{1}{3}\right)$, $P(X = 3) = \left(\dfrac{2}{3}\right)^2\left(\dfrac{1}{3}\right)$,

    $P(X = 4) = \left(\dfrac{2}{3}\right)^3\left(\dfrac{1}{3}\right)$, $P(X = 5) = \left(\dfrac{2}{3}\right)^4\left(\dfrac{1}{3}\right)$, . . . , and we see that the distribution has its maximum value at $X = 1$.

18. **(B)** The chi-square tests all involve counts, and comparing means doesn't make sense in this context.

19. **(A)** The median of X is -2, and this is also true of distribution 1 (note that a horizontal line from 0.5 strikes curve 1 above -2 on the x-axis). Y has a smaller standard deviation than Z (tighter clustering around the mean), so Y must correspond to distribution 2, which shows almost all values are between -1 and 1.

20. **(E)**

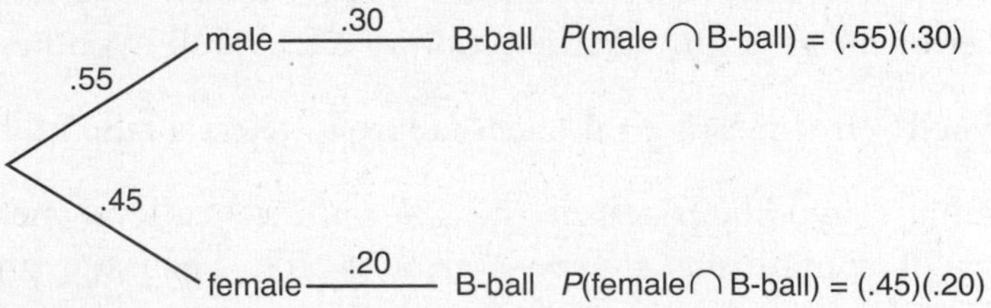

    .55 male —.30— B-ball $P(\text{male} \cap \text{B-ball}) = (.55)(.30)$

    .45 female —.20— B-ball $P(\text{female} \cap \text{B-ball}) = (.45)(.20)$

    $\text{P(B-ball)} = (.55)(.30) + (.45)(.20)$

    $\text{P(male}\,|\,\text{B-ball)} = \dfrac{(.55)(.30)}{(.55)(.30) + (.45)(.20)}$

21. **(C)** Adding the same constant to every value in a set adds the same constant to the mean but leaves the standard deviation unchanged. Multiplying every value in a set by the same constant multiplies the mean and standard deviation by that constant. So the new mean is $5/9 \times (78.35 - 32) + 273 = 298.75$, and the new standard deviation is $5/9 \times 6.3 = 3.5$.

22. **(B)** Design 2 is an example of a matched pairs design, a special case of a block design; here, each subject is compared to itself with respect to the two treatments. Both designs definitely use randomization with regard to assignment of treatments, but since they do not use randomization in selecting subjects from the general population, care must be taken in generalizing any conclusions. It's not clear whether or not the researchers who do the observations and measurements know which treatment individual cows are receiving, so there is no way to conclude if there is or is not blinding. The two sources of BVH are different treatments, and so they are not being confounded. In both designs treatments are randomly applied, so neither is an observational study.

23. **(A)** The linear regression *t*-test has null hypothesis H_0: $\beta = 0$ that there is no linear relationship; if the *P*-value is small enough, then there is evidence of a linear association, that is, there is evidence that $\beta \neq 0$.

24. **(A)** We have a binomial distribution with mean $= np = 12\left(\frac{1}{6}\right) = 2$. Answer (A) is the only reasonable choice.

25. **(B)** The size of the sample always matters; the larger the sample, the greater the power of statistical tests. One percent of a large population is large. Larger samples are better, but if the sample is greater than 10% of the population, the best statistical techniques are not those covered in the AP curriculum.

26. **(E)** The points on the scatterplot all fall on the straight line:

$$\textit{Female length} = \textit{Male length} + 0.5$$

27. **(A)** $\chi^2 = \sum \frac{(observed - expected)^2}{expected}$ and expected are found by multiplying the proportions times the sample size of 100.

28. **(D)** The distributions in (A), (B), and (C) appear roughly symmetric, so the mean and median will be roughly the same. The distribution in (D) is skewed to the right, so the mean will be greater than the median, while the distribution in (E) is skewed to the left, so the mean will be less than the median.

29. **(C)** This is a binomial with $n = 10$ and $p = .38$, so the mean is $np = 10(.38) = 3.8$.

30. **(E)** Answers (A), (B), and (C) are common misconceptions. Since the 95% confidence interval contains 80, a two-sided test would not be significant at the 5% significance level or lower. The interval can be expressed as 77.5 ± 3, that is, we are 95% confident that the true mean fastball speed is within 3 mph of 77.5 mph.

31. **(A)** Residual = Observed – Predicted, so 1.0 = 11 – Predicted and Predicted =10.

32. **(B)** Stratified sampling is when the population is divided into homogeneous groups (the three Divisions in this example), and a random sample of individuals is chosen from each group.

33. **(D)** There are 8 outcomes {TTT, TTH, THT, THH, HTT, HTH, HHT, HHH}, so P(0 heads) = 1/8, P(1 head) = 3/8, P(2 heads) = 3/8, and P(3 heads) = 1/8. Thus, we assign one digit to the results of 0 heads and 3 heads and 3 digits to the results of 1 head and 2 heads and ignore the other 2 digits (of the 10 available digits).

34. **(A)** The margin of error, $\pm t^* \frac{s}{\sqrt{n}}$, depends on the sample size, not the population size.

35. **(B)** We have $(Q_1 + Q_3)/2 = 20$ or $Q_3 + Q_1 = 40$, and $Q_3 - Q_1 = 20$, which algebraically gives $Q_1 = 10$ and $Q_3 = 30$ [add the equations to obtain $2Q_3 = 60$ so $Q_3 = 30$; then plugging into either equation and solving for Q_1 gives $Q_1 = 10$.]

36. **(E)** The larger the sample size, the closer the sample distribution is to the population distribution. The central limit theorem roughly says that if multiple samples of size *n* are drawn randomly and independently from a population, then the histogram of the means of those samples will be approximately normal. Statistics have probability distributions called sampling distributions. The standard error is based on the spread of the population and on the sample size. The central limit theorem does not apply to all statistics as it does to sample means. Many sampling distributions are not normal; for example, the sampling distribution of the sample max is not a normal distribution. An estimator of a parameter is unbiased if we have a method that, through repeated samples, is on average the same value as the parameter.

37. **(A)** With 0.068 in a tail, the confidence interval with 34 at one end would have a confidence level of 1 – 2(0.068) = .864, so anything higher than 86.4% confidence will contain 34.

38. **(C)** From a boxplot there is no way of telling if a distribution is bell-shaped (very different distributions can have the same five-number summary). Distribution I appears strongly skewed right, and so its mean is probably much greater than its median, while distribution II appears roughly symmetric, and so its mean is probably close to its median. The interquartile range, not the range, in I is 13.

39. **(C)** To make money, there must be more wins than losses, so with 50 plays, we need to calculate $P(X > 25)$. We have a binomial distribution with $n = 50$ and probability of success $p = 18/38$. On a calculator such as the TI-84 we find $P(X > 25) = 1 - P(X \leq 25) =$ `1-binomcdf(50,18/38,25)` $= 0.303$. [or on the Nspire: `binomcdf(50,18/38,26,50)`].

40. **(B)** $\chi^2 = \sum \frac{(observed - expected)^2}{expected}$ and cell calculations [expected value of a cell equals (row total)(column total)/(table total)] or χ^2-test on a calculator such as the TI-84 will yield expected cells of 29, 36, 35, 29, 36, 35.

Section II

1. (a) There are $2 \times 2 \times 2 = 8$ different treatments:
   Two-day, aerobic, with adjuvant
   Two-day, aerobic, without adjuvant
   Two-day, anaerobic, with adjuvant
   Two-day, anaerobic, without adjuvant
   Five-day, aerobic, with adjuvant
   Five-day, aerobic, without adjuvant
   Five-day, anaerobic, with adjuvant
   Five-day, anaerobic, without adjuvant

   (b) We must randomly assign the treatment combinations to the beds. (Roses have already been randomly assigned to the beds.) With 8 treatments and 16 beds, each treatment should be assigned to 2 beds. For example, give each bed a random number between 1 and 16 (no repeats), and then assign the first treatment in the above list to the beds with the numbers 1 and 2, assign the second treatment in the above list to the beds with the numbers 3 and 4, and so on.

   (c) Using only mini-pink roses in this experiment gives reduced variability and increases the likelihood of determining differences among the treatments.

   (d) Using only mini-pink roses in this experiment limits the scope and makes it difficult to generalize the results to other species of roses.

Scoring

Part (a) is essentially correct for correctly listing all eight treatment combinations and is incorrect otherwise.

Part (b) is essentially correct if each treatment combination is randomly assigned to two beds of roses. Part (b) is partially correct if each treatment is randomly assigned to two beds but the method is unclear, or if a method of randomization is correctly described but the method may not assure that each treatment is assigned to *two* beds.

Part (c) is essentially correct for noting *reduced variability* and for explaining that this increases likelihood of determining differences among the treatments. Part (c) is partially correct for only one of these two components.

Part (d) is essentially correct noting *limited scope* and for explaining that this makes generalization to other species difficult. Part (d) is partially correct for only one of these two components.

Count partially correct answers as one-half an essentially correct answer.

4 Complete Answer	Four essentially correct answers.
3 Substantial Answer	Three essentially correct answers.
2 Developing Answer	Two essentially correct answers.
1 Minimal Answer	One essentially correct answer.

Use a holistic approach to decide a score totaling between two numbers.

TIP

Graders want to give you credit. Help them! Make them understand *what* you are doing, *why* you are doing it, and *how* you are doing it. Don't make the reader guess at what you are doing. **Communication** is just as important as statistical knowledge!

2. Part 1: State the correct hypotheses.

$H_0 : \mu = 82$ and $H_a : \mu > 82$

Part 2: Identify the correct test and check assumptions.

One-sample t-test with $t = \dfrac{\overline{x} - \mu_0}{\frac{s}{\sqrt{n}}}$

Assumptions: Random sample (given) and either a normal population distribution or a large sample (in this case, the sample size, $n = 47$, is large).

Part 3: Correctly calculate the test statistic t, the degrees of freedom df, and the p-value.

$$t = \frac{83.5 - 82}{\frac{3.4}{\sqrt{47}}} = 3.02,\ df = 47 - 1 = 46, \text{ and } p = .002$$

Part 4: Using the *p*-value, give a correct conclusion in context. With this small a *p*-value, $p = .002$, there is strong evidence to reject H_0 and conclude that there is strong evidence that his speed with the new racquet is an improvement over the old.

Scoring

Part 1 either is essentially correct or is incorrect.

Part 2 is essentially correct if the test is correctly identified by name or formula and the assumptions are checked. Part 2 is partially correct if only one of these two elements is correct.

Part 3 is essentially correct if the *t*-value is calculated and both *df* and *p* are stated. Part 3 is partially correct if only one of these two elements is correct.

Part 4 is essentially correct if the correct conclusion is given in context and the conclusion is linked to the *p*-value. Part 4 is partially correct if the correct conclusion is given in context but there is no linkage to the *p*-value.

Count partially correct answers as one-half an essentially correct answer.

4 Complete Answer	Four essentially correct answers.
3 Substantial Answer	Three essentially correct answers.
2 Developing Answer	Two essentially correct answers.
1 Minimal Answer	One essentially correct answer.

Use a holistic approach to decide a score totaling between two numbers.

3. (a) Chi-square goodness-of-fit test

(b) H_0: The new freshman class is distributed 2.7% non-Hispanic Black, 3.7% Asian or Pacific Islander, 4.0% Hispanic, 80.0% non-Hispanic White, and 9.6% other/unknown.
H_a: The new freshman class has a distribution different from that in October, 2008.
Randomization is given and the expected cell frequencies—$2.7\% \times 200 = 5.4$, $3.7\% \times 200 = 7.4$, $4.0\% \times 200 = 8.0$, $80.0\% \times 200 = 160.0$, and $9.6\% \times 200 = 19.2$—are all greater than 5.

(c) $\chi^2 = \sum \frac{(obs - \exp)^2}{\exp} = \frac{(3-5.4)^2}{5.4} + \frac{(4-7.4)^2}{7.4} + \frac{(14-8)^2}{8} + \frac{(150-160)^2}{160} + \frac{(29-19.2)^2}{19.2} = 12.76$

with $df = 5 - 1 = 4$, and a *p*-value of .013. With such a small *p*-value, there is evidence to reject H_0 and conclude that there is statistical evidence of a change in ethnic/racial composition.

(d) No, this test/procedure targeted students visiting the campus, and such students might be different from the targeted population of students making up the new freshman class. For example, some students who eventually make up the freshman class might not have the funds or the time to visit the campus. Or it can be argued that even if all potential students do visit the campus, there is no reason to conclude that the distribution of visiting students is the same as the distribution of students who both are accepted and decide to attend this college.

Scoring

Part (a–b) is essentially correct if the correct test is named, the hypotheses are correctly stated, and the assumption of all expected cell frequencies being greater than five is checked. Part (a–b) is partially correct if only two of these three elements are correct. Part (a–b) is incorrect if only one of these three elements is correct.

Part (c1) is essentially correct if the chi-square value is calculated and both *df* and *p* are stated. Part (c1) is partially correct if only one of these two elements is correct.

Part (c2) is essentially correct if the correct conclusion is given in context and the conclusion is linked to the *p*-value. Part (c2) is partially correct if the correct conclusion is given in context but there is no linkage to the *p*-value.

Part (d) is essentially correct or incorrect. It is essentially correct for both stating that the intended population is not targeted and giving a clear argument for this answer.

A partially correct answer counts half of an essentially correct answer.

4 Complete Answer	Four essentially correct answers.
3 Substantial Answer	Three essentially correct answers.
2 Developing Answer	Two essentially correct answers.
1 Minimal Answer	One essentially correct answer.

Use a holistic approach to decide a score totaling between two numbers.

4. (a) $P(x > 2400) = P\left(z > \frac{2400-2317}{128}\right) = P(z > 0.6484) = .258$

(b) $P(\textit{at least } 3 \textit{ out of } 5 \textit{ are} > 2400) = 10(.258)^3(.742)^2 + 5(.258)^4(.742) + (.258)^5 = .112$

[On the TI-84, 1 – binomcdf(5,.258,2) = .112]

(c) The distribution of $\bar{x}$ is normal with mean $\mu_{\bar{x}} = 2317$ and standard deviation

$\sigma_{\bar{x}} = \frac{128}{\sqrt{5}} = 57.24$. So

$$P(\bar{x} > 2400) = P\left(z > \frac{2400 - 2317}{57.24}\right) = P(z > 1.450) = .0735$$

Scoring

Part (a) is essentially correct if the correct probability is calculated and the derivation is clear. Simply writing normalcdf(2400,∞,2317,128) = .258 is a partially correct response.

Part (b) is essentially correct if the correct probability is calculated and the derivation is clear. Part (b) is partially correct for indicating a binomial with $n = 5$ and p = answer from (a) but calculating incorrectly. Simply writing 1-binomcdf(5,.258,2) = .112 is also a partially correct response.

Part (c) is essentially correct for specifying both $\mu_{\bar{x}}$ and $\sigma_{\bar{x}}$ and correctly calculating the probability. Part (c) is partially correct for specifying both $\mu_{\bar{x}}$ and $\sigma_{\bar{x}}$ but incorrectly calculating the probability, or for failing to specify both $\mu_{\bar{x}}$ and $\sigma_{\bar{x}}$ but correctly calculating the probability.

4 Complete Answer	All three parts essentially correct.
3 Substantial Answer	Two parts essentially correct and one part partially correct.
2 Developing Answer	Two parts essentially correct OR one part essentially correct and one or two parts partially correct OR all three parts partially correct.
1 Minimal Answer	One part essentially correct OR two parts partially correct.

5. (a) $\Sigma\, xP(x) = 135p + (-35)(1 - p) = -35 + 170p$

   (b) $\Sigma\, xP(x) = (-5)p + 25(1 - p) = 25 - 30p$

   (c) $-35 + 170p = 25 - 30p$ gives $p = .3$. When $p > .3$, the expected return for oil paintings is greater than that for finger paintings, and when $p < .3$, the expected return for finger paintings is greater than for oil paintings. (These statements follow from the positive slope of $R = -35 + 170p$ and the negative slope of $R = 25 - 30p$.)

(d) First, name the confidence interval: a 95% confidence interval for the proportion p of similar establishments with tourists who were primarily art collectors.

Second, check conditions: We are given that this is a random sample, it is reasonable to assume that the sample is less than 10 percent of all similar establishments, and the sample size is large enough ($np = 33$ and $n(1 - p) = 117$ are both greater than 10).

Third, demonstrate correct mechanics: With $\hat{p} = \frac{33}{150} = .22$, we have

$$\hat{p} \pm z\sqrt{\frac{\hat{p}(1-\hat{p})}{n}} = .22 \pm 1.96\sqrt{\frac{(.22)(.78)}{150}} = .22 \pm .066.$$

Fourth, interpret in context: We are 95% confident that the true proportion of similar establishments with tourists who were primarily art collectors is between .154 and .286.

(e) The entire interval in (d) is below $p = .3$, so based on (c), the expected return for finger paintings is greater than for oil paintings for all p in this interval.

Scoring

Parts (a) and (b) are scored together. They are essentially correct if both answers are correct and partially correct if one answer is correct.

Part (c) is essentially correct for a correct calculation of the intersection together with a correct conclusion of what it means when p is greater or less than .3. Part (c) is partially correct if the intersection is not correctly calculated, but the conclusions are correct based on the incorrect intersection value.

Part (d) is essentially correct if steps 2, 3, and 4 are correct. (Step 1 is only a restatement from the question.) Part 3 is partially correct if only two of these three steps are correct.

Part (e) is essentially correct if the correct conclusion is given with clear linkage to the results from both (c) and (d). Part (e) is partially correct if the explanation of the linkage is present but weak.

Give 1 point for each essentially correct part and ½ point for each partially correct part.

4 Complete Answer	4 points
3 Substantial Answer	3 points
2 Developing Answer	2 points
1 Minimal Answer	1 point

Use a holistic approach to decide a score totaling between two numbers.

6. (a) Salary and years of experience exhibit an approximately linear relationship. As years of experience increase, so does the median salary. The third-quartile, Q_3, salary also increases with years, and so does the Q_1 salary with the exception of one year. As years of experience increase, there are only minor differences in measures of variability in the salaries, with roughly the same range (except for the last two) and fairly consistent interquartile ranges.

   (b) First, the boxplots indicate that an overall scatterplot pattern would be roughly linear. Second, the residual plot shows no pattern. Third, the histogram of residuals appears roughly normal (unimodal, symmetric, and without clear skewness or outliers).

   (c) With $df = 98$ and .025 in each tail, the critical t-values are ±1.984. $b_1 \pm ts_{b_1} = 0.910 \pm 1.984(0.03273) = 0.910 \pm .065$. We are 95% confident that for each additional year of experience, the average increase in salary is between \$845 and \$975.

   (d) The sum and thus the mean of the residuals is always 0. The standard deviation of the residuals is σ, which can be estimated with $s = 0.9402$. With a roughly normal distribution, we have $P(X \geq 1) = .14$.

TIP

Read carefully and recognize that sometimes very different tests are required in different parts of the same problem.

Scoring

Part (a) is essentially correct for correctly noting a linear relationship, noting that as years of experience increase so does the median salary (or Q_3 or generally Q_1), and noting that measures of variability (range or IQR) stay roughly the same. Part (a) is partially correct for correctly noting two of the three features.

Part (b) is essentially correct for noting the three conditions (roughly linear scatterplot, no pattern in the residual plot, and roughly normal histogram of residuals). Part (b) is partially correct for correctly noting two of the three conditions.

Part (c) is essentially correct for both a correct calculation of the confidence interval and a correct interpretation in context. Part (c) is partially correct for a correct calculation without the interpretation in context, or for a correct interpretation based on an incorrect calculation.

Part (d) is essentially correct noting that the distribution of residuals is roughly normal with mean 0 and standard deviation 0.9402, and then using this to correctly calculate the probability. Part (d) is partially correct for correctly noting the distribution of residuals but incorrectly calculating the probability, or for making a calculation based on a normal distribution with mean 0 but using an incorrect standard deviation.

Count partially correct answers as one-half an essentially correct answer.

4 Complete Answer	Four essentially correct answers.
3 Substantial Answer	Three essentially correct answers.
2 Developing Answer	Two essentially correct answers.
1 Minimal Answer	One essentially correct answer.

Use a holistic approach to decide a score totaling between two numbers.

AP SCORE FOR THE DIAGNOSTIC EXAM

Multiple-Choice section (40 questions)

Number correct × 1.25 = __________

Free-Response section (5 open-ended questions plus an investigative task)

Question 1 __________ (out of 4) × 1.875 = __________

Question 2 __________ (out of 4) × 1.875 = __________

Question 3 __________ (out of 4) × 1.875 = __________

Question 4 __________ (out of 4) × 1.875 = __________

Question 5 __________ (out of 4) × 1.875 = __________

Question 6 __________ (out of 4) × 3.125 = __________

Total points from Multiple-Choice and Free-Response sections = __________

Conversion chart based on a recent AP exam:

Total points	AP Score
70–100	5
57–69	4
44–56	3
33–43	2
0–32	1

STUDY GUIDE FOR THE DIAGNOSTIC TEST MULTIPLE-CHOICE QUESTIONS

Note in which Themes your missed questions fall. Then give special note to the Topics corresponding to the missed questions. Additionally, whenever you wish to test yourself on a particular Theme, go back to the designated questions.

Theme One: Exploratory Analysis

Question 1	Topic Five: Exploring Categorical Data
Question 6	Topic Four: Exploring Bivariate Data
Question 10	Topic Two: Summarizing Distributions
Question 12	Topic Two: Summarizing Distributions
Question 19	Topic One: Graphical Displays
	Topic Eleven: The Normal Distribution
Question 21	Topic Two: Summarizing Distributions
Question 26	Topic Four: Exploring Bivariate Data
Question 28	Topic One: Graphical Displays
	Topic Two: Summarizing Distributions
Question 31	Topic Four: Exploring Bivariate Data
Question 35	Topic Two: Summarizing Distributions
Question 38	Topic One: Graphical Displays
	Topic Two: Summarizing Distributions

Theme Two: Planning a Study

Question 5	Topic Seven: Planning and Conducting Surveys
Question 14	Topic Seven: Planning and Conducting Surveys
Question 22	Topic Eight: Planning and Conducting Experiments
Question 25	Topic Seven: Planning and Conducting Surveys
Question 32	Topic Seven: Planning and Conducting Surveys

Theme Three: Probability

Question 3	Topic Nine: Probability as Relative Frequency
Question 9	Topic Nine: Probability as Relative Frequency
	Topic Fourteen: Tests of Significance—Proportions and Means
Question 11	Topic Twelve: Sampling Distributions
Question 15	Topic Ten: Combining Independent Random Variables
	Topic Eleven: The Normal Distribution
Question 17	Topic Nine: Probability as Relative Frequency
Question 19	Topic Eleven: The Normal Distribution
	Topic One: Graphical Displays
Question 20	Topic Nine: Probability as Relative Frequency
Question 24	Topic Nine: Probability as Relative Frequency
Question 29	Topic Nine: Probability as Relative Frequency
Question 33	Topic Nine: Probability as Relative Frequency
Question 36	Topic Twelve: Sampling Distributions
Question 39	Topic Nine: Probability as Relative Frequency

Theme Four: Statistical Inference

Question 2	Topic Thirteen: Confidence Intervals
Question 4	Topic Fourteen: Tests of Significance—Proportions and Means
Question 7	Topic Fourteen: Tests of Significance—Proportions and Means
Question 8	Topic Fourteen: Tests of Significance—Proportions and Means
Question 9	Topic Fourteen: Tests of Significance—Proportions and Means Topic Nine: Probability as Relative Frequency
Question 13	Topic Thirteen: Confidence Intervals
Question 16	Topic Thirteen: Confidence Intervals
Question 18	Topic Fifteen: Tests of Significance—Chi-Square and Slope of Least Squares Line
Question 23	Topic Fifteen: Tests of Significance—Chi-Square and Slope of Least Squares Line
Question 27	Topic Fifteen: Tests of Significance—Chi-Square and Slope of Least Squares Line
Question 30	Topic Thirteen: Confidence Intervals Topic Fourteen: Tests of Significance—Proportions and Means
Question 34	Topic Thirteen: Confidence Intervals
Question 37	Topic Thirteen: Confidence Intervals Topic Fourteen: Tests of Significance—Proportions and Means
Question 40	Topic Fifteen: Tests of Significance—Chi-Square and Slope of Least Squares Line

THEME ONE: EXPLORATORY ANALYSIS

TOPIC 1

Graphical Displays

- Dotplots
- Bar Charts
- Histograms
- Stemplots
- Center and Spread
- Clusters and Gaps
- Outliers
- Modes
- Shape
- Cumulative Relative Frequency Plots
- Skewness

There are a variety of ways to organize and arrange data. Much information can be put into tables, but these arrays of bare figures tend to be spiritless and sometimes even forbidding. Some form of graphical display is often best for seeing patterns and shapes and for presenting an immediate impression of everything about the data. Among the most common visual representations of data are dotplots, bar charts, histograms, and stemplots. It is important to remember that all graphical displays should be clearly labeled, leaving no doubt what the picture represents—**AP Statistics scoring guides harshly penalize the lack of titles and labels!**

TIP

The first thing to do with data is to draw a picture—always.

DOTPLOTS

Dotplots and bar charts are particularly useful with regard to *categorical* (or *qualitative*) *variables*, that is, variables that note the category to which each individual belongs. This is in contrast to *quantitative variables*, which take on numerical values.

TIP

Just because a variable has numerical values doesn't necessarily mean that it's quantitative.

EXAMPLE 1.1

Suppose that in a class of 35 students, 10 choose basketball as their favorite sport while 7 pick baseball, 6 pick football, 5 pick tennis, 5 pick soccer, and 2 pick hockey. These data can be displayed in the following *dotplot*:

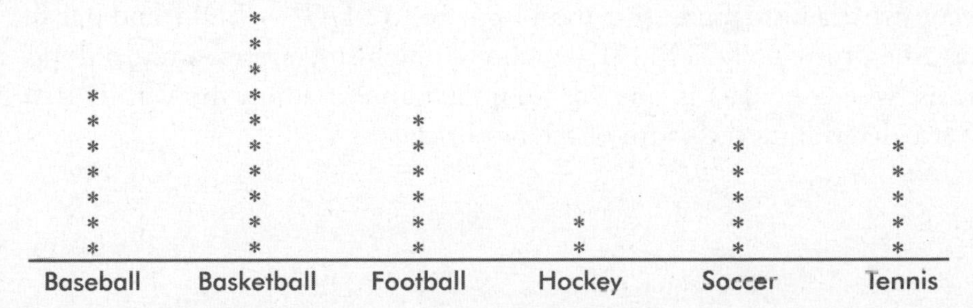

The *frequency* of each result is indicated by the *number of dots* representing that result.

The dotplot can also be drawn with a vertical axis and horizontal rows of dots. In fact, in almost all displays, vertical and horizontal can be switched depending upon which picture seems easier to read or simply which better fits the page.

BAR CHARTS

A common visual display to compare the sizes of categories or groups is the *bar chart.* Sizes can be measured as frequencies or as percents.

EXAMPLE 1.2

In a survey taken during the first week of January 1999, 55% of those surveyed wanted the Clinton impeachment trial to end immediately, 15% wanted it to continue with witnesses, 25% wanted the trial to continue without calling witnesses, and 5% expressed no opinion. These data can be displayed in the following bar chart (or *bar graph*):

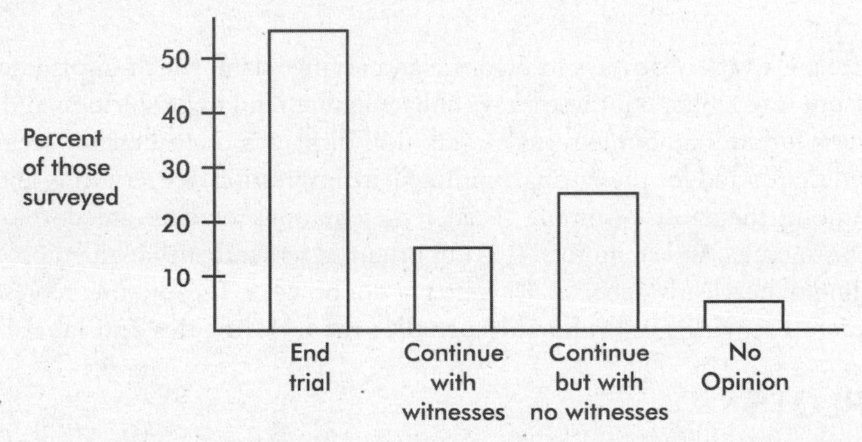

The relative frequencies of different results are indicated by the heights of the bars representing these results.

HISTOGRAMS

Histograms, useful for large data sets involving quantitative variables, show counts or percents falling either at certain values or between certain values. While the AP Statistics Exam does not stress construction of histograms, there are often questions on interpreting given histograms.

To construct a histogram using the TI-84, go to STAT → EDIT and put the data in a list, then turn a STAT PLOT on, choose the histogram icon under Type, specify the list where the data is, and use ZoomStat and/or adjust the WINDOW. Note that XSCL determines the width of the bin or class.

EXAMPLE 1.3

Suppose there are 2000 families in a small town and the distribution of children among them is as follows: 300 families are childless, 400 have one child, 700 have two children, 300 have three, 100 have four, 100 have five, and 100 have six. These data can be displayed in the following histogram:

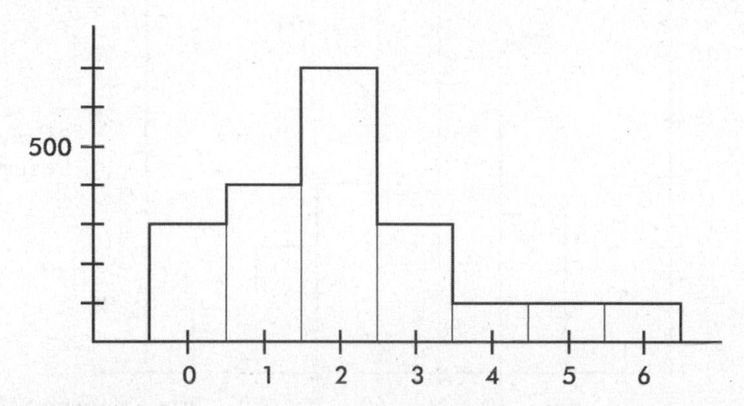

Sometimes, instead of labeling the vertical axis with frequencies, it is more convenient or more meaningful to use *relative frequencies*, that is, frequencies divided by the total number in the population.

Number of Children	Frequency	Relative Frequency
0	300	300/2000 = .150
1	400	400/2000 = .200
2	700	700/2000 = .350
3	300	300/2000 = .150
4	100	100/2000 = .050
5	100	100/2000 = .050
6	100	100/2000 = .050

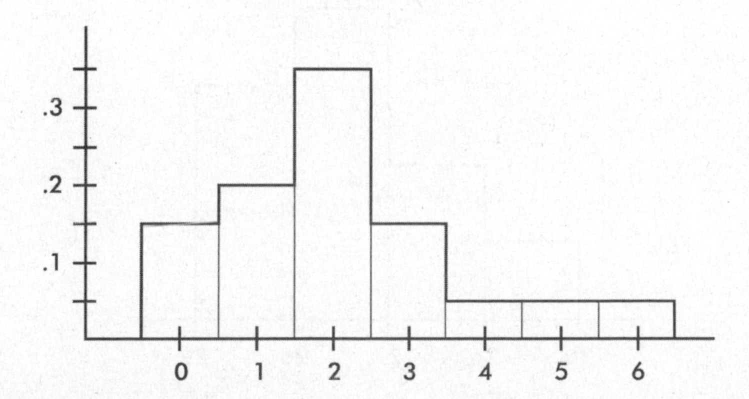

Note that the shape of the histogram is the same whether the vertical axis is labeled with frequencies or with relative frequencies. Sometimes we show both frequencies and relative frequencies on the same graph.

EXAMPLE 1.4

Consider the following histogram displaying 40 salaries paid to the top-level executives of a large company.

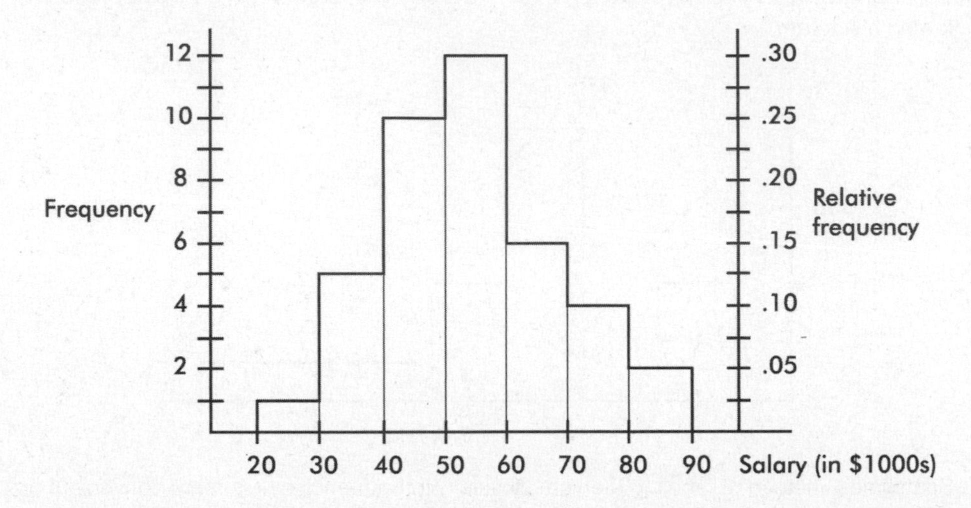

What can we learn from this histogram? For example, none of the executives earned more than $90,000 or less than $20,000. Twelve earned between $50,000 and $60,000. Twenty-five percent earned between $40,000 and $50,000. Note how this histogram shows the number of items (salaries) falling *between* certain values, whereas the preceding histogram showed the number of items (families) falling *at* each value. For example, in Example 1.4 we see that ten salaries fell somewhere between $40,000 and $50,000, while in Example 1.3 we see that 700 families had exactly two children.

EXAMPLE 1.5

Consider the following histogram where the vertical axis has not been labeled. What can we learn from this histogram?

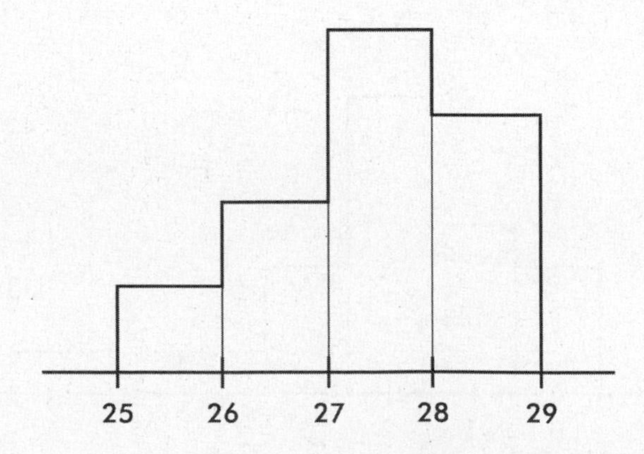

(continued)

Answer: It is impossible to determine the actual frequencies; however, we can determine the relative frequencies by noting the fraction of the total *area* that is over any interval:

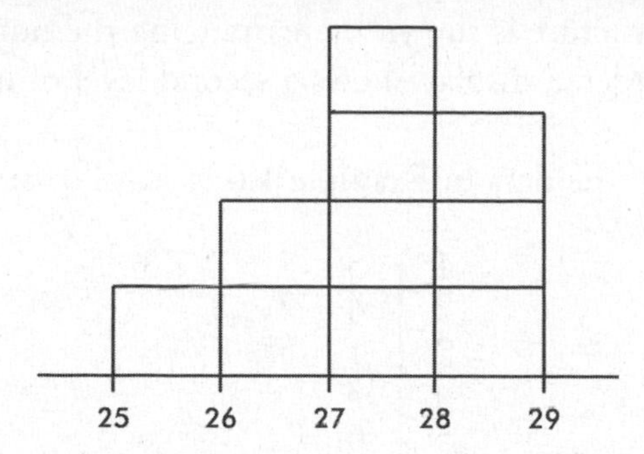

We can divide the area into ten equal portions, and then note that $\frac{1}{10}$ or 10% of the area is above 25–26, 20% is above 26–27, 40% is above 27–28, and 30% is above 28–29.

Although it is usually not possible to divide histograms so nicely into ten equal areas, the principle of relative frequencies corresponding to relative areas still applies.

Relative frequencies are the usual choice when comparing distributions of different size populations.

STEMPLOTS

Although a histogram may show how many scores fall into each grouping or interval, the exact values of individual scores are often lost. An alternative pictorial display, called a stemplot, retains this individual information.

EXAMPLE 1.6

Consider the set {17, 17, 18, 13, 28, 38, 31, 27, 35, 50, 43, 37, 24} of percentages of three-point shots made by Michael Jordan during his 13 years with the Bulls. Let 1, 2, 3, 4, and 5 be placeholders for 10, 20, 30, 40, and 50. List the last digit of each value from the original set after the appropriate placeholder.

The result is a *stemplot* (also called a *stem and leaf display*) of these data:

Stems	**Leaves**
1	7 7 8 3
2	8 7 4
3	8 1 5 7
4	3
5	0

Drawing a continuous line around the leaves would result in a horizontal histogram:

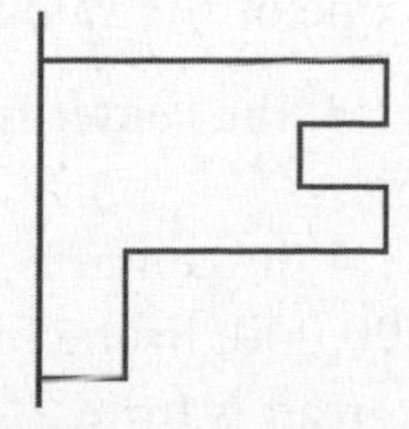

Note that the stemplot gives the shape of the histogram and, unlike the histogram, indicates the values of the original data.

Sometimes further structure is shown by rearranging the numbers in each row in ascending order. This ordered display shows a second level of information from the original stemplot.

The revised display of the data in Example 1.6 is as follows:

```
1 | 3 7 7 8
2 | 4 7 8
3 | 1 5 7 8
4 | 3
5 | 0
```

EXAMPLE 1.7

```
40 | 7
41 |
42 |
43 |
44 |
45 |
46 |
47 |
48 | 8
49 |
50 |
51 | 0
52 | 6799
53 | 04469
54 | 2467
55 | 03578
56 | 1235
57 | 59
58 | 56
```

(40 | 7 means 4.07 gm/cm^3)

Using a "torsion balance," Henry Cavendish (in 1798) made 29 measurements of Earth's density, obtaining values of 5.5, 5.57, 5.42, 5.61, 5.53, 5.47, 4.88, 5.62, 5.63, 4.07, 5.29, 5.34, 5.26, 5.44, 5.46, 5.55, 5.34, 5.3, 5.36, 5.79, 5.75, 5.29, 5.1, 5.86, 5.58, 5.27, 5.85, 5.65, and 5.39 gm/cm^3.

To the left is a stemplot of this data.

Note that the scale is such that one must multiply each value in the dataset by 0.01 to return the original value. For example, $407 \times 0.01 = 4.07$.

CENTER AND SPREAD

TIP

Center and spread should always be described together.

Looking at a graphical display, we see that two important aspects of the overall pattern are

1. the *center*, which separates the values (or area under the curve in the case of a histogram) roughly in half, and
2. the *spread*, that is, the scope of the values from smallest to largest.

In the histogram of Example 1.3, the center is 2 children while the spread is from 0 to 6 children.

In the histogram of Example 1.4 the center is between \$50,000 and \$60,000, and the spread is from \$20,000 to \$90,000; in the histogram of Example 1.5, the center is between 27 and 28, and the spread is from 25 to 29.

In the stemplot of Example 1.6, the center is 28%, and the spread is from 13% to 50%; in the stemplot of Example 1.7, the center is 5.46, and the spread is from 4.07 to 5.86.

CLUSTERS AND GAPS

Other important aspects of the overall pattern are

1. *clusters,* which show natural subgroups into which the values fall (for example, the salaries of teachers in Ithaca, NY, fall into three overlapping clusters, one for public school teachers, a higher one for Ithaca College professors, and an even higher one for Cornell University professors), and
2. *gaps,* which show holes where no values fall (for example, the Office of the Dean sends letters to students being put on the honor roll and to those being put on academic warning for low grades; thus the GPA distribution of students receiving letters from the Dean has a huge middle gap).

EXAMPLE 1.8

Consider the following histogram:

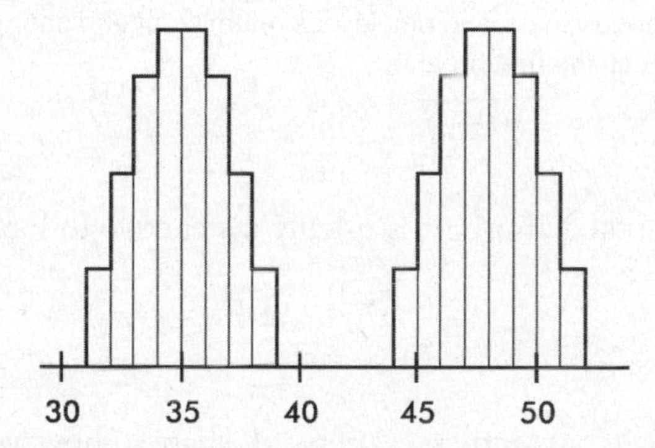

Simply saying that the center of the distribution is around 42 and the spread is from 31 to 52 clearly misses something. The values fall into two distinct clusters with a gap between.

OUTLIERS

TIP

Pay attention to outliers!

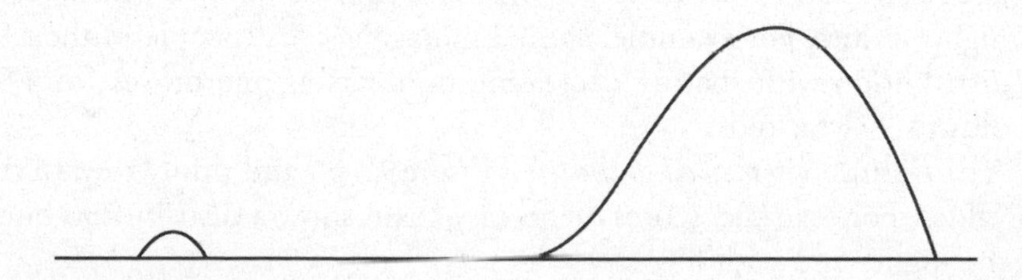

Extreme values, called *outliers,* are found in many distributions. Sometimes they are the result of errors in measurements and deserve scrutiny; however, outliers can also be the result of natural chance variation. Outliers may occur on one side or both sides of a distribution. In the stemplot of Example 1.7, 4.07 is clearly an outlier.

MODES

Some distributions have one or more major peaks, called *modes.* (The values with the peaks above them are the modes.) With exactly one or two such peaks, the distribution is said to be *unimodal* or *bimodal,* respectively. But every little bump in the data is not a mode! You should always look at the big picture and decide whether or not two (or more) phenomena are affecting the histogram.

EXAMPLE 1.9

The histogram below shows employee computer usage (number accessing the Internet) at given times at a company main office.

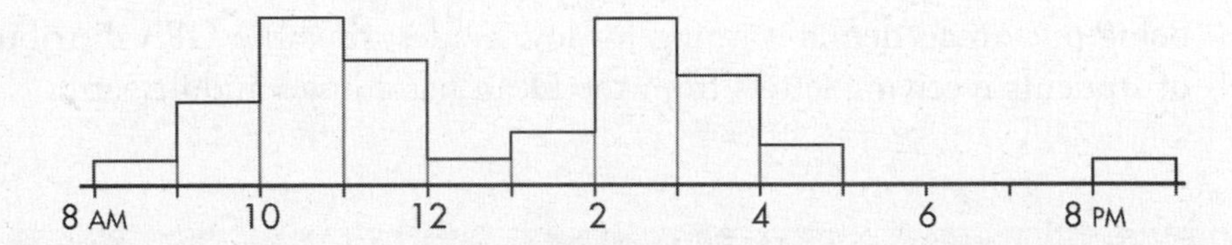

Note that this is a *bimodal* distribution. Computer usage at this company appears heaviest at midmorning and midafternoon, with a dip in usage during the noon lunch hour. There is an evening outlier possibly indicating employees returning after dinner (or perhaps custodial cleanup crews taking an Internet break!).

Note that, as illustrated above, it is usually instructive to look for reasons behind outliers and modes.

SHAPE

Distributions come in an endless variety of shapes; however, certain common patterns are worth special mention:

1. A *symmetric* distribution is one in which the two halves are mirror images of each other. For example, the weights of all people in some organizations fall into symmetric distributions with two mirror-image bumps, one for men's weights and one for women's weights.
2. A distribution is *skewed to the right* if it spreads far and thinly toward the higher values. For example, ages of nonagenarians (people in their 90s) is a distribution with sharply decreasing numbers as one moves from 90-year-olds to 99-year-olds.
3. A distribution is *skewed to the left* if it spreads far and thinly toward the lower values. For example, scores on an easy exam show a distribution bunched at the higher end with few low values.
4. A *bell-shaped* distribution is symmetric with a center mound and two sloping tails. For example, the distribution of IQ scores across the general population is roughly symmetric with a center mound at 100 and two sloping tails.

5. A distribution is *uniform* if its histogram is a horizontal line. For example, tossing a fair die and noting how many spots (pips) appear on top yields a uniform distribution with 1 through 6 all equally likely.

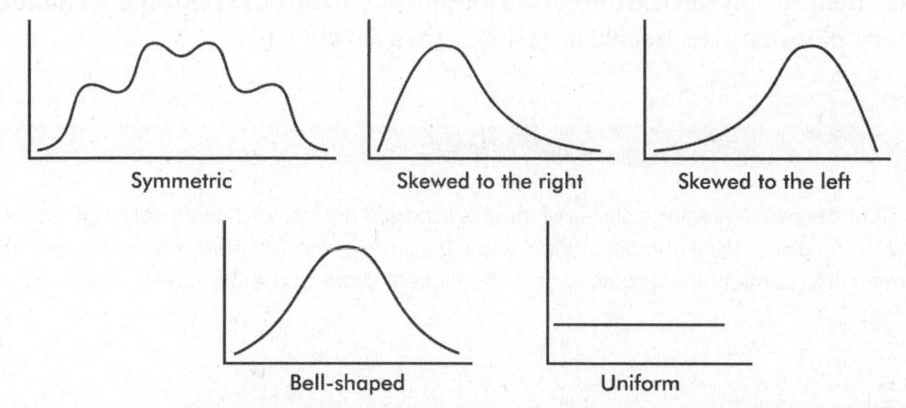

Even when a basic shape is noted, it is important also to note if some of the data deviate from this shape. For example, in the stemplot of Example 1.8, there is an outlier at 4.07 and a value at 4.88, with the remaining values showing a roughly bell-shaped distribution.

CUMULATIVE RELATIVE FREQUENCY PLOTS

Sometimes we sum frequencies and show the result visually in a *cumulative relative frequency plot* (also known as an *ogive*).

EXAMPLE 1.10

The following graph shows 1996 school enrollment in the United States by age.

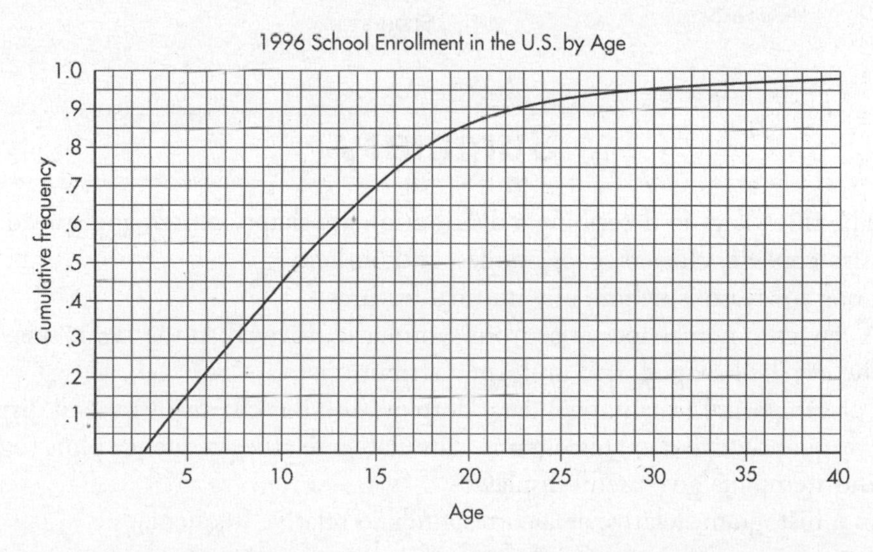

What can we learn from this cumulative relative frequency plot? For example, going up to the graph from age 5, we see that .15 or 15% of school enrollment is below age 5. Going over to the graph from .5 on the vertical axis, we see that 50% of the school enrollment is below and 50% is above a middle age of 11. Going up from age 30, we see that .95 or 95% of the enrollment is below age 30, and thus 5% is above age 30. Going over from .25 and .75 on the vertical axis, we see that the middle 50% of school enrollment is between ages 6 and 7 at the lower end and age 16 at the upper end.

CUMULATIVE RELATIVE FREQUENCY AND SKEWNESS

A distribution skewed to the left has a cumulative frequency plot that rises slowly at first and then steeply later, while a distribution skewed to the right has a cumulative frequency plot that rises steeply at first and then slowly later.

EXAMPLE 1.11

Consider the essay grading policies of three teachers, Abrams, who gives very high scores, Brown, who gives equal numbers of low and high scores, and Connors, who gives very low scores. Histograms of the grades (with 1 the highest score and 4 the lowest score) are as follows:

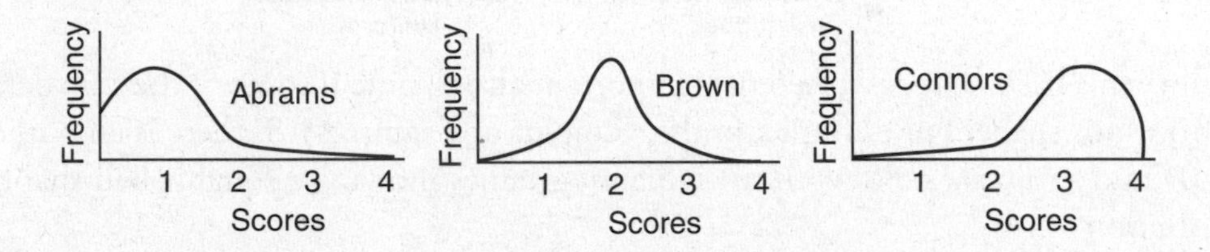

These translate into the following cumulative frequency plots:

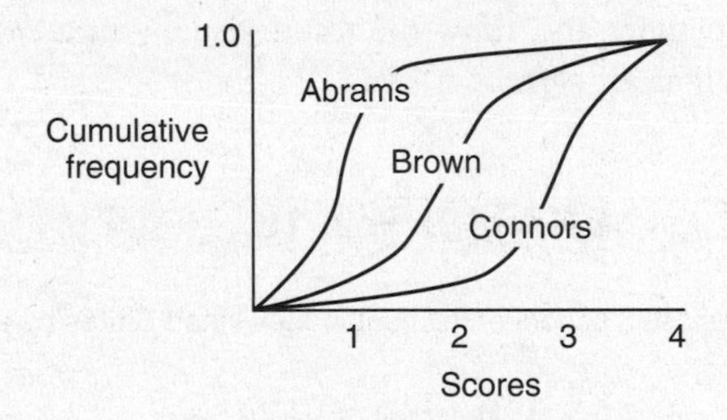

Summary

- The three keys to describing a distribution are shape, center, and spread.
- Also consider clusters, gaps, modes, and outliers.
- Look for reasons behind any unusual features.
- A few common shapes arise from symmetric, skewed to the right, skewed to the left, bell-shaped, and uniform distributions.
- For categorical (qualitative) data, dotplots and bar charts give useful displays.
- For quantitative data, histograms, cumulative relative frequency plots (ogives), and stemplots give useful displays.
- In a histogram, relative area corresponds to relative frequency.

Questions on Topic One: Graphical Displays

Multiple-Choice Questions

Directions: The questions or incomplete statements that follow are each followed by five suggested answers or completions. Choose the response that best answers the question or completes the statement.

1. The stemplot below shows ages of CEOs of a select group of corporations.

2	28
3	36
4	27
5	0258
6	3379
7	48
8	3
9	0

Which of the following is not a correct statement about this distribution?

(A) The distribution is bell-shaped.
(B) The distribution is skewed left and right.
(C) The center is around 60.
(D) The spread is from 28 to 90.
(E) There are no outliers.

2. Which of the following statements are true?

I. Stemplots are useful both for quantitative and categorical data sets.
II. Stemplots are equally useful for small and very large data sets.
III. Stemplots can show symmetry, gaps, clusters, and outliers.

(A) I only
(B) II only
(C) III only
(D) I and II
(E) I and III

3. Consider the following picture:

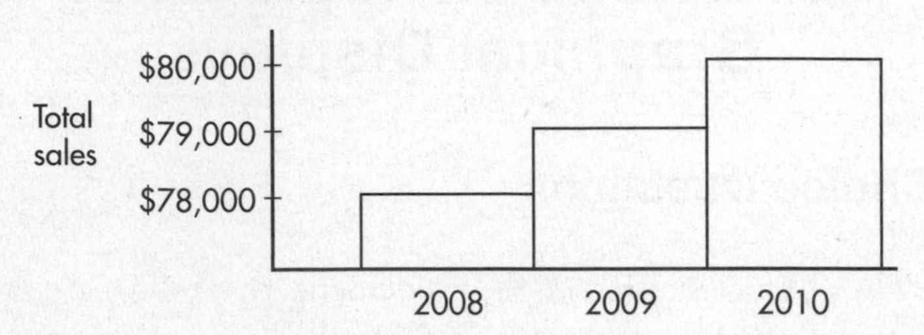

Which of the following statements are true?

I. Total sales in 2009 were two times total sales in 2008, while total sales in 2010 were three times the 2008 total.
II. The choice of labeling for the vertical axis results in a misleading sales picture.
III. A histogram showing the same information, but this time with a vertical axis starting at $78,000, would be less misleading.

(A) I only
(B) II only
(C) III only
(D) II and III
(E) None of the above gives the complete set of true responses.

4. Consider the following histogram:

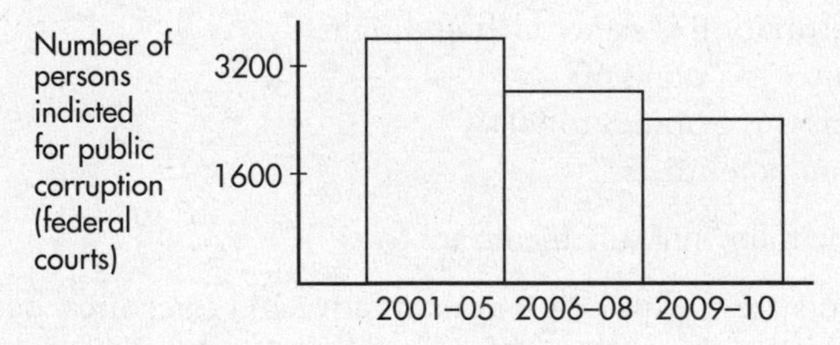

Which of the following statements are true?

I. Each year from 2001 to 2010 the number of indictments has steadily decreased.
II. While the number of indictments has decreased each year, the amount of decrease has lessened.
III. The labeling of the horizontal axis has resulted in a misleading picture.

(A) I only
(B) II only
(C) III only
(D) I and II
(E) None of the above gives the complete set of true responses.

5. Which of the following are true statements?

   I. Both dotplots and stemplots can show symmetry, gaps, clusters, and outliers.
   II. In histograms, relative areas correspond to relative frequencies.
   III. In histograms, frequencies can be determined from relative heights.

   (A) II only
   (B) I and II
   (C) I and III
   (D) II and III
   (E) I, II, and III

6. Which of the following are true statements?

   I. All symmetric histograms have single peaks.
   II. All symmetric bell-shaped curves are normal.
   III. All normal curves are bell-shaped and symmetric.

   (A) I only
   (B) II only
   (C) III only
   (D) I and II
   (E) None of the above gives the complete set of true responses.

7. Which of the following distributions are more likely to be skewed to the right than skewed to the left?

   I. Household incomes
   II. Home prices
   III. Ages of teenage drivers

   (A) II only
   (B) I and II
   (C) I and III
   (D) II and III
   (E) I, II, and III

8. Which of the following are true statements?

   I. Two students working with the same set of data may come up with histograms that look different.
   II. Displaying outliers is less problematic when using histograms than when using stemplots.
   III. Histograms are more widely used than stemplots or dotplots because histograms display the values of individual observations.

   (A) I only
   (B) II only
   (C) III only
   (D) I and II
   (E) II and III

9. Following is a histogram of test scores.

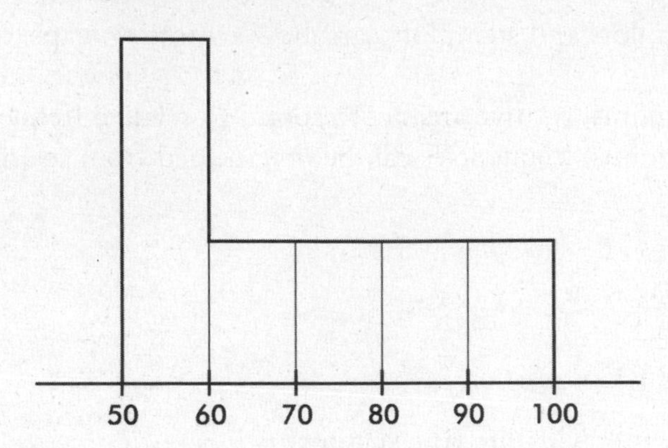

Which of the following statements are true?

I. The middle (median) score was 75.
II. If the passing score was 60, most students failed.
III. More students scored between 50 and 60 than between 90 and 100.

(A) I only
(B) II only
(C) III only
(D) II and III
(E) I, II, and III

Questions 10–14 refer to the following five cumulative relative frequency plots:

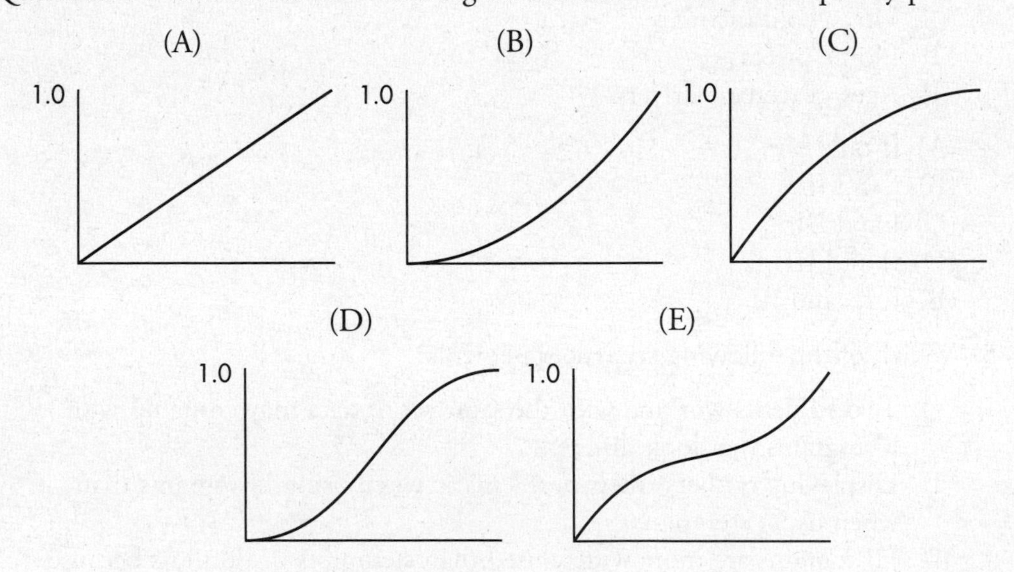

10. To which of the above cumulative relative frequency plots does the following histogram correspond?

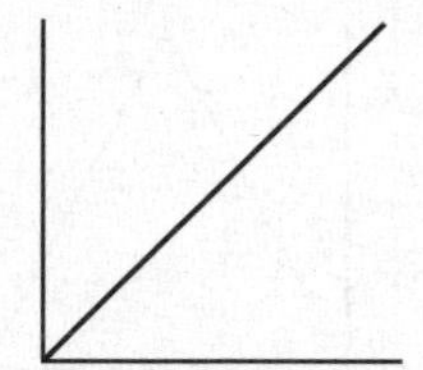

(A) A
(B) B
(C) C
(D) D
(E) E

11. To which of the above cumulative relative frequency plots does the following histogram correspond?

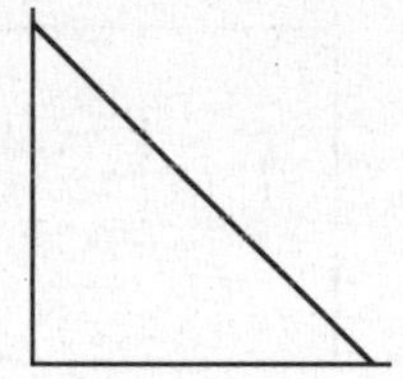

(A) A
(B) B
(C) C
(D) D
(E) E

12. To which of the above cumulative relative frequency plots does the following histogram correspond?

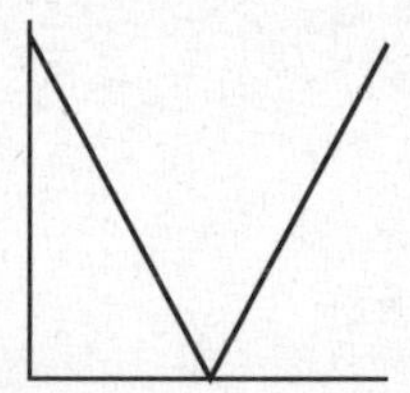

(A) A
(B) B
(C) C
(D) D
(E) E

13. To which of the above cumulative relative frequency plots does the following histogram correspond?

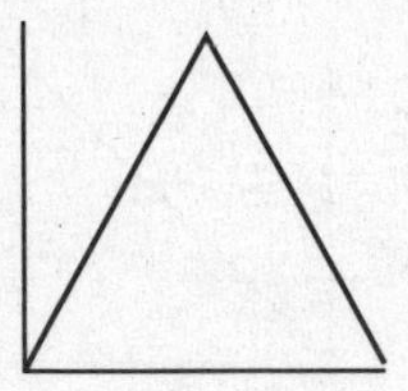

(A) A
(B) B
(C) C
(D) D
(E) E

14. To which of the above cumulative relative frequency plots does the following histogram correspond?

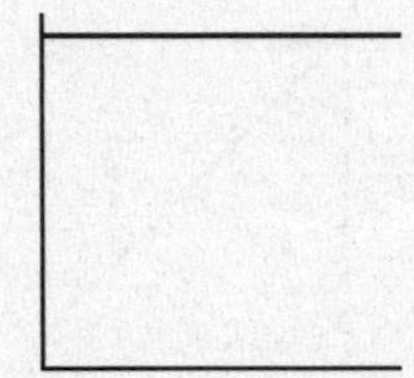

(A) A
(B) B
(C) C
(D) D
(E) E

Free-Response Questions

Directions: You must show all work and indicate the methods you use. You will be graded on the correctness of your methods and on the accuracy of your final answers.

Three Open-Ended Questions

1. The dotplot below shows the numbers of goals scored by the 20 teams playing in a city's high school soccer games on a particular day.

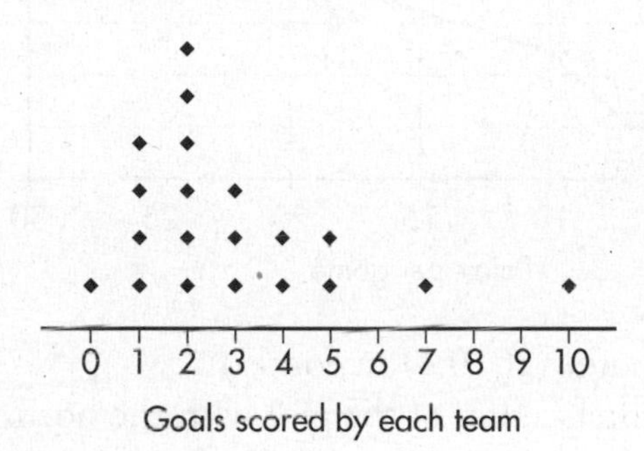

   (a) Describe the distribution.
   (b) One superstar scored six goals, but his team still lost. What are all possible final scores for that game? Explain.
   (c) Is it possible that all the teams scoring exactly two goals won their games? Explain.

2. The winning percentages for a major league baseball team over the past 22 years are shown in the following stemplot:

Stem	Leaf
46	08
47	1479
48	5889
49	347
50	
51	
52	58
53	256
54	489
55	6

(55 | 6 means 55.6%)

   (a) Interpret the lowest value.
   (b) Describe the distribution.
   (c) Give a reason that one might argue that the team is more likely to lose a given game than win it.
   (d) Give a reason that one might argue that the team is more likely to win a given game than lose it.

3. A college basketball team keeps records of career average points per game of players playing at least 75% of team games during their college careers. The cumulative relative frequency plot below summarizes statistics of players graduating over the past 10 years.

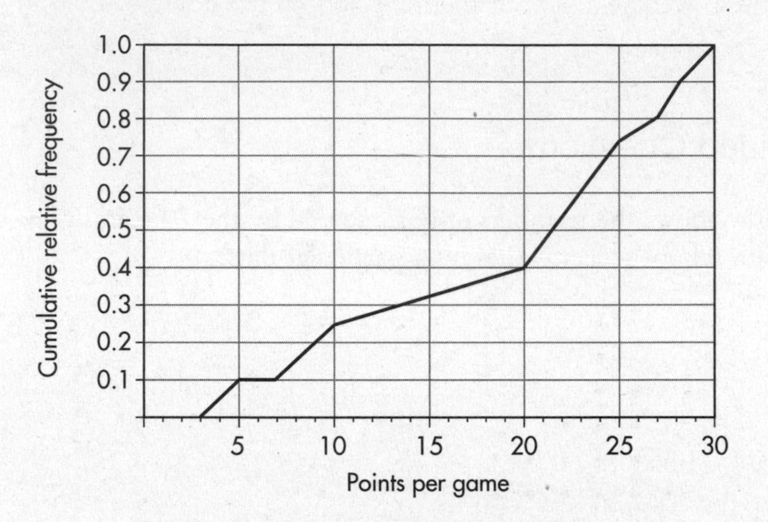

(a) Interpret the point (20, 0.4) in context.
(b) Interpret the intersection of the plot with the horizontal axis in context.
(c) Interpret the horizontal section of plot from 5 to 7 points per game in context.
(d) The players with the top 10% of the career average points per game achievements will be listed on a plaque. What is the cutoff score for being included on the plaque?
(e) What proportion of the players averaged between 10 and 20 points per game?

Answer Key

1. **B**	4. **C**	7. **B**	10. **B**	13. **D**
2. **C**	5. **B**	8. **A**	11. **C**	14. **A**
3. **B**	6. **C**	9. **C**	12. **E**	

Answers Explained

Multiple-Choice

1. **(B)** There is no such thing as being skewed both left and right.

2. **(C)** Stemplots are not used for categorical data sets and are too unwieldy to be used for very large data sets.

3. **(B)** The vertical axis, starting at $77,000, results in a misleading sales picture. It would be better to start at $0, not $78,000.

4. **(C)** Labeling the horizontal axis with different year spans results in a misleading picture. The number of indictments per year is actually increasing.

5. **(B)** In general, histograms give information about relative frequencies, not actual frequencies.

6. **(C)** Symmetric histograms can have any number of peaks. All normal curves are bell-shaped and symmetric, but not all symmetric bell-shaped curves are normal.

7. **(B)** Incomes and home prices tend to have a few very high scores that make the distributions skewed to the right. Teenage drivers mostly have ages in the last teenage years with a scattering of younger drivers, and thus a distribution skewed to the left.

8. **(A)** Choice of width and number of classes changes the appearance of a histogram. Displaying outliers is *more* problematic with histograms. Histograms do not show individual observations.

9. **(C)** The median score splits the area in half, and so the median is not 75. The area between 50 and 60 is greater than the area between 90 and 100 but is less than the area between 60 and 100.

10. **(B)** A histogram with little area under the curve early and much greater area later results in a cumulative relative frequency plot which rises slowly at first and then at a much faster rate later.

11. **(C)** A histogram with large area under the curve early and much less area later results in a cumulative relative frequency plot which rises quickly at first and then at a much slower rate later.

12. **(E)** A histogram with little area under the curve in the middle and much greater area on both ends results in a cumulative relative frequency plot which rises quickly at first, then almost levels off, and finally rises quickly at the end.

13. **(D)** A histogram with little area under the curve on the ends and much greater area in the middle results in a cumulative relative frequency plot which rises slowly at first, then quickly in the middle, and finally slowly again at the end.

14. **(A)** Uniform distributions result in cumulative relative frequency plots which rise at constant rates, thus linear.

Free-Response

1. (a) A complete answer considers shape, center, and spread.
   Shape: unimodal, skewed right, outlier at 10
   Center: around 2 or 3
   Spread: from 0 to 10

   (b) If the player scored six goals, his/her team must have scored either 7 or 10, but they lost, so they scored 7, and the only possible final score is that they lost by a score of 10 to 7.

   (c) No, there were six teams that scored exactly two goals, but there were only five teams that scored less than two goals, so not all the two-goal teams could have won.

2. (a) The lowest winning percentage over the past 22 years is 46.0%.

   (b) A complete answer considers shape, center, and spread.
   Shape: two clusters, each somewhat bell-shaped
   Center: around 50%
   Spread: from 46.0 to 55.6%

   (c) The team had more losing seasons (13) than winning seasons (9).

   (d) The cluster of winning percentages is further above 50% than the cluster of losing percentages is below 50%.

3. (a) 40% of the players averaged fewer than 20 points per game.

   (b) All the players averaged at least 3 points per game.

   (c) No players averaged between 5 and 7 points per game because the cumulative relative frequency was 10% for both 5 and 7 points.

   (d) Go over to the plot from 0.9 on the vertical axis, and then down to the horizontal axis to result in 28 points per game.

   (e) Reading up to the plot and then over from 10 and from 20 shows that 0.25 of the players averaged under 10 points per game and 0.4 of the players averaged under 20 points per game. Thus, 0.4 – 0.25 = 0.15 gives the proportion of players who averaged between 10 and 20 points per game.

Summarizing Distributions

TOPIC 2

- Measuring the Center
- Measuring Spread
- Measuring Position
- Empirical Rule
- Histograms
- Boxplots
- Changing Units

Given a raw set of data, often we can detect no overall pattern. Perhaps some values occur more frequently, a few extreme values may stand out, and the range of values is usually apparent. The presentation of data, including summarizations and descriptions, and involving such concepts as representative or average values, measures of dispersion, positions of various values, and the shape of a distribution, falls under the broad topic of *descriptive statistics.* This aspect of statistics is in contrast to *statistical analysis,* the process of drawing inferences from limited data, a subject discussed in later topics.

MEASURING THE CENTER: MEDIAN AND MEAN

The word *average* is used in phrases common to everyday conversation. People speak of bowling and batting averages or the average life expectancy of a battery or a human being. Actually the word *average* is derived from the French *avarie,* which refers to the money that shippers contributed to help compensate for losses suffered by other shippers whose cargo did not arrive safely (i.e., the losses were shared, with everyone contributing an average amount). In common usage *average* has come to mean a representative score or a typical value or the center of a distribution. Mathematically, there are a variety of ways to define the average of a set of data. In practice, we use whichever method is most appropriate for the particular case under consideration. However, beware of a headline with the word *average;* the writer has probably chosen the method that emphasizes the point he or she wishes to make.

In the following paragraphs we consider the two primary ways of denoting an average:

1. The *median,* which is the middle number of a set of numbers arranged in numerical order.
2. The *mean,* which is found by summing items in a set and dividing by the number of items.

EXAMPLE 2.1

Consider the following set of home run distances (in feet) to center field in 13 ballparks: {387, 400, 400, 410, 410, 410, 414, 415, 420, 420, 421, 457, 461}. What is the average?

Answer: The median is 414 (there are six values below 414 and six values above), while the mean is

$$\frac{387 + 400 + 400 + 410 + 410 + \cdots + 457 + 461}{13} = 417.3 \text{ feet}$$

TIP

Don't forget to put the data in order before finding the median.

Median

The word *median* is derived from the Latin *medius* which means "middle." The values under consideration are arranged in ascending or descending order. If there is an odd number of values, the median is the middle one. If there is an even number, the median is found by adding the two middle values and dividing by 2. Thus the median of a set has the same number of elements above it as below it.

The median is not affected by exactly how large the larger values are or by exactly how small the smaller values are. Thus it is a particularly useful measurement when the extreme values, called *outliers*, are in some way suspicious or when we want do diminish their effect. For example, if ten mice try to solve a maze, and nine succeed in less than 15 minutes while one is still trying after 24 hours, the most representative value is the median (not the mean, which is over 2 hours). Similarly, if the salaries of four executives are each between $240,000 and $245,000 while a fifth is paid less than $20,000, again the most representative value is the median (the mean is under $200,000). It is often said that the median is "resistant" to extreme values.

In certain situations the median offers the most economical and quickest way to calculate an average. For example, suppose 10,000 lightbulbs of a particular brand are installed in a factory. An average life expectancy for the bulbs can most easily be found by noting how much time passes before exactly one-half of them have to be replaced. The median is also useful in certain kinds of medical research. For example, to compare the relative strengths of different poisons, a scientist notes what dosage of each poison will result in the death of exactly one-half the test animals. If one of the animals proves especially susceptible to a particular poison, the median lethal dose is not affected.

Mean

While the median is often useful in descriptive statistics, the *mean,* or more accurately, the *arithmetic mean,* is most important for statistical inference and analysis. Also, for the layperson, the average is usually understood to be the mean.

The mean of a *whole population* (the complete set of items of interest) is often denoted by the Greek letter μ (mu), while the mean of a *sample* (a part of a population) is often denoted by $\bar{x}$. For example, the mean value of the set of all houses in the United States might be μ = \$56,400, while the mean value of 100 randomly chosen houses might be $\bar{x}$ = \$52,100 or perhaps $\bar{x}$ = \$63,800 or even $\bar{x}$ = \$124,000.

In statistics we learn how to estimate a population mean from a sample mean. Throughout this book, the word *sample* often implies a *simple random sample* (SRS), that is, a sample selected in such a way that every possible sample of the desired size has an equal chance of being included. (It is also true that each element of the population will have an equal chance of being included.) In the real world, this process of random selection is often very difficult to achieve, and so we proceed, with caution, as long as we have good reason to believe that our sample is representative of the population.

Mathematically, the mean = $\frac{\Sigma x}{n}$, where Σx represents the sum of all the elements of the set under consideration and n is the actual number of elements. Σ is the uppercase Greek letter sigma.

EXAMPLE 2.2

Suppose that the numbers of unnecessary procedures recommended by five doctors in a 1-month period are given by the set {2, 2, 8, 20, 33}. Note that the median is 8 and the mean is $\frac{2+2+8+20+33}{5} = 13$. If it is discovered that the fifth doctor also recommended an additional 25 unnecessary procedures, how will the median and mean be affected?

Answer: The set is now {2, 2, 8, 20, 58}. The median is still 8; however, the mean changes to $\frac{2+2+8+20+58}{5} = 18$.

The above example illustrates how the mean, unlike the median, is sensitive to a change in any value.

EXAMPLE 2.3

Suppose the salaries of six employees are $3000, $7000, $15,000, $22,000, $23,000, and $38,000, respectively.

a. What is the mean salary?

Answer:

$$\frac{3000 + 7000 + 15{,}000 + 22{,}000 + 23{,}000 + 38{,}000}{6} = \$18{,}000$$

b. What will the new mean salary be if everyone receives a $3000 increase?

Answer:

$$\frac{6000 + 10{,}000 + 18{,}000 + 25{,}000 + 26{,}000 + 41{,}000}{6} = \$21{,}000$$

Note that $18,000 + $3000 = $21,000.

c. What if everyone receives a 10% raise?

Answer:

$$\frac{3300 + 7700 + 16{,}500 + 24{,}200 + 25{,}300 + 41{,}800}{6} = \$19{,}800$$

Note that 110% of $18,000 is $19,800.

The above example illustrates how adding the same constant to each value increases the mean (and median) by a like amount. Similarly, multiplying each value by the same constant multiplies the mean (and median) by a like amount.

MEASURING SPREAD: RANGE, INTERQUARTILE RANGE, VARIANCE, AND STANDARD DEVIATION

TIP

Understanding variation is the key to understanding statistics.

In describing a set of numbers, not only is it useful to designate an average value but it is also important to be able to indicate the *variability* or the *dispersion* of the measurements. A producer of time bombs aims for small variability—it would not be good for his 30-minute fuses actually to have a range of 10–50 minutes before detonation. On the other hand, a teacher interested in distinguishing better students from poorer students aims to design exams with large variability in results—it would not be helpful if all her students scored exactly the same. The players on two basketball teams may have the same average height, but this observation doesn't tell the whole story. If the dispersions are quite different, one team may have a 7-foot player, whereas the other has no one over 6 feet tall. Two Mediterranean holiday cruises may advertise the same average age for their passengers. One, however, may have only passengers between 20 and 25 years old, while the other has only middle-aged parents in their forties together with their children under age 10.

There are four primary ways of describing variability or dispersion:

1. The *range*, which is the difference between the largest and smallest values
2. The *interquartile range*, IQR, which is the difference between the largest and smallest values after removing the lower and upper quarters (i.e., IQR is the range of the middle 50%); that is, IQR = $Q_3 - Q_1$ = 75th percentile minus 25th percentile
3. The *variance*, which is determined by averaging the squared differences of all the values from the mean
4. The *standard deviation*, which is the square root of the variance.

EXAMPLE 2.4

The monthly rainfall in Monrovia, Liberia, where May through October is the rainy season and November through April the dry season, is as follows:

Month:	Jan	Feb	Mar	Apr	May	June	July	Aug	Sept	Oct	Nov	Dec
Rain (in.):	1	2	4	6	18	37	31	16	28	24	9	4

The mean is

$$\frac{1+2+4+6+18+37+31+16+28+24+9+4}{12} = 15 \text{ inches}$$

What are the measures of variability?

Answer: Range: The maximum is 37 inches (June), and the minimum is 1 inch (January). Thus the range is 37 – 1 = 36 inches of rain.

Interquartile range: Removing the lower and upper quarters leaves 4, 6, 9, 16, 18, and 24. Thus the interquartile range is 24 – 4 = 20. [The interquartile range is sometimes calculated as follows: The median of the lower half is $Q_1 = \frac{4+4}{2} = 4$, the median of the upper half is

(continued)

$Q_3 = \frac{24+28}{2} = 26$, and the interquartile range is $Q_3 - Q_1 = 22$. When there is a large number of values in the set, the two methods give the same answer.]

Variance:

$$\frac{14^2 + 13^2 + 11^2 + 9^2 + 3^2 + 22^2 + 16^2 + 1^2 + 13^2 + 9^2 + 6^2 + 11^2}{12} = 143.7$$

Standard deviation: $\sqrt{143.7} = 12.0$ inches

Range

The simplest, most easily calculated measure of variability is the *range*. The difference between the largest and smallest values can be noted quickly, and the range gives some impression of the dispersion. However, it is entirely dependent on the two extreme values and is insensitive to the ones in the middle.

One use of the range is to evaluate samples with very few items. For example, some quality control techniques involve taking periodic small samples and basing further action on the range found in several such samples.

Interquartile Range

Finding the *interquartile range* is one method of removing the influence of extreme values on the range. It is calculated by arranging the data in numerical order, removing the upper and lower quarters of the values, and noting the range of the remaining values. That is, it is the range of the middle 50% of the values.

A numerical rule sometimes used for designating outliers is to calculate 1.5 times the interquartile range (IQR) and then call a value an outlier if it is more than $1.5 \times$ IQR below the first quartile or $1.5 \times$ IQR above the third quartile.

EXAMPLE 2.5

Suppose that farm sizes in a small community have the following characteristics: the smallest value is 16.6 acres, 10% of the values are below 23.5 acres, 25% are below 41.1 acres, the median is 57.6 acres, 60% are below 87.2 acres, 75% are below 101.9 acres, 90% are below 124.0 acres, and the top value is 201.7 acres.

a. What is the range?
Answer: The range is 201.7 – 16.6 = 185.1 acres.

b. What is the interquartile range?
Answer: The interquartile range, with the highest and lowest quarters of the values removed, is 101.9 – 41.1 = 60.8 acres. Thus, while the largest farm is 185 acres more than the smallest, the middle 50% of the farm sizes range over a 61-acre interval.

(continued)

c. When the numerical rule is used for outliers, should either the smallest or largest value be called an outlier?
Answer: $1.5 \times IQR = 1.5 \times 60.8 = 91.2$. If a number is more than 91.2 below the first quartile, 41.1, or more than 91.2 above the third quartile, 101.9, then it will be called an outlier. Since the largest value, 201.7, is greater than $101.9 + 91.2 = 193.1$, it is considered an outlier by the numerical rule.

Variance

Dispersion is often the result of various chance happenings. For example, consider the motion of microscopic particles suspended in a liquid. The unpredictable motion of any particle is the result of many small movements in various directions caused by random bumps from other particles. If we average the total displacements of all the particles from their starting points, the result will not increase in direct proportion to time. If, however, we average the *squares* of the total displacements of all the particles, this result will increase in direct proportion to time.

The same holds true for the movement of paramecia. Their seemingly random motions as seen under a microscope can be described by the observation that the average of the squares of the displacements from their starting points is directly proportional to time. Also, consider ping-pong balls dropped straight down from a high tower and subjected to chance buffeting in the air. We can measure the deviations from a center spot on the ground to the spots where the balls actually strike. As the height of the tower is increased, the average of the squared deviations increases proportionately.

In a wide variety of cases we are in effect trying to measure dispersion from the mean due to a multitude of chance effects. The proper tool in these cases is the average of the squared deviations from the mean; it is called the *variance* and is denoted by σ^2 (σ is the lowercase Greek letter sigma):

$$\sigma^2 = \frac{\Sigma(x-\mu)^2}{n}$$

For circumstances specified later, the variance of a sample, denoted by s^2, is calculated as

$$s^2 = \frac{\Sigma(x-\bar{x})^2}{n-1}$$

EXAMPLE 2.6

During the years 1929–39 of the Great Depression, the weekly average hours worked in manufacturing jobs were 45, 43, 41, 39, 39, 35, 37, 40, 39, 36, and 37, respectively. What is the variance?

(continued)

Answer: The variance can be quickly found on any calculator with a simple statistical package, or it can be found as follows:

$$\mu = \frac{45+43+41+39+39+35+37+40+39+36+37}{11} = 39.2 \text{ hours}$$

$$\sigma^2 = \frac{(45-39.2)^2+(43-39.2)^2+\cdots+(36-39.2)^2+(37-39.2)^2}{11} = 8.1$$

EXAMPLE 2.7

Let $X = \{3, 7, 15, 23\}$. What is the variance?

Answer: Again, use the statistical package on your calculator, or

$$\Sigma x = 3+7+15+23 = 48$$

$$\mu = \frac{\Sigma x}{n} = \frac{48}{4} = 12$$

The variance can be calculated from its definition:

$$\sigma^2 = \frac{\Sigma(x-\mu)^2}{n} = \frac{(3-12)^2+(7-12)^2+(15-12)^2+(23-12)^2}{4} = 59$$

Standard Deviation

Suppose we wish to pick a representative value for the variability of a certain population. The preceding discussions indicate that a natural choice is the value whose square is the average of the squared deviations from the mean. Thus we are led to consider the square root of the variance. This value is called the *standard deviation*, is denoted by σ, and is calculated on your calculator or as follows:

$$\sigma = \sqrt{\frac{\Sigma(x-\mu)^2}{n}}$$

Similarly, the standard deviation of a sample is denoted by s and is calculated on your calculator or as follows:

$$s = \sqrt{\frac{\Sigma(x-\bar{x})^2}{n-1}}$$

While variance is measured in square units, standard deviation is measured in the same units as are the data.

For the various x-values, the deviations $x - \bar{x}$ are called *residuals*, and s is a "typical value" for the residuals. While s is not the average of the residuals (the average of the residuals is always 0), s does give a measure of the spread of the x-values around the sample mean.

EXAMPLE 2.8

Putting the data {1, 6, 3, 8} into a calculator such as the TI-84 gives what value for the standard deviation?

Answer: The TI-84 gives

```
1-Var Stats
x̄=4.5
∑x=18
∑x²=110
Sx=3.109126351
σx=2.692582404
```

Thus, if the data are a population, the standard deviation is $\sigma = 2.693$, while if the data are a sample, the standard deviation is $s = 3.109$.

MEASURING POSITION: SIMPLE RANKING, PERCENTILE RANKING, AND Z-SCORE

We have seen several ways of choosing a value to represent the center of a distribution. We also need to be able to talk about the *position* of any other values. In some situations, such as wine tasting, simple rankings are of interest. Other cases, for example, evaluating college applications, may involve positioning according to percentile rankings. There are also situations in which position can be specified by making use of measurements of both central tendency and variability.

There are three important, recognized procedures for designating position:

1. *Simple ranking,* which involves arranging the elements in some order and noting where in that order a particular value falls
2. *Percentile ranking,* which indicates what percentage of all values fall below the value under consideration
3. The *z-score,* which states very specifically by how many standard deviations a particular value varies from the mean.

EXAMPLE 2.9

The water capacities (in gallons) of the 57 major solid-fuel boilers sold in the United States are 6.3, 7.4, 8.6, 10, 12.1, 50, 8.2, 9.8, 11.4, 12.9, 14.5, 16.1, 26, 21, 27, 40, 55, 30, 35, 55, 65, 18.8, 23.3, 28.3, 33.8, 26.4, 33, 50, 35, 21, 21, 18.5, 26.4, 37, 12, 12, 50, 65, 65, 56, 66, 60, 70, 27.7, 34.3, 42, 46.2, 33, 25, 29, 40, 19, 9, 12.5, 15.8, 24.5, and 16.5, respectively (John W. Bartok, *Solid-Fuel Furnaces and Boilers,* Garden Way, Charlotte, Vermont). What is the position of the Passat HO-45, which has a capacity of 46.2 gallons?

Answer: Since there are 12 boilers with higher capacities on the list, the Passat has a simple ranking of thirteenth (out of 57). Forty-four boilers have lower capacities, and so the Passat has a percentile ranking of $\frac{44}{57} = 77.2\%$. The above list has a mean of 30.3 and a standard deviation of 17.8, so the Passat has a *z*-score of $\frac{46.2-30.3}{17.8} = 0.89$.

Simple Ranking

Simple ranking is easily calculated and easily understood. We know what it means for someone to graduate second in a class of 435, or for a player from a team of size 30 to have the seventh-best batting average. Simple ranking is useful even when no numerical values are associated with the elements. For example, detergents can be ranked according to relative cleansing ability without any numerical measurements of strength.

Percentile Ranking

Percentile ranking, another readily understood measurement of position, is helpful in comparing positions with different bases. We can more easily compare a rank of 176 out of 704 with a rank of 187 out of 935 by noting that the first has a rank of 75%, and the second, a rank of 80%. Percentile rank is also useful when the exact population size is not known or is irrelevant. For example, it is more meaningful to say that Jennifer scored in the 90th percentile on a national exam rather than trying to determine her exact ranking among some large number of test takers.

The *quartiles*, Q_1 and Q_3, lie one-quarter and three-quarters of the way up a list, respectively. Their percentile ranks are 25% and 75%, respectively. The interquartile range defined earlier can also be defined to be $Q_3 - Q_1$. The *deciles* lie one-tenth and nine-tenths of the way up a list, respectively, and have percentile ranks of 10% and 90%.

z-Score

The *z-score* is a measure of position that takes into account both the center and the dispersion of the distribution. More specifically, the *z*-score of a value tells how many standard deviations the value is from the mean. Mathematically, $x - \mu$ gives the raw distance from μ to x; dividing by σ converts this to number of standard deviations. Thus $z = \frac{x-\mu}{\sigma}$, where x is the raw score, μ is the mean, and σ is the standard deviation. If the score x is greater than the mean μ, then z is positive; if x is less than μ, then z is negative.

Given a *z*-score, we can reverse the procedure and find the corresponding raw score. Solving for x gives $x = \mu + z\sigma$.

EXAMPLE 2.10

Suppose the average (mean) price of gasoline in a large city is \$1.80 per gallon with a standard deviation of \$0.05. Then \$1.90 has a *z*-score of $\frac{1.90-1.80}{0.05} = +2$, while \$1.65 has a *z*-score of $\frac{1.65-1.80}{0.05} = -3$. Alternatively, a *z*-score of +2.2 corresponds to a raw score of 1.80 + 2.2(0.05) = 1.80 + 0.11 = 1.91, while a *z*-score of –1.6 corresponds to 1.80 – 1.6(0.05) = 1.72.

It is often useful to portray integer *z*-scores and the corresponding raw scores as follows:

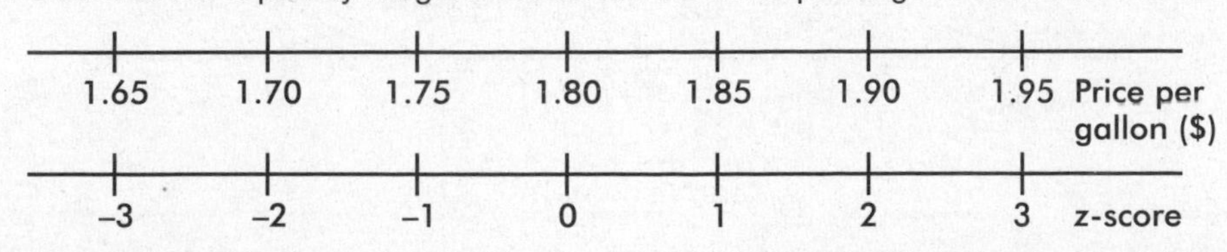

EXAMPLE 2.11

Suppose the attendance at a movie theater averages 780 with a standard deviation of 40. Adding multiples of 40 to and subtracting multiples of 40 from the mean 780 gives

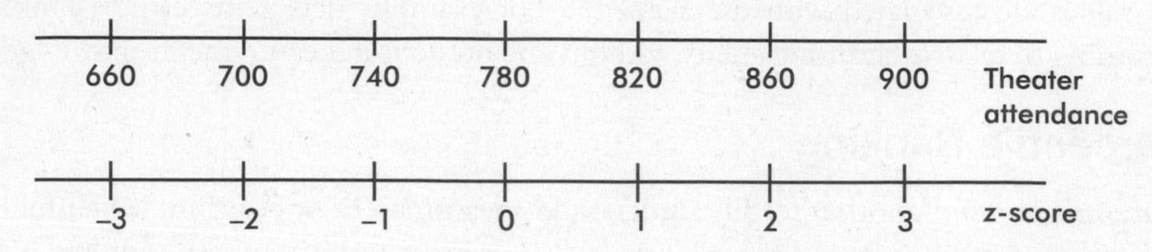

A theater attendance of 835 is converted to a z-score as follows: $\frac{835-780}{40} = \frac{55}{40} = 1.375$. A z-score of -2.15 is converted to a theater attendance as follows: $780 - 2.15(40) = 694$.

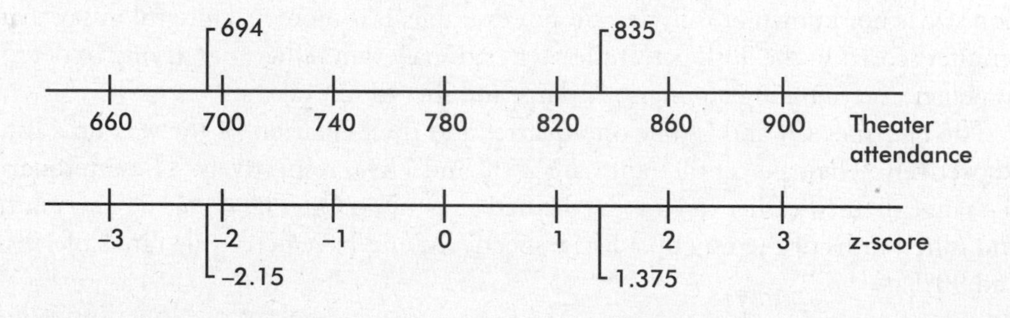

EMPIRICAL RULE

The *empirical rule* (also called the *68-95-99.7 rule*) applies specifically to symmetric bell-shaped data (not to skewed data!). In this case, about 68% of the values lie within 1 standard deviation of the mean, about 95% of the values lie within 2 standard deviations of the mean, and more than 99% of the values lie within 3 standard deviations of the mean.

In the following figure the horizontal axis shows z-scores:

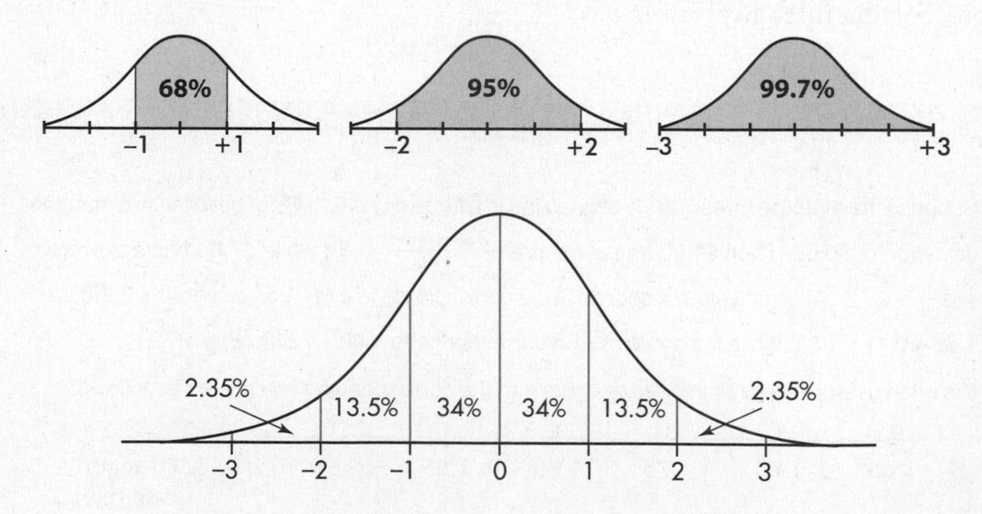

EXAMPLE 2.12

Suppose that taxicabs in New York City are driven an average of 75,000 miles per year with a standard deviation of 12,000 miles. What information does the empirical rule give us?

Answer: Assuming that the distribution is bell-shaped, we can conclude that approximately 68% of the taxis are driven between 63,000 and 87,000 miles per year, approximately 95% are driven between 51,000 and 99,000 miles, and virtually all are driven between 39,000 and 111,000 miles.

The empirical rule also gives a useful quick estimate of the standard deviation in terms of the range. We can see in the figure above that 95% of the data fall within a span of 4 standard deviations (from –2 to +2 on the *z*-score line) and 99.7% of the data fall within 6 standard deviations (from –3 to +3 on the *z*-score line). It is therefore reasonable to conclude that for these data the standard deviation is roughly between one-fourth and one-sixth of the range. Since we can find the range of a set almost immediately, the empirical rule technique for estimating the standard deviation is often helpful in pointing out gross arithmetic errors.

EXAMPLE 2.13

If the range of a data set is 60, what is an estimate for the standard deviation?

Answer: By the empirical rule, the standard deviation is expected to be between $\left(\frac{1}{6}\right)60 = 10$ and $\left(\frac{1}{4}\right)60 = 15$. If the standard deviation is calculated to be 0.32 or 87, there is probably an arithmetic error; a calculation of 12, however, is reasonable.

However, it must be stressed that the above use of the range is not intended to provide an accurate value for the standard deviation. It is simply a tool for pointing out unreasonable answers rather than, for example, blindly accepting computer outputs.

HISTOGRAMS AND MEASURES OF CENTRAL TENDENCY

Suppose we have a detailed histogram such as

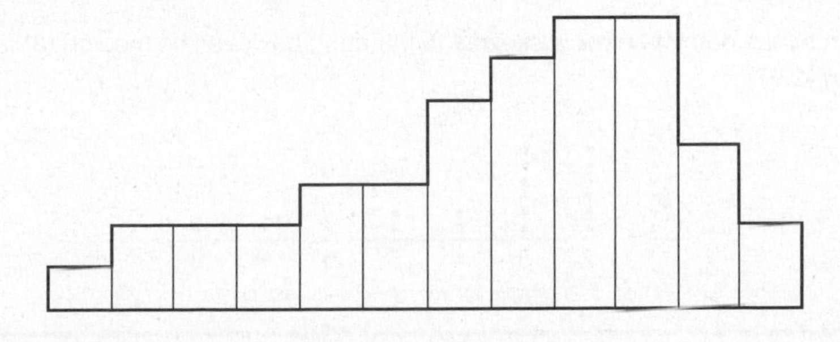

Our measures of central tendency fit naturally into such a diagram.

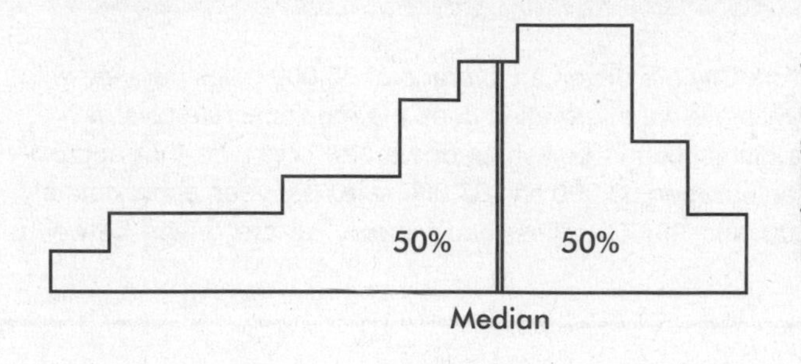

The *median* divides a distribution in half, so it is represented by a line that divides the area of the histogram in half.

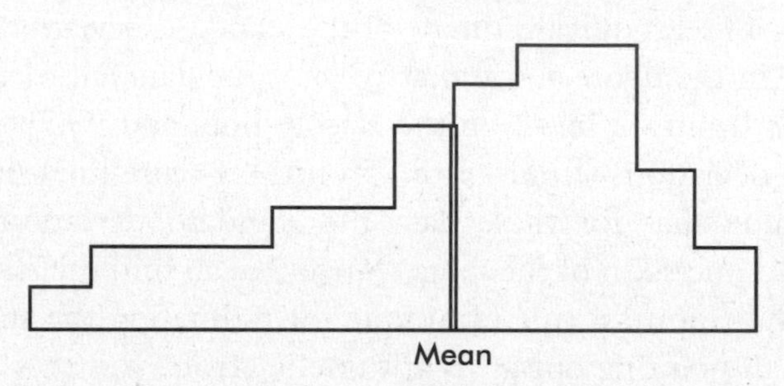

The *mean* is affected by the spacing of all the values. Therefore, if the histogram is considered to be a solid region, the mean corresponds to a line passing through the center of gravity, or balance point.

The above distribution, spread thinly far to the low side, is said to be *skewed to the left*. Note that in this case the mean is usually less than the median. Similarly, a distribution spread far to the high side is *skewed to the right*, and its mean is usually greater than its median.

EXAMPLE 2.14

Suppose that the faculty salaries at a college have a median of \$32,500 and a mean of \$38,700. What does this indicate about the shape of the distribution of the salaries?

Answer: The median is less than the mean, and so the salaries are probably skewed to the right. There are a few highly paid professors, with the bulk of the faculty at the lower end of the pay scale.

It should be noted that the above principle is a useful, but not hard-and-fast, rule.

EXAMPLE 2.15

The set given by the dotplot below is skewed to the right; however, its median (3) is greater than its mean (2.97).

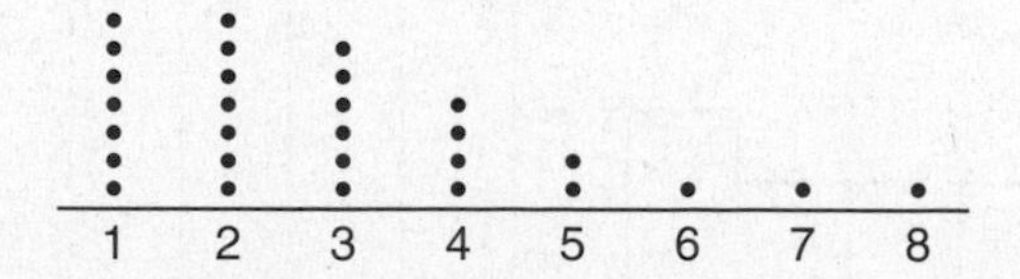

HISTOGRAMS, *Z*-SCORES, AND PERCENTILE RANKINGS

We have seen that relative frequencies are represented by relative areas, and so labeling the vertical axis is not crucial. If we know the standard deviation, the horizontal axis can be labeled in terms of *z*-scores. In fact, if we are given the percentile rankings of various *z*-scores, we can construct a histogram.

EXAMPLE 2.16

Suppose we are asked to construct a histogram from these data:

z-score:	−2	−1	0	1	2
Percentile ranking:	0	20	60	70	100

We note that the entire area is less than *z*-score +2 and greater than *z*-score −2. Also, 20% of the area is between *z*-scores −2 and −1, 40% is between −1 and 0, 10% is between 0 and 1, and 30% is between 1 and 2. Thus the histogram is as follows:

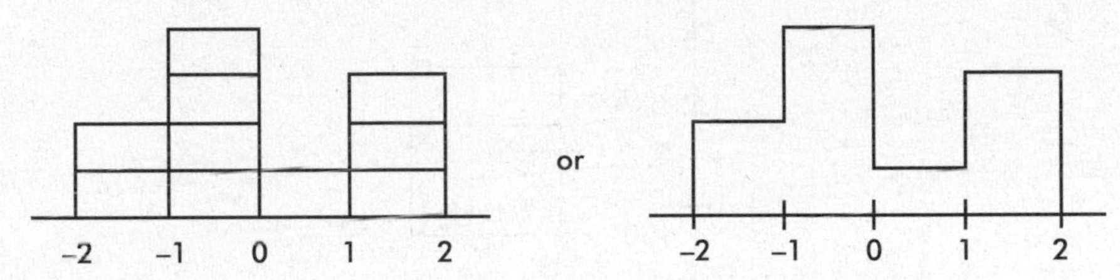

Now suppose we are given four in-between *z*-scores as well:

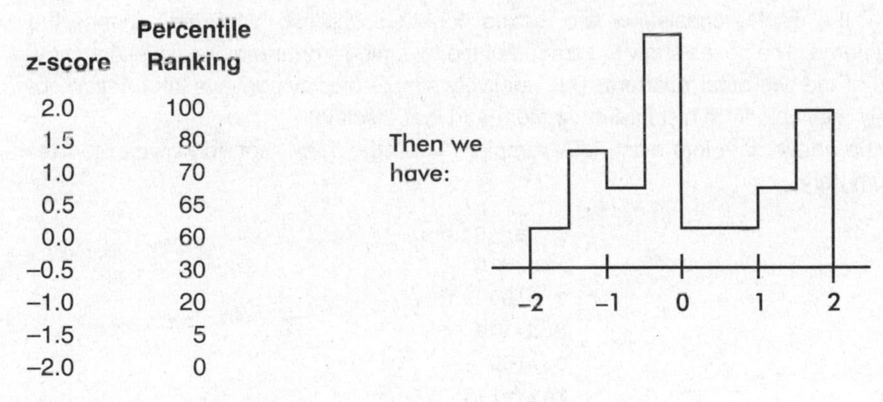

z-score	Percentile Ranking
2.0	100
1.5	80
1.0	70
0.5	65
0.0	60
−0.5	30
−1.0	20
−1.5	5
−2.0	0

With 1000 *z*-scores perhaps the histogram would look like

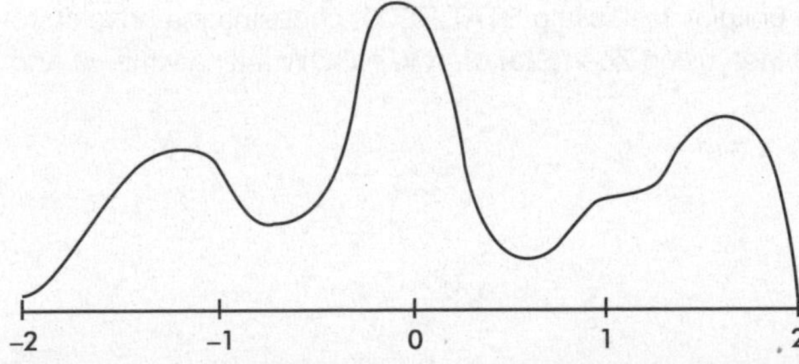

The height at any point is meaningless; what is important is relative areas. For example, in the final diagram above, what percentage of the area is between *z*-scores of +1 and +2?

Answer: Still 30%.

What percent is to the left of 0?

Answer: Still 60%.

BOXPLOTS

A *boxplot* (also called a *box and whisker display*) is a visual representation of dispersion that shows the smallest value, the largest value, the middle (median), the middle of the bottom half of the set (Q_1), and the middle of the top half of the set (Q_3).

EXAMPLE 2.17

The total farm product indexes for the years 1919–45 (with 1910–14 as 100) are 215, 210, 130, 140, 150, 150, 160, 150, 140, 150, 150, 125, 85, 70, 75, 90, 115, 120, 125, 100, 95, 100, 130, 160, 200, 200, 210, respectively. (Note the instability of prices received by farmers!) The largest value is 215, the smallest is 70, the middle is 140, the middle of the top half is 160, and the middle of the bottom half is 100. A boxplot of these five numbers is

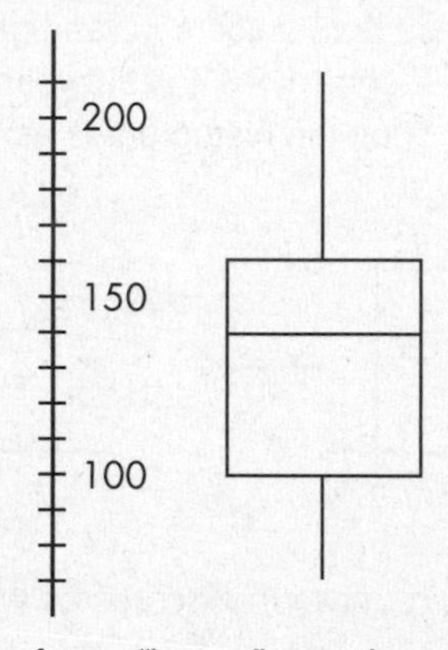

Note that the display consists of two "boxes" together with two "whiskers"—hence the alternative name. The boxes show the spread of the two middle quarters; the whiskers show the spread of the two outer quarters. This relatively simple display conveys information not immediately available from histograms or stem and leaf displays.

Putting the above data into a list, for example, L1, on the TI-84, not only gives the five-number summary

```
1-Var Stats
minX=70
Q1=100
Med=140
Q3=160
MaxX=215
```

but also gives the boxplot itself using STAT PLOT, choosing the boxplot from among the six type choices, and then using ZoomStat or in WINDOW letting Xmin=0 and Xmax=225.

On the TI-Nspire the data can be put in a list (here called *index*), and then a simultaneous multiple view is possible.

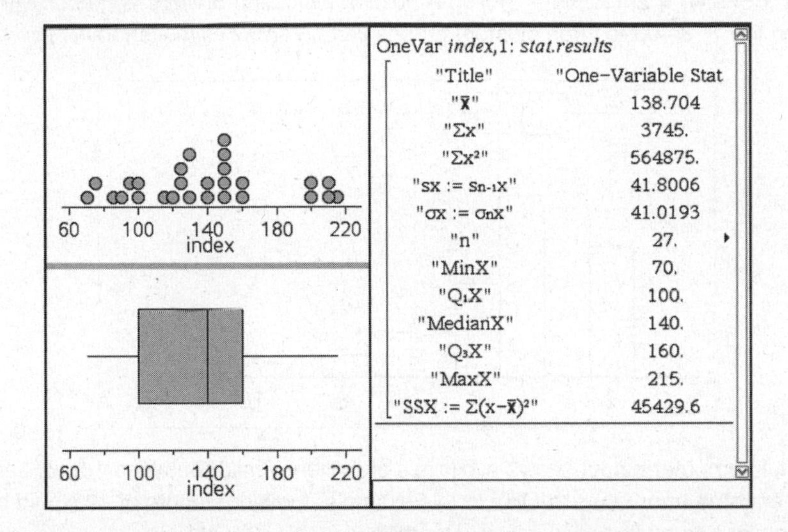

When a distribution is strongly skewed, or when it has pronounced outliers, drawing a boxplot with its five-number summary including median, quartiles, and extremes, gives a more useful description than calculating a mean and a standard deviation.

Sometimes values more than 1.5 × IQR (1.5 times the interquartile range) outside the two boxes are plotted separately as possible outliers. (The TI-84 has a modified boxplot option. Note the two options in the second row of Type in StatPlot.)

EXAMPLE 2.18

Inputting the lengths of words in a selection of Shakespeare's plays results in a calculator output of

```
1-Var Stats
minX=1
Q1=3
Med=4
Q3=5
maxX=12
```

Outliers consist of any word lengths less than $Q_1 - 1.5(IQR) = 3 - 1.5(5 - 3) = 0$ or greater than $Q_3 + 1.5(IQR) = 5 + 1.5(5 - 3) = 8$. A boxplot indicating outliers, together with a histogram (on the TI-84 up to three different graphs can be shown simultaneously) is

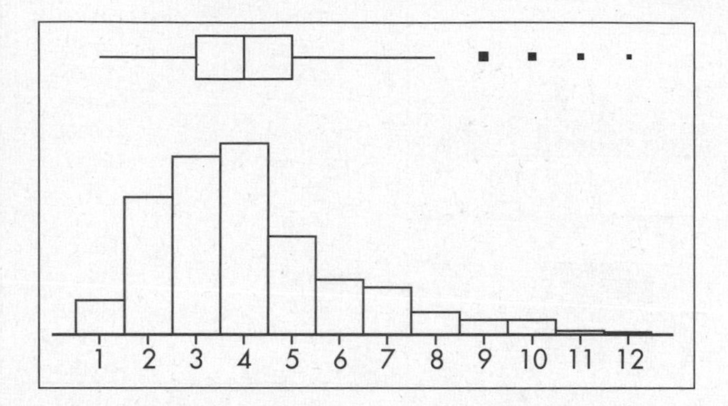

Note: Some computer output shows *two* levels of outliers—mild (between 1.5 IQR and 3 IQR) and extreme (more than 3 IQR). In this example, the word length of 12 would be considered an extreme outlier since it is greater than $5 + 3(5 - 3) = 11$.

It should be noted that two sets can have the same five-number summary and thus the same boxplots but have dramatically different distributions.

EXAMPLE 2.19

Let $A = \{0, 5, 10, 15, 25, 30, 35, 40, 45, 50, 71, 72, 73, 74, 75, 76, 77, 78, 100\}$ and $B = \{0, 22, 23, 24, 25, 26, 27, 28, 29, 50, 55, 60, 65, 70, 75, 85, 90, 95, 100\}$. Simple inspection indicates very different distributions, however the TI-84 gives identical boxplots with Min = 0, Q_1 = 25, Med = 50, Q_3 = 75, and Max = 100 for each.

EFFECT OF CHANGING UNITS

Changing units, for example, from dollars to rubles or from miles to kilometers, is common in a world that seems to become smaller all the time. It is instructive to note how measures of center and spread are affected by such changes.

Adding the same constant to every value increases the mean and median by that same constant; however, the distances between the increased values stay the same, and so the range and standard deviation are unchanged.

EXAMPLE 2.20

A set of experimental measurements of the freezing point of an unknown liquid yield a mean of 25.32 degrees Celsius with a standard deviation of 1.47 degrees Celsius. If all the measurements are converted to the Kelvin scale, what are the new mean and standard deviation?

Answer: Kelvins are equivalent to degrees Celsius plus 273.16. The new mean is thus 25.32 + 273.16 = 298.48 kelvins. However, the standard deviation remains numerically the same, 1.47 kelvins. Graphically, you should picture the whole distribution moving over by the constant 273.16; the mean moves, but the standard deviation (which measures spread) doesn't change.

Multiplying every value by the same constant multiplies the mean, median, range, and standard deviation all by that constant.

EXAMPLE 2.21

Measurements of the sizes of farms in an upstate New York county yield a mean of 59.2 hectares with a standard deviation of 11.2 hectares. If all the measurements are converted from hectares (metric system) to acres (one acre was originally the area a yoke of oxen could plow in one day), what are the new mean and standard deviation?

Answer: One hectare is equivalent to 2.471 acres. The new mean is thus $2.471 \times 59.2 = 146.3$ acres with a standard deviation of $2.471 \times 11.2 = 27.7$ acres. Graphically, multiplying each value by the constant 2.471 both moves and spreads out the distribution.

Summary

- The two principle measurements of the center of a distribution are the mean and the median.
- The principle measurements of the spread of a distribution are the range (maximum value minus minimum value), the interquartile range ($IQR = Q_3 - Q_1$), the variance, and the standard deviation.
- Adding the same constant to every value in a set adds the same constant to the mean and median but leaves all the above measures of spread unchanged.
- Multiplying every value in a set by the same constant multiplies the mean, median, range, IQR, and standard deviation by that constant.
- The mean, range, variance, and standard deviation are sensitive to extreme values, while the median and interquartile range are not.
- The principle measurements of position are simple ranking, percentile ranking, and the *z*-score (which measures the number of standard deviations from the mean).
- The empirical rule (the 68-95-99.7 rule) applies specifically to symmetric bell-shaped data.
- In skewed left data, the mean is usually less than the median, while in skewed right data, the mean is usually greater than the median.
- Boxplots visually show the five-number summary: the minimum value, the first quartile (Q_1), the median, the third quartile (Q_3), and the maximum value; and usually indicate outliers as distinct points.
- Note that two sets can have the same five-number summary and thus the same boxplots but have dramatically different distributions.

Questions on Topic Two: Summarizing Distributions

Multiple-Choice Questions

Directions: The questions or incomplete statements that follow are each followed by five suggested answers or completions. Choose the response that best answers the question or completes the statement.

1. The graph below shows household income in Laguna Woods, California.

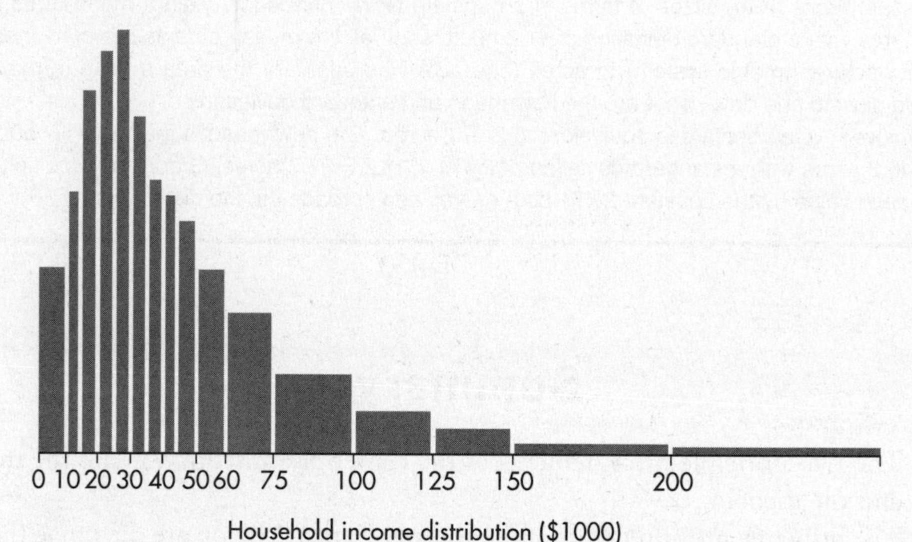

What can be said about the ratio $\dfrac{\text{Mean family income}}{\text{Median family income}}$?

(A) Approximately zero
(B) Less than one, but definitely above zero
(C) Approximately one
(D) Greater than one
(E) Cannot be answered without knowing the standard deviation

2. Which of the following are true statements?

I. The range of the sample data set is never greater than the range of the population.
II. The interquartile range is half the distance between the first quartile and the third quartile.
III. While the range is affected by outliers, the interquartile range is not.

(A) I only
(B) II only
(C) III only
(D) I and II
(E) I and III

3. Dieticians are concerned about sugar consumption in teenagers' diets (a 12-ounce can of soft drink typically has 10 teaspoons of sugar). In a random sample of 55 students, the number of teaspoons of sugar consumed for each student on a randomly selected day is tabulated. Summary statistics are noted below:

   Min = 10   Max = 60   First quartile = 25   Third quartile = 38
   Median = 31   Mean = 31.4 $n = 55$ $s = 11.6$

   Which of the following is a true statement?

   (A) None of the values are outliers.
   (B) The value 10 is an outlier, and there can be no others.
   (C) The value 60 is an outlier, and there can be no others.
   (D) Both 10 and 60 are outliers, and there may be others.
   (E) The value 60 is an outlier, and there may be others at the high end of the data set.

Problems 4–6 refer to the following five boxplots.

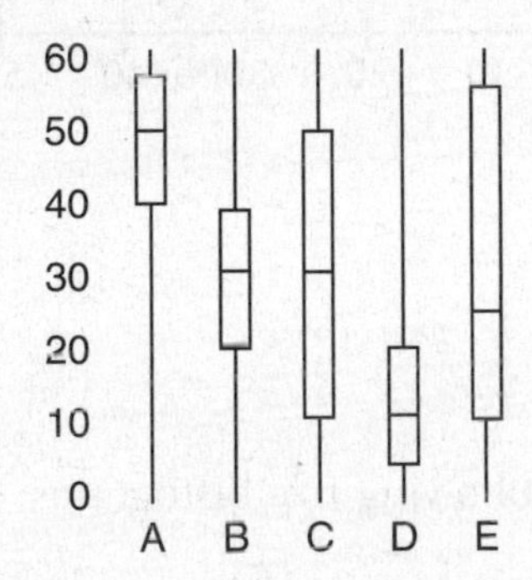

4. To which of the above boxplots does the following histogram correspond?

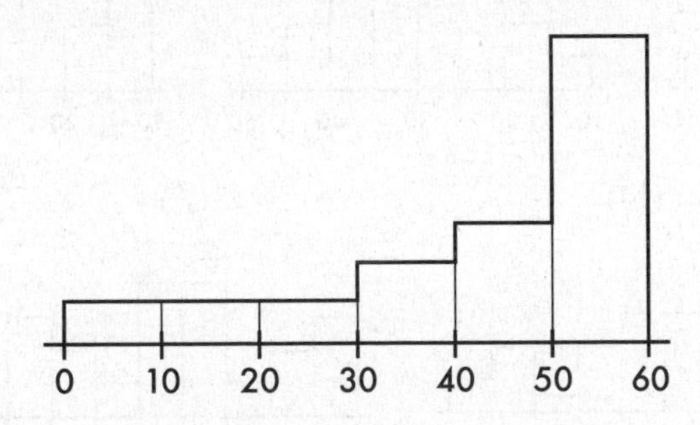

   (A) A
   (B) B
   (C) C
   (D) D
   (E) E

5. To which of the above boxplots does the following histogram correspond?

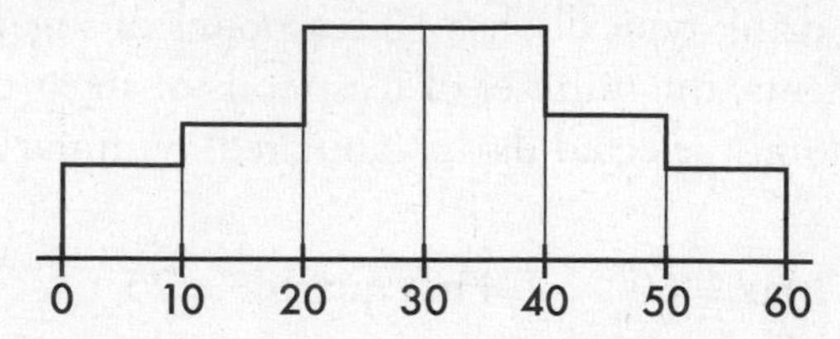

(A) A
(B) B
(C) C
(D) D
(E) E

6. To which of the above boxplots does the following histogram correspond?

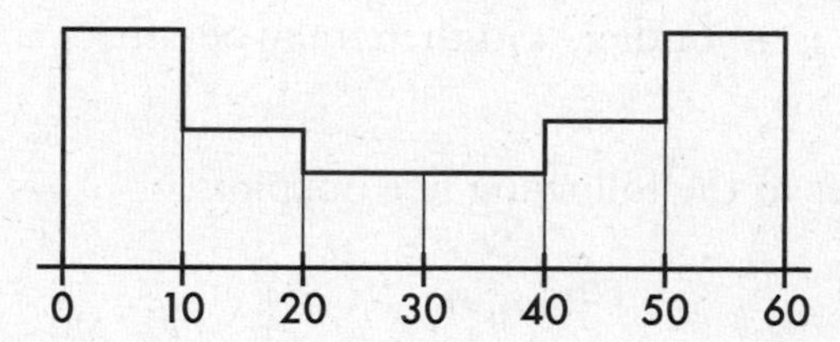

(A) A
(B) B
(C) C
(D) D
(E) E

Problems 7–9 refer to the following five histograms:

(A)

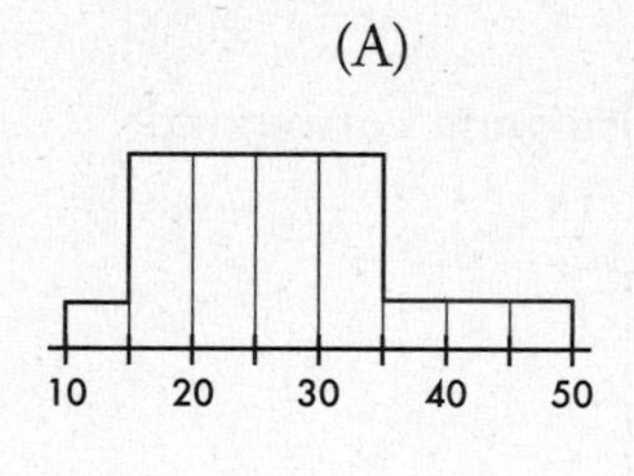

(B)

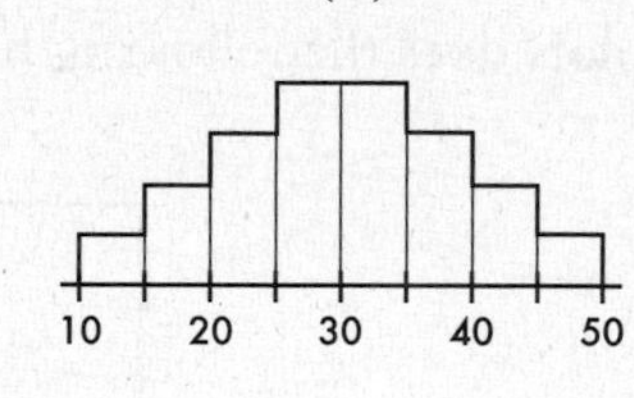

(C)

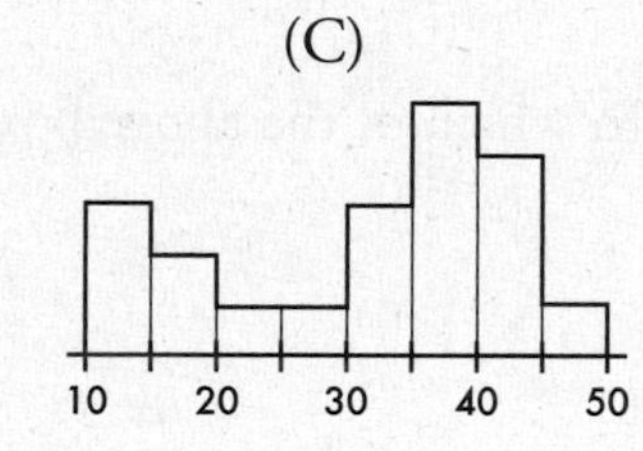

(D)

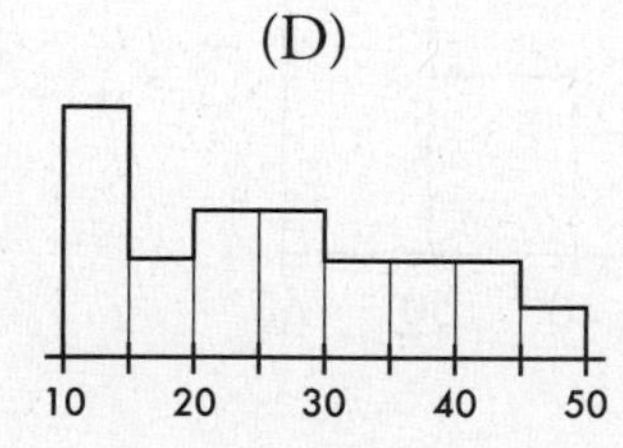

(E)

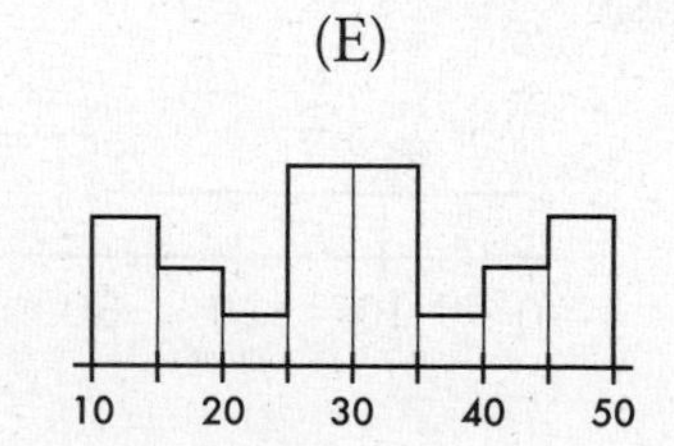

7. To which of the above histograms does the following boxplot correspond?

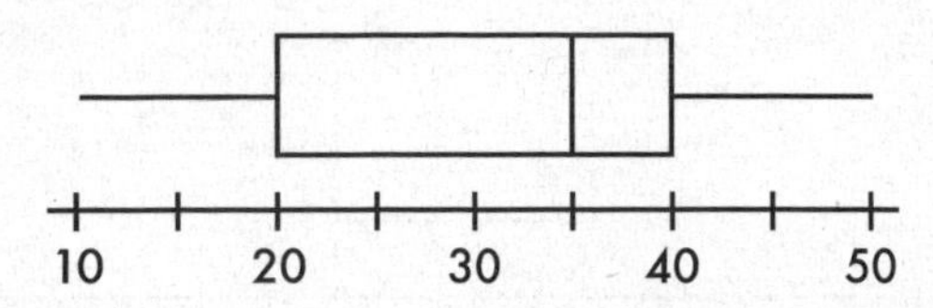

(A) A
(B) B
(C) C
(D) D
(E) E

8. To which of the above histograms does the following boxplot correspond?

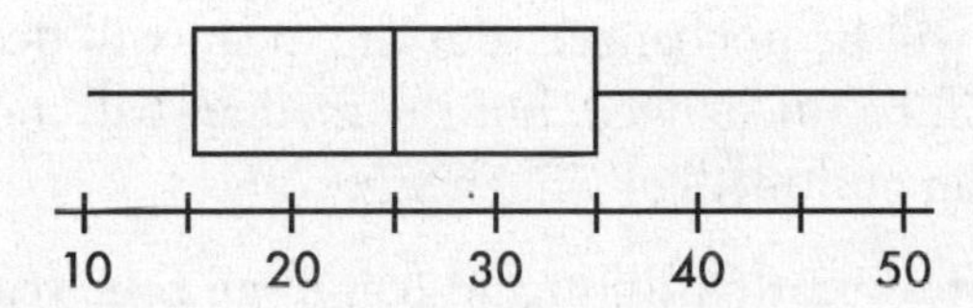

(A) A
(B) B
(C) C
(D) D
(E) E

9. To which of the above histograms does the following boxplot correspond?

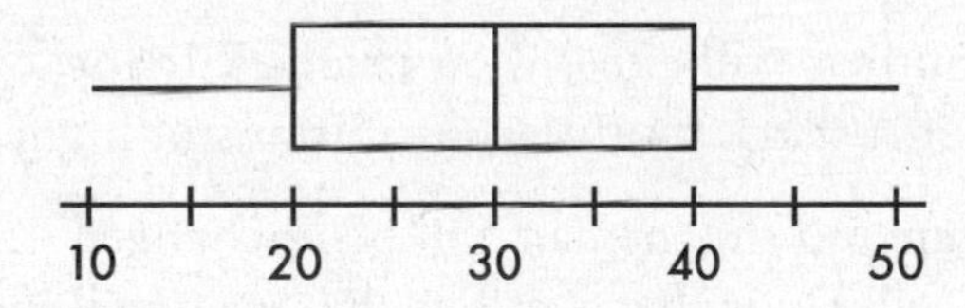

(A) A
(B) B
(C) C
(D) D
(E) E

10. Below is a boxplot of CO_2 levels (in grams per kilometer) for a sampling of 2008 vehicles.

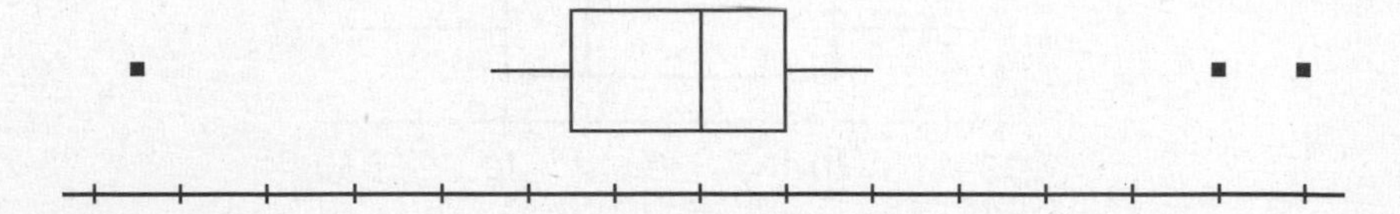

Suppose follow-up testing determines that the low outlier should be 10 grams per kilometer less and the two high outliers should each be 5 grams per kilometer greater. What effect, if any, will these changes have on the mean and median CO_2 levels?

(A) Both the mean and median will be unchanged.
(B) The median will be unchanged, but the mean will increase.
(C) The median will be unchanged, but the mean will decrease.
(D) The mean will be unchanged, but the median will increase.
(E) Both the mean and median will change.

11. Below is a boxplot of yearly tuition and fees of all four year colleges and universities in a Western state. The low outlier is from a private university that gives full scholarships to all accepted students, while the high outlier is from a private college catering to the very rich.

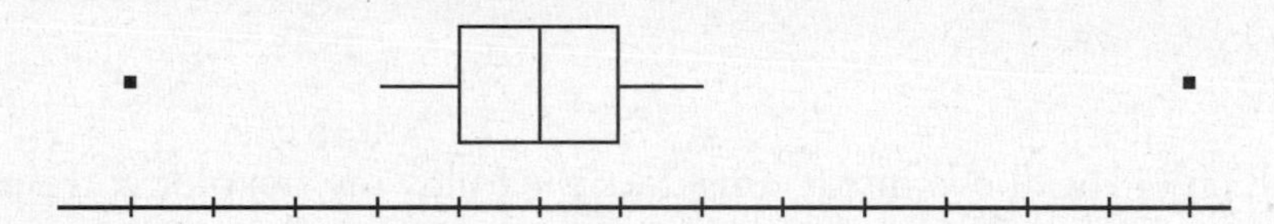

Removing both outliers will effect what changes, if any, on the mean and median costs for this state's four year institutions of higher learning?

(A) Both the mean and the median will be unchanged.
(B) The median will be unchanged, but the mean will increase.
(C) The median will be unchanged, but the mean will decrease.
(D) The mean will be unchanged, but the median will increase.
(E) Both the mean and median will change.

12. Suppose the average score on a national test is 500 with a standard deviation of 100. If each score is increased by 25, what are the new mean and standard deviation?

(A) 500, 100
(B) 500, 125
(C) 525, 100
(D) 525, 105
(E) 525, 125

13. Suppose the average score on a national test is 500 with a standard deviation of 100. If each score is increased by 25%, what are the new mean and standard deviation?

    (A) 500, 100
    (B) 525, 100
    (C) 625, 100
    (D) 625, 105
    (E) 625, 125

14. If quartiles $Q_1 = 20$ and $Q_3 = 30$, which of the following must be true?

    I. The median is 25.
    II. The mean is between 20 and 30.
    III. The standard deviation is at most 10.

    (A) I only
    (B) II only
    (C) III only
    (D) All are true.
    (E) None are true.

15. A 1995 poll by the Program for International Policy asked respondents what percentage of the U.S. budget they thought went to foreign aid. The mean response was 18%, and the median was 15%. (The actual amount is less than 1%.) What do these responses indicate about the likely shape of the distribution of all the responses?

    (A) The distribution is skewed to the left.
    (B) The distribution is skewed to the right.
    (C) The distribution is symmetric around 16.5%.
    (D) The distribution is bell-shaped with a standard deviation of 3%.
    (E) The distribution is uniform between 15% and 18%.

16. Assuming that batting averages have a bell-shaped distribution, arrange in ascending order:

    I. An average with a *z*-score of -1.
    II. An average with a percentile rank of 20%.
    III. An average at the first quartile, Q_1.

    (A) I, II, III
    (B) III, I, II
    (C) II, I, III
    (D) II, III, I
    (E) III, II, I

17. Which of the following are true statements?

    I. If the sample has variance zero, the variance of the population is also zero.
    II. If the population has variance zero, the variance of the sample is also zero.
    III. If the sample has variance zero, the sample mean and the sample median are equal.

    (A) I and II
    (B) I and III
    (C) II and III
    (D) I, II, and III
    (E) None of the above gives the complete set of true responses.

18. When there are multiple gaps and clusters, which of the following is the best choice to give an overall picture of a distribution?

    (A) Mean and standard deviation
    (B) Median and interquartile range
    (C) Boxplot with its five-number summary
    (D) Stemplot or histogram
    (E) None of the above are really helpful in showing gaps and clusters.

19. Suppose the starting salaries of a graduating class are as follows:

Number of Students	Starting Salary ($)
10	15,000
17	20,000
25	25,000
38	30,000
27	35,000
21	40,000
12	45,000

What is the mean starting salary?

(A) $30,000
(B) $30,533
(C) $32,500
(D) $32,533
(E) $35,000

20. When a set of data has suspect outliers, which of the following are preferred measures of central tendency and of variability?

    (A) mean and standard deviation
    (B) mean and variance
    (C) mean and range
    (D) median and range
    (E) median and interquartile range

21. If the standard deviation of a set of observations is 0, you can conclude

    (A) that there is no relationship between the observations.
    (B) that the average value is 0.
    (C) that all observations are the same value.
    (D) that a mistake in arithmetic has been made.
    (E) none of the above.

22. A teacher is teaching two AP Statistics classes. On the final exam, the 20 students in the first class averaged 92 while the 25 students in the second class averaged only 83. If the teacher combines the classes, what will the average final exam score be?

    (A) 87
    (B) 87.5
    (C) 88
    (D) None of the above
    (E) More information is needed to make this calculation.

23. The 60 longest rivers in the world have lengths distributed as follows:

Length (mi):	1000–1499	1500–1999	2000–2499	2500 2999	3000 3499	3500–3999	4000–4499
Number of rivers:	21	22	4	8	2	2	1

(The Nile is the longest with a length of 4145 miles, and the Amazon is the second longest at 3900 miles.)

Which of the following best describes these data?

(A) Skewed distribution, mean greater than median
(B) Skewed distribution, median greater than mean
(C) Symmetric distribution, mean greater than median
(D) Symmetric distribution, median greater than mean
(E) Symmetric distribution with outliers on high end

24. In 1993 the seven states with the fewest business bankruptcies were Vermont (900), Alaska (1000), North Dakota (1100), Wyoming (1300), South Dakota (1400), Hawaii (1500), and Delaware (1600).

    Which of the following are reasonable conclusions about the distribution of bankruptcies throughout the 50 states in 1993?

    I. The total number of bankruptcies in the United States in 1993 was approximately 50(1300) = 65,000.
    II. Because of these low values, the distribution was skewed to the left.
    III. The range of the distribution was approximately 50(1600 – 1000) = 30,000.

    (A) I only
    (B) II only
    (C) III only
    (D) All are reasonable.
    (E) None are reasonable.

25. Suppose 10% of a data set lie between 40 and 60. If 5 is first added to each value in the set and then each result is doubled, which of the following is true?

    (A) 10% of the resulting data will lie between 85 and 125.
    (B) 10% of the resulting data will lie between 90 and 130.
    (C) 15% of the resulting data will lie between 80 and 120.
    (D) 20% of the resulting data will lie between 45 and 65.
    (E) 30% of the resulting data will lie between 85 and 125.

26. A stemplot for the 1988 per capita personal income (in hundreds of dollars) for the 50 states is

Stem	Leaves
11	0 7
12	6 2 7 8 2 7 5 7 8 5 0
13	3 7 7
14	9 7 8 1 6
15	0 9 0 5 2 5 0 4 4
16	4 5 9 4 8 2 8 6
17	7 6 4 6
18	9
19	5 3 0 3
20	7
21	9
22	8

    Which of the following best describes these data?

    (A) Skewed distribution, mean greater than median
    (B) Skewed distribution, median greater than mean
    (C) Symmetric distribution, mean greater than median
    (D) Symmetric distribution, median greater than mean
    (E) Symmetric distribution with outliers on high end

27. Which of the following statements are true?

    I. If the right and left sides of a histogram are mirror images of each other, the distribution is symmetric.
    II. A distribution spread far to the right side is said to be skewed to the right.
    III. If a distribution is skewed to the right, its mean is often greater than its median.

    (A) I only
    (B) I and II
    (C) I and III
    (D) II and III
    (E) None of the above gives the complete set of true responses.

28. Which of the following statements are true?

    I. In a stemplot the number of leaves equals the size of the set of data.
    II. Both the dotplot and the stemplot are useful in identifying outliers.
    III. Histograms do not retain the identity of individual scores; however, dotplots, stemplots, and boxplots all do.

    (A) I and II
    (B) I and III
    (C) II and III
    (D) I, II, and III
    (E) None of the above gives the complete set of true responses.

29. The 70 highest dams in the world have an average height of 206 meters with a standard deviation of 35 meters. The Hoover and Grand Coulee dams have heights of 221 and 168 meters, respectively. The Russian dams, the Nurek and Charvak, have heights with *z*-scores of +2.69 and –1.13, respectively. List the dams in order of ascending size.

    (A) Charvak, Grand Coulee, Hoover, Nurek
    (B) Charvak, Grand Coulee, Nurek, Hoover
    (C) Grand Coulee, Charvak, Hoover, Nurek
    (D) Grand Coulee, Charvak, Nurek, Hoover
    (E) Grand Coulee, Hoover, Charvak, Nurek

30. The first 115 Kentucky Derby winners by color of horse were as follows: roan, 1; gray, 4; chestnut, 36; bay, 53; dark bay, 17; and black, 4. (You should "bet on the bay!") Which of the following visual displays is most appropriate?

    (A) Bar chart
    (B) Histogram
    (C) Stemplot
    (D) Boxplot
    (E) Time plot

For Questions 31 and 32 consider the following: The graph below shows cumulative proportions plotted against grade point averages for a large public high school.

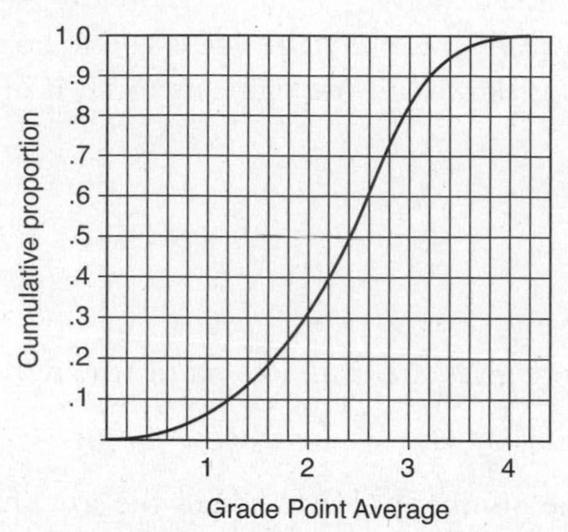

31. What is the median grade point average?

(A) 0.8
(B) 2.0
(C) 2.4
(D) 2.5
(E) 2.6

32. What is the interquartile range?

(A) 1.0
(B) 1.8
(C) 2.4
(D) 2.8
(E) 4.0

33. The following dotplot shows the speeds (in mph) of 100 fastballs thrown by a major league pitcher.

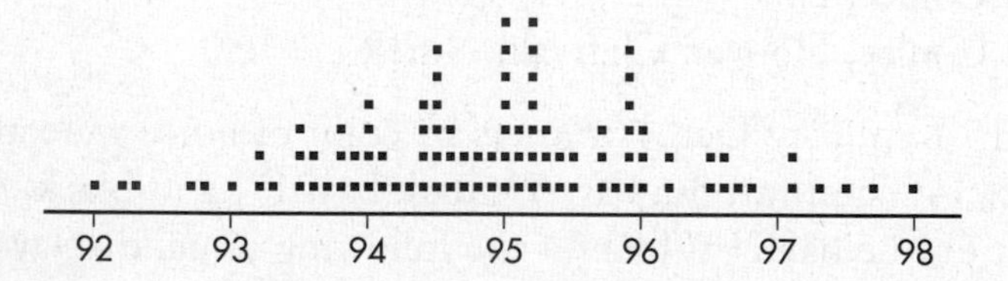

Which of the following is the best estimate of the standard deviation of these speeds?

(A) 0.5 mph
(B) 1.1 mph
(C) 1.6 mph
(D) 2.2 mph
(E) 6.0 mph

Free-Reponse Questions

Directions: You must show all work and indicate the methods you use. You will be graded on the correctness of your methods and on the accuracy of your final answers.

Four Open-Ended Questions

1. Victims spend from 5 to 5840 hours repairing the damage caused by identity theft with a mean of 330 hours and a standard deviation of 245 hours.

   (a) What would be the mean, range, standard deviation, and variance for hours spent repairing the damage caused by identity theft if each of the victims spent an additional 10 hours?

   (b) What would be the mean, range, standard deviation, and variance for hours spent repairing the damage caused by identity theft if each of the victims' hours spent increased by 10%?

2. In a study of all school districts in a state, the median 4-year graduation rate was 78.0% with Q_1 = 60.4% and Q_3 = 82.6%. The only rates below Q_1 or above Q_3 were 26.4%, 32.2%, 49.0%, 57.9%, 88.3%, and 98.1%.

   (a) Draw a boxplot.

   (b) Describe the distribution.

   (c) Is the mean 4-year graduation rate probably close to, below, or above 78.0%? Explain.

   (d) Would a stemplot give more, less, or basically the same information?

3. The Children's Health Insurance Program (CHIP) provides health benefits to children from families whose incomes exceed the eligibility for Medicaid. Each state sets its own eligibility criteria. The following boxplot shows recent yearly expenditures on this program by state.

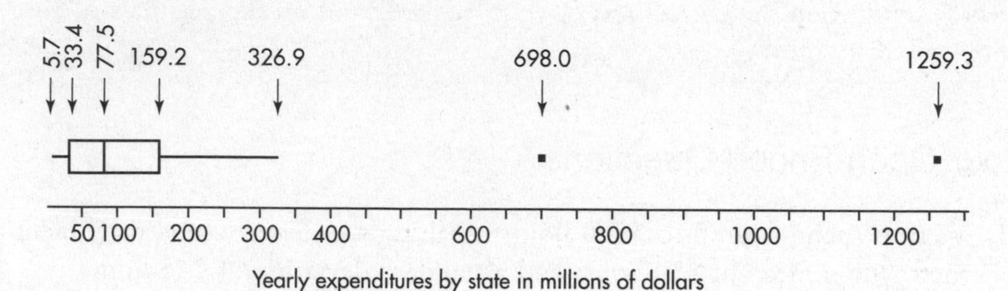

(a) What are the median and interquartile range of the distribution of yearly state expenditures in the CHIP program?
(b) Suppose the federal government takes over three million dollars of administrative costs from the state CHIP expenditures. What are the median and interquartile range of the new reduced expenditure distribution?
(c) Suppose instead the federal government picks up the tab for half of all state CHIP expenditures. What are the median and interquartile range of this new reduced expenditure distribution?
(d) Based on the above boxplot, which of the following is the most reasonable value for the mean state expenditure (in millions of dollars): 78, 135, 325, 630, or 750? Explain.

4. Suppose a distribution has mean 300 and standard deviation 25. If the *z*-score of Q_1 is –0.7 and the *z*-score of Q_3 is 0.7, what values would be considered to be outliers?

Answer Key

1. **D**	8. **D**	15. **B**	22. **A**	29. **A**
2. **E**	9. **E**	16. **A**	23. **A**	30. **A**
3. **E**	10. **A**	17. **C**	24. **E**	31. **C**
4. **A**	11. **C**	18. **D**	25. **B**	32. **A**
5. **B**	12. **C**	19. **B**	26. **A**	33. **B**
6. **C**	13. **E**	20. **E**	27. **E**	
7. **C**	14. **E**	21. **C**	28. **A**	

Answers Explained

Multiple-Choice

1. **(D)** The distribution is clearly skewed right, so the mean is greater than the median, and the ratio is greater than one.

2. **(E)** All elements of the sample are taken from the population, and so the smallest value in the sample cannot be less than the smallest value in the population; similarly, the largest value in the sample cannot be greater than the largest value in the population. The interquartile range is the full distance between the first quartile and the third quartile. Outliers are extreme values, and while they may affect the range, they do not affect the interquartile range when the lower and upper quarters have been removed before calculation.

3. **(E)** Outliers are any values below $Q_1 - 1.5(IQR) = 5.5$ or above $Q_3 + 1.5(IQR) = 57.5$.

4. **(A)** The value 50 seems to split the area under the histogram in two, so the median is about 50. Furthermore, the histogram is skewed to the left with a tail from 0 to 30.

5. **(B)** Looking at areas under the curve, Q_1 appears to be around 20, the median is around 30, and Q_3 is about 40.

6. **(C)** Looking at areas under the curve, Q_1 appears to be around 10, the median is around 30, and Q_3 is about 50.

7. **(C)** The boxplot indicates that 25% of the data lie in each of the intervals 10–20, 20–35, 35–40, and 40–50. Counting boxes, only histogram C has this distribution.

8. **(D)** The boxplot indicates that 25% of the data lie in each of the intervals 10–15, 15–25, 25–35, and 35–50. Counting boxes, only histogram D has this distribution.

9. **(E)** The boxplot indicates that 25% of the data lie in each of the intervals 10–20, 20–30, 30–40, and 40–50. Counting boxes, only histogram E has this distribution.

10. **(A)** Subtracting 10 from one value and adding 5 to two values leaves the sum of the values unchanged, so the mean will be unchanged. Exactly what values the outliers take will not change what value is in the middle, so the median will be unchanged.

11. **(C)** The high outlier is further from the mean than is the low outlier, so removing both will decrease the mean. However, removing the lowest and highest values will not change what value is in the middle, so the median will be unchanged.

12. **(C)** Adding the same constant to every value increases the mean by that same constant; however, the distances between the increased values and the increased mean stay the same, and so the standard deviation is unchanged. Graphically, you should picture the whole distribution as moving over by a constant; the mean moves, but the standard deviation (which measures spread) doesn't change.

13. **(E)** Multiplying every value by the same constant multiplies both the mean and the standard deviation by that constant. Graphically, increasing each value by 25% (multiplying by 1.25) both moves and spreads out the distribution.

14. **(E)** The median is somewhere between 20 and 30, but not necessarily at 25. Even a single very large score can result in a mean over 30 and a standard deviation over 10.

15. **(B)** The median is less than the mean, and so the responses are probably skewed to the right; there are a few high guesses, with most of the responses on the lower end of the scale.

16. **(A)** Given that the empirical rule applies, a *z*-score of -1 has a percentile rank of about 16%. The first quartile Q_1 has a percentile rank of 25%.

17. **(C)** If the variance of a set is zero, all the values in the set are equal. If all the values of the population are equal, the same holds true for any subset; however, if all the values of a subset are the same, this may not be true of the whole population. If all the values in a set are equal, the mean and the median both equal this common value and so equal each other.

18. **(D)** Stemplots and histograms can show gaps and clusters that are hidden when one simply looks at calculations such as mean, median, standard deviation, quartiles, and extremes.

19. **(B)** There are a total of $10 + 17 + 25 + 38 + 27 + 21 + 12 = 150$ students. Their total salary is $10(15{,}000) + 17(20{,}000) + 25(25{,}000) + 38(30{,}000) + 27(35{,}000) + 21(40{,}000) + 12(45{,}000) = \$4{,}580{,}000$. The mean is $\frac{4{,}580{,}000}{150} = \$30{,}533$.

20. **(E)** The mean, standard deviation, variance, and range are all affected by outliers; the median and interquartile range are not.

21. **(C)** Because of the squaring operation in the definition, the standard deviation (and also the variance) can be zero only if all the values in the set are equal.

22. **(A)** The sum of the scores in one class is $20 \times 92 = 1840$, while the sum in the other is $25 \times 83 = 2075$. The total sum is $1840 + 2075 = 3915$. There are $20 + 25 = 45$ students, and so the average score is $\frac{3915}{45} = 87$.

23. **(A)** A distribution spread thinly on the high end is a skewed distribution with the mean greater than the median.

24. **(E)** None are reasonable because we are not looking at a random sample of states but rather only at the seven lowest values, that is, at the very low end of the tail of the whole distribution. (The distribution was actually skewed to the right, with California having 159,700 bankruptcies, followed by New York with 51,300, and all other states below 50,000.)

25. **(B)** Increasing every value by 5 gives 10% between 45 and 65, and then doubling gives 10% between 90 and 130.

26. **(A)** A distribution spread thinly on the high end is a skewed distribution with the mean greater than the median.

27. **(E)** All three statements are true.

28. **(A)** Dotplots and stemplots retain the identity of individual scores; however, histograms and boxplots do not.

29. **(A)** $206 + 2.69(35) = 300$; $206 - 1.13(35) = 166$.

30. **(A)** Bar charts are used for categorical variables.

31. **(C)** The median corresponds to the 0.5 cumulative proportion.

32. **(A)** The 0.25 and 0.75 cumulative proportions correspond to $Q_1 = 1.8$ and $Q_3 = 2.8$, respectively, and so the interquartile range is $2.8 - 1.8 = 1.0$.

33. **(B)** With bell-shaped data the empirical rule applies, giving that the spread from 92 to 98 is roughly 6 standard deviations, and so one SD is about 1.

Free-Response

1. (a) Adding 10 to each value increases the mean by 10, but leaves measures of variability unchanged, so the new mean is 340 hours while the range stays at 5835 hours, the standard deviation remains at 245 hours, and the variance remains at $245^2 = 60{,}025$ hr^2.

(b) Increasing each value by 10% (multiplying by 1.10) will increase the mean to 1.1(330) = 363 hours, the range to 1.1(5835) = 6418.5 hours, the standard deviation to 1.1(245) = 269.5 hours, and the variance to $(269.5)^2$ = 72,630.25 hr^2. (Note that the variance increases by a multiple of $(1.1)^2$ not by a multiple of 1.1.)

2. (a) Check for outliers: IQR = 82.6 – 60.4 = 22.2. Q_1 – 1.5(IQR) = 27.1 while Q_3 + 1.5(IQR) = 115.9, so the only outlier is 26.4.

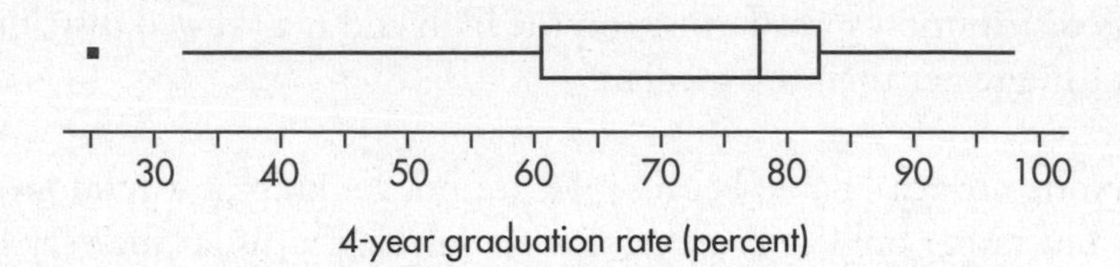

(b) A complete answer considers shape, center, and spread.

Shape: appears skewed left with an outlier at 26.4
Center: median is 78.0
Spread: from 26.4 to 98.1

(c) When the distribution is skewed left, the mean is usually less than the median.

(d) A stemplot would show more information because it shows all the original data, not just the few values given above; a stemplot can show clusters and gaps which are hidden by a boxplot.

3. (a) The median is 77.5 (millions of dollars), and the IQR = $Q_3 - Q_1$ = 159.2 – 33.4 = 125.8 (millions of dollars).

(b) Reducing every value by 3 will reduce the median by 3 but leave measures of variability unchanged, so the new mean is 77.5 – 3 = 74.5 (millions of dollars), and the IQR will still be 125.8 (millions of dollars).

(c) Reducing every value by 50% reduces the median to (0.5)(77.5) = 38.75 (millions of dollars) and reduces the IQR to (0.5)(125.8) = 62.9 (millions of dollars).

(d) The boxplot indicates that the distribution is skewed right, so the mean will be greater than the median. It is unlikely that the two outliers will pull the mean out as far as 325, so the most reasonable value for the mean is 135 (millions of dollars).

4. *Z*-scores give the number of standard deviations from the mean, so Q_1 = 300 – 0.7(25) = 282.5 and Q_3 = 300 + 0.7(25) = 317.5.
The interquartile range is IQR = 317.5 – 282.5 = 35, and 1.5(IQR) = 1.5(35) = 52.5.
The standard definition of outliers encompasses all values less than Q_1 – 52.5 = 230 and all values greater than Q_3 + 52.5 = 370.

Comparing Distributions

TOPIC 3

- Dotplots
- Double Bar Charts
- Back-to-back Stemplots
- Parallel Boxplots
- Cumulative Frequency Plots

Many real-life applications of statistics involve comparisons of *two* populations. Such comparisons can involve modifications of graphical displays such as dotplots, bar charts, stemplots, boxplots, and cumulative frequency plots to portray both sets simultaneously.

TIP

When asked for a comparison, simply making two separate lists is not enough and will be penalized.

DOTPLOTS

EXAMPLE 3.1

The caloric intakes of 25 people on each of two weight loss programs are recorded as follows:

Program A: 1000, 1000, 1100, 1100, 1100, 1200, 1200, 1200, 1200, 1300, 1300, 1300, 1300, 1300, 1400, 1400, 1400, 1400, 1400, 1500, 1500, 1600, 1600, 1700, 1900

Program B: 1000, 1100, 1100, 1200, 1200, 1200, 1300, 1300, 1300, 1400, 1400, 1400, 1400, 1500, 1500, 1500, 1500, 1500, 1600, 1600, 1600, 1700, 1700, 1800, 1800

These data can be compared with dotplots, one above the other, using the same horizontal scale.

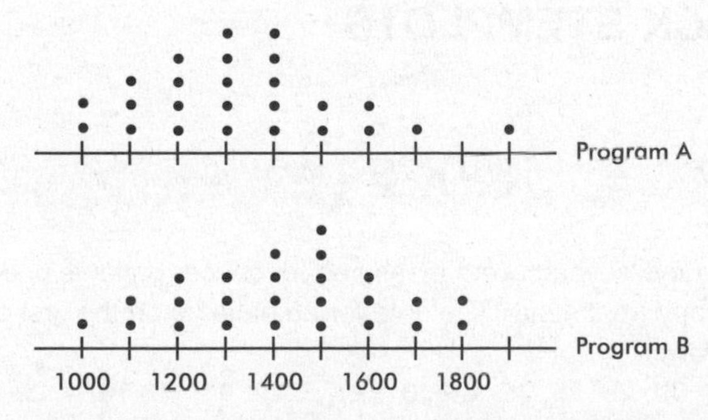

Program A appears to be associated with a lower average caloric intake than Program B. Comparing shape, center, and spread, we have:

Shape: We see that both sets of data are roughly bell-shaped (the empirical rule applies), and Program A has an outlier at 1900 calories (while Program B has no outliers).

(continued)

Center: Visually, or by counting dots, the centers of the two distributions are 1300 and 1400 calories, for Programs A and B respectively. (A calculator gives means of $\bar{x}_A = 1336$ and $\bar{x}_B = 1424$.) By any method, the center for Program B is higher.

Spread: The spreads are approximately the same, 1000 to 1900 calories for Program A and 1000 to 1800 calories for Program B. (A calculator gives standard deviations of $s_A = 218$ and $s_B = 218$.)

DOUBLE BAR CHARTS

EXAMPLE 3.2

A study tabulated the percentages of young adults who recognized various photographs as follows: Joe Stalin (10%), Joe Camel (95%), Senator Simpson of Wyoming (5%), Bart Simpson (80%), Al Gore (30%), Al Bundy (60%), Mickey Mantle (20%), Mickey Mouse (100%), Charlie Chaplin (25%), Charlie the Tuna (90%). These data can be illustrated with a bar chart appropriately displayed in pairs of bars.

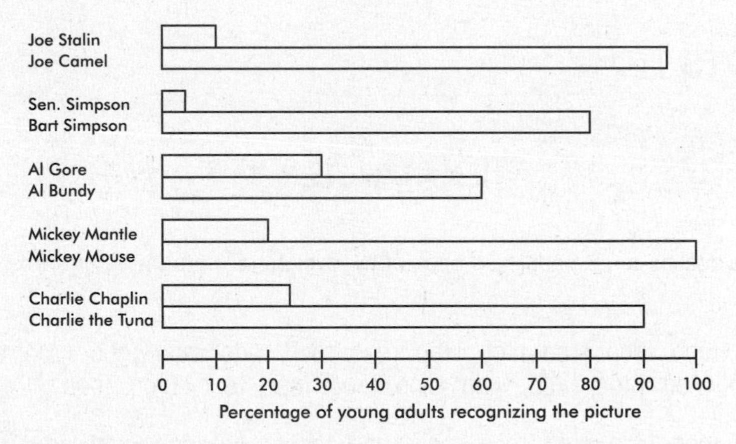

The pairs of bar graphs visually indicate the reason for concern felt by some educators.

BACK-TO-BACK STEMPLOTS

EXAMPLE 3.3

In a 40-year study, survival years were measured for cancer patients undergoing one of two different chemotherapy treatments. The data for 25 patients on the first drug and 30 on the second were as follows:

Drug A: 5, 10, 17, 39, 29, 25, 20, 4, 8, 31, 21, 3, 12, 11, 19, 10, 4, 22, 17, 18, 13, 28, 11, 14, 21

Drug B: 19, 12, 20, 28, 22, 35, 1, 21, 21, 26, 18, 28, 29, 20, 15, 32, 31, 24, 22, 26, 18, 20, 22, 35, 30, 18, 25, 24, 19, 21

(continued)

In drawing a back-to-back stemplot of the above data, we place a vertical line on each side of the column of stems and then arrange one set of leaves extending out to the right while the other extends out to the left.

Drug B		Drug A
1	0	3 4 4 5 8
9 9 8 8 8 5 2	1	0 0 1 1 2 3 4 7 7 8 9
9 8 8 6 6 5 4 4 2 2 2 1 1 1 0 0 0	2	0 1 1 2 5 8 9
5 5 2 1 0	3	1 9

(Stem unit = 10)

Note that even though drug A showed the longest-surviving patient (39 years) and drug B showed the shortest-surviving patient (1 year), the back-to-back stemplot indicates that the bulk of patients on drug B survived longer than the bulk of patients on drug A.

Comparing shape, center, and spread, we have:

Shape: Both distributions are roughly bell-shaped (the empirical rule applies). The drug A distribution appears to have a high outlier at 39, while the drug B distribution appears to have a low outlier at 1.

Center: Visually, or by counting values, the centers of the two distributions are 17 and 22, respectively. (A calculator gives means of $\bar{x}_A = 16.48$ and $\bar{x}_B = 22.73$.) By either method, the drug B distribution has a greater center.

Spread: The spreads are 3 to 39 survival years for drug A and 1 to 35 survival years for drug B. (A calculator gives standard deviations of $s_A = 9.25$ and $s_B = 6.97$.) By either method, the drug A distribution has a greater spread than the drug B distribution.

PARALLEL BOXPLOTS

EXAMPLE 3.4

Mail-order labs and 1-hour minilabs were compared with regard to price for developing and printing one 24-exposure roll of 35-millimeter color-print film. Prices included shipping and handling charges where applicable. Following is a computer output describing the results:

For mail-order labs:

```
Mean = 5.37    Standard deviation = 1.92    Min = 3.51
Max = 8.00    N = 18    Median = 4.77
Quartiles = 3.92, 6.45
```

For 1-hour minilabs:

```
Mean = 10.11    Standard deviation = 1.32    Min = 8.58
Max = 11.95    N = 15    Median = 10.08
Quartiles = 8.97, 11.51
```

(continued)

In drawing parallel boxplots (also called side-by-side boxplots) of the above data, we place both on the same diagram:

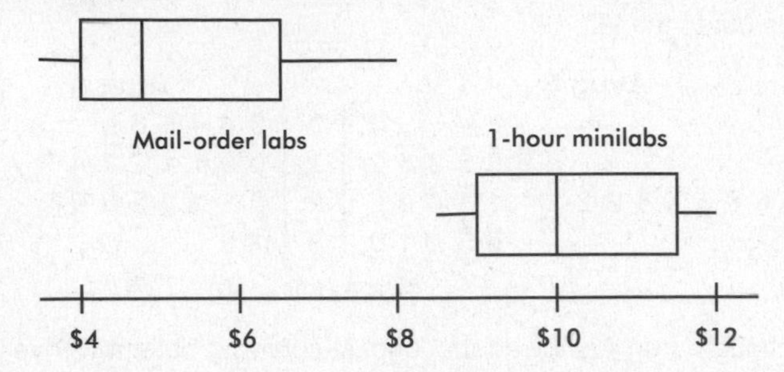

Boxplots show the minimum, maximum, median, and quartile values. The distribution of mail-order lab prices is lower and more spread out than that of prices of 1-hour minilabs. Both are slightly skewed toward the upper end (the skewness can also be noted from the computer output showing the mean to be greater than the median in both cases.)

Parallel boxplots are useful in presenting a picture of the comparison of several distributions.

EXAMPLE 3.5

Following are parallel boxplots showing the daily price fluctuations of a certain common stock over the course of 5 years. What trends do the boxplots show?

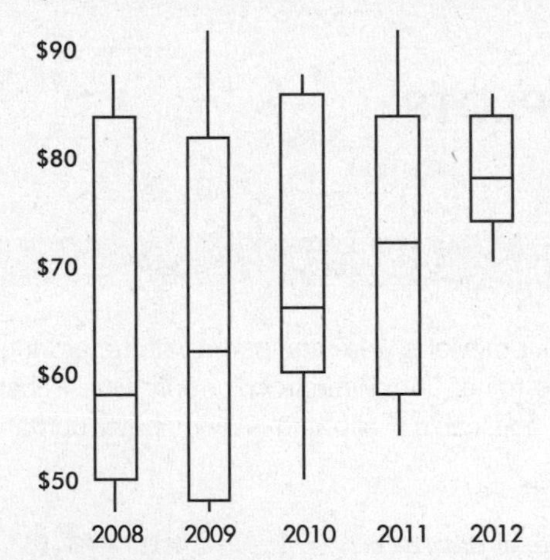

The parallel boxplots show that from year to year the median daily stock price has steadily risen 20 points from about $58 to about $78, the third quartile value has been roughly stable at about $84, the yearly low has never decreased from that of the previous year, and the interquartile range has never increased from one year to the next.

CUMULATIVE FREQUENCY PLOTS

EXAMPLE 3.6

The graph below compares cumulative frequency plotted against age for the U.S. population in 1860 and in 1980.

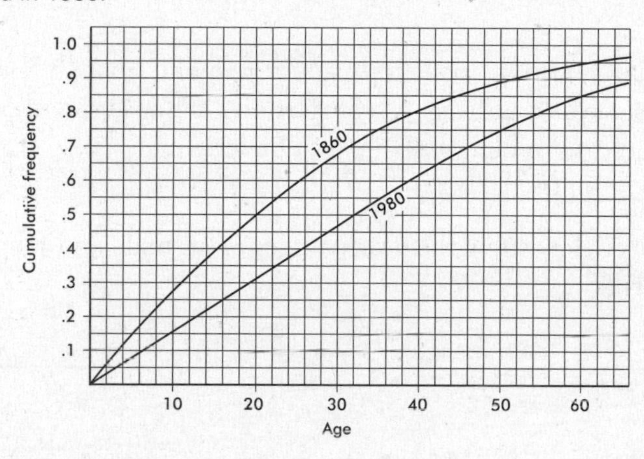

How do the medians and interquartile ranges compare?

Answer: Looking across from .5 on the vertical axis, we see that in 1860 half the population was under the age of 20, while in 1980 all the way up to age 32 must be included to encompass half the population. Looking across from .25 and .75 on the vertical axis, we see that for 1860, $Q_1 = 9$ and $Q_3 = 35$ and so the interquartile range is $35 - 9 = 26$ years, while for 1980, $Q_1 = 16$ and $Q_3 = 50$ and so the interquartile range is $50 - 16 = 34$ years.

Summary

To visually compare two or more distributions use:

- Dotplots, either one above the other or side-by-side
- Double bar charts
- Histograms, either one above the other or side-by-side
- Back-to-back stemplots
- Parallel boxplots
- Cumulative frequency plots on the same grid
- For all the above, make note of any similarities and differences in shape, center, and spread.

Questions on Topic Three: Comparing Distributions

Multiple-Choice Questions

Directions: The questions or incomplete statements that follow are each followed by five suggested answers or completions. Choose the response that best answers the question or completes the statement.

1. The dotplots below show the yearly wages of all male and female executives at a large firm.

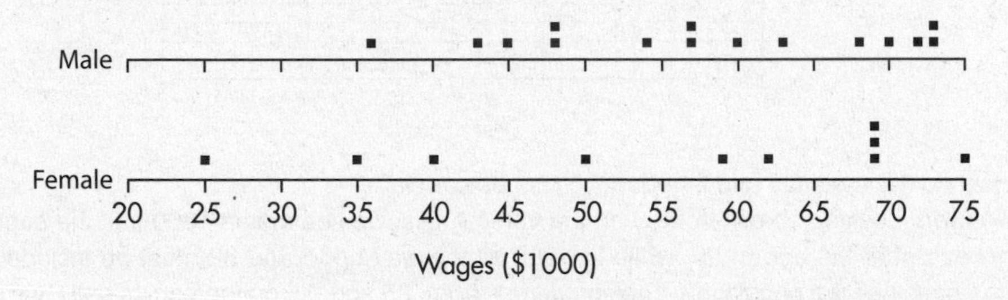

Which of the following conclusions cannot be drawn from the plots?

(A) A greater proportion of male employees than female employees are executives at this firm.
(B) No executive receives a salary less than $25,000.
(C) The median salary paid to male executives is less than the median salary paid to female executives.
(D) The range of salaries paid to male executives is less than the range of salaries paid to female executives.
(E) More male than female executives have salaries over $70,000.

For Questions 2 and 3 consider the following two histograms:

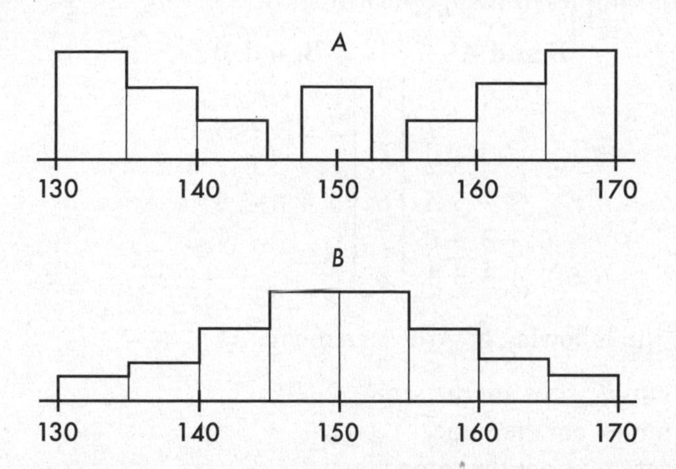

2. Which of the following statements are true?

   I. Both sets have the same mean.
   II. Both sets have the same range.
   III. Both sets have the same variance.

   (A) I only
   (B) I and II
   (C) I and III
   (D) I, II, and III
   (E) None of the above gives the complete set of true responses.

3. Which of the following statements are true?

   I. The empirical rule applies only to set *A*.
   II. You can be sure that the standard deviation of set *A* is greater than 5.
   III. You can be sure that the standard deviation of set *B* is greater than 5.

   (A) I only
   (B) II only
   (C) I and II
   (D) I and III
   (E) II and III

4. Consider the following back-to-back stemplots comparing car battery lives (in months) of samples of two popular brands.

Brand A		Brand B
	3	7
7	4	2 3 4 8 8
3 2	5	1 4 5 6 7 8 9 9
8 7 5 4	6	3 4 6 6 8
9 6 5 3 3 0	7	6
6 5 4 3 3 3 1	8	

Which of the following are true statements?

I. The sample sizes are the same.
II. The ranges are the same.
III. The variances are the same.
IV. The means are the same.
V. The medians are the same.

(A) I and II
(B) I and IV
(C) II and V
(D) III and V
(E) I, II, and III

5. The following boxplots were constructed from SAT math scores of boys and girls at a high school:

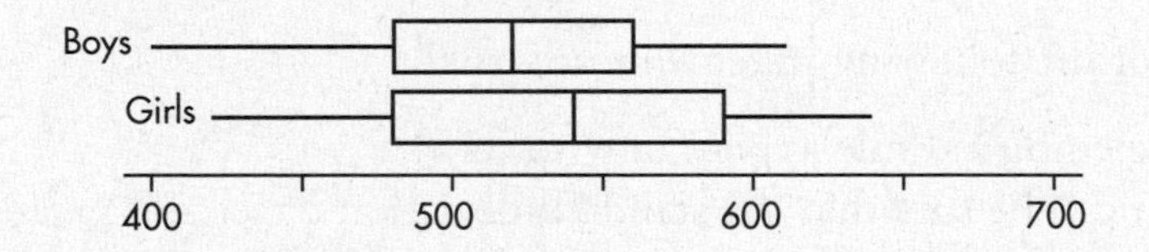

Which of the following is a possible boxplot for the combined scores of all the students?

(A)
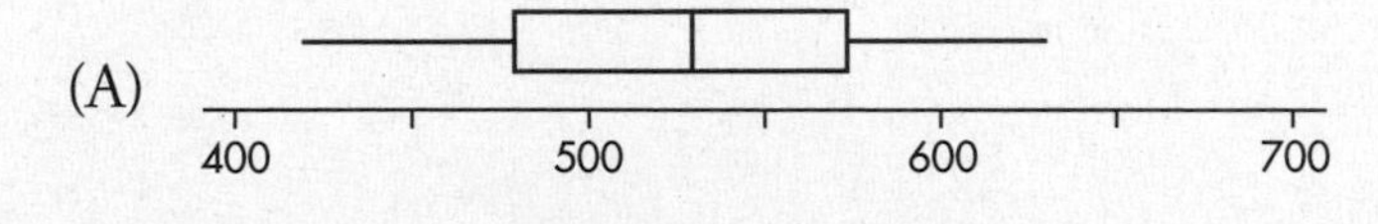

(B)
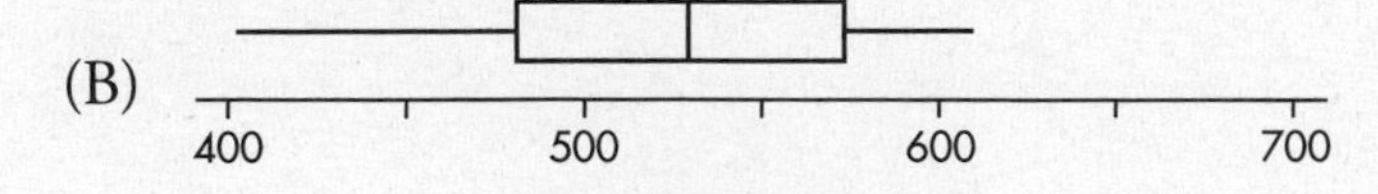

(C)
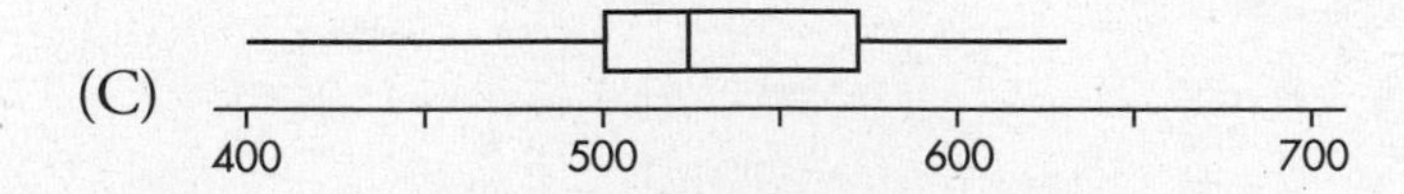

(D)
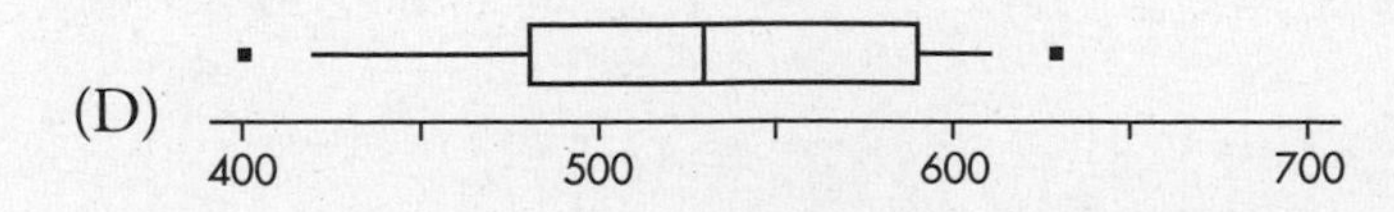

(E)
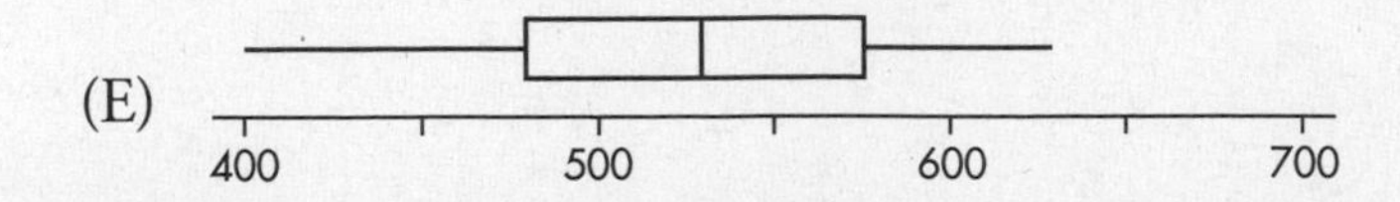

Questions 6–8 refer to the following population pyramids (source: U.S. Census Bureau).

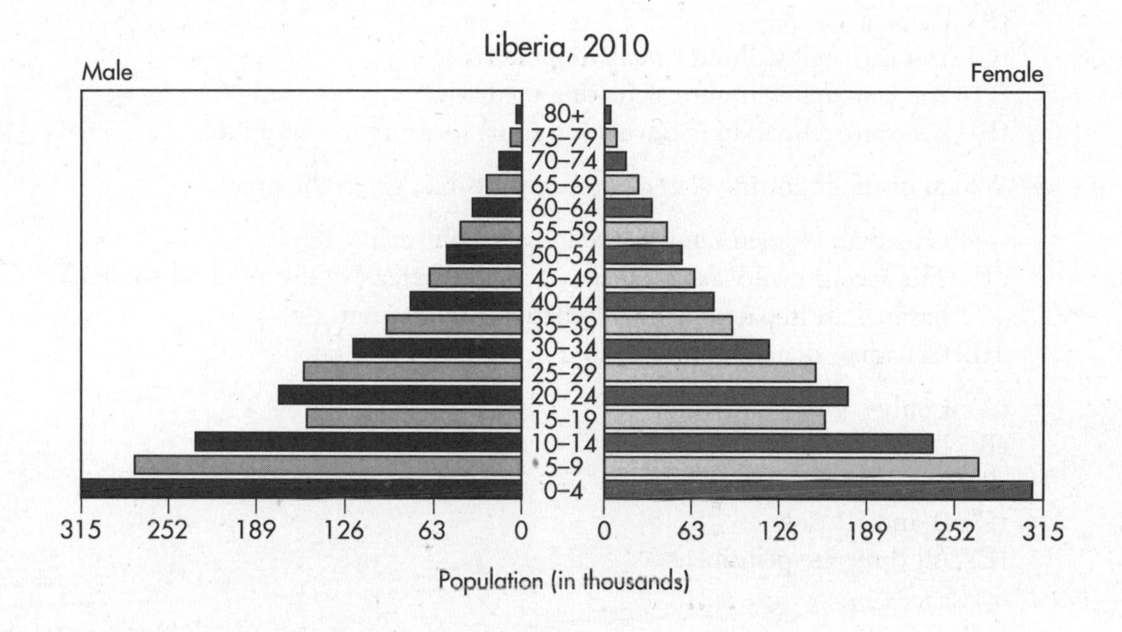

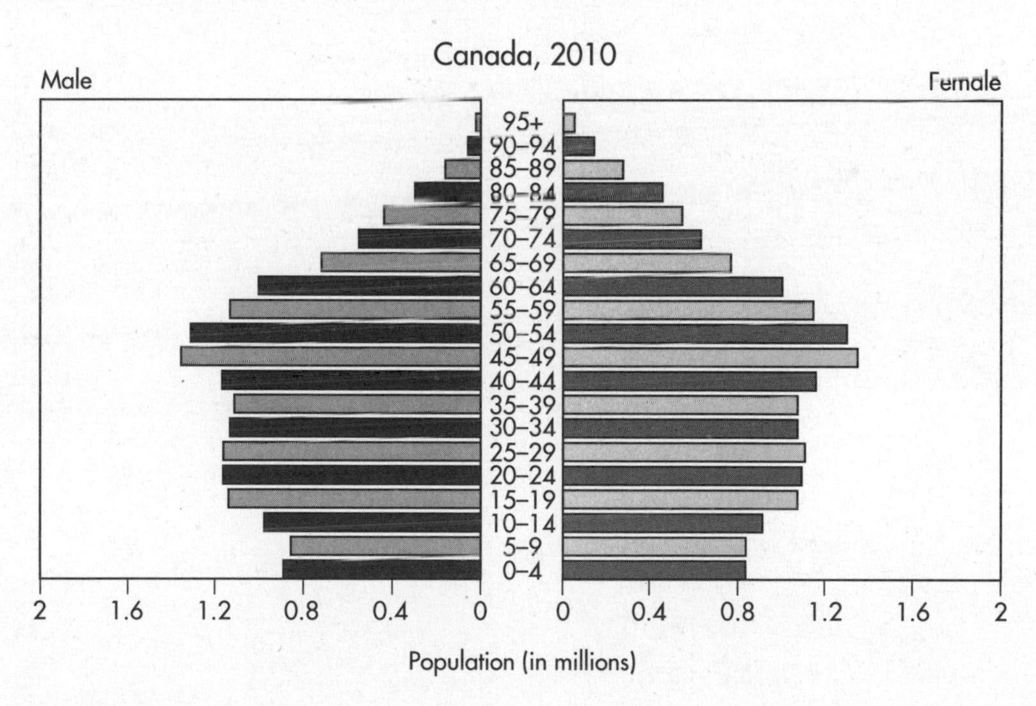

6. What is the approximate median age of the Liberian population?

   (A) 0–4
   (B) 15–19
   (C) 30–34
   (D) 40–44
   (E) There is insufficient information to approximate the median.

7. Which country has more children younger than 10 years of age?

   (A) Liberia
   (B) Canada
   (C) You can't tell without calculating means.
   (D) You can't tell without calculating medians.
   (E) You can't tell without calculating some measure of variability.

8. Which of the following statements are plausible, given the graphs?

   I. Canadian women tend to live longer than men.
   II. The recent civil war in Liberia, with the extensive use of child soldiers, has had an impact on the population age distribution.
   III. Canadian demographics show a decreasing birth rate.

   (A) I only
   (B) I and II only
   (C) I and III only
   (D) II and III only
   (E) All three are plausible.

Free-Response Questions

Directions: You must show all work and indicate the methods you use. You will be graded on the correctness of your methods and on the accuracy of your final answers.

Four Open-Ended Questions

1. Census data was used to compare the marital status versus age of residents of a small town in rural upstate New York. Resulting histograms are given below.

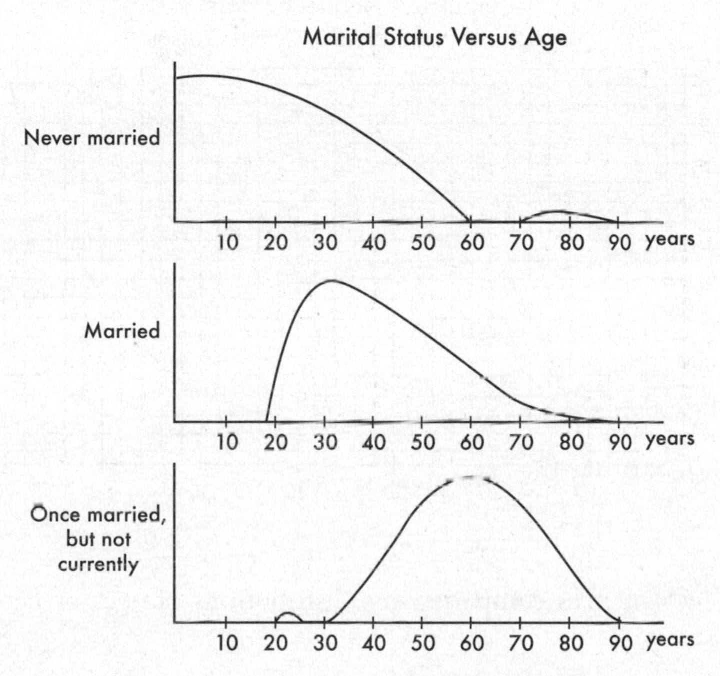

Write a few sentences comparing the distributions of ages of people in this town who have never been married, who are currently married, and who are no longer married.

2. During one season, 20 top NBA rebounding leaders averaged the following numbers of rebounds per game: 9.4, 9.6, 9.7, 10.3, 10.4, 10.6, 10.6, 10.6, 10.8, 10.8, 10.9, 10.9, 10.9, 11.0, 11.1, 11.4, 12.6, 13.4, 14.1, and 16.8, while 20 top assist leaders averaged the following number of assists per game: 6.1, 6.2, 6.4, 6.9, 7.1, 7.2, 7.4, 7.5, 7.6, 7.7, 7.7, 7.9, 8.2, 8.3, 8.7, 8.8, 9.3, 9.4, 10.2, and 12.3. Is it easier for the best NBA players to get a rebound or an assist? Answer the question and compare these data using back-to-back stemplots.

3. In 1980 automobile registrations by states ranged from a low of 262,000 (Alaska) to a high of 16,873,000 (California) with a median of 2,329,000 (Iowa), a 25th percentile of 834,000, and a 75th percentile of 3,749,000. In 1990 the registrations ranged from 462,000 (Vermont) to 21,926,000 (California) with a median of 2,649,000 (Oklahoma), a 25th percentile of 1,054,000, and a 75th percentile of 4,444,000. Draw parallel boxplots and compare the distributions. Comment on whether or not any of the six named states are outliers in their respective distributions.

4. Cumulative frequency graphs of the ages of people on three different Caribbean cruises (A, B, and C) are given below:

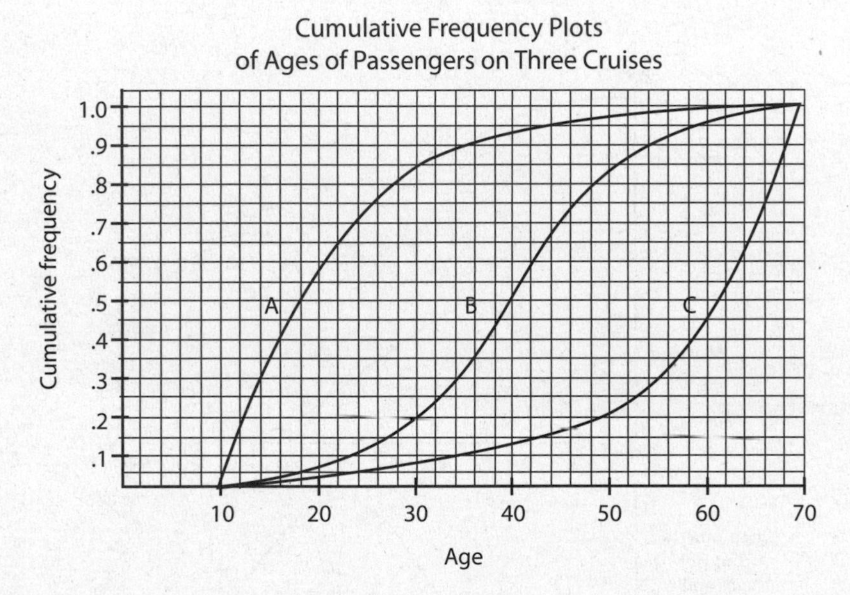

Write a few sentences comparing the distributions of ages of people on the three cruises.

Answer Key

1. **A** 3. **E** 5. **E** 7. **B**
2. **B** 4. **A** 6. **B** 8. **E**

Answers Explained

Multiple-Choice

1. **(A)** The numbers of male and female employees are not given so proportions who are executives cannot be determined.

2. **(B)** Both sets are symmetric about 150 and so have the same mean. Both sets have the range 170–130 = 40. Set *A* is much more spread out than set *B*, and so set *A* has the greater variance.

3. **(E)** The empirical rule applies to bell-shaped data like those found in set *B*, not in set *A*. For bell-shaped data, 95% of the values fall within two standard deviations of the mean, and 99.7% within three. However, in the histogram for set *B* one sees that 95% of the data are not between 140 and 160 and 99.7% are not between 135 and 165. Thus the standard deviation for set *B* must be greater than 5. The standard deviation for set *A* is even larger, and so it too must be greater than 5.

4. **(A)** Both sets have 20 elements. The ranges, 76–37 = 39 and 86–47 = 39, are equal. Brand A clearly has the larger mean and median, and with its skewness it also has the larger variance.

5. **(E)** The minimum of the combined set of scores must be the min of the boys since it is lower; the maximum of the combined set of scores must be the max of the girls since it is higher; the first quartile must be the same as the identical first quartiles of the two original distributions. There are no outliers (scores more than 1.5(IQR) from the first and third quartiles).

6. **(B)** Roughly 50% of total bar length is above and below the 15–19 interval.

7. **(B)** There are about 1.2 million younger than the age of 10 in Liberia (boys and girls) and roughly 3.5 million in Canada.

8. **(E)** In the Canadian graph, all higher age groups show greater numbers of women than men. In the Liberian graph, the smaller 15–19 age group shows a definite break with the overall pattern (a great number of child soldiers died in the fighting). In the Canadian graph, the narrowing base indicates a decreasing birth rate.

Free-Response

1. A complete answer compares shape, center, and spread.

Shape: The "never married" and "married" distributions are skewed right (toward the higher ages), while the "once married" distribution appears more bell-shaped and symmetric. The "never married" has a gap between 60 and 70 with a cluster between 70 and 90, while the "once married" has a small gap between 25 and 30 with perhaps outliers between 20 and 25.

Center: Considering the center to be a value separating the area under the histogram roughly in half, the center of the "never married" distribution is the least (at approximately 20), and the center of the "once married" is the greatest (at approximately 60).

Spread: The spread of the "married" distribution is the least (with a range of approximately 85 – 18 = 67), and the spread of the "never married" distribution is the greatest (with a range of approximately 90 – 0 = 90).

2.

Assists per game		Rebounds per game
9 4 2 1	6	
9 7 7 6 5 3 2 1	7	
8 7 3 2	8	
4 3	9	4 6 7
2	10	3 4 6 6 6 8 8 9 9 9
	11	0 1 4
3	12	6
	13	4
	14	1
	15	
	16	8

(Stem unit is 1.0, leaf unit is 0.1)

Clearly the number of rebounds that the best rebounders get per game is greater than the number of assists that the best assist leaders get per game. A complete answer compares shape, center, and spread.

Shape: Both distributions are skewed right (toward higher values). The lower 16 values of each distribution look symmetric and bell-shaped. Each distribution has a clear outlier on the upper end (summary statistics were not asked for, but would actually have shown four upper-end outliers in the "rebound" distribution).

Center: The center of the "assists" distribution (at a little under 8) is less than the spread of the "rebounds" distribution (at a little under 11).

Spread: The spread of the "assists" distribution (with a range of 12.3 – 6.1 = 6.2) is less than the spread of the "rebounds" distribution (with a range of 16.8 – 9.4 = 7.4).

3.

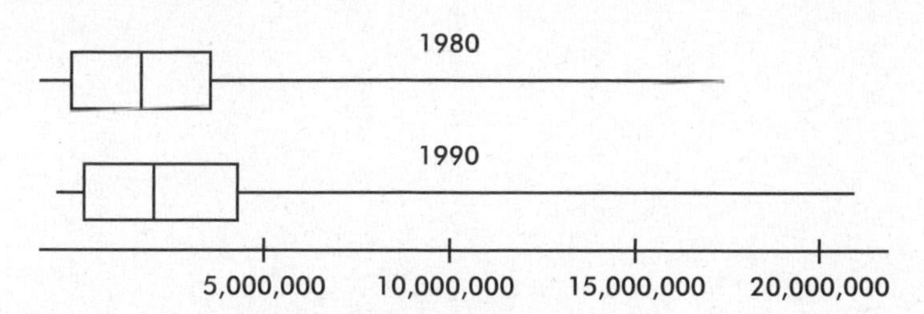

A complete answer compares shape, center, and spread, and notes outliers.

Shape: The most striking aspect is how skewed to the right (towards higher values) both sets of data appear to be.

Center: The median value changed very little from 1980 to 1990 (rising slightly from 2,329,000 to 2,649,000).

Spread: The difference between the lowest value and the median changed very little between 1980 and 1990 (2,329,000 – 262,000 = 2,067,000 in 1980, and 2,649,000 – 462,000 = 2,187,000 in 1990), while the upper-half spread increased substantially (from 16,873,000 – 2,329,000 = 14,544,000 in 1980 to 21,926,000 – 2,649,000 = 19,277,000 in 1990).

Outliers: For 1980, IQR = $Q_3 - Q_1$ = 3,749,000 – 834,000 = 2,915,000, and outliers are any values less than $Q_1 - 1.5(IQR)$ = –3,538,500 or greater than $Q_3 + 1.5(IQR)$ = 8,121,500, so California does represent an outlier. For 1990, IQR = $Q_3 - Q_1$ = 4,444,000 – 1,054,000 = 3,390,000, and outliers are any values less than $Q_1 - 1.5(IQR)$ = –4,031,000 or greater than $Q_3 + 1.5(IQR)$ = 9,529,000, so again California represents an outlier.

4. A complete answer compares shape, center, and spread.

Shape: Cruise A, for which the cumulative frequency plot rises steeply at first, has more younger passengers, and thus a distribution skewed to the right (towards the higher ages). Cruise C, for which the cumulative frequency plot rises slowly at first and then steeply towards the end, has more older passengers, and thus a distribution skewed to the left (towards the younger ages). Cruise B, for which the cumulative frequency plot rises slowly at each end and steeply in the middle, has a more bell-shaped distribution.

Center: Considering the center to be a value separating the area under the histogram roughly in half, the centers will correspond to a cumulative frequency of 0.5. Reading across from 0.5 to the intersection of each graph, and then down to the *x*-axis, shows centers of approximately 18, 40, and 61 years, respectively. Thus, the center of distribution A is the least, and the center of distribution C is the greatest.

Spread: The spreads of the age distributions of all three cruises are the same: from 10 to 70 years.

Exploring Bivariate Data

TOPIC 4

- Scatterplots
- Correlation and Linearity
- Least Squares Regression Line
- Residual Plots
- Outliers and Influential Points
- Transformations to Achieve Linearity

Our studies so far have been concerned with measurements of a single variable. However, many important applications of statistics involve examining whether two or more variables are related to one another. For example, is there a relationship between the smoking histories of pregnant women and the birth weights of their children? Between SAT scores and success in college? Between amount of fertilizer used and amount of crop harvested?

Two questions immediately arise. First, how can the strength of an apparent relationship be measured? Second, how can an observed relationship be put into functional terms? For example, a real estate broker might not only wish to determine whether a relationship exists between the prime rate and the number of new homes sold in a month but might also find useful an expression with which to predict the number of home sales given a particular value of the prime rate.

A graphical display, called a *scatterplot*, gives an immediate visual impression of a possible relationship between two variables, while a numerical measurement, called a *correlation coefficient*, is often used as a quantitative value of the strength of a linear relationship. In either case, evidence of a relationship is not evidence of causation.

SCATTERPLOTS

Suppose a relationship is perceived between two quantitative variables X and Y, and we graph the pairs (x, y). We are interested in the strength of this relationship, the scatterplot arising from the relationship, and any deviation from the basic pattern of this relationship. In this topic we examine whether the relationship can be reasonably explained in terms of a linear function, that is, one whose graph is a straight line.

TIP

Recognize explanatory (x) and response (y) variables in context.

For example, we might be looking at a plot such as

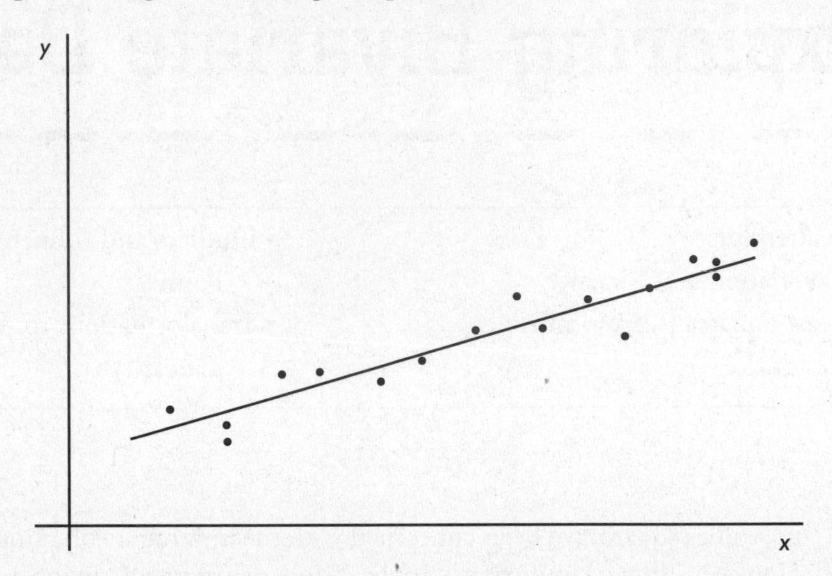

We need to know what the term *best-fitting straight line* means and how we can find this line. Furthermore, we want to be able to gauge whether the relationship between the variables is strong enough so that finding and making use of this straight line is meaningful.

Patterns in Scatterplots

When larger values of one variable are associated with larger values of a second variable, the variables are called *positively associated.* When larger values of one are associated with smaller values of the other, the variables are called *negatively associated.*

EXAMPLE 4.1

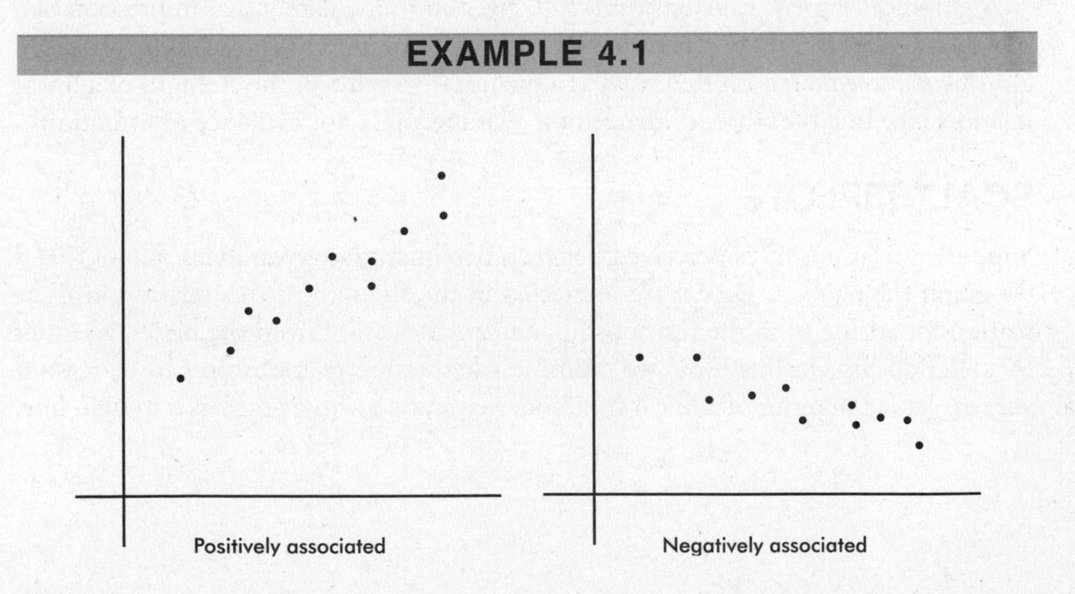

The strength of the association is gauged by how close the plotted points are to a straight line.

EXAMPLE 4.2

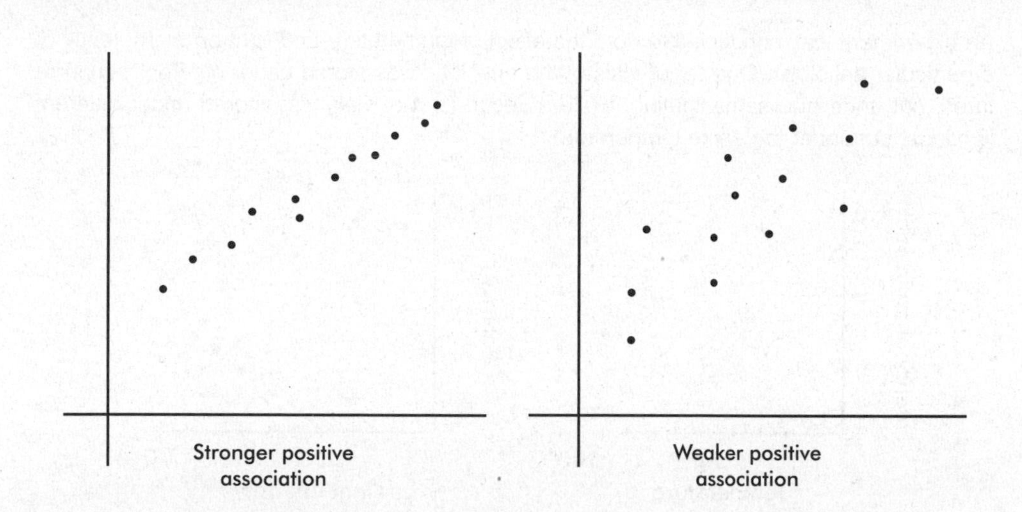

Sometimes different dots in a scatterplot are labeled with different symbols or different colors to show a categorical variable. The resulting *labeled scatterplot* might distinguish between men and women, between stocks and bonds, and so on.

EXAMPLE 4.3

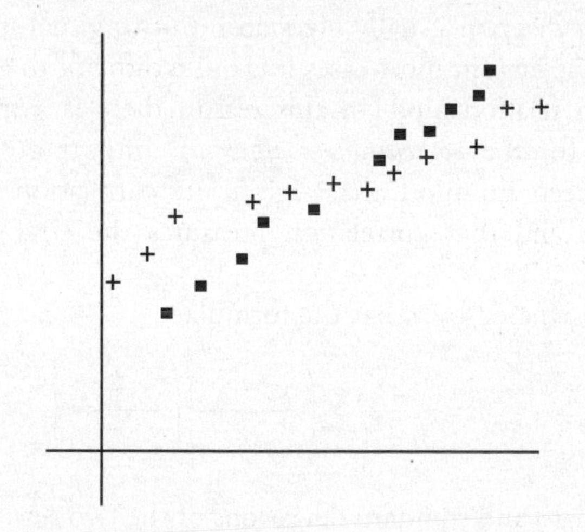

The above diagram is a labeled scatterplot distinguishing men with plus signs and women with square dots to show a categorical variable.

When analyzing the overall pattern in a scatterplot, it is also important to note *clusters* and *outliers.*

EXAMPLE 4.4

An experiment was conducted to note the effect of temperature and light on the potency of a particular antibiotic. One set of vials of the antibiotic was stored under different temperatures, but under the same lighting, while a second set of vials was stored under different lightings, but under the same temperature.

TIP

Note when the data falls in distinct groups.

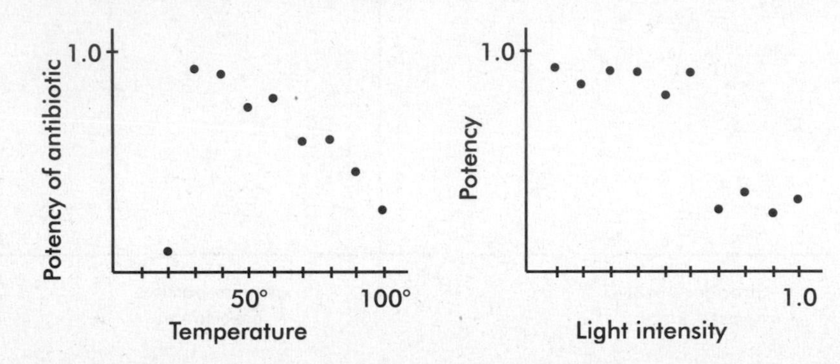

In the first scatterplot note the linear pattern with one outlier far outside this pattern. A possible explanation is that the antibiotic is more potent at lower temperatures, but only down to a certain temperature at which it drastically loses potency.

In the second histogram note the two clusters. It appears that below a certain light intensity the potency is one value, while above that intensity it is another value. In each cluster there seems to be no association between intensity and potency.

CORRELATION AND LINEARITY

TIP

Correlation does not imply causation!

Although a scatter diagram usually gives an intuitive visual indication when a linear relationship is strong, in most cases it is quite difficult to visually judge the specific strength of a relationship. For this reason there is a mathematical measure called *correlation* (or the *correlation coefficient*). Important as correlation is, we always need to keep in mind that significant correlation does not necessarily indicate *causation* and that correlation measures the strength only of a *linear* relationship.

Correlation, designated by r, has the formula

$$r = \frac{1}{n-1}\sum\left(\frac{x_i - \bar{x}}{s_x}\right)\left(\frac{y_i - \bar{y}}{s_y}\right)$$

TIP

With standardized data (z-scores), r is the slope of the regression line.

in terms of the means and standard deviations of the two sets. We note that the formula is actually the sum of the products of the corresponding z-scores divided by 1 less than the sample size. However, you should be able to quickly calculate correlation using the statistical package on your calculator. (Examining the formula helps you understand where correlation is coming from, but you will NOT have to use the formula to calculate r.)

Note from the formula that correlation does not distinguish between which variable is called x and which is called y. The formula is also based on standardized scores (z-scores), and so changing units does not change the correlation. Finally, since means and standard deviations can be strongly influenced by outliers, correlation is also strongly affected by extreme values.

The value of r always falls between -1 and $+1$, with -1 indicating perfect negative correlation and $+1$ indicating perfect positive correlation. It should be stressed that a correlation at or near zero doesn't mean there isn't a relationship between the variables; there may still be a strong *nonlinear* relationship.

EXAMPLE 4.5

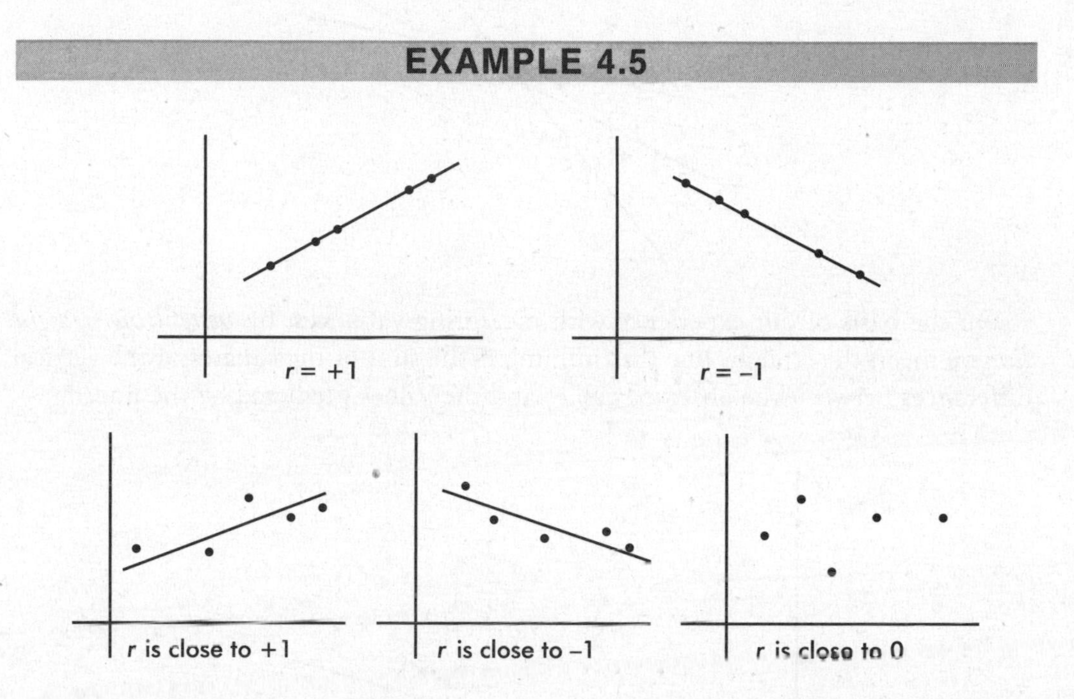

It can be shown that r^2, called the *coefficient of determination*, is the ratio of the variance of the predicted values $\hat{y}$ to the variance of the observed values y. That is, there is a partition of the y-variance, and r^2 is the proportion of this variance that is predictable from a knowledge of x. Alternatively, we can say that r^2 gives the percentage of variation in y that is explained by the variation in x. In either case, always interpret r^2 in context of the problem. Remember when calculating r from r^2 that r may be positive or negative.

While the correlation r is given as a decimal between -1.0 and 1.0, the coefficient of determination r^2 is usually given as a percentage. An r^2 of 100% is a perfect fit, with all the variation in y explained by variation in x. How large a value of r^2 is desirable depends on the application under consideration. While scientific experiments often aim for an r^2 in the 90% or above range, observational studies with r^2 of 10% to 20% might be considered informative. Note that while a correlation of .6 is twice a correlation of .3, the corresponding r^2 of 36% is *four* times the corresponding r^2 of 9%.

LEAST SQUARES REGRESSION LINE

> **TIP**
>
> If a scatterplot indicates a nonlinear relationship, don't try to force a straight-line fit.

What is the best-fitting straight line that can be drawn through a set of points?

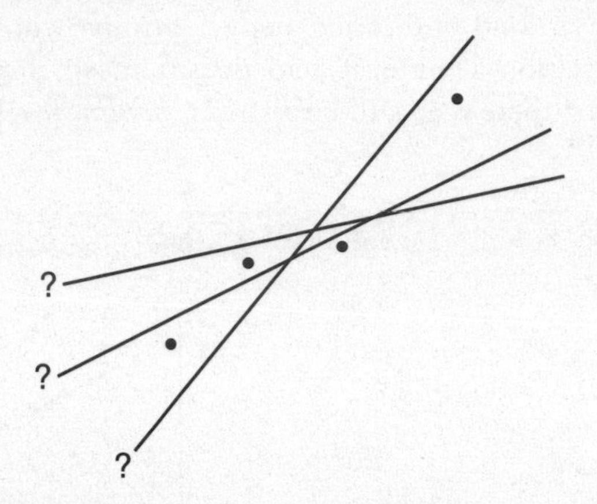

On the basis of our experience with measuring variances, by *best-fitting straight line* we mean the straight line that minimizes the sum of the squares of the vertical differences between the observed values and the values predicted by the line.

> **TIP**
>
> A hat over a variable means it is a predicted version of the variable.

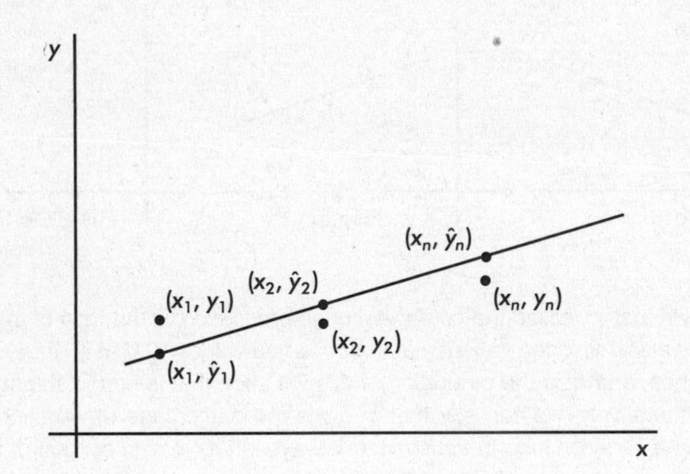

That is, in the above figure, we wish to minimize

$$(y_1 - \hat{y}_1)^2 + (y_2 - \hat{y}_2)^2 + \cdots + (y_n - \hat{y}_n)^2$$

It is reasonable, intuitive, and correct that the best-fitting line will pass through $(\bar{x}, \bar{y})$ where $\bar{x}$ and $\bar{y}$ are the means of the variables X and Y. Then, from the basic expression for a line with a given slope through a given point, we have

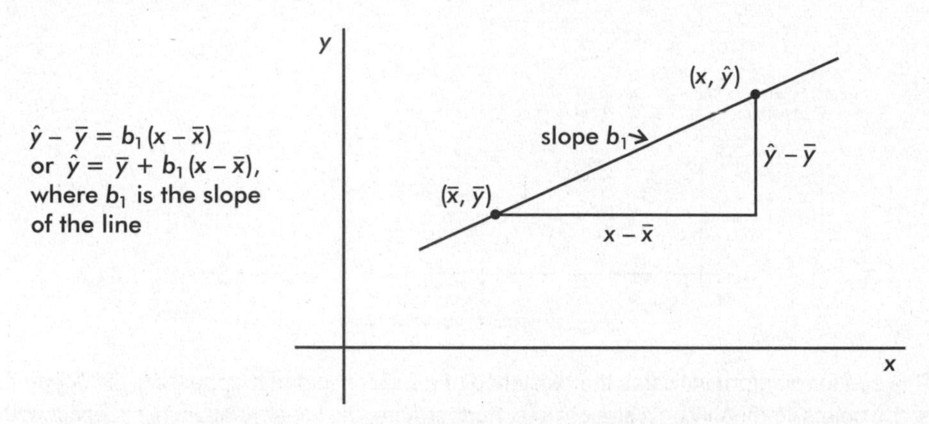

The slope b_1 can be determined from the formula

$$b_1 = r\frac{s_y}{s_x}$$

where r is the correlation and s_x and s_y are the standard deviations of the two sets. That is, each standard deviation change in x results in a change of r standard deviations in y. If you graph z-scores for the y-variable against z-scores for the x-variable, the slope of the regression line is precisely r, and, in fact, the linear equation becomes $z_y = rz_x$.

This best-fitting straight line, that is, the line that minimizes the sum of the squares of the differences between the observed values and the values predicted by the line, is called the *least squares regression line* or simply the *regression line*. It can be calculated directly by entering the two data sets and using the statistics package on your calculator.

> **TIP**
>
> Just because we can calculate a regression line doesn't mean it is useful.

EXAMPLE 4.6

An insurance company conducts a survey of 15 of its life insurance agents. The average number of minutes spent with each potential customer and the number of policies sold in a week are noted for each agent. Letting X and Y represent the average number of minutes and the number of sales, respectively, we have

X: 25 23 30 25 20 33 18 21 22 30 26 26 27 29 20
Y: 10 11 14 12 8 18 9 10 10 15 11 15 12 14 11

Find the equation of the best-fitting straight line for the data.

(continued)

Answer: Plotting the 15 points (25, 10), (23, 11), . . . , (20, 11) gives an intuitive visual impression of the relationship:

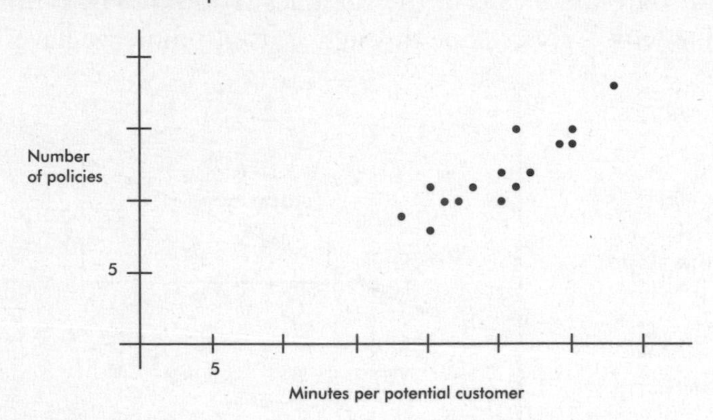

TIP

Be sure to **label axes** and **show number scales** whenever possible.

This scatter diagram indicates the existence of a relationship that appears to be *linear*, that is, the points lie roughly on a straight line. Furthermore, the linear relationship is *positive*; that is, as one variable increases, so does the other (the straight line slopes upward).

Using a calculator, we find the correlation to be $r = .8836$, the coefficient of determination to be $r^2 = .78$ (indicating that 78% of the variation in the number of policies is explained by the variation in the number of minutes spent), and the regression line to be

$$\begin{aligned} \hat{y} &= \bar{y} + b_1(x - \bar{x}) \\ &= 12 + 0.5492(x - 25) \\ &= -1.73 + 0.5492x \end{aligned}$$

We also write: $\widehat{\text{Policies}} = -1.73 + 0.5492$ Minutes

Adding this to our scatterplot yields

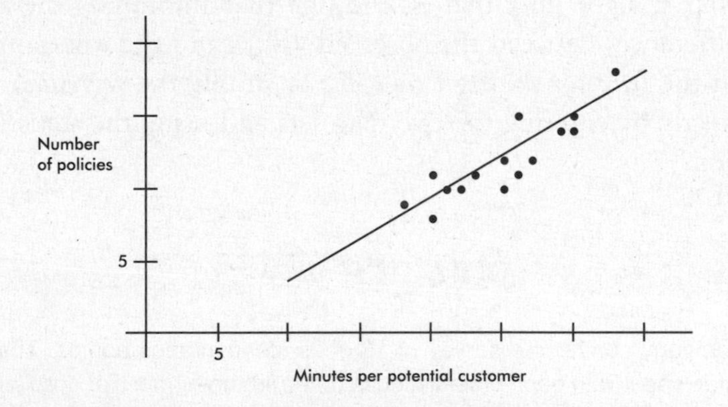

Thus, for example, we might predict that agents who average 24 minutes per customer will average $0.5492(24) - 1.73 = 11.45$ sales per week. We also note that each additional minute spent seems to produce an average 0.5492 extra sale.

EXAMPLE 4.7

Following are advertising expenditures and total sales for six detergent products:

Advertising ($1000) ($x$):	2.3	5.7	4.8	7.3	5.9	6.2
Total sales ($1000) ($y$):	77	105	96	118	102	95

Predict the total sales if $5000 is spent on advertising and interpret the slope of the regression line. What if $100,000 is spent on advertising?

Answer: With your calculator, the equation of the regression line is found to be

$$\hat{y} = 98.833 + 7.293(x - 5.367) = 59.691 + 7.293x$$

(A calculator like the TI-84, with less round-off error, directly gives $\hat{y} = 59.683 + 7.295x$.)

It is also worthwhile to replace the x and y with more appropriately named variables, resulting, for example, in

$$\widehat{\text{Sales}} = 59.691 + 7.293 \text{ Advcost}$$

TIP

Use meaningful variable names.

The regression line predicts that if $5000 is spent on advertising, the resulting total sales will be 7.293(5) + 59.691 = 96.156 thousands of dollars ($96,156).

The slope of the regression line indicates that every extra $1000 spent on advertising will result in an average of $7293 in added sales.

If $100,000 is spent on advertising, we calculate 7.293(100) + 59.691 = 788.991 thousands of dollars (≈$789,000). How much confidence should we have in this answer? Not much! We are trying to use the regression line to predict a value far outside the range of the data values. This procedure is called *extrapolation* and must be used with great care.

TIP

Be careful about extrapolation beyond the observed x-values.

It should be noted that when we use the regression line to predict a y-value for a given x-value, we are actually predicting the *mean* y-value for that given x-value. For any given x-value, there are many possible y-values, and we are predicting their mean. So if $5000 is spent many times on advertising, various resulting total sales figures may result, but their predicted average is $96,156.

The TI-Nspire gives

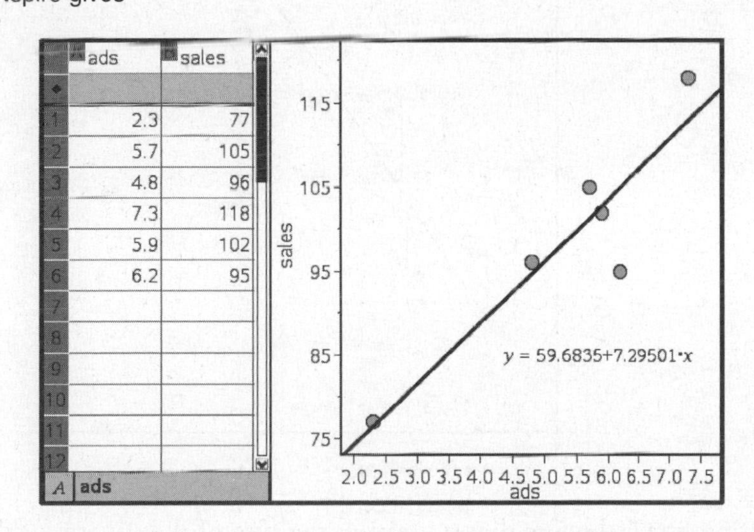

EXAMPLE 4.8

A random sample of 30 U.S. farm regions surveyed during the summer of 2003 produced the following statistics:

Average temperature (°F) during growing season: $\bar{x} = 81$, $s_x = 3$
Average corn yield per acre (bu.): $\bar{y} = 131$, $s_y = 5$
Correlation $r = .32$

Based on this study, what is the mean predicted corn yield for a region where the average growing season temperature is 76.5°F?

Answer: 76.5 is $\frac{76.5-81}{3} = -1.5$ standard deviations below the average temperature reported in the study, so with a correlation of $r = .32$, the predicted corn yield is $.32(-1.5) = -0.48$ standard deviations below the average corn yield, or $131 - 0.48(5) = 128.6$ bushels per acre.

Alternatively, we could have found the linear regression equation relating these variables: $slope = r\frac{s_y}{s_x} = .32\left(\frac{5}{3}\right) \approx 0.533$, intercept $\approx 131 - 81(0.533) \approx 87.8$, and thus $\widehat{Yield} = 0.533$ Temp + 87.8. Then $0.533(76.5) + 87.8 \approx 128.6$.

RESIDUAL PLOTS

The difference between an observed and predicted value is called the *residual.* When the regression line is graphed on the scatterplot, the residual of a point is the vertical distance the point is from the regression line.

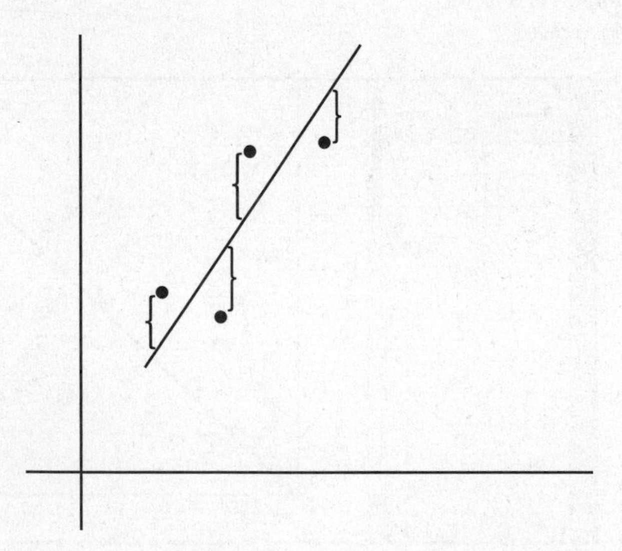

TIP

Be able to calculate the residual for each data value.

The regression line is the line that minimizes the sum of the squares of the residuals.

EXAMPLE 4.9

We calculate the predicted values from the regression line in Example 4.7 and subtract from the observed values to obtain the residuals:

x	2.3	5.7	4.8	7.3	5.9	6.2
y	77	105	96	118	102	95
$\hat{y}$	76.5	101.3	94.7	112.9	102.7	104.9
$y-\hat{y}$	0.5	3.7	1.3	5.1	–0.7	–9.9

Note that the sum of the residuals is

$$0.5 + 3.7 + 1.3 + 5.1 - 0.7 - 9.9 = 0.0$$

The TI-Nspire easily shows the "squares of the residuals," the sum of which is minimized by the regression line.

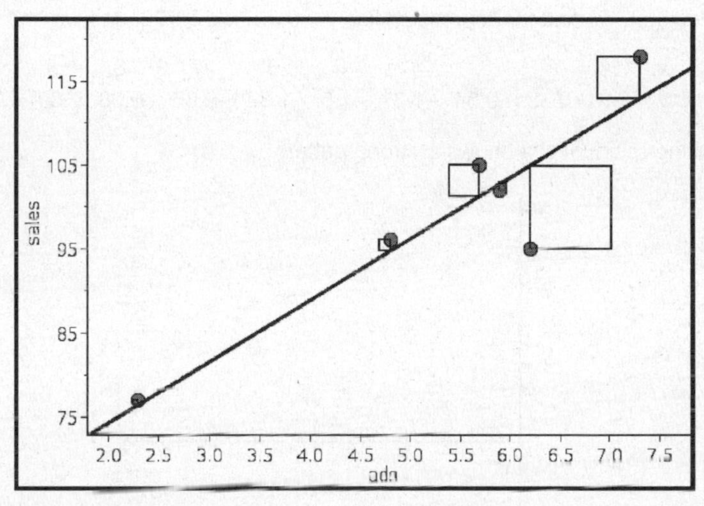

The above equation is true in general; that is, *the sum and thus the mean of the residuals is always zero.*

The notation for residuals is $\hat{e}_i = y_i - \hat{y}_i$ and so $\sum_{i=1}^{n} \hat{e}_i = 0$. The standard deviation of the residuals is calculated as follows:

$$s_e = \sqrt{\frac{\sum e_i^2}{n-2}} = \sqrt{\frac{\sum (y_i - \hat{y}_i)^2}{n-2}}$$

s_e gives a measure of how the points are spread around the regression line.

Plotting the residuals gives further information. In particular, a residual plot with a definite pattern is an indication that a nonlinear model will show a better fit to the data than the straight regression line. In addition to whether or not the residuals are randomly distributed, one should look at the balance between positive and negative residuals and also the size of the residuals in comparison to the associated y-values.

The residuals can be plotted against either the x-values or the $\hat{y}$-values (since $\hat{y}$ is a linear transformation of x, the plots are identical except for scale and a left-right reversal when the slope is negative).

It is also important to understand that a linear model may be appropriate, but weak, with a low correlation. And, alternatively, a linear model may not be the best model (as evidenced by the residual plot), but it still might be a very good model with high r^2.

EXAMPLE 4.10

Suppose the drying time of a paint product varies depending on the amount of a certain additive it contains.

Additive (oz), x:	1	2	3	4	5	6	7	8	9	10
Drying time (hr), y:	4	2.1	1.5	1	1.2	1.7	2.5	3.6	4.9	6.1

Using a calculator, we find the regression line $\hat{y} = 0.327x + 1.062$, and we find the residuals:

x:	1	2	3	4	5	6	7	8	9	10
$y - \hat{y}$:	2.61	0.38	–0.54	–1.37	–1.5	–1.32	–0.85	–0.08	0.89	1.77

The resulting residual plot shows a strong pattern:

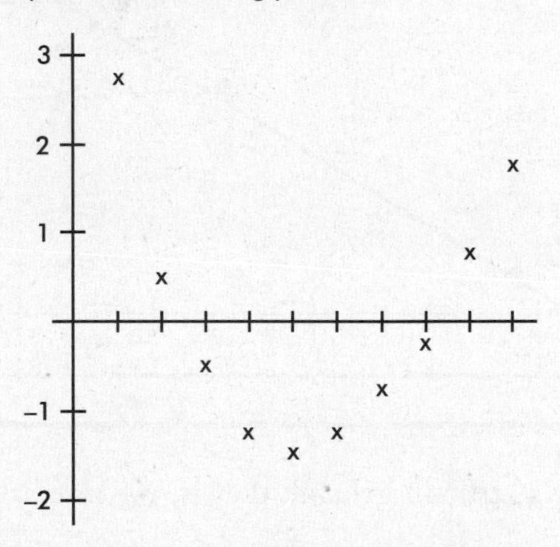

This pattern indicates that a nonlinear model will be a better fit than a straight-line model. A scatterplot of the original data shows clearly what is happening:

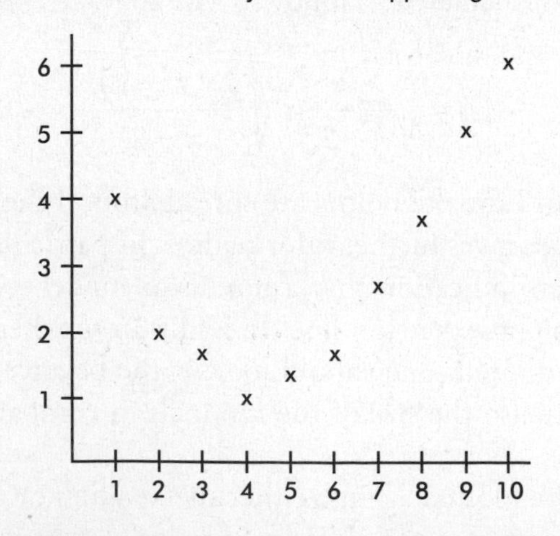

The ability to interpret computer output is important not only to do well on the AP Statistics exam, but also to understand statistical reports in the business and scientific world.

EXAMPLE 4.11

Miles per gallon versus speed for a new model automobile is fitted with a least squares regression line. The graph of the residuals and some computer output for the regression are as follows:

Regression Analysis: MPG Versus Speed

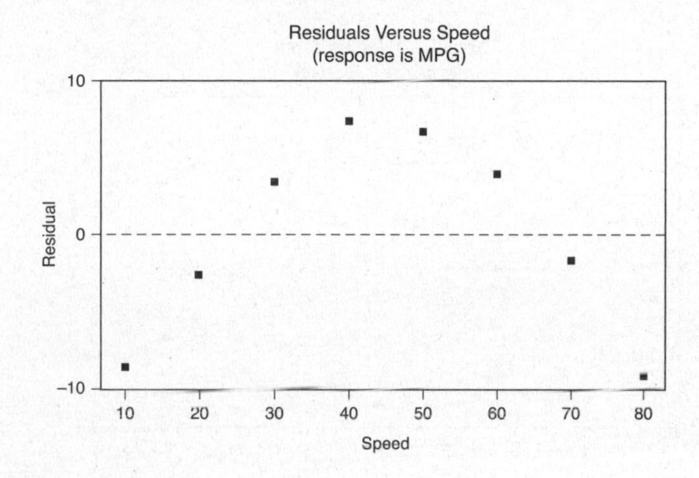


```
The regression equation is
MPG = 38.9 – 0.218 Speed

Predictor        Coef     SE Coef         T          P
Constant       38.929       5.651      6.89      0.000
Speed         –0.2179      0.1119     –1.95      0.099

S = 7.252   R–Sq = 38.7%   R–Sq(adj) = 28.5%

Analysis of Variance

Source            DF         SS          MS          F      P
Regression         1     199.34      199.34       3.79  0.099
Residual Error     6     315.54       52.59
Total              7     514.88
```

a. Interpret the slope of the regression line in context.
Answer: The slope of the regression line is –0.2179, indicating that, on average, the MPG drops by 0.2179 for every increase of one mile per hour in speed.

b. What is the mean predicted MPG at a speed of 30 mph?
Answer: At 30 mph, the mean predicted MPG is –0.2179(30) + 38.929, or about 32.4 MPG.

c. What was the actual MPG at a speed of 30 mph?
Answer: The residual for 30 mph is about +3.5, and since residual = actual – predicted, we estimate the actual MPG to be 32.4 + 3.5, or about 36 MPG.

d. Is a line the most appropriate model? Explain.
Answer: The fact that the residuals show such a strong curved pattern indicates that a nonlinear model would be more appropriate.

e. What does "S = 7.252" refer to?
Answer: The standard deviation of the residuals, $s_e = 7.252$, is a "typical value" of the residuals and gives a measure of how the points are spread around the regression line.

EXAMPLE 4.12

The number of youngsters playing Little League baseball in Ithaca, New York, during the years 1995–2003 is fitted with a least squares regression line. The graph of the residuals and some computer output for their regression are as follows:

Regression Analysis: Number of Players Versus Years Since 1995

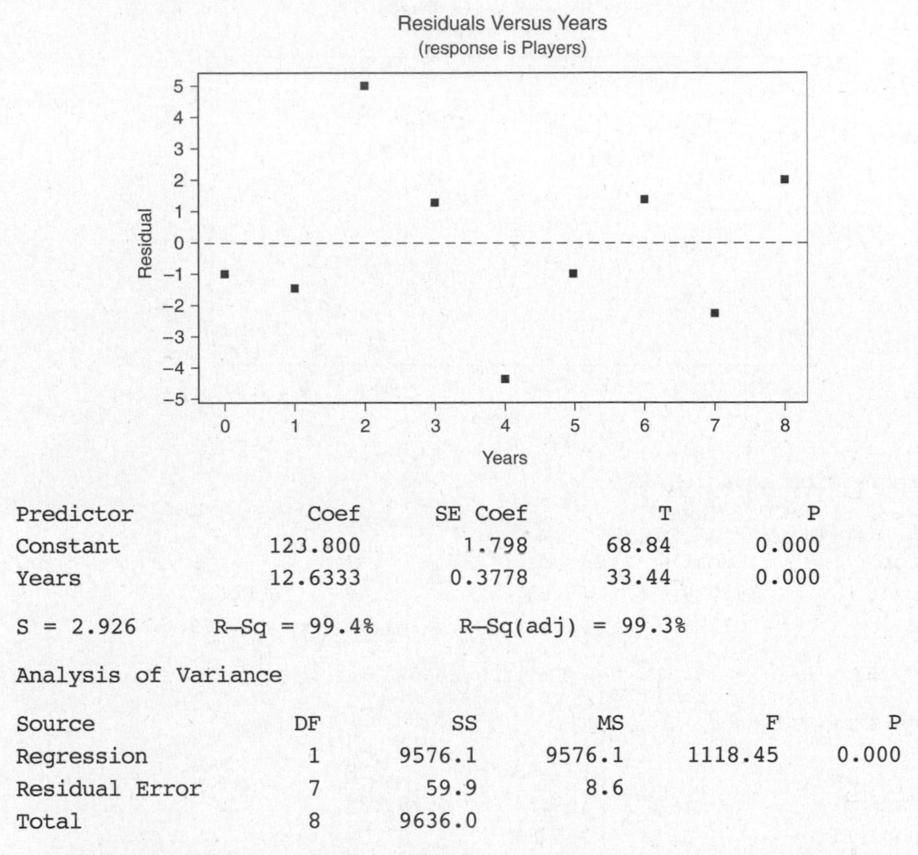

Predictor	Coef	SE Coef	T	P
Constant	123.800	1.798	68.84	0.000
Years	12.6333	0.3778	33.44	0.000

S = 2.926   R–Sq = 99.4%   R–Sq(adj) = 99.3%

Analysis of Variance

Source	DF	SS	MS	F	P
Regression	1	9576.1	9576.1	1118.45	0.000
Residual Error	7	59.9	8.6		
Total	8	9636.0			

a. Does it appear that a line is an appropriate model for the data? Explain.
Answer: Yes. R–Sq = 99.4% is large, and the residual plot shows no pattern. Thus a linear model is appropriate.

b. What is the equation of the regression line (in context)?
Answer: Predicted # of players = 123.8 + 12.6 (years since 1995)

c. Interpret the slope of the regression line in the context of the problem.
Answer: The slope of the regression line is 12.6, indicating that, on average, the predicted number of children playing Little League baseball in Ithaca increased by 12 or 13 players per year during the 1995–2003 time period.

d. Interpret the *y*-intercept of the regression line in the context of the problem.
Answer: The *y*-intercept, 123.8, refers to the year 1995. Thus the number of players in Little League in Ithaca in 1995 was predicted to be around 124.

e. What is the predicted number of players in 1997?
Answer: For 1997, $x = 2$, so the predicted number of players is 12.6(2) + 123.8 = 149.

f. What was the actual number of players in 1997?
Answer: The residual for 1997 ($x = 2$) from the residual plot is +5, so actual − predicted = 5, and thus the actual number of players in 1997 must have been 5 + 149 = 154.

(continued)

TIP

Simply using a calculator to find a regression line is not enough; you must understand it (for example, be able to interpret the slope and intercepts in context).

g. What years, if any, did the number of players decrease from the previous year? Explain.
Answer: The number would decrease if one residual were more than 12.6 greater than the next residual. This never happens, so the number of players never decreased.

OUTLIERS AND INFLUENTIAL POINTS

In a scatterplot, regression outliers arc indicated by points falling far away from the overall pattern. That is, a point is an outlier if its residual is an outlier in the set of residuals.

EXAMPLE 4.13

A scatterplot of grade point average (GPA) versus weekly television time for a group of high school seniors is as follows:

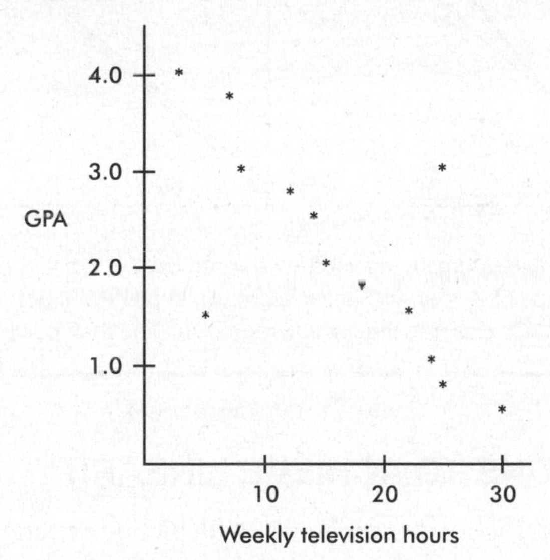

By direct observation of the scatterplot, we note that there are two outliers: one person who watches 5 hours of television weekly yet has only a 1.5 GPA, and another person who watches 25 hours weekly yet has a 3.0 GPA. Note also that while the value of 30 weekly hours of television may be considered an outlier for the television hours variable and the 0.5 GPA may be considered an outlier for the GPA variable, the point (30, 0.5) is *not* an outlier in the regression context because it does not fall off the straight-line pattern.

Scores whose removal would sharply change the regression line are called *influential scores.* Sometimes this description is restricted to points with extreme x-values. An influential score may have a small residual but still have a greater effect on the regression line than scores with possibly larger residuals but average x-values.

EXAMPLE 4.14

Consider the following scatterplot of six points and the regression line:

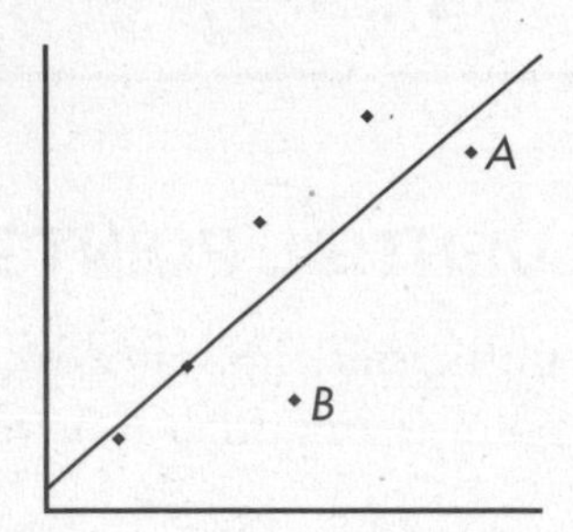

The heavy line in the scatterplot on the left below shows what happens when point *A* is removed, and the heavy line in the scatterplot on the right below shows what happens when point *B* is removed.

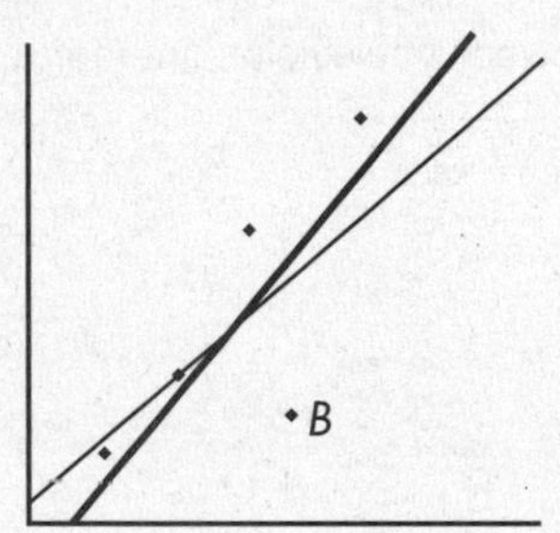

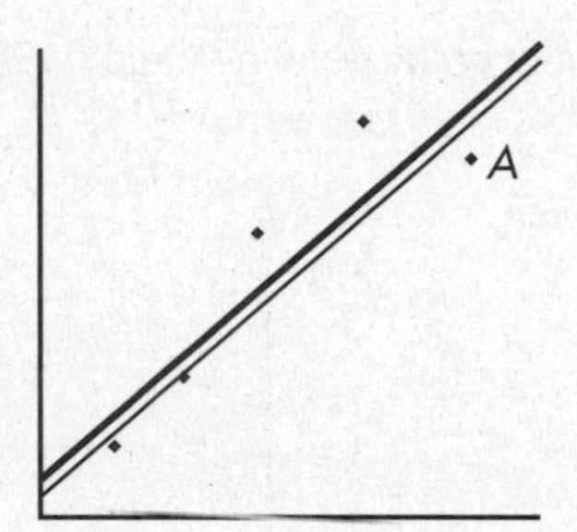

Note that the regression line is greatly affected by the removal of point *A* but not by the removal of point *B*. Thus point *A* is an *influential score*, while point *B* is not. This is true in spite of the fact that point *A* is closer to the original regression line than point *B*.

TRANSFORMATIONS TO ACHIEVE LINEARITY

Often a straight-line pattern is not the best model for depicting a relationship between two variables. A clear indication of this problem is when the scatterplot shows a distinctive curved pattern. Another indication is when the residuals show a distinctive pattern rather than a random scattering. In such a case, the nonlinear model can sometimes be revealed by transforming one or both of the variables and then noting a linear relationship. Useful transformations often result from using the *log* or *ln* buttons on your calculator to create new variables.

EXAMPLE 4.15

Consider the following years and corresponding populations:

Year, x:	1950	1960	1970	1980	1990
Population (1000s), y:	50	67	91	122	165

The scatterplot and residual plot indicate that a nonlinear relationship would be an even stronger model.

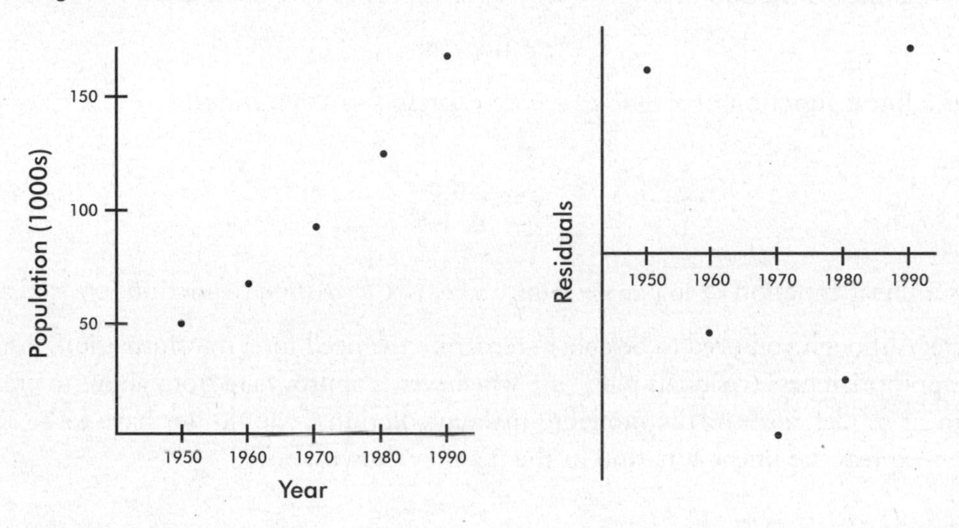

Letting log y be a new variable, we obtain

x:	1950	1960	1970	1980	1990
log y:	1.70	1.83	1.96	2.09	2.22

The scatterplot and residual plot now indicate a stronger linear relationship.

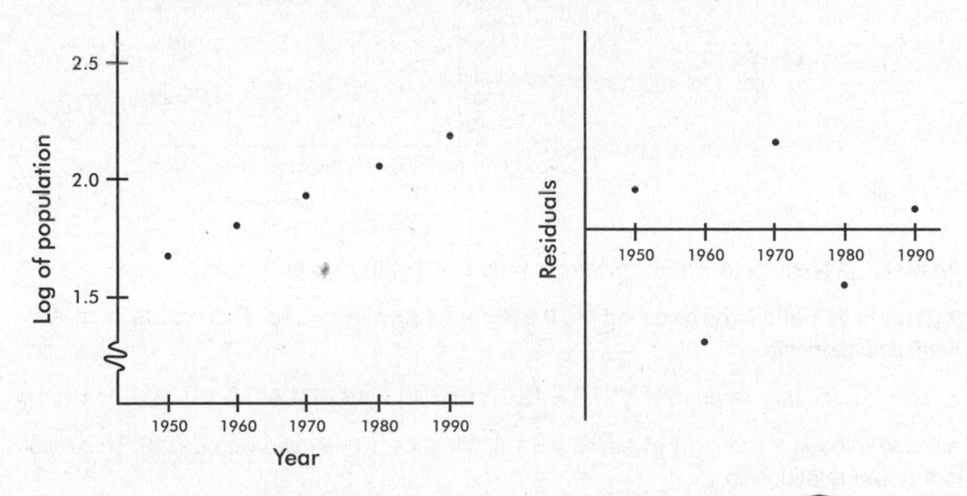

A regression analysis yields $\widehat{\log y} = 0.013x - 23.65$. In context we have: $\widehat{\log(\text{Pop})} = -23.65 + 0.013(\text{Year})$. So, for example, the population predicted for the year 2000 would be calculated $\widehat{\log(\text{Pop})} = 0.013(2000) - 23.65 = 2.35$, and so Pop $= 10^{2.35} \approx 224$ thousand, or 224,000. The "linear" equation $\widehat{\log y} = 0.013x - 23.65$ can be re-expressed as $\hat{y} = 10^{0.013x-23.65}$ [$= 10^{-23.65} \times (10^{0.013})^x = 2.2387E{-24} \times 1.0304^x$].

There are many useful transformations. For example:

Log y as a linear function of x, $\log y = ax + b$, re-expresses as an *exponential*:

$$y = 10^{ax+b} \text{ or } y = b_1 10^{ax} \text{ where } b_1 = 10^b$$

Log y as a linear function of log x, $\log y = a \log x + b$, re-expresses as a *power*:

$$y = 10^{a \log x + b} \text{ or } y = b_1 x^a \text{ where } b_1 = 10^b$$

$\sqrt{y}$ as a linear function of x, $\sqrt{y} = ax + b$, re-expresses as a *quadratic*:

$$y = (ax + b)^2$$

$\frac{1}{y}$ as a linear function of x, $\frac{1}{y} = ax + b$, re-expresses as a *reciprocal*:

$$y = \frac{1}{ax + b}$$

y as a linear function of log x, $y = a \log x + b$, is a *logarithmic* function.

Note: Although you need to be able to recognize the need for a transformation, justify its appropriateness (residuals plot), use whichever is appropriate from above to create a linear model, and use the model to make predictions, you do not have to be able to re-express the linear equation in the manner shown above.

EXAMPLE 4.16

What are possible models for the following data?

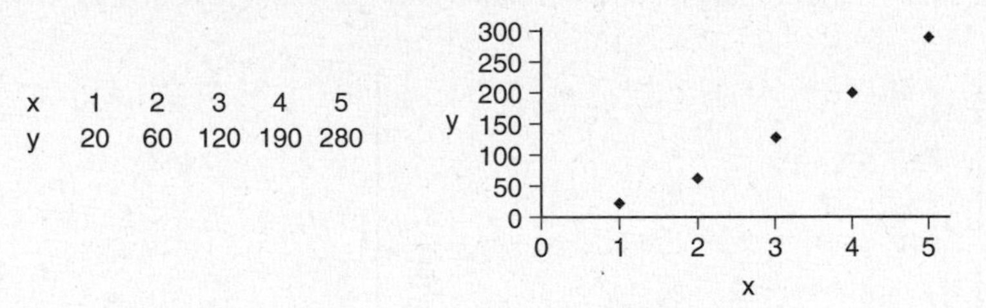

x	1	2	3	4	5
y	20	60	120	190	280

Answer: A linear fit to x and y gives $\hat{y} = 65x - 61$ with $r = .99$.

A linear fit to x and log y gives $\log y = 0.279x + 1.139$ with $r = .98$. This results in an *exponential* relationship:

$$\hat{y} = 10^{0.279x+1.139} = 13.77(10^{0.279x}) = 13.77(1.901^x)$$

A linear fit to log x and log y gives $\log y = 1.639 \log x + 1.295$, also with $r = .99$. This results in a *power* relationship:

$$\hat{y} = 10^{1.639 \log x + 1.295} = 19.72(x^{1.639})$$

(continued)

All three models give high correlation and are reasonable fits. Further analysis can be done by examining the residual plots:

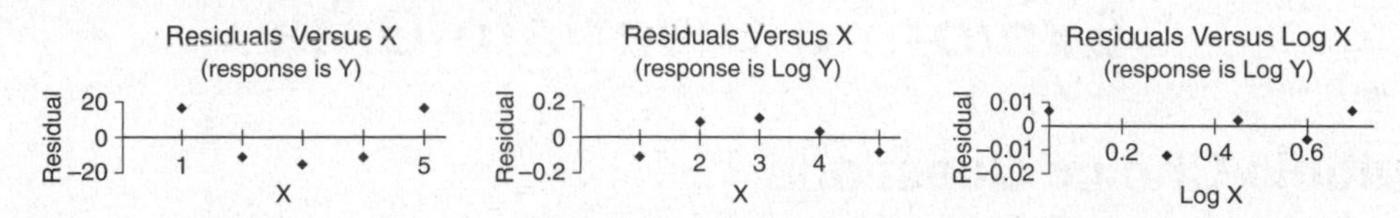

The first two residual plots have distinct curved patterns. The third residual plot illustrates both a more random pattern and smaller residuals. (One can also create residual plots using the derived curved models, but we choose to restrict our attention to residual plots of the linear models.) Among the above three models, the power model $\hat{y} = 19.72(x^{1.639})$, appears to be best.

Summary

- A scatterplot gives an immediate indication of the shape (linear or not), strength, and direction (positive or negative) of a possible relationship between two variables.
- If the relationship appears roughly linear, then the correlation coefficient, r, is a useful measurement.
- The value of r is always between –1 and +1, with positive values indicating positive association and negative values indicating negative association; and values close to –1 or +1 indicating a stronger linear association than values close to 0, which indicate a weaker linear association.
- Evidence of an association is not evidence of a cause-and-effect relationship!
- Correlation is not affected by which variable is called x and which y or by changing units.
- Correlation can be strongly affected by extreme values.
- The differences between the observed and predicted values are called residuals.
- The best-fitting straight line, called the regression line, minimizes the sum of the squares of the residuals.
- For the linear regression model, the mean of the residuals is always 0.
- A definite pattern in the residual plot indicates that a nonlinear model may fit the data better than the straight regression line.
- The coefficient of determination, r^2, gives the percentage of variation in y that is accounted for by the variation in x.
- Influential scores are scores whose removal would sharply change the regression line.
- Nonlinear models can sometimes be studied by transforming one or both variables and then noting a linear relationship.
- It is very important to be able to interpret generic computer output.

Questions on Topic Four: Exploring Bivariate Data

Multiple-Choice Questions

Directions: The questions or incomplete statements that follow are each followed by five suggested answers or completions. Choose the response that best answers the question or completes the statement.

1. A study collects data on average combined SAT scores (math, critical reading, and writing) and percentage of students who took the exam at 100 randomly selected high schools. Following is part of the computer printout for regression:

Variable	**Coefficient**	**s.e. of coeff**	**t-ratio**	**prob**
Constant	1576.32	12.65	124.6	≤ 0.0001
SAT	-2.84276	0.2461	-11.55	≤ 0.0001

R-squared = 76.5%   R-squared (adj) = 76.1%

Which of the following is a correct conclusion?

(A) SAT in the variable column indicates that SAT is the dependent (response) variable.
(B) The correlation is ±0.875, but the sign cannot be determined.
(C) The *y*-intercept indicates the mean combined SAT score if percent of students taking the exam has no effect on combined SAT scores.
(D) The R^2 value indicates that the residual plot does not show a strong pattern.
(E) Schools with lower percentages of students taking the exam tend to have higher average combined SAT scores.

2. A simple random sample of 35 world-ranked chess players provides the following statistics:

Number of hours of study per day: $\bar{x} = 6.2$, $s_x = 1.3$
Yearly winnings: $\bar{y} = \$208,000$, $s_y = \$42,000$
Correlation $r = .15$

Based on this data, what is the resulting linear regression equation?

(A) $\widehat{\text{Winnings}}$ = 178,000 + 4850 Hours
(B) $\widehat{\text{Winnings}}$ = 169,000 + 6300 Hours
(C) $\widehat{\text{Winnings}}$ = 14,550 + 31,200 Hours
(D) $\widehat{\text{Winnings}}$ = 7750 + 32,300 Hours
(E) $\widehat{\text{Winnings}}$ = –52,400 + 42,000 Hours

3. A rural college is considering constructing a windmill to generate electricity but is concerned over noise levels. A study is performed measuring noise levels (in decibels) at various distances (in feet) from the campus library, and a least squares regression line is calculated with a correlation of 0.74. Which of the following is a proper and most informative conclusion for an observation with a negative residual?

   (A) The measured noise level is 0.74 times the predicted noise level.
   (B) The predicted noise level is 0.74 times the measured noise level.
   (C) The measured noise level is greater than the predicted noise level.
   (D) The predicted noise level is greater than the measured noise level.
   (E) The slope of the regression line at that point must also be negative.

4. Consider the following three scatterplots:

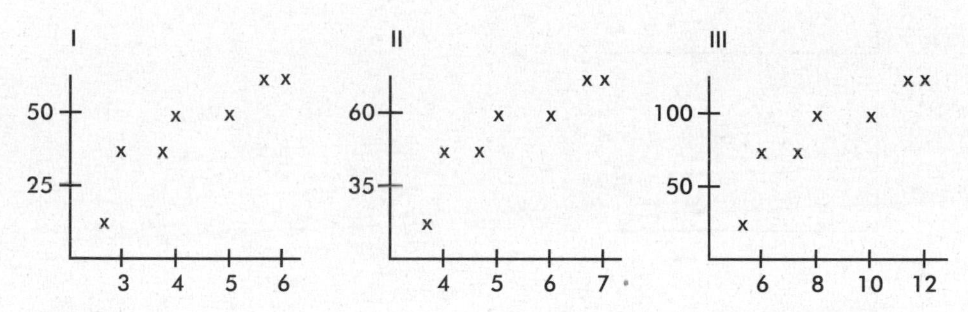

   Which has the greatest correlation coefficient?

   (A) I
   (B) II
   (C) III
   (D) They all have the same correlation coefficient.
   (E) This question cannot be answered without additional information.

5. Suppose the correlation is negative. Given two points from the scatterplot, which of the following is possible?

   I. The first point has a larger x-value and a smaller y-value than the second point.
   II. The first point has a larger x-value and a larger y-value than the second point.
   III. The first point has a smaller x-value and a larger y-value than the second point.

   (A) I only
   (B) II only
   (C) III only
   (D) I and III
   (E) I, II, and III

6. Consider the following residual plot:

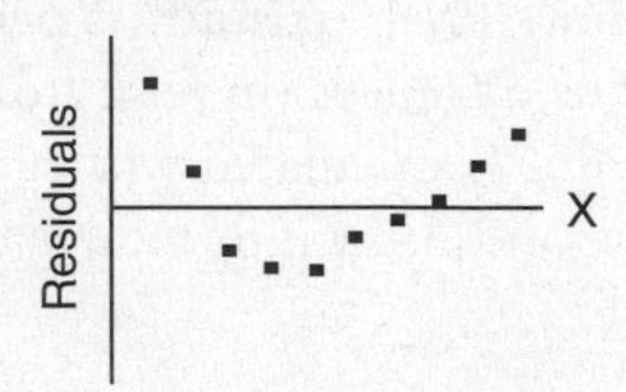

Which of the following scatterplots could have resulted in the above residual plot? (The *y*-axis scales are not the same in the scatterplots as in the residual plot.)

(A) 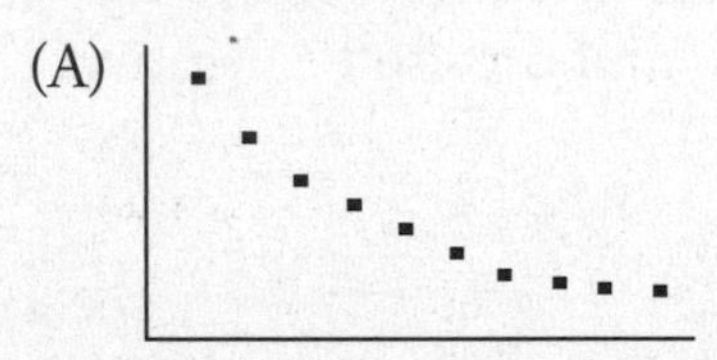

(B) 

(C) 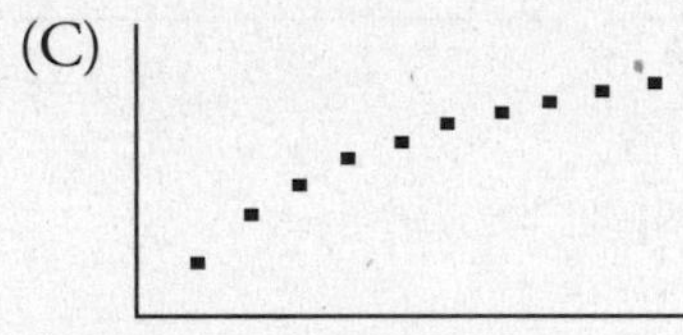

(D) 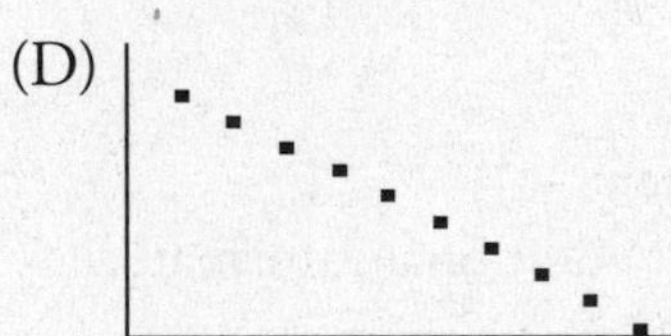

(E) None of these could result in the given residual plot.

7. Suppose the regression line for a set of data, $\hat{y} = 3x + b$, passes through the point (2, 5). If $\bar{x}$ and $\bar{y}$ are the sample means of the *x*- and *y*-values, respectively, then $\bar{y}$ =

(A) $\bar{x}$.
(B) $\bar{x} - 2$.
(C) $\bar{x} + 5$.
(D) $3\bar{x}$.
(E) $3\bar{x} - 1$.

8. Suppose a study finds that the correlation coefficient relating family income to SAT scores is $r = +1$. Which of the following are proper conclusions?

   I. Poverty causes low SAT scores.
   II. Wealth causes high SAT scores.
   III. There is a very strong association between family income and SAT scores.

   (A) I only
   (B) II only
   (C) III only
   (D) I and II
   (E) I, II, and III

9. A study of department chairperson ratings and student ratings of the performance of high school statistics teachers reports a correlation of $r = 1.15$ between the two ratings. From this information we can conclude that

   (A) chairpersons and students tend to agree on who is a good teacher.
   (B) chairpersons and students tend to disagree on who is a good teacher.
   (C) there is little relationship between chairperson and student ratings of teachers.
   (D) there is strong association between chairperson and student ratings of teachers, but it would be incorrect to infer causation.
   (E) a mistake in arithmetic has been made.

10. Which of the following statements about correlation r is true?

   (A) A correlation of .2 means that 20% of the points are highly correlated.
   (B) Perfect correlation, that is, when the points lie exactly on a straight line, results in $r = 0$.
   (C) Correlation is not affected by which variable is called x and which is called y.
   (D) Correlation is not affected by extreme values.
   (E) A correlation of .75 indicates a relationship that is 3 times as linear as one for which the correlation is only .25.

Questions 11–13 refer to the following:

The relationship between winning game proportions when facing the sun and when the sun is on one's back is analyzed for a random sample of 10 professional players. The computer printout for regression is below:

```
Predictor        Coef     SE Coef         T        P
Constant      0.05590     0.02368      2.36    0.046
Facing        0.92003     0.03902     23.58    0.000

S = 0.0242922    R-Sq = 98.6%    R-Sq(adj) = 98.4%

Analysis of Variance

Source            DF         SS         MS        F        P
Regression         1    0.32812    0.32812   556.03    0.000
Residual Error     8    0.00472    0.00059
Total              9    0.33284
```

11. What is the equation of the regression line, where *face* and *back* are the winning game proportions when facing the sun and with back to the sun, respectively?

(A) $\widehat{face} = 0.056 + 0.920\ back$

(B) $\widehat{back} = 0.056 + 0.920\ facing$

(C) $\widehat{face} = 0.920 + 0.056\ back$

(D) $\widehat{back} = 0.920 + 0.056\ facing$

(E) $\widehat{face} = 0.024 + 0.039\ back$

12. What is the *correlation*?

(A) –.984
(B) –.986
(C) .984
(D) .986
(E) .993

13. For one player, the winning game proportions were 0.55 and 0.59 for *facing* and *back*, respectively. What was the associated *residual*?

(A) –0.028
(B) 0.028
(C) –0.0488
(D) 0.0488
(E) 0.3608

14. Which of the following statements about residuals are true?

I. The mean of the residuals is always zero.
II. The regression line for a residual plot is a horizontal line.
III. A definite pattern in the residual plot is an indication that a nonlinear model will show a better fit to the data than the straight regression line.

(A) I and II
(B) I and III
(C) II and III
(D) I, II, and III
(E) None of the above gives the complete set of true responses.

15. Data are obtained for a group of college freshmen examining their SAT scores (math plus writing plus critical reading) from their senior year of high school and their GPAs during their first year of college. The resulting regression equation is

$$\widehat{GPA} = 0.55 + 0.00161 \text{ (SAT total)} \quad \text{with} \quad r = .632$$

What percentage of the variation in GPAs can be explained by looking at SAT scores?

(A) 0.161%
(B) 16.1%
(C) 39.9%
(D) 63.2%
(E) This value cannot be computed from the information given.

16. In a study of whether the structure of the adult human brain changes when a new skill is learned, the gray matter volume of four individuals was measured before and after learning a new cognitive skill. The resulting scatterplot was:

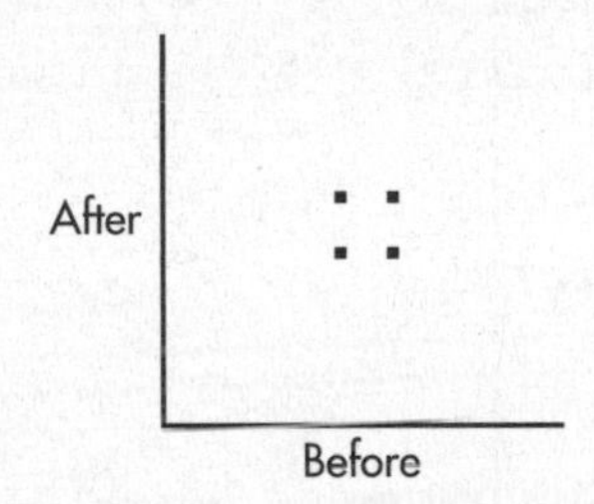

The correlation above is 0. Three researchers each run the experiment on a new subject and each obtain an additional data point:

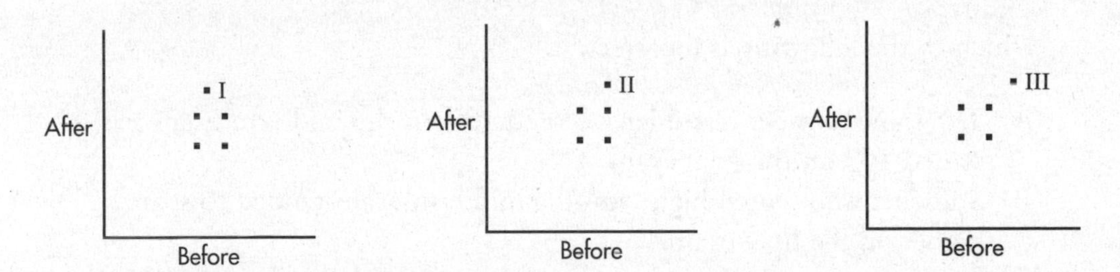

Match the above scatterplots with their new correlations.

(A) I: –0.33 II: 0 III: 0.33
(B) I: 0 II: 0.33 III: 0.64
(C) I: 0 II: 0.33 III: 1.0
(D) I: –0.33 II: 0 III: 1.0
(E) I: 0 II: 0.50 III: 1.0

17. In a study of winning percentage in home games versus average home attendance for professional baseball teams, the resulting regression line is:

$$\widehat{\text{Winning percentage}} = 44 + 0.0003$$

What is the residual if a team has a winning percentage of 55% with an average attendance of 34,000?

(A) –11.0
(B) –0.8
(C) 0.8
(D) 11.0
(E) 23.0

18. Consider the following scatterplot of midterm and final exam scores for a class of 15 students.

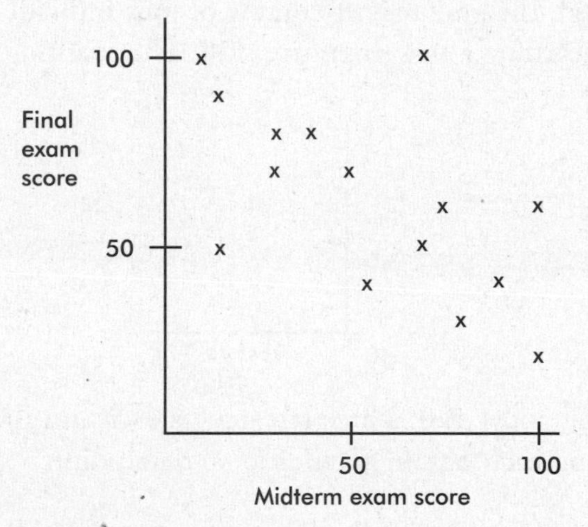

Which of the following is incorrect?

(A) The same number of students scored 100 on the midterm exam as scored 100 on the final exam.
(B) Students who scored higher on the midterm exam tended to score higher on the final exam.
(C) The scatterplot shows a moderate negative correlation between midterm and final exam scores.
(D) The coefficient of determination here is positive.
(E) No one scored 90 or above on both exams.

19. If every woman married a man who was exactly 2 inches taller than she, what would the correlation between the heights of married men and women be?

(A) Somewhat negative
(B) 0
(C) Somewhat positive
(D) Nearly 1
(E) 1

20. Suppose the correlation between two variables is $r = .23$. What will the new correlation be if .14 is added to all values of the x-variable, every value of the y-variable is doubled, and the two variables are interchanged?

    (A) .23
    (B) .37
    (C) .74
    (D) −.23
    (E) −.74

21. Suppose the correlation between two variables is −.57. If each of the y-scores is multiplied by −1, which of the following is true about the new scatterplot?

    (A) It slopes up to the right, and the correlation is −.57.
    (B) It slopes up to the right, and the correlation is +.57.
    (C) It slopes down to the right, and the correlation is −.57.
    (D) It slopes down to the right, and the correlation is +.57.
    (E) None of the above is true.

22. Consider the set of points {(2, 5), (3, 7), (4, 9), (5, 12), (10, n)}. What should n be so that the correlation between the x- and y-values is 1?

    (A) 21
    (B) 24
    (C) 25
    (D) A value different from any of the above.
    (E) No value for n can make $r = 1$.

23. Consider the following three scatterplots:

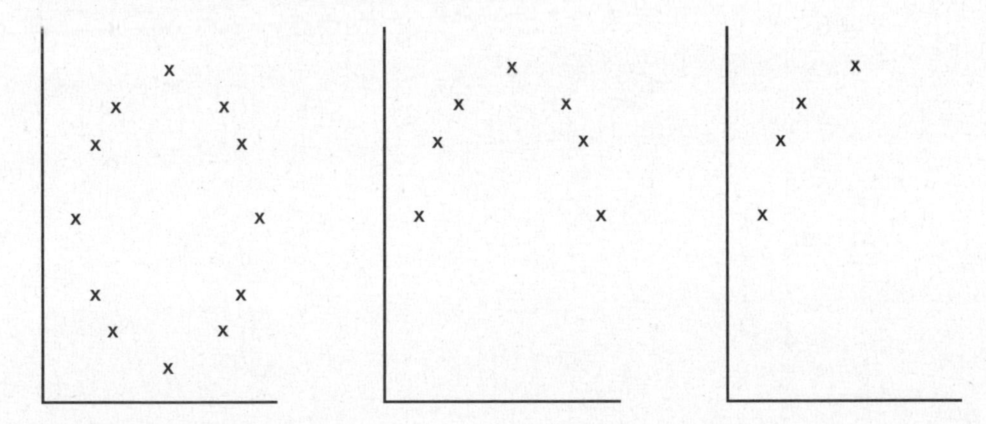

Which of the following is a true statement about the correlations for the three scatterplots?

(A) None are 0.
(B) One is 0, one is negative, and one is positive.
(C) One is 0, and both of the others are positive.
(D) Two are 0, and the other is 1.
(E) Two are 0, and the other is close to 1.

24. Consider the three points (2, 11), (3, 17), and (4, 29). Given any straight line, we can calculate the sum of the squares of the three vertical distances from these points to the line. What is the smallest possible value this sum can be?

    (A) 6
    (B) 9
    (C) 29
    (D) 57
    (E) None of these values

25. Suppose that the scatterplot of log X and log Y shows a strong positive correlation close to 1. Which of the following is true?

    (A) The variables X and Y also have a correlation close to 1.
    (B) A scatterplot of the variables X and Y shows a strong nonlinear pattern.
    (C) The residual plot of the variables X and Y shows a random pattern.
    (D) A scatterplot of X and log Y shows a strong linear pattern.
    (E) A cause-and-effect relationship can be concluded between log X and log Y.

26. Consider n pairs of numbers. Suppose $\bar{x} = 2$, $s_x = 3$, $\bar{y} = 4$, and $s_y = 5$. Of the following, which could be the least squares line?

    (A) $\hat{y} = -2 + x$
    (B) $\hat{y} = 2x$
    (C) $\hat{y} = -2 + 3x$
    (D) $\hat{y} = \frac{5}{3} - x$
    (E) $\hat{y} = 6 - x$

Free-Response Questions

Directions: You must show all work and indicate the methods you use. You will be graded on the correctness of your methods and on the accuracy of your final answers.

Ten Open-Ended Questions

1. Average home attendance and number of home wins for the 2009–10 NBA Pacific Division teams were as follows:

	Lakers	Suns	Clippers	Warriors	Kings
Average attendance	18,997	17,648	16,343	18,027	13,254
Home wins	34	32	21	18	18

   (a) Does a winning team bring out the fans? Can average attendance be predicted from number of wins? Find the equation of the best-fitting straight line.
   (b) Interpret the slope.
   (c) Predict the average attendance for a team with 25 home wins.
   (d) Predict the number of home wins which will bring out an average of at least 17,000 fans.
   (e) What is the residual for the Lakers average attendance?

2. The shoe sizes and the number of ties owned by ten corporate vice presidents are as follows.

Shoe size, x:	8	9.5	9	11	9	9.5	8.5	9	9	9.5
Number of ties, y:	10	10	8	15	12	13	16	7	12	4

   (a) Draw a scatterplot for these data.
   (b) Find the correlation r.
   (c) Can we find the best-fitting straight-line approximation to the above data? Does it make sense to use this equation to predict the number of ties owned by a corporate executive who wears size 10 shoes? Explain.

3. Following is a scatterplot of the average life expectancies and per capita incomes (in thousands of dollars) for people in a sample of 50 countries.

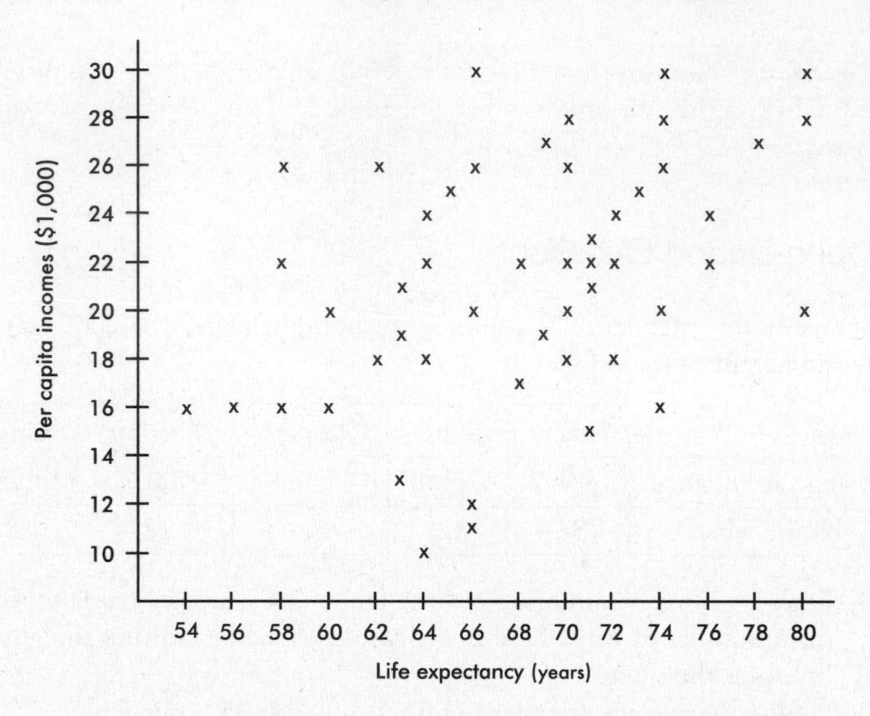

(a) Estimate the mean for the set of 50 life expectancies and for the set of 50 per capita incomes.
(b) Estimate the standard deviation for the set of life expectancies and for the set of per capita incomes. Explain your reasoning.
(c) Does the scatterplot show a correlation between per capita income and life expectancy? Is it positive or negative? Is it weak or strong?

4. (a) Find the correlation r for each of the three sets:

A = {(5, 5), (5, 10), (10, 5), (10, 10)}
B = {(50, 50), (50, 55), (55, 50), (55, 55)}
C = {(90, 90), (90, 95), (95, 90), (95, 95)}

(b) Find the correlation for the set consisting of the 12 scores from A, B, and C.
(c) Comment on the above results.

5. An outlier can have a striking effect on the correlation r. For example, comment on the following three scatterplots:

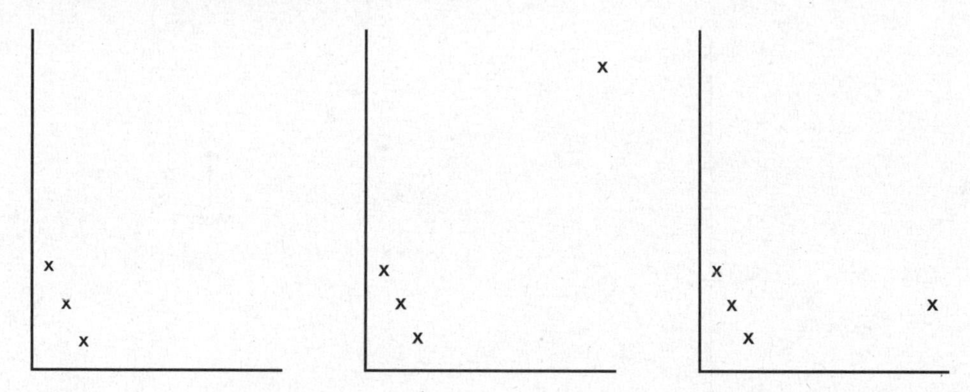

6. Fuel economy y (in miles per gallon) is tabulated for various speeds x (in miles per hour) for a certain car model. A linear regression model gives Predicted fuel economy = 34.8 – 0.16 (Speed) with the following residual plot:

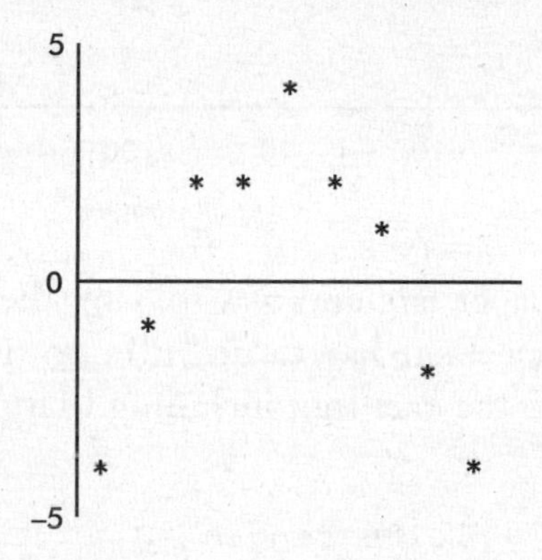

A quadratic regression model gives $\hat{y} = -0.0032x^2 + 0.26x + 23.8$ with the following residual plot:

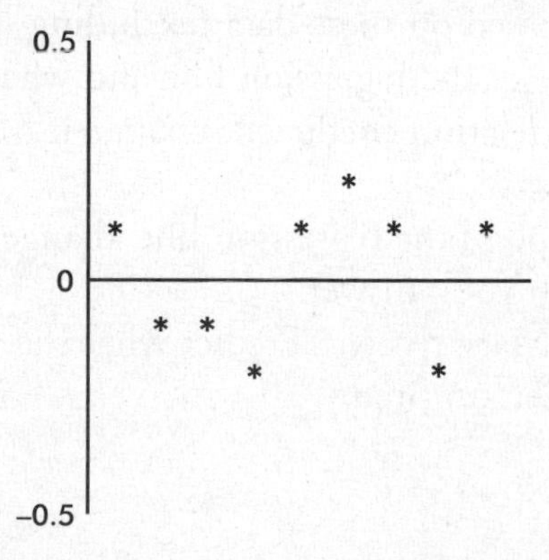

(a) What does each model predict for fuel economy at 50 miles per hour?
(b) Which model is a better fit? Explain.

7. The following scatterplot shows the grades for research papers for a sociology professor's class plotted against the lengths of the papers (in pages).

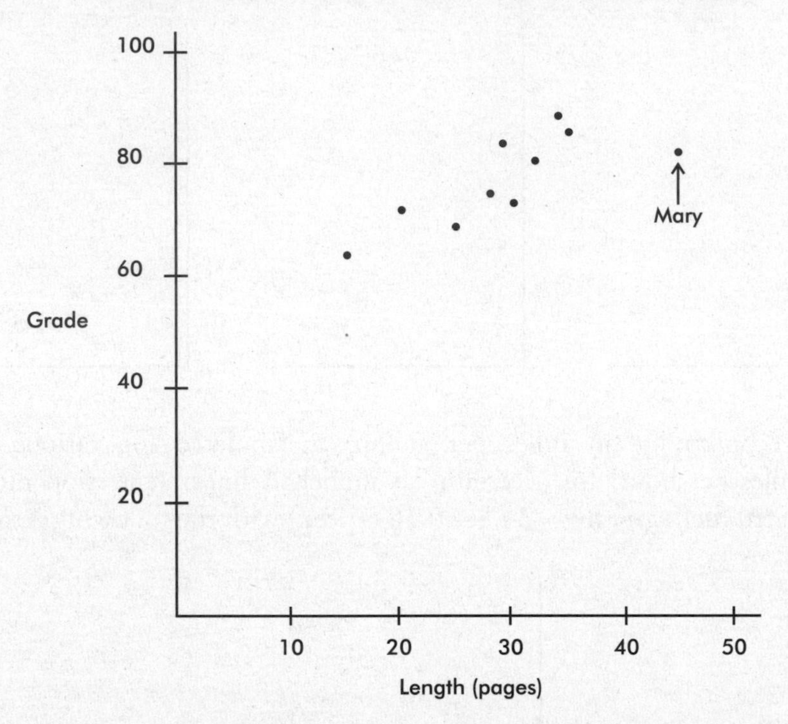

Mary turned in her paper late and was told by the professor that her grade would have been higher if she had turned it in on time. A computer printout fitting a straight line to the data (not including Mary's score) by the method of least squares gives

```
Grade = 46.51 + 1.106 Length
R–sq = 74.6%
```

(a) Find the correlation coefficient for the relationship between grade and length of paper based on these data (excluding Mary's paper).
(b) What is the slope of the regression line and what does it signify?
(c) How will the correlation coefficient change if Mary's paper is included? Explain your answer.
(d) How will the slope of the regression line change if Mary's paper is included? Explain your answer.
(e) What grade did Mary receive? Predict what she would have received if her paper had been on time.

8. Data show a trend in winning long jump distances for an international competition over the years 1972–92. With jumps recorded in inches and dates in years since 1900, a least squares regression line is fit to the data. The computer output and a graph of the residuals are as follows:

```
R squared = 92.1%
Variable    Coefficient    SE of Coeff    t-ratio    Prob
Constant      256.576         11.59         22.1     0.0001
Year          0.95893         0.141         6.81     0.0024
```

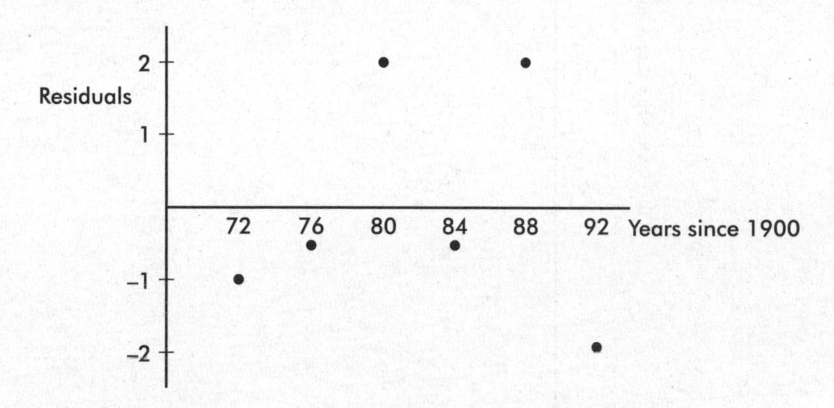

(a) Does a line appear to be an appropriate model? Explain.
(b) What is the slope of the least squares line? Give an interpretation of the slope.
(c) What is the correlation?
(d) What is the predicted winning distance for the 1980 competition?
(e) What was the actual winning distance in 1980?

9. A scatterplot of the number of accidents per day on a particular interstate highway during a 30-day month is as follows:

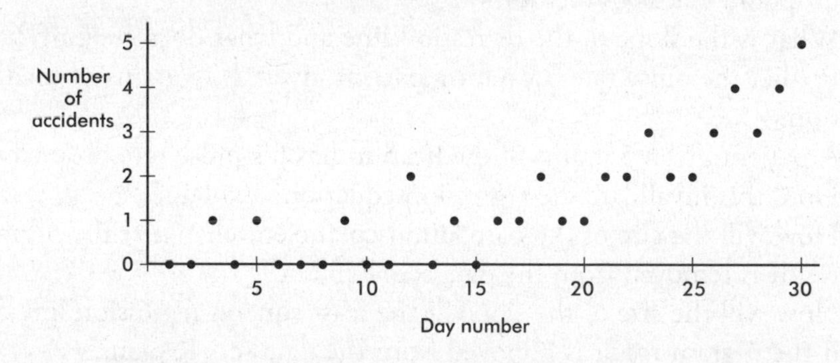

(a) Draw a histogram of the frequencies of the number of accidents.
(b) Draw a boxplot of the number of accidents.
(c) Name a feature apparent in the scatterplot but not in the histogram or boxplot.
(d) Name a feature clearly shown by the histogram and boxplot but not as obvious in the scatterplot.

10. The following scatterplot shows the pulse rate drop (in beats per minute) plotted against the amount of medication (in grams) of an experimental drug being field-tested in several hospitals.

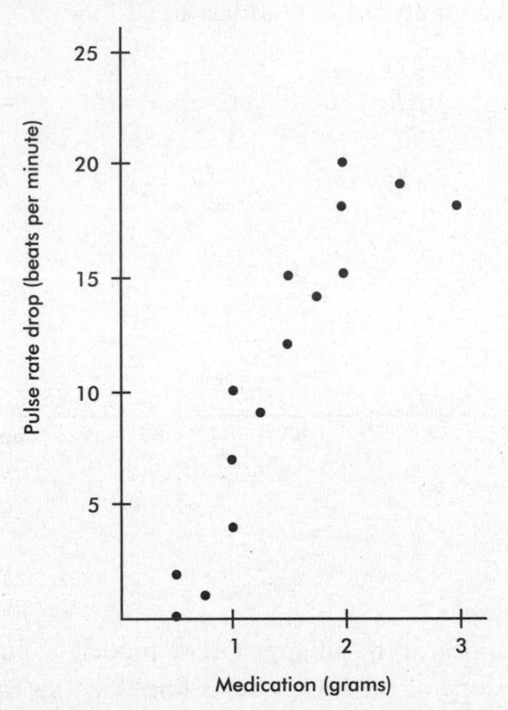

A computer printout showing the results of fitting a straight line to the data by the method of least squares gives

```
PulseRateDrop = −1.68 + 8.5 Grams
R−sq = 81.9%
```

(a) Find the correlation coefficient for the relationship between pulse rate drop and grams of medication.
(b) What is the slope of the regression line and what does it signify?
(c) Predict the pulse rate drop for a patient given 2.25 grams of medication.
(d) A patient given 5 grams of medication has his pulse rate drop to zero. Does this invalidate the regression equation? Explain.
(e) How will the size of the correlation coefficient change if the 3-gram result is removed from the data set? Explain.
(f) How will the size of the slope of the least squares regression line change if the 3-gram result is removed from the data set? Explain.

Answer Key

1. **E**	7. **E**	12. **E**	17. **C**	22. **E**
2. **A**	8. **C**	13. **B**	18. **B**	23. **E**
3. **D**	9. **E**	14. **D**	19. **E**	24. **A**
4. **D**	10. **C**	15. **C**	20. **A**	25. **B**
5. **E**	11. **B**	16. **B**	21. **B**	26. **E**
6. **A**				

Answers Explained

Multiple-Choice

1. **(E)** The variable column indicates the independent (explanatory) variable. The sign of the correlation is the same as the sign of the slope (negative here). In this example, the *y*-intercept is meaningless (predicted SAT result if no students take the exam). There can be a strong linear relation, with high R^2 value, but still a distinct pattern in the residual plot indicating that a non-linear fit may be even stronger. The negative value of the slope (–2.84276) gives that the predicted combined SAT score of a school is 2.84 points lower for each one unit higher in the percentage of students taking the exam, on average.

2. **(A)** Slope = $.15(\frac{42,000}{1.3}) \approx 4850$ and intercept = $208,000 - 4850(6.2) \approx 178,000$.

3. **(D)** Residual = Measured – Predicted, so if the residual is negative, the predicted must be greater than the measured (observed).

4. **(D)** The correlation coefficient is not changed by adding the same number to each value of one of the variables or by multiplying each value of one of the variables by the same positive number.

5. **(E)** A negative correlation shows a tendency for higher values of one variable to be associated with lower values of the other; however, given any two points, anything is possible.

6. **(A)** This is the only scatterplot in which the residuals go from positive to negative and back to positive.

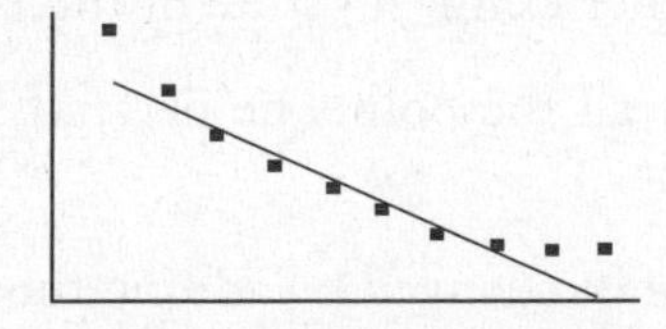

7. **(E)** Since (2, 5) is on the line $y = 3x + b$, we have $5 = 6 + b$ and $b = -1$. Thus the regression line is $y = 3x - 1$. The point $(\bar{x}, \bar{y})$ is always on the regression line, and so we have $\bar{y} = 3\bar{x} - 1$.

8. **(C)** The correlation r measures association, not causation.

9. **(E)** The correlation r cannot take a value greater than 1.

10. **(C)** If the points lie on a straight line, $r = \pm 1$. Correlation has the formula $r = \frac{1}{n-1}\sum\left(\frac{x_i - \bar{x}}{s_x}\right)\left(\frac{y_i - \bar{y}}{s_y}\right)$ so x and y are interchangeable, and r does not depend on which variable is called x or y. However, since means and standard deviations can be strongly influenced by outliers, r too can be strongly affected by extreme values. While $r = .75$ indicates a better fit with a linear model than $r = .25$ does, we cannot say that the linearity is threefold.

11. **(B)** The "Predictor" column indicates the independent variable with its coefficient to the right.

12. **(E)** $r = \sqrt{.986} = .993$

13. **(B)** $\widehat{back} = 0.056 + 0.920\,(0.55) = 0.562$ and so the *residual* = 0.59 – 0.562 = 0.028

14. **(D)** The sum and thus the mean of the residuals are always zero. In a good straight-line fit, the residuals show a random pattern.

15. **(C)** The coefficient of determination r^2 gives the proportion of the y-variance that is predictable from a knowledge of x. In this case $r^2 = (.632)^2 = .399$ or 39.9%.

16. **(B)** The point I doesn't contribute to a line with negative or positive slope. In none of the scatterplots do the points fall on a straight line, so none of them have correlation 1.0.

17. **(C)** Predicted winning percentage = 44 + 0.0003(34,000) = 54.2, and Residual = Observed – Predicted = 55 – 54.2 = 0.8.

18. **(B)** On each exam, two students had scores of 100. There is a general negative slope to the data showing a moderate negative correlation. The coefficient of determination, r^2, is always ≥ 0. While several students scored 90 or above on one or the other exam, no student did so on both exams.

19. **(E)** On the scatterplot all the points lie perfectly on a line sloping up to the right, and so $r = 1$.

20. **(A)** The correlation is not changed by adding the same number to every value of one of the variables, by multiplying every value of one of the variables by the same positive number, or by interchanging the x- and y-variables.

21. **(B)** The slope and the correlation coefficient have the same sign. Multiplying every y-value by –1 changes this sign.

22. **(E)** A scatterplot readily shows that while the first three points lie on a straight line, the fourth point does not lie on this line. Thus no matter what the fifth point is, all the points cannot lie on a straight line, and so r cannot be 1.

23. **(E)** All three scatterplots show very strong nonlinear patterns; however, the correlation r measures the strength of only a linear association. Thus $r = 0$ in the first two scatterplots and is close to 1 in the third.

24. **(A)** Using your calculator, find the regression line to be $\hat{y} = 9x - 8$. The regression line, also called the least squares regression line, minimizes the sum of the squares of the vertical distances between the points and the line. In this case (2, 10), (3, 19), and (4, 28) are on the line, and so the minimum sum is $(10 - 11)^2 + (19 - 17)^2 + (28 - 29)^2 = 6$.

25. **(B)** When transforming the variables leads to a linear relationship, the original variables have a nonlinear relationship, their correlation (which measures linearity) is not close to 1, and the residuals do not show a random pattern. While r close to 1 indicates strong association, it does not indicate cause and effect.

26. **(E)** The least squares line passes through $(\bar{x}, \bar{y}) = (2, 4)$, and the slope b satisfies $b = r\frac{s_y}{s_x} = \frac{5r}{3}$. Since $-1 \le r \le 1$, we have $-\frac{5}{3} \le b \le \frac{5}{3}$.

Free-Response

1. (a) A calculator gives $\widehat{\text{Attendance}} = 12{,}416 + 180.4\,(\text{Wins})$.
   (b) Each additional home win raises the average attendance by about 180 people, on average.
   (c) 12,416 + 180.4(25) = 16,926
   (d) 17,000 = 12,416 + 180.4(Wins) gives Wins = 25.4 so 26 wins needed to average at least 17,000 average attendance.
   (e) With 34 wins, the predicted average attendance is 12,416 + 180.4(34) = 18,550 so the residual is 18,997 – 18,550 = 447.

2. (a)

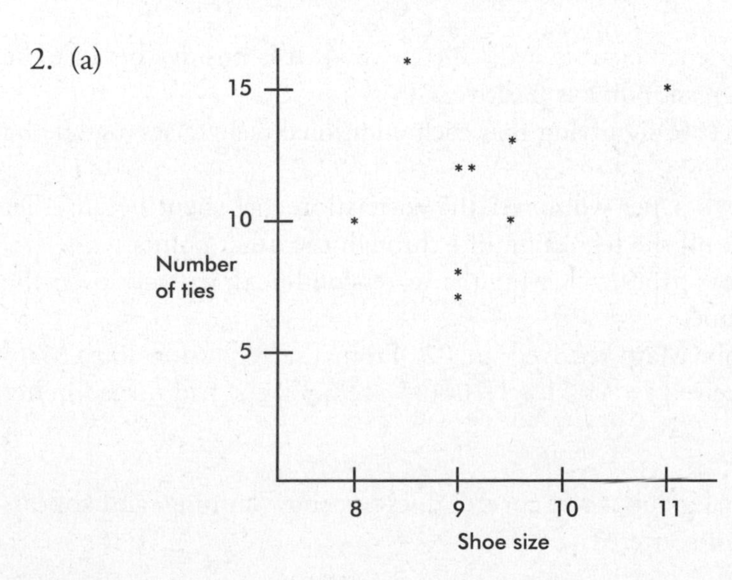

(b) A calculator gives $r = .1568$.

(c) The correlation r is low for this number of data scores, and the scatterplot shows no linear pattern whatsoever. Although theoretically we could use our techniques to find the best-fitting straight-line approximation, the result would be meaningless and should not be used for predictions.

3. (a) By visual inspection $\bar{x} \approx 68$ and $\bar{y} \approx 21$.

   (b) The range of the life expectancies is $80 - 54 = 26$, and so the standard deviation is roughly $\frac{26}{4} = 6.5$. Similarly the standard deviation of the per capita incomes is roughly $\frac{30-10}{4} = 5$.

   (c) While the points generally fall from the lower left to the upper right, they are still widely scattered. Thus the scatterplot shows a weak positive correlation between per capita income and life expectancy.

4. (a) The correlation for each of the three sets is 0.

   (b) The correlation for the set consisting of all 12 scores is .9948.

   (c) The data from each set taken separately show no linear pattern. However, together they show a strong linear fit. Note the positions of the data from the separate sets in the complete scatterplot.

5. In the first scatterplot, the points fall exactly on a downward sloping straight line, so $r = -1$. In the second scatterplot, the isolated point is an influential point, and r is close to $+1$. In the third scatterplot, the isolated point is also influential, and r is close to 0.

6. (a) $\hat{y} = -0.16(50) + 34.8 = 26.8$ miles per gallon, and $\hat{y} = -0.0032(50)^2 + 0.258(50) + 23.8 = 28.7$ miles per gallon.

   (b) Model 2 is the better fit. First, the residuals are much smaller for model 2, indicating that this model gives values much closer to the observed values. Second, a curved residual pattern like that in model 1 indicates that a non-linear model would be better. A more uniform residual scatter as in model 2 indicates a better fit.

7. (a) The correlation coefficient is $r = \sqrt{.746} = .864$. It is positive because the slope of the regression line is positive.

   (b) The slope is 1.106, signifying that each additional page raises a grade by 1.106.

   (c) Including Mary's paper will lower the correlation coefficient because her result seems far off the regression line through the other points.

   (d) Including Mary's paper will swing the regression line down and lower the value of the slope.

   (e) From the graph, Mary received an 82. From the regression line, Mary would have received $\hat{y} = 46.51 + 1.106(45) = 96.3$ if she had turned in her paper on time.

8. (a) Yes. The residual graph is not curved, does not show fanning, and appears to be random or scattered.

(b) The slope is 0.95893, indicating that the winning jump improves 0.95893 inches per year on average or about 3.8 inches every four years on average.
(c) With r^2 = .921, the correlation r is .96.
(d) 0.95893(80) + 256.576 ≈ 333.3 inches
(e) The residual for 1980 is +2, and so the actual winning distance must have been 333.3 + 2 = 335.3 inches.

9. 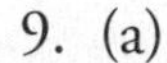
(a)

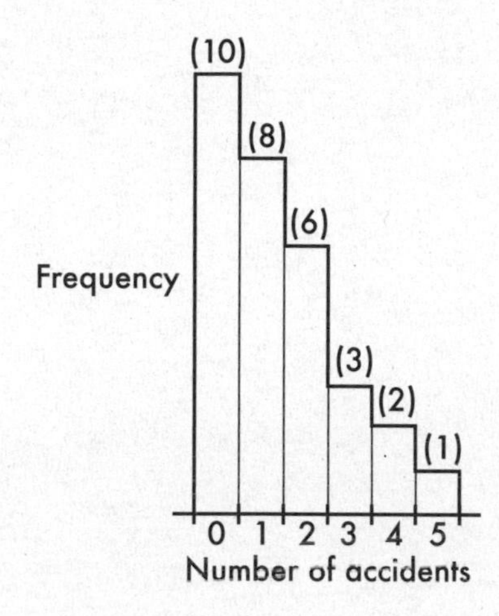

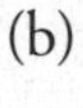
(b)

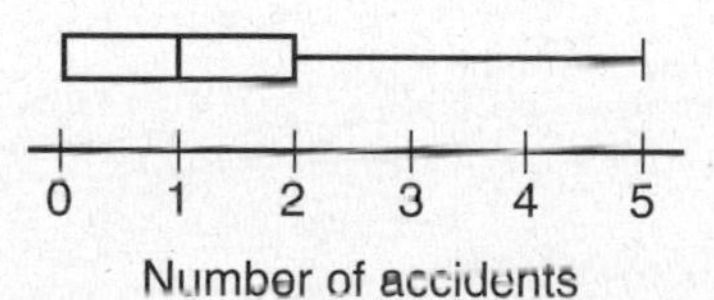

(c) There is a roughly linear trend with daily accidents increasing during the month.
(d) The daily number of accidents is strongly skewed to the right.

10. (a) The correlation coefficient $r = \sqrt{.819} = .905$. It is positive because the slope of the regression line is positive.
(b) The slope is 8.5, signifying that each gram of medication lowers the pulse rate by 8.5 beats per minute.
(c) $\hat{y} = -1.68 + 8.5(2.25) = 17.4$ beats per minute.
(d) There is always danger in using a regression line to extrapolate beyond the values of x contained in the data. In this case, the 5 grams was an overdose, the patient died, and the regression line cannot be used for such values beyond the data set.
(e) Removing the 3-gram result from the data set will increase the correlation coefficient because the 3-gram result appears to be far off a regression line through the remaining points.
(f) Removing the 3-gram result from the data set will swing the regression line upward so that the slope will increase.

Exploring Categorical Data: Frequency Tables

TOPIC 5

- Marginal Frequencies for Two-Way Tables
- Conditional Relative Frequencies and Association

While many variables such as age, income, and years of education are quantitative or numerical in nature, others such as gender, race, brand preference, mode of transportation, and type of occupation are qualitative or categorical. Quantitative variables, too, are sometimes grouped into categorical classes.

MARGINAL FREQUENCIES FOR TWO-WAY TABLES

Qualitative data often encompass two categorical variables that may or may not have a dependent relationship. These data can be displayed in a *two-way contingency table.*

EXAMPLE 5.1

A 4-year study, reported in *The New York Times*, on men more than 70 years old analyzed blood cholesterol and noted how many men with different cholesterol levels suffered nonfatal or fatal heart attacks.

	Low cholesterol	Medium cholesterol	High cholesterol
Nonfatal heart attacks	29	17	18
Fatal heart attacks	19	20	9

Severity of heart attacks is the *row variable*, while cholesterol level is the *column variable*.

One method of analyzing these data involves first calculating the totals for each row and each column:

	Low cholesterol	Medium cholesterol	High cholesterol	Total
Nonfatal heart attacks	29	17	18	64
Fatal heart attacks	19	20	9	48
Total	48	37	27	112

These totals are placed in the right and bottom margins of the table and thus are called *marginal frequencies.*

These marginal frequencies are often put in the form of proportions or percentages. The *marginal distribution* of the cholesterol level is

Low: $\frac{48}{112} = .429 = 42.9\%$

Medium: $\frac{37}{112} = .330 = 33.0\%$

High: $\frac{27}{112} = .241 = 24.1\%.$

This distribution can also be displayed in a bar graph as follows:

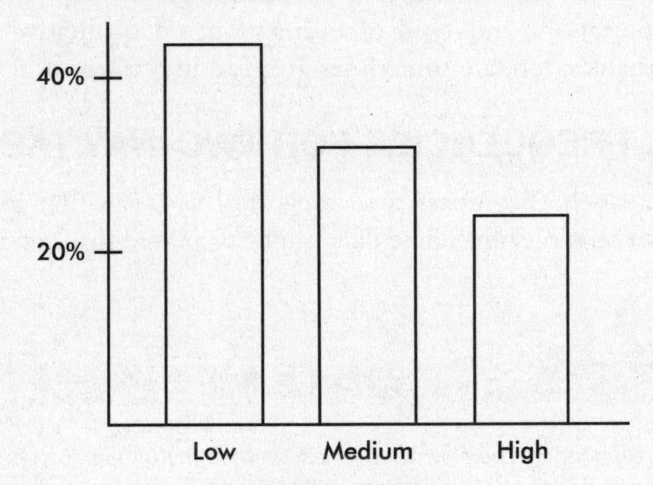

Cholesterol levels
(of elderly men who suffered heart attacks)

Similarly we can determine the marginal distribution for the severity of heart attacks:

Nonfatal: $\frac{64}{112} = .571 = 57.1\%$

Fatal: $\frac{48}{112} = .429 = 42.9\%.$

(continued)

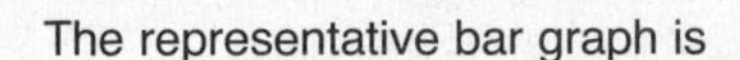

The representative bar graph is

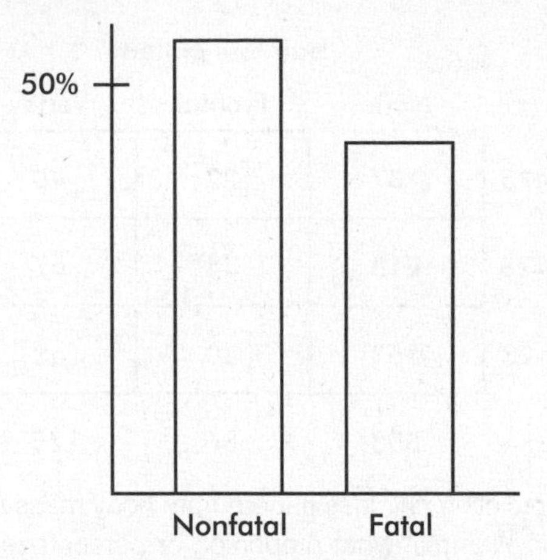

Severity of heart attacks
(of elderly men who suffered heart attacks)

CONDITIONAL RELATIVE FREQUENCIES AND ASSOCIATION

The marginal distributions described and calculated above do not describe or measure the relationship between the two categorical variables. For this we must consider the information in the body of the table, not just the sums in the margins.

EXAMPLE 5.2

Is hair loss pattern related to body mass index? One study (*Journal of the American Medical Association*, February 24, 1993, page 1000) of 769 men showed the following numbers:

		Hair loss pattern		
		None	Frontal	Vertex
Body mass index	<25	137	22	40
	25–28	218	34	67
	>28	153	30	68

(continued)

The analysis first involves finding the row and column totals as we did before.

		Hair loss pattern			
		None	Frontal	Vertex	
Body mass index	<25	137	22	40	199
	25–28	218	34	67	319
	>28	153	30	68	251
		508	86	175	769

We are interested in predicting hair loss pattern from body mass index, and so we look at each row separately. For example, what proportion or percentage of the 199 men with a body mass index less than 25 have each of the hair loss patterns?

None: $\frac{137}{199} = .688 = 68.8\%$

Frontal: $\frac{22}{199} = .111 = 11.1\%$

Vertex: $\frac{40}{199} = .201 = 20.1\%$.

These *conditional relative frequencies* can be displayed either with groupings of bars or by a segmented bar chart where each segment has a length corresponding to its relative frequency:

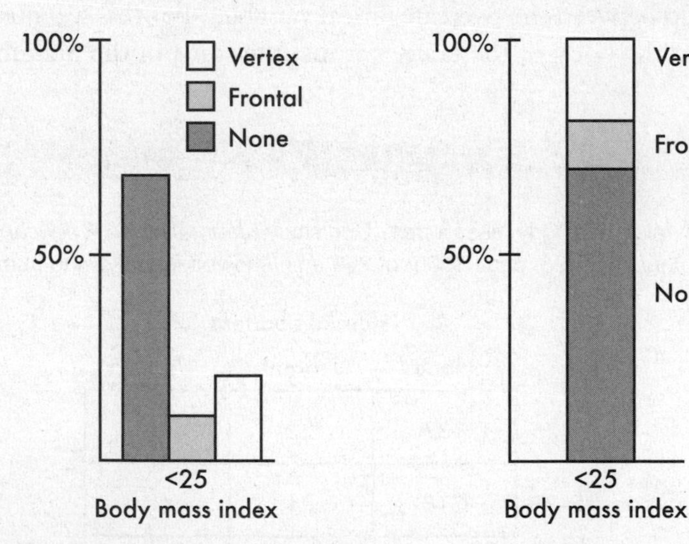

Similarly, the conditional relative frequencies for the 319 men with a body mass index between 25 and 28 are

None: $\frac{218}{319} = .683 = 68.3\%$

Frontal: $\frac{34}{319} = .107 = 10.7\%$

Vertex: $\frac{67}{319} = .210 = 21.0\%$.

(continued)

For the 251 men with a body mass index of more than 28 we have

None: $\frac{153}{251} = .610 = 61.0\%$

Frontal: $\frac{30}{251} = .120 = 12.0\%$

Vertex: $\frac{68}{251} = .271 = 27.1\%$.

Both of the following bar charts give good visual pictures:

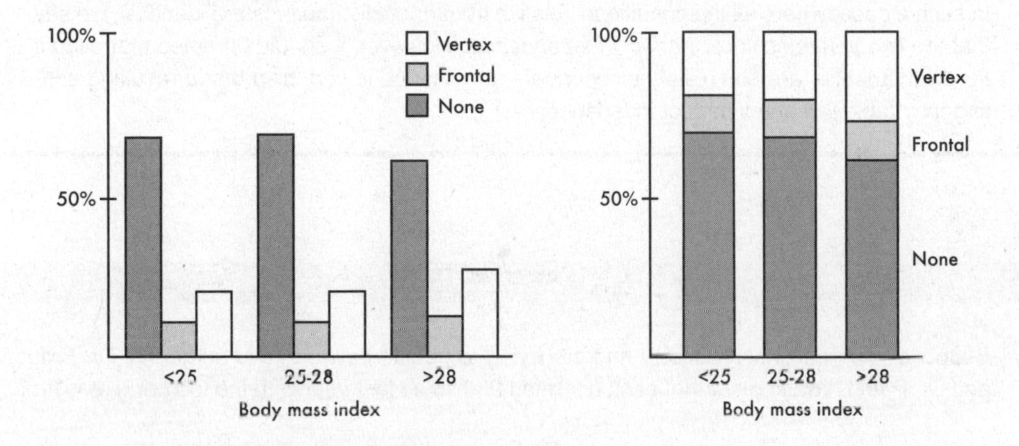

Segmented bar charts indicate a slight relationship between higher vertex pattern baldness and a body mass index of more than 28.

EXAMPLE 5.3

A study was made to compare year in high school with preference for vanilla or chocolate ice cream with the following results:

	Vanilla	Chocolate
Freshman	20	10
Sophomore	24	12
Junior	18	9
Senior	22	11

(continued)

What are the conditional relative frequencies for each class?

Freshmen: $\frac{20}{30} = .667$ prefer vanilla and $\frac{10}{30} = .333$ prefer chocolate.

Sophomores: $\frac{24}{36} = .667$ prefer vanilla and $\frac{12}{36} = .333$ prefer chocolate.

Juniors: $\frac{18}{27} = .667$ prefer vanilla and $\frac{9}{27} = .333$ prefer chocolate.

Seniors: $\frac{22}{33} = .667$ prefer vanilla and $\frac{11}{33} = .333$ prefer chocolate.

In such a case, where all the conditional relative frequency distributions are identical, we say that the two variables show *perfect independence.* (However, it should be noted that even if the two variables are completely independent, the chance is very slim that a resulting contingency table will show perfect independence.)

EXAMPLE 5.4

Suppose you need heart surgery and are trying to decide between two surgeons, Dr. Fixit and Dr. Patch. You find out that each operated 250 times last year with the following results:

	Dr. F	Dr. P
Died	60	50
Survived	190	200

Whom should you go to? Among Dr. Fixit's 250 patients 190 survived, for a survival rate of $\frac{190}{250} = .76$ or 76%, while among Dr. Patch's 250 patients 200 survived, for a survival rate of $\frac{200}{250} = .80$ or 80%. Your choice seems clear.

However, everything may not be so clear-cut. Suppose that on further investigation you determine that the surgeons operated on patients who were in either good or poor condition with the following results:

Good condition

	Dr. F	Dr. P
Died	8	17
Survived	60	120

Poor condition

	Dr. F	Dr. P
Died	52	33
Survived	130	80

Note that adding corresponding boxes from these two tables gives the original table above.

How do the surgeons compare when operating on patients in good health? Dr. Fixit's 68 patients in good condition have a survival rate of $\frac{60}{68} = .882$ or 88.2%, while Dr. Patch's 137 patients in good condition have a survival rate of $\frac{120}{137} = .876$ or 87.6%. Similarly, we note that Dr. Fixit's 182 patients in poor condition have a survival rate of $\frac{130}{182} = .714$ or 71.4%, while Dr. Patch's 113 patients in poor condition have a survival rate of $\frac{80}{113} = .708$ or 70.8%.

Thus Dr. Fixit does better with patients in good condition (88.2% versus Dr. Patch's 87.6%) and also does better with patients in poor condition (71.4% versus Dr. P's 70.8%). However, Dr. Fixit has a lower overall patient survival rate (76% versus Dr. Patch's 80%)! How can this be?

This problem is an example *of Simpson's paradox*, where a comparison can be reversed when more than one group is combined to form a single group. The effect of another variable, sometimes called a *lurking variable*, is masked when the groups are combined. In this particular example, closer scrutiny reveals that Dr. Fixit operates on many more patients in poor condition than Dr. Patch, and these patients in poor condition are precisely the ones with lower survival rates. Thus even though Dr. Fixit does better with all patients, his overall rating is lower. Our original table hid the effect of the lurking variable related to the condition of the patients.

Summary

- Two-way contingency tables are useful in showing relationships between two categorical variables.
- The row and column totals lead to calculations of the marginal distributions.
- Focusing on single rows or columns leads to calculations of conditional distributions.
- Segmented bar charts are a useful visual tool to show conditional distributions.
- Simpson's paradox occurs when the results from a combined grouping seem to contradict the results from the individual groups.

Questions on Topic Five: Exploring Categorical Data

Multiple-Choice Questions

Directions: The questions or incomplete statements that follow are each followed by five suggested answers or completions. Choose the response that best answers the question or completes the statement.

Questions 1–5 are based on the following: To study the relationship between party affiliation and support for a balanced budget amendment, 500 registered voters were surveyed with the following results:

	For	Against	No opinion
Democrat	50	150	50
Republican	125	50	25
Independent	15	10	25

1. What percentage of those surveyed were Democrats?

   (A) 10%
   (B) 20%
   (C) 30%
   (D) 40%
   (E) 50%

2. What percentage of those surveyed were for the amendment and were Republicans?

   (A) 25%
   (B) 38%
   (C) 40%
   (D) 62.5%
   (E) 65.8%

3. What percentage of Independents had no opinion?

   (A) 5%
   (B) 10%
   (C) 20%
   (D) 25%
   (E) 50%

4. What percentage of those against the amendment were Democrats?

(A) 30%
(B) 42%
(C) 50%
(D) 60%
(E) 71.4%

5. Voters of which affiliation were most likely to have no opinion about the amendment?

(A) Democrat
(B) Republican
(C) Independent
(D) Republican and Independent, equally
(E) Democrat, Republican, and Independent, equally

Questions 6–10 are based on the following: A study of music preferences in three geographic locations resulted in the following segmented bar chart:

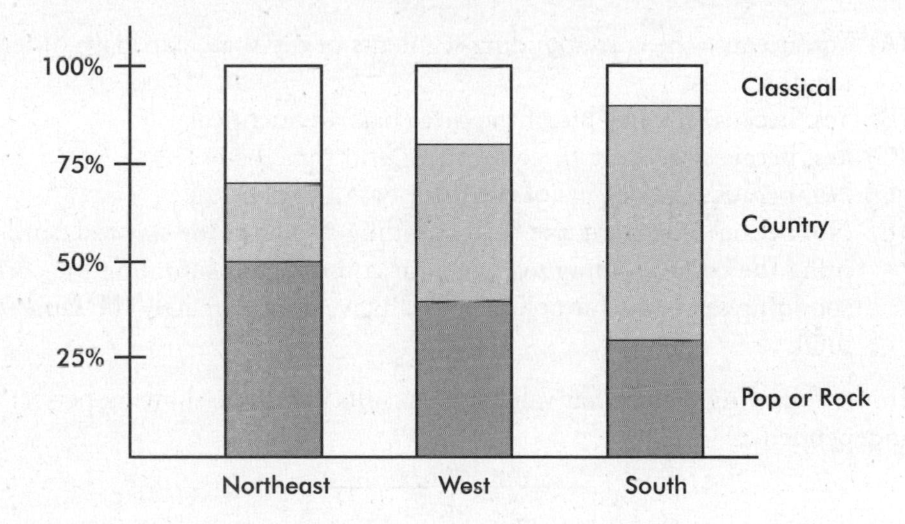

6. What percentage of those surveyed from the Northeast prefer country music?

(A) 20%
(B) 30%
(C) 40%
(D) 50%
(E) 70%

7. Which of the following is greatest?

(A) The percentage of those from the Northeast who prefer classical.
(B) The percentage of those from the West who prefer country.
(C) The percentage of those from the South who prefer pop or rock.
(D) The above are all equal.
(E) It is impossible to determine the answer without knowing the actual numbers of people involved.

8. Which of the following is greatest?

   (A) The number of people in the Northeast who prefer pop or rock.
   (B) The number of people in the West who prefer classical.
   (C) The number of people in the South who prefer country.
   (D) The above are all equal.
   (E) It is impossible to determine the answer without knowing the actual numbers of people involved.

9. All three bars have a height of 100%.

   (A) This is a coincidence.
   (B) This happened because each bar shows a complete distribution.
   (C) This happened because there are three bars each divided into three segments.
   (D) This happened because of the nature of musical patterns.
   (E) None of the above is true.

10. Based on the given segmented bar chart, does there seem to be a relationship between geographic location and music preference?

    (A) Yes, because the corresponding segments of the three bars have different lengths.
    (B) Yes, because the heights of the three bars are identical.
    (C) Yes, because there are three segments and three bars.
    (D) No, because the heights of the three bars are identical.
    (E) No, because summing the corresponding segments for classical, summing the corresponding segments for country, and summing the corresponding segments for pop or rock all give approximately the same total.

11. In the following table, what value for n results in a table showing perfect independence?

20	50
30	n

    (A) 10
    (B) 40
    (C) 60
    (D) 75
    (E) 100

12. A company employs both men and women in its secretarial and executive positions. In reports filed with the government, the company shows that the percentage of female employees who receive raises is higher than the percentage of male employees who receive raises. A government investigator claims that the percentage of male secretaries who receive raises is higher than the percentage of female secretaries who receive raises, and that the percentage of male executives who receive raises is higher than the percentage of female executives who receive raises. Is this possible?
    (A) No, either the company report is wrong or the investigator's claim is wrong.
    (B) No, if the company report is correct, then either a greater percentage of female secretaries than of male secretaries receive raises or a greater percentage of female executives than of male executives receive raises.
    (C) No, if the investigator is correct, then by summation of the corresponding numbers, the total percentage of male employees who receive raises would have to be greater than the total percentage of female employees who receive raises.
    (D) All of the above are true.
    (E) It is possible for both the company report to be true and the investigator's claim to be correct.

Free-Response Questions

Directions: You must show all work and indicate the methods you use. You will be graded on the correctness of your methods and on the accuracy of your final answers.

Two Open-Ended Questions

1. The following table gives the numbers (in thousands) of officers and enlisted personnel by military branch in the U.S. armed forces.

	Army	Navy	Marine Corps	Air Force
Officers	88	52	20	65
Enlisted	452	276	178	258

(a) Calculate the percentage

i. of military men and women who are enlisted.
ii. of military men and women who are not Marine Corps officers.
iii. of officers who are in the Navy.

(b) Construct a graphical display showing the association between career path (officer vs. enlisted) and military branch.
(c) Summarize what the graphical display illustrates about the association between career path (officer vs. enlisted) and military branch.

2. The graduate school at the University of California at Berkeley reported that in 1973 they accepted 44% of 8442 male applicants and 35% of 4321 female applicants. Concerned that one of their programs was guilty of gender bias, the graduate school analyzed admissions to the six largest graduate programs and obtained the following results:

Program	Men Accepted	Men Rejected	Women Accepted	Women Rejected
A	511	314	89	19
B	352	208	17	8
C	120	205	202	391
D	137	270	132	243
E	53	138	95	298
F	22	351	24	317

(a) Find the percentage of men and the percentage of women accepted by each program. Comment on any pattern or bias you see.

(b) Find the percentage of men and the percentage of women accepted overall by these six programs. Does this appear to contradict the results from part (a)?

(c) If you worked in the Graduate Admissions Office, what would you say to an inquiring reporter who is investigating gender bias in graduate admissions?

Answer Key

1. **E**	4. **E**	7. **B**	10. **A**
2. **A**	5. **C**	8. **E**	11. **D**
3. **E**	6. **A**	9. **B**	12. **E**

Answers Explained

Multiple-Choice

1. **(E)** Of the 500 people surveyed, 50 + 150 + 50 = 250 were Democrats, and $\frac{250}{500} = .5$ or 50%.

2. **(A)** Of the 500 people surveyed, 125 were both for the amendment and Republicans, and $\frac{125}{500} = .25$ or 25%.

3. **(E)** There were 15 + 10 + 25 = 50 Independents; 25 of them had no opinion, and $\frac{25}{50} = .5$ or 50%.

4. **(E)** There were 150 + 50 + 10 = 210 people against the amendment; 150 of them were Democrats, and $\frac{150}{210} = .714$ or 71.4%.

5. **(C)** The percentages of Democrats, Republicans, and Independents with no opinion are 20%, 12.5%, and 50%, respectively.

6. **(A)** In the bar corresponding to the Northeast, the segment corresponding to country music stretches from the 50% level to the 70% level, indicating a length of 20%.

7. **(B)** Based on lengths of indicated segments, the percentage from the West who prefer country is the greatest.

8. **(E)** The given bar chart shows percentages, not actual numbers.

9. **(B)** In a complete distribution, the probabilities sum to 1, and the relative frequencies total 100%.

10. **(A)** The different lengths of corresponding segments show that in different geographic regions different percentages of people prefer each of the music categories.

11. **(D)** Relative frequencies must be equal. Either looking at rows gives $\frac{20}{70} = \frac{30}{30+n}$ or looking at columns gives $\frac{20}{50} = \frac{50}{50+n}$. We could also set up a proportion $\frac{n}{30} = \frac{50}{20}$ or $\frac{n}{50} = \frac{30}{20}$. Solving any of these equations gives $n = 75$.

12. **(E)** It is possible for both to be correct, for example, if there were 11 secretaries (10 women, 3 of whom receive raises, and 1 man who receives a raise) and 11 executives (10 men, 1 of whom receives a raise, and 1 woman who does not receive a raise). Then 100% of the male secretaries receive raises while only 30% of the female secretaries do; and 10% of the male executives receive raises

while 0% of the female executives do. However, overall 3 out of 11 women receive raises, while only 2 out of 11 men receive raises. This is an example of Simpson's paradox.

Free-Response

1. (a) i. $\frac{1164}{1389} \approx 83.80\%$

   ii. $1 - \frac{20}{1389} \approx 98.56\%$

   iii. $\frac{52}{225} \approx 23.11\%$

   (b) Calculate row or column totals, and then show either a side-by-side bar graph or a segmented bar graph, showing percentages, and conditioned on either career path (officer vs. enlisted) or military branch:

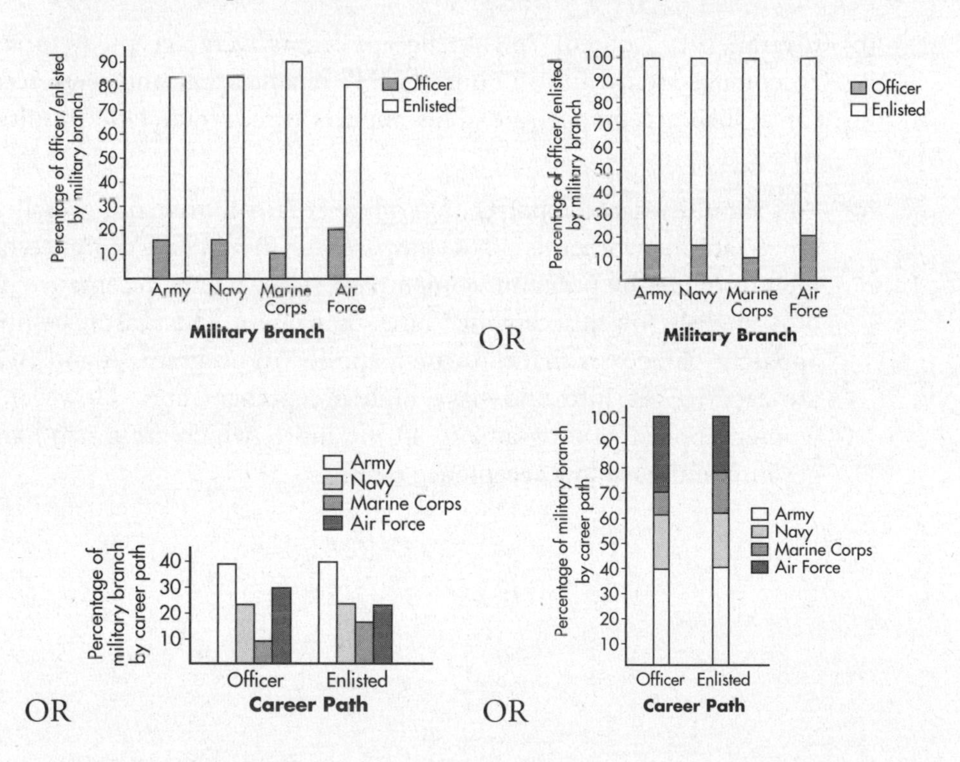

   (c) The Army and the Navy have about the same percentage of officers (16%), while the Air Force has a higher percentage of officers (20%), and the Marine Corps has a lower percentage of officers (10%).

   OR

   Among the officers and the enlisted career paths there are about the same percentage Army (39%), and about the same percentage Navy (23%), while the officers have a lower percentage Marine Corps than the enlisted (9% vs. 16%) and the officers have a higher percentage Air Force than the enlisted (29% vs. 22%).

2. (a)

Program	Percentage of Men Accepted (%)	Percentage of Women Accepted (%)
A	62	82
B	63	68
C	37	34
D	33	35
E	28	24
F	6	7

There doesn't appear to be any real pattern; however, women seem to be favored in four of the programs, while men seem to be slightly favored in the other two programs.

(b) Overall, 1195 out of 2681 male applicants were accepted, for a 45% acceptance rate, while 559 out of 1835 female applicants were accepted, for a 30% acceptance rate. This appears to contradict the results from part *a*.

(c) You should tell the reporter that while it is true that the overall acceptance rate for women is 30% compared to the 44% acceptance rate for men, program by program women have either higher acceptance rates or only slightly lower acceptance rates than men. The reason behind this apparent paradox is that most men applied to programs A and B, which are easy to get into and have high acceptance rates. However, most women applied to programs C, D, E, and F, which are much harder to get into and have low acceptance rates.

THEME TWO: PLANNING A STUDY

Overview of Methods of Data Collection

• Census	• Experiment
• Sample Survey	• Observational Study

In the real world, time and cost considerations usually make it impossible to analyze an entire population. Does the government question you and your parents before announcing the monthly unemployment rates? Does a television producer check every household's viewing preferences before deciding whether a pilot program will be continued? In studying statistics we learn how to estimate *population* characteristics by considering a *sample*. For example, later in this book we will see how to estimate population means and proportions by looking at sample means and proportions.

To derive conclusions about the larger population, we need to be confident that the sample we have chosen represents that population fairly. Analyzing the data with computers is often easier than gathering the data, but the frequently quoted "Garbage in, garbage out" applies here. Nothing can help if the data are badly collected. Unfortunately, many of the statistics with which we are bombarded by newspapers, radio, and television are based on poorly designed data collection procedures.

CENSUS

A *census* is a complete enumeration of an entire population. In common use, it is often thought of as an official attempt to contact every member of the population, usually with details regarding age, marital status, race, gender, occupation, income, years of school completed, and so on. Every 10 years the U.S. Bureau of the Census divides the nation into nine regions and attempts to gather information about everyone in the country. A massive amount of data is obtained, but even with the resources of the U.S. government, the census is not complete. For example, many homeless people are always missed, or counted at two temporary residences, and there are always households that do not respond even after repeated requests for information. It was estimated that the 2000 census missed about 3.3 million people (1.2% of the population).

In most studies, both in the private and public sectors, a complete census is unreasonable because of time and cost involved. Furthermore, attempts to gather complete data have been known to lead to carelessness. Finally, and most important, a well-designed, well-conducted sample survey is far superior to a poorly designed study involving a complete census. For example, a poorly worded question might give meaningless data even if everyone in the population answers.

SAMPLE SURVEY

The census tries to count everyone; it is not a sample. A sample survey aims to obtain information about a whole population by studying a part of it, that is, a sample. The goal is to gather information without disturbing or changing the population. Numerous procedures are used to collect data through sampling, and much of the statistical information distributed to us comes from sample surveys. Often, controlled experiments are later undertaken to demonstrate relationships suggested by sample surveys.

However, the one thing that most quickly invalidates a sample and makes useful information impossible to obtain is *bias.* A sample is biased if in some critical way it does not represent the population. The main technique to avoid bias is to incorporate *randomness* into the selection process. Randomization protects us from effects and influences, both known and unknown. Finally, the larger the sample, the better the results, but what is critical is the sample size, not the percentage or fraction of the population. That is, a random sample of size 500 from a population of size 100,000 is just as representative as a random sample of size 500 from a population of size 1,000,000.

EXPERIMENT

In a controlled study, called an *experiment,* the researcher should randomly divide subjects into appropriate groups. Some action is taken on one or more of the groups, and the response is observed. For example, patients may be randomly given unmarked capsules of either aspirin or acetaminophen and the effects of the medication measured. Experiments often have a *treatment group* and a *control group*; in the ideal situation, neither the subjects nor the researcher knows which group is which. The Salk vaccine experiment of the 1950s, in which half the children received the vaccine and half were given a placebo, with not even their doctors knowing who received what, is a classic example of this *double-blind* approach. Controlled experiments can indicate cause-and-effect relationships.

The critical principles behind good experimental design include *control* (outside of who receives what treatments, conditions should be as similar as possible for all involved groups), *blocking* (the subjects can be divided into representative groups to bring certain differences directly into the picture), *randomization* (unknown and uncontrollable differences are handled by randomizing who receives what treatments), *replication* (treatments need to be repeated on a sufficient number of subjects), and *generalizability* (ability to repeat an experiment in a variety of settings).

OBSERVATIONAL STUDY

Sample surveys are one example of what are called *observational studies*. In observational studies there is no choice in regard to who goes into the treatment and control groups. For example, a researcher cannot ethically tell 100 people to smoke three packs of cigarettes a day and 100 others to smoke only one pack per day; he can only observe people who habitually smoke these amounts. In observational studies the researcher strives to determine which variables affect the noted response. While results may suggest relationships, it is difficult to conclude cause and effect.

Observational studies are primary, vital sources of data; however, they are a poor method of measuring the effect of change. To evaluate responses to change, one must impose change, that is, perform an experiment. Furthermore, observational studies on the impact of some variable on another variable often fail because explanatory variables are *confounded* with other variables.

Summary

- A complete census is usually unreasonable because of time and cost constraints.
- Estimate population characteristics (called parameters) by considering statistics from a sample.
- Analysis of badly gathered sample data is usually a meaningless exercise.
- A sample is biased if in some critical way it does not represent the population.
- The main technique to avoid bias is to incorporate randomness into the selection process.
- Experiments involve applying a treatment to one or more groups and observing the responses.
- Observational studies involve observing responses to choices people make.

Questions on Topic Six: Overview of Methods of Data Collection

Multiple-Choice Questions

Directions: The questions or incomplete statements that follow are each followed by five suggested answers or completions. Choose the response that best answers the question or completes the statement.

1. When travelers change airlines during connecting flights, each airline receives a portion of the fare. Several years ago, the major airlines used a sample trial period to determine what percentage of certain fares each should collect. Using these statistical results to determine fare splits, the airlines now claim huge savings over previous clerical costs. Which of the following is true?

    I. The airlines ran an experiment using a trial period for the control group.
    II. The airlines ran an observational study using the calculations from a trial period as a sample.
    III. The airlines feel that any monetary error in fare splitting resulting from using a statistical sample is smaller than the previous clerical costs necessary to calculate exact fare splits.

    (A) I only
    (B) II only
    (C) III only
    (D) I and III
    (E) II and III

2. Which of the following are true statements?

    I. In an experiment some treatment is intentionally forced on one group to note the response.
    II. In an observational study information is gathered on an already existing situation.
    III. Sample surveys are observational studies, not experiments.

    (A) I and II
    (B) I and III
    (C) II and III
    (D) I, II, and III
    (E) None of the above gives the complete set of true responses.

3. Which of the following are true statements?

    I. In an experiment researchers decide how people are placed in different groups.
    II. In an observational study, the participants select which group they are in.
    III. A control group is most often a self-selected grouping in an experiment.

    (A) I and II
    (B) I and III
    (C) II and III
    (D) I, II, and III
    (E) None of the above gives the complete set of true responses.

4. In one study on the effect of niacin on cholesterol level, 100 subjects who acknowledged being long-time niacin takers had their cholesterol levels compared with those of 100 people who had never taken niacin. In a second study, 50 subjects were randomly chosen to receive niacin and 50 were chosen to receive a placebo.

    (A) The first study was a controlled experiment, while the second was an observational study.
    (B) The first study was an observational study, while the second was a controlled experiment.
    (C) Both studies were controlled experiments.
    (D) Both studies were observational studies.
    (E) Each study was part controlled experiment and part observational study.

5. In one study subjects were randomly given either 500 or 1000 milligrams of vitamin C daily, and the number of colds they came down with during a winter season was noted. In a second study people responded to a questionnaire asking about the average number of hours they sleep per night and the number of colds they came down with during a winter season.

    (A) The first study was an experiment without a control group, while the second was an observational study.
    (B) The first study was an observational study, while the second was a controlled experiment.
    (C) Both studies were controlled experiments.
    (D) Both studies were observational studies.
    (E) None of the above is a correct statement.

6. In a 1992 London study, 12 out of 20 migraine sufferers were given chocolate whose flavor was masked by peppermint, while the remaining eight sufferers received a similar-looking, similar-tasting tablet that had no chocolate. Within 1 day, five of those receiving chocolate complained of migraines, while no complaints were made by any of those who did not receive chocolate. Which of the following is a true statement?

   (A) This study was an observational study of 20 migraine sufferers in which it was noted how many came down with migraines after eating chocolate.
   (B) This study was a sample survey in which 12 out of 20 migraine sufferers were picked to receive peppermint-flavored chocolate.
   (C) A census of 20 migraine sufferers was taken, noting how many were given chocolate and how many developed migraines.
   (D) A study was performed using chocolate as a placebo to study one cause of migraines.
   (E) An experiment was performed comparing a treatment group that was given chocolate to a control group that was not.

7. Suppose you wish to compare the average class size of mathematics classes to the average class size of English classes in your high school. Which is the most appropriate technique for gathering the needed data?

   (A) Census
   (B) Sample survey
   (C) Experiment
   (D) Observational study
   (E) None of these methods is appropriate.

8. Which of the following are true statements?

   I. Based on careful use of control groups, experiments can often indicate cause-and-effect relationships.
   II. While observational studies may suggest relationships, great care must be taken in concluding that there is cause and effect because of the lack of control over lurking variables.
   III. A complete census is the only way to establish a cause-and-effect relationship absolutely.

   (A) I and II
   (B) I and III
   (C) II and III
   (D) I, II, and III
   (E) None of the above gives the complete set of true responses.

9. Two studies are run to compare the experiences of families living in high-rise public housing to those of families living in townhouse subsidized rentals. The first study interviews 25 families who have been in each government program for at least 1 year, while the second randomly assigns 25 families to each program and interviews them after 1 year. Which of the following is a true statement?

   (A) Both studies are observational studies because of the time period involved.
   (B) Both studies are observational studies because there are no control groups.
   (C) The first study is an observational study, while the second is an experiment.
   (D) The first study is an experiment, while the second is an observational study.
   (E) Both studies are experiments.

10. Two studies are run to determine the effect of low levels of wine consumption on cholesterol level. The first study measures the cholesterol levels of 100 volunteers who have not consumed alcohol in the past year and compares these values with their cholesterol levels after 1 year, during which time each volunteer drinks one glass of wine daily. The second study measures the cholesterol levels of 100 volunteers who have not consumed alcohol in the past year, randomly picks half the group to drink one glass of wine daily for a year while the others drink no alcohol for the year, and finally measures their levels again. Which of the following is a true statement?

    (A) The first study is an observational study, while the second is an experiment.
    (B) The first study is an experiment, while the second is an observational study.
    (C) Both studies are observational studies, but only one uses both randomization and a control group.
    (D) The first study is a census of 100 volunteers, while the second study is an experiment.
    (E) Both studies are experiments.

Answer Key

1. **E**	3. **A**	5. **A**	7. **A**	9. **C**
2. **D**	4. **B**	6. **E**	8. **A**	10. **E**

Answers Explained

Multiple-Choice

1. **(E)** This study is not an experiment in which responses are being compared. It is an observational study in which the airlines use split fare calculations from a trial period as a sample to indicate the pattern of all split fare transactions. They claim that this leads to "huge savings."

2. **(D)** I and II can be considered part of the definitions of *experiment* and *observational study.* A sample survey does not impose any treatment; it simply counts a certain outcome, and so it is an observational study, not an experiment.

3. **(A)** In an experiment a control group is the untreated group picked by the researchers. It is usually best when the selection process involves chance.

4. **(B)** The first study was observational because the subjects were not chosen for treatment.

5. **(A)** The first study was an experiment with two treatment groups and no control group. The second study was observational; the researcher did not randomly divide the subjects into groups and have each group sleep a designated number of hours per night.

6. **(E)** This study was an experiment in which the researchers divided the subjects into treatment and control groups. A census would involve a study of all migraine sufferers, not a sample of 20. The response of the treatment group receiving chocolate was compared to the response of the control group receiving a placebo. The peppermint tablet with no chocolate was the placebo.

7. **(A)** The main office at your school should be able to give you the class sizes of every math and English class. If need be, you can check with every math and English teacher.

8. **(A)** A complete census can provide much information about a population, but it doesn't necessarily establish a cause-and-effect relationship among seemingly related population parameters.

9. **(C)** In the first study the families were already in the housing units, while in the second study one of two treatments was applied to each family.

10. **(E)** Both studies apply treatments and measure responses, and so both are experiments.

TOPIC 7

Planning and Conducting Surveys

- Simple Random Sampling
- Characteristics of a Well-Designed, Well-Conducted Survey
- Sampling Error
- Sources of Bias
- Other Sampling Methods

Most data collection involves observational studies, not controlled experiments. Furthermore, while most data collection has some purpose, many studies come to mind after the data have been assembled and examined. For data collection to be useful, the resulting sample must be representative of the population under consideration.

SIMPLE RANDOM SAMPLING

How can a good, that is, a representative, sample be chosen? The most accurate technique would be to write the name of each member of the population on a card, mix the cards thoroughly in a large box, and pull out a specified number of cards. This method would give everyone in the population an equal chance of being selected as part of the sample. Unfortunately, this method is usually too time-consuming and too costly, and bias might still creep in if the mixing is not thorough. A *simple random sample*, that is, **one in which every possible sample of the desired size has an equal chance of being selected**, can more easily be obtained by assigning a number to everyone in the population and using a random number table or having a computer generate random numbers to indicate choices.

EXAMPLE 7.1

Suppose 80 students are taking an AP Statistics course and the teacher wants to randomly pick out a sample of 10 students to try out a practice exam. She first assigns the students numbers 01, 02, 03, . . . , 80. Reading off two digits at a time from a random number table, she ignores any over 80 and ignores repeats, stopping when she has a set of ten. If the table began 75425 56573 90420 48642 27537 61036 15074 84675, she would choose the students numbered 75, 42, 55, 65, 73, 04, 27, 53, 76, and 10. Note that 90 and 86 are ignored because they are over 80, and the second and third occurrences of 42 are ignored because they are repeats.

CHARACTERISTICS OF A WELL-DESIGNED, WELL-CONDUCTED SURVEY

A well-designed survey always incorporates chance, such as using random numbers from a table or a computer. However, the use of probability techniques is not enough to ensure a representative sample. Often we don't have a complete listing of the population, and so we have to be careful about exactly how we are applying "chance." Even when subjects are picked by chance, they may choose not to respond to the survey or they may not be available to respond, thus calling into question how representative the final sample really is. The wording of the questions must be neutral—subjects give different answers depending on the phrasing.

EXAMPLE 7.2

Suppose we are interested in determining the percentage of adults in a small town who eat a nutritious breakfast. How about randomly selecting 100 numbers out of the telephone book, calling each one, and asking whether the respondent is intelligent enough to eat a nutritious breakfast every morning?

Answer: Random selection is good, but a number of questions should be addressed. For example, are there many people in the town without telephones or with unlisted numbers? How will the time of day the calls are made affect whether the selected people are reachable? If people are unreachable, will replacements be randomly chosen in the same way or will this lead to a certain class of people being underrepresented? Finally, even if these issues are satisfactorily addressed, the wording of the question is clearly not neutral—unless the phrase *intelligent enough* is dropped, answers will be almost meaningless.

SAMPLING ERROR: THE VARIATION INHERENT IN A SURVEY

No matter how well-designed and well-conducted a survey is, it still gives a sample *statistic* as an estimate for a population *parameter*. Different samples give different sample statistics, all of which are estimates for the same population parameter, and so error, called *sampling error*, is naturally present. This error can be described using probability; that is, we can say how likely we are to have a certain size error. Generally, the chance of this error occurring is smaller when the sample size is larger. However, the way the data are obtained is crucial—a large sample size cannot make up for a poor survey design or faulty collection techniques.

EXAMPLE 7.3

Each of four major news organizations surveys likely voters and separately reports that the percentage favoring the incumbent candidate is 53.4%, 54.1%, 52.0%, and 54.2%, respectively. What is the correct percentage? Did three or more of the news organizations make a mistake?

Answer: There is no way of knowing the correct population percentage from the information given. The four surveys led to four statistics, each an estimate of the population parameter. No one made a mistake unless there was a bad survey, for example, one without the use of chance, or not representative of the population, or with poor wording of the question. Sampling differences are natural.

SOURCES OF BIAS IN SURVEYS

Poorly designed sampling techniques result in *bias,* that is, in a tendency to favor the selection of certain members of a population. If a study is biased, size doesn't help—a large sample size will simply result in a large worthless study. Think about bias before running a study, because once all the data comes in, there is no way to recover if the sample was biased. Sometimes pilot testing with a small sample will show bias that can be corrected before a larger sample is obtained. Although each of the following sources of bias is defined separately, there is overlap, and many if not most examples of bias involve more than one of the following.

TIP

Sampling error is to be expected, while bias is to be avoided.

Household bias: When a sample includes only one member of any given household, members of large households are underrepresented. To respond to this, pollsters sometimes give greater weight to members of larger households.

Nonresponse bias: A good example is that of most mailed questionnaires, as they tend to have very low response percentages, and it is often unclear which part of the population is responding. Sometimes people chosen for a survey simply refuse to respond or are unreachable or too difficult to contact. Answering machines and caller ID prevent easy contacts. To maximize response rates, one can use multiple follow-up contacts and cash or other incentives. Also, short, easily understood surveys generally have higher response rates.

Quota sampling bias: This results when interviewers are given free choice in picking people, for example, to obtain a particular percentage men, a particular percentage Catholic, or a particular percentage African-American. This flawed technique resulted in misleading polls leading to the *Chicago Tribune* making an early incorrect call of Thomas E. Dewey as the winner over Harry S. Truman in the 1948 presidential election.

Response bias: The very question itself can lead to misleading results. People often don't want to be perceived as having unpopular or unsavory views and so may respond untruthfully when face to face with an interviewer or when filling out a questionnaire that is not anonymous. Patients may lie about following doctors' orders, dieters may be dishonest about how strictly they've followed a weight loss program, students may shade the truth about how many hours they've studied for exams, and viewers may not want to admit they they watch certain television programs.

TIP

Think about potential bias before collecting data.

Selection bias: An often-cited example is the *Literary Digest* opinion poll that predicted a landslide victory for Alfred Landon over Franklin D. Roosevelt in the 1936 presidential election. The *Digest* surveyed people with cars and telephones, but in 1936 only the wealthy minority, who mainly voted Republican, had cars and telephones. In spite of obtaining more than two million responses, the *Digest* picked a landslide for the wrong man!

Size bias: Throwing darts at a map to decide in which states to sample would bias in favor of geographically large states. Interviewing people checking out of the hospital would bias in favor of patients with short stays, since due to costs, more people today have shorter stays. Having each student pick one coin out of a bag of 1000 coins to help estimate the total monetary value of the coins in the bag would bias in favor of large coins, for example, quarters over dimes.

Undercoverage bias: This happens when there is inadequate representation. For example, telephone surveys simply ignore all those possible subjects who don't have telephones. In the 2008 presidential election surveys, phone surveys went only to land line phones, leaving out many young adults who have only cell phones. Another example is *convenience samples,* like interviews at shopping malls, which are based on choosing individuals who are easy to reach. These interviews tend to produce data highly unrepresentative of the entire population. Door-to-door household surveys typically miss college students and prison inmates, as well as the homeless.

Voluntary response bias: Samples based on individuals who offer to participate typically give too much emphasis to people with strong opinions. For example, radio call-in programs about controversial topics such as gun control, abortion, and school segregation do not produce meaningful data on what proportion of the population favor or oppose related issues. Online surveys posted to websites are a modern source of voluntary response bias.

Wording bias: Nonneutral or poorly worded questions may lead to answers that are very unrepresentative of the population. To avoid such bias, do not use *leading* questions, and write questions that are clear and relatively short. Also be careful of sequences of questions that lead respondents toward certain answers.

Note: Again, it should be understood that there is considerable overlap among the above classifications. For example, a nonneutral question may be said to have both *response* bias and *wording* bias. *Selection* bias and *undercoverage* bias often go hand in hand. *Voluntary response* bias and *nonresponse* bias are clearly related.

OTHER SAMPLING METHODS

Time- and cost-saving modifications are often used to implement sampling procedures other than simple random samples.

Systematic sampling involves listing the population in some order (for example, alphabetically), choosing a random point to start, and then picking every tenth (or hundredth, or thousandth, or *k*th) person from the list. This gives a reasonable sample as long as the original order of the list is not in any way related to the variables under consideration.

In *stratified sampling* the population is divided into *homogeneous* groups called *strata*, and random samples of persons from all strata are chosen. For example, we can stratify by age or gender or income level or race and pick a sample of people from each stratum. Note that all individuals in a given stratum have a characteristic in common. We could further do *proportional sampling*, where the sizes of the random samples from each stratum depend on the proportion of the total population represented by the stratum.

TIP

Know the difference between strata and clusters.

In *cluster sampling* the population is divided into *heterogeneous* groups called *clusters*, and we then take a random sample of clusters from among all the clusters. For example, to survey high school seniors we could randomly pick several senior class homerooms in which to conduct our study. Note that each cluster should resemble the entire population.

Multistage sampling refers to a procedure involving two or more steps, each of which could involve any of the various sampling techniques. The Gallup organization often follows a procedure in which nationwide locations are randomly selected, then neighborhoods are randomly selected in each of these locations, and finally households are randomly selected in each of these neighborhoods.

EXAMPLE 7.4

Suppose a sample of 100 high school students from a school of size 5000 is to be chosen to determine their views on the death penalty. One method would be to have each student write his or her name on a slip of paper, put the papers in a box, and have the principal reach in and pull out 100 of the papers. However, questions could arise regarding how well the papers are mixed up in the box. For example, how might the outcome be affected if all students in one homeroom toss in their names at the same time so that their papers are clumped together? Another method would be to assign each student a number from 1 to 5000 and then use a random number table, picking out four digits at a time and tossing out repeats and numbers over 5000 (simple random sampling). What are alternative procedures?

Answer: From a list of the students, the surveyor could simply note every fiftieth name (systematic sampling). Since students in each class have certain characteristics in common, the surveyor could use a random selection method to pick 25 students from each of the separate lists of freshmen, sophomores, juniors, and seniors (stratified sampling). The researcher could separate the homerooms by classes; then randomly pick five freshmen homerooms, five sophomore homerooms, five junior homerooms, and five senior homerooms (cluster sampling); and then randomly pick five students from each of the homerooms (multistage sampling). The surveyor could separately pick random samples of males and females (stratified sampling), the size of each of the two samples chosen according to the proportion of male and female students attending the school (proportional sampling).

It should be noted that none of the alternative procedures in the above example result in a *simple random* sample because every possible sample of size 100 does not have an equal chance of being selected.

Summary

- A simple random sample (SRS) is one in which every possible sample of the desired size has an equal chance of being selected.
- Sampling error is not an error, but rather refers to the natural variability between samples.
- Bias is the tendency to favor the selection of certain members of a population.
- Nonresponse bias occurs when a large fraction of those sampled do not respond (most mailed questionnaires are good examples).
- Response bias happens when the question itself leads to misleading results (for example, people don't want to be perceived as having unpopular, unsavory, or illegal views).
- Undercoverage bias occurs when part of the population is ignored (for example, telephone surveys miss all those without phones).
- Voluntary response bias occurs when individuals choose whether to respond (for example, radio call-in surveys).
- Systematic sampling involves listing the population, choosing a random point to start, and then picking every nth person for some n.
- Stratified sampling involves dividing the population into homogeneous groups called strata and then picking random samples from each of the strata.
- Cluster sampling involves dividing the population into heterogeneous groups called clusters, and then picking everyone in a random sample of the clusters.
- Multistage sampling refers to procedures involving two or more steps, each of which could involve any of the sampling techniques.

Questions on Topic Seven: Planning and Conducting Surveys

Multiple-Choice Questions

Directions: The questions or incomplete statements that follow are each followed by five suggested answers or completions. Choose the response that best answers the question or completes the statement.

1. Ann Landers, who wrote a daily advice column appearing in newspapers across the country, once asked her readers, "If you had it to do over again, would you have children?" Of the more than 10,000 readers who responded, 70% said no. What does this show?

   (A) The survey is meaningless because of voluntary response bias.
   (B) No meaningful conclusion is possible without knowing something more about the characteristics of her readers.
   (C) The survey would have been more meaningful if she had picked a random sample of the 10,000 readers who responded.
   (D) The survey would have been more meaningful if she had used a control group.
   (E) This was a legitimate sample, randomly drawn from her readers and of sufficient size to allow the conclusion that most of her readers who are parents would have second thoughts about having children.

2. Which of the following is a true statement?

   (A) If bias is present in a sampling procedure, it can be overcome by dramatically increasing the sample size.
   (B) There is no such thing as a "bad sample."
   (C) Sampling techniques that use probability techniques effectively eliminate bias.
   (D) Convenience samples often lead to undercoverage bias.
   (E) Voluntary response samples often underrepresent people with strong opinions.

3. Two possible wordings for a questionnaire on gun control are as follows:

   I. The United States has the highest rate of murder by handguns among all countries. Most of these murders are known to be crimes of passion or crimes provoked by anger between acquaintances. Are you in favor of a 7-day cooling-off period between the filing of an application to purchase a handgun and the resulting sale?

   II. The United States has one of the highest violent crime rates among all countries. Many people want to keep handguns in their homes for self-protection. Fortunately, U.S. citizens are guaranteed the right to bear arms by the Constitution. Are you in favor of a 7-day waiting period between the filing of an application to purchase a needed handgun and the resulting sale?

   One of these questions showed that 25% of the population favored a 7-day waiting period between application for purchase of a handgun and the resulting sale, while the other question showed that 70% of the population favored the waiting period. Which produced which result and why?

   (A) The first question probably showed 70% and the second question 25% because of the lack of randomization in the choice of pro-gun and anti-gun subjects as evidenced by the wording of the questions.
   (B) The first question probably showed 25% and the second question 70% because of a placebo effect due to the wording of the questions.
   (C) The first question probably showed 70% and the second question 25% because of the lack of a control group.
   (D) The first question probably showed 25% and the second question 70% because of response bias due to the wording of the questions.
   (E) The first question probably showed 70% and the second question 25% because of response bias due to the wording of the questions.

4. Each of the 29 NBA teams has 12 players. A sample of 58 players is to be chosen as follows. Each team will be asked to place 12 cards with its players' names into a hat and randomly draw out two names. The two names from each team will be combined to make up the sample. Will this method result in a simple random sample of the 348 basketball players?

   (A) Yes, because each player has the same chance of being selected.
   (B) Yes, because each team is equally represented.
   (C) Yes, because this is an example of stratified sampling, which is a special case of simple random sampling.
   (D) No, because the teams are not chosen randomly.
   (E) No, because not each group of 58 players has the same chance of being selected.

5. To survey the opinions of bleacher fans at Wrigley Field, a surveyor plans to select every one-hundredth fan entering the bleachers one afternoon. Will this result in a simple random sample of Cub fans who sit in the bleachers?

   (A) Yes, because each bleacher fan has the same chance of being selected.
   (B) Yes, but only if there is a single entrance to the bleachers.
   (C) Yes, because the 99 out of 100 bleacher fans who are not selected will form a control group.
   (D) Yes, because this is an example of systematic sampling, which is a special case of simple random sampling.
   (E) No, because not every sample of the intended size has an equal chance of being selected.

6. Which of the following is a true statement about sampling error?

   (A) Sampling error can be eliminated only if a survey is both extremely well designed and extremely well conducted.
   (B) Sampling error concerns natural variation between samples, is always present, and can be described using probability.
   (C) Sampling error is generally larger when the sample size is larger.
   (D) Sampling error implies an error, possibly very small, but still an error, on the part of the surveyor.
   (E) Sampling error is higher when bias is present.

7. What fault do all these sampling designs have in common?

   I. The *Wall Street Journal* plans to make a prediction for a presidential election based on a survey of its readers.
   II. A radio talk show asks people to phone in their views on whether the United States should pay off its huge debt to the United Nations.
   III. A police detective, interested in determining the extent of drug use by teenagers, randomly picks a sample of high school students and interviews each one about any illegal drug use by the student during the past year.

   (A) All the designs make improper use of stratification.
   (B) All the designs have errors that can lead to strong bias.
   (C) All the designs confuse *association* with *cause and effect.*
   (D) None of the designs satisfactorily controls for sampling error.
   (E) None of the designs makes use of chance in selecting a sample.

8. A state auditor is given an assignment to choose and audit 26 companies. She lists all companies whose name begins with A, assigns each a number, and uses a random number table to pick one of these numbers and thus one company. She proceeds to use the same procedure for each letter of the alphabet and then combines the 26 results into a group for auditing. Which of the following is a true statement?

   (A) Each company has an equal probability of being audited.
   (B) Each set of 26 companies has an equal chance of being selected.
   (C) Her procedure results in a simple random sample.
   (D) Her procedure doesn't truly make use of chance.
   (E) She could have used a calculator random number generator in place of using a random number table to achieve similar results.

9. A researcher planning a survey of heads of households in a particular state has census lists for each of the 23 counties in that state. The procedure will be to obtain a random sample of heads of households from each of the counties rather than grouping all the census lists together and obtaining a sample from the entire group. Which of the following is an *incorrect* statement about the resulting stratified sample?

   (A) It is not a simple random sample.
   (B) It is easier and less costly to obtain than a simple random sample.
   (C) It gives comparative information that a simple random sample wouldn't give.
   (D) A cluster sample would have been more appropriate.
   (E) Differences in county sizes could be taken into account by making the size of the random sample from each county depend on the proportion of the total population represented by the county.

10. To find out the average occupancy size of student-rented apartments, a researcher picks a simple random sample of 100 such apartments. Even after one follow-up visit, the interviewer is unable to make contact with anyone in 27 of these apartments. Concerned about nonresponse bias, the researcher chooses another simple random sample and instructs the interviewer to continue this procedure until contact is made with someone in a total of 100 apartments. The average occupancy size in the final 100-apartment sample is 2.78. Is this estimate probably too low or too high?

   (A) Too low, because of undercoverage bias.
   (B) Too low, because convenience samples overestimate average results.
   (C) Too high, because of undercoverage bias.
   (D) Too high, because convenience samples overestimate average results.
   (E) Too high, because voluntary response samples overestimate average results.

11. To conduct a survey of long-distance calling patterns, a researcher opens a telephone book to a random page, closes his eyes, puts his finger down on the page, and then reads off the next 50 names. Which of the following is *incorrect*?

    (A) The survey design incorporates chance.
    (B) Assuming the page and starting point on the page are randomly selected, each person in the phone book has an equal chance of being selected.
    (C) The procedure could easily result in selection bias.
    (D) The procedure does not result in a simple random sample.
    (E) This is the typical methodology of a systematic sample.

12. Consider the following three events:

    I. Although 18% of the student body are minorities, in a random sample of 20 students, 5 are minorities.
    II. In a survey about sexual habits, an embarrassed student deliberately gives the wrong answers.
    III. A surveyor mistakenly records answers to one question in the wrong space.

    Which of the following correctly characterizes the above?

    (A) I, sampling error; II, response bias; III, human mistake
    (B) I, sampling error; II, nonresponse bias; III, hidden error
    (C) I, hidden bias; II, voluntary sample bias; III, sampling error
    (D) I, undercoverage error; II, voluntary error; III, unintentional error
    (E) I, small sample error; II, deliberate error; III, mistaken error

13. A researcher plans a study to examine the depth of belief in God among the adult population. He obtains a simple random sample of 100 adults as they leave church one Sunday morning. All but one of them agree to participate in the survey. Which of the following is a true statement?

    (A) Proper use of chance as evidenced by the simple random sample makes this a well-designed survey.
    (B) The high response rate makes this a well-designed survey.
    (C) Selection bias makes this a poorly designed survey.
    (D) The validity of this survey depends on whether or not the adults attending this church are representative of all churches.
    (E) The validity of this survey depends upon whether or not similar numbers of those surveyed are male and female.

Free-Response Questions

Directions: You must show all work and indicate the methods you use. You will be graded on the correctness of your methods and on the accuracy of your final answers.

Seven Open-Ended Questions

1. Cell phones emit a form of electromagnetic radiation, and there is a concern on how this affects the human body. A World Health Organization (WHO) study of 12,000 people found no connection between moderate cell phone use and brain cancer, although the report does mention a higher incidence of brain cancer for heavy users (defined as those who used their phone for at least half an hour a day). A study of 420,095 persons in Denmark found no correlation between length of cell phone subscriptions (in years) and brain tumor incidence.
   (a) Were these studies observational studies or experiments or one of each? Explain.
   (b) Does the WHO study definition of "heavy users" seem reasonable? Explain.
   (c) Neither study tries to distinguish between voice or text messaging use of the cell phone. Should this have any affect on conclusions? Explain.
   (d) What is a weakness in the Denmark study that the WHO study does take into account?

2. A questionnaire is being designed to determine whether most people are or are not in favor of legislation protecting the habitat of the spotted owl. Give two examples of poorly worded questions, one biased toward each response.

3. To obtain a sample of 25 students from among the 500 students present in school one day, a surveyor decides to pick every twentieth student waiting in line to attend a required assembly in the gym.
   (a) Explain why this procedure will not result in a simple random sample of the students present that day.
   (b) Describe a procedure that will result in a simple random sample of the students present that day.

4. A hot topic in government these days is welfare reform. Suppose a congresswoman wishes to survey her constituents concerning their opinions on whether the federal government should turn welfare over to the states. Discuss possible sources of bias with regard to the following four options: (1) conducting a survey via random telephone dialing into her district, (2) sending out a mailing using a registered voter list, (3) having a pollster interview everyone who walks past her downtown office, and (4) broadcasting a radio appeal urging interested citizens in her district to call in their opinions to her office.

5. You and nine friends go to a restaurant and check your coats. You all forget to pick up the ticket stubs, and so when you are ready to leave, Hilda, the hat-check girl, randomly gives each of you one of the ten coats. You are surprised that one person actually receives the correct coat. You would like to explore this further and decide to use a random number table to simulate the situation. Describe how the random number table can be used to simulate one trial of the coat episode. Explain what each of the digits 0 through 9 will represent.

6.

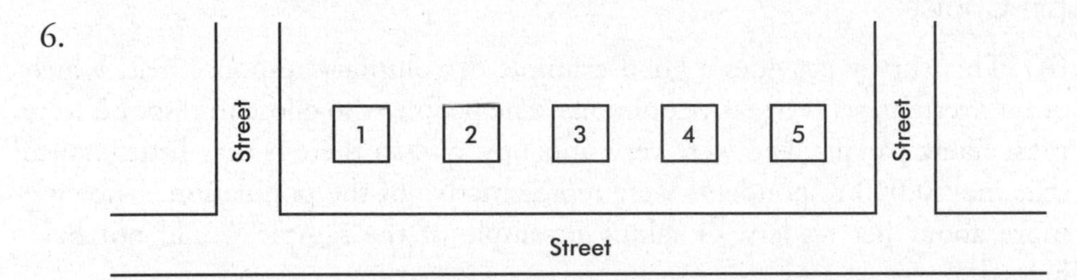

You are supposed to interview the residents of two of the above five houses.

(a) How would you choose which houses to interview?
(b) You plan to visit the homes at 9 a.m. If someone isn't home, explain the reasons for and against substituting another house.
(c) Are there any differences you might expect to find among the residents based on the above sketch?

7. A cable company plans to survey potential customers in a small city currently served by satellite dishes. Two sampling methods are being considered. Method A is to randomly select a sample of 25 city blocks and survey every family living on those blocks. Method B is to randomly select a sample of families from each of the five natural neighborhoods making up the city.

(a) What is the statistical name for the sampling technique used in Method A, and what is a possible reason for using it rather than an SRS?
(b) What is the statistical name for the sampling technique used in Method B, and what is a possible reason for using it rather than an SRS?

Answer Key

1. **A**	4. **E**	7. **B**	10. **C**	13. **C**
2. **D**	5. **E**	8. **E**	11. **E**	
3. **E**	6. **B**	9. **D**	12. **A**	

Answers Explained

Multiple-Choice

1. **(A)** This survey provides a good example of voluntary response bias, which often overrepresents negative opinions. The people who chose to respond were most likely parents who were very unhappy, and so there is very little chance that the 10,000 respondents were representative of the population. Knowing more about her readers, or taking a sample of the sample would not have helped.

2. **(D)** If there is bias, taking a larger sample just magnifies the bias on a larger scale. If there is enough bias, the sample can be worthless. Even when the subjects are chosen randomly, there can be bias due, for example, to non-response or to the wording of the questions. Convenience samples, like shopping mall surveys, are based on choosing individuals who are easy to reach, and they typically miss a large segment of the population. Voluntary response samples, like radio call-in surveys, are based on individuals who offer to participate, and they typically overrepresent persons with strong opinions.

3. **(E)** The wording of the questions can lead to response bias. The neutral way of asking this question would simply have been: Are you in favor of a 7-day waiting period between the filing of an application to purchase a handgun and the resulting sale?

4. **(E)** In a simple random sample, every possible group of the given size has to be equally likely to be selected, and this is not true here. For example, with this procedure it will be impossible for all the Bulls to be together in the final sample. This procedure is an example of stratified sampling, but stratified sampling does not result in simple random samples.

5. **(E)** In a simple random sample, every possible group of the given size has to be equally likely to be selected, and this is not true here. For example, with this procedure it will be impossible for all the early arrivals to be together in the final sample. This procedure is an example of systematic sampling, but systematic sampling does not result in simple random samples.

6. **(B)** Different samples give different sample statistics, all of which are estimates of a population parameter. Sampling error relates to natural variation between samples, can never be eliminated, can be described using probability, and is generally smaller if the sample size is larger.

7. **(B)** The *Wall Street Journal* survey has strong selection bias; that is, people who read the *Journal* are not very representative of the general population. The talk show survey results in a *voluntary response sample*, which typically gives too much emphasis to persons with strong opinions. The police detective's survey has strong response bias in that students may not give truthful responses to a police detective about their illegal drug use.

8. **(E)** While the auditor does use chance, each company will have the same chance of being audited only if the same number of companies have names starting with each letter of the alphabet. This will not result in a simple random sample because each possible set of 26 companies does not have the same chance of being picked as the sample. For example, a group of companies whose names all start with *A* will not be chosen. Calculator random number generators and random number tables have similar uses and results.

9. **(D)** This is not a simple random sample because all possible sets of the required size do not have the same chance of being picked. For example, a set of households all from just half the counties has no chance of being picked to be the sample. Stratified samples are often easier and less costly to obtain and also make comparative data available. In this case responses can be compared among various counties. There is no reason to assume that each county has heads of households with the same characteristics and opinions as the state as a whole, so cluster sampling is not appropriate. When conducting stratified sampling, proportional sampling is used when one wants to take into account the different sizes of the strata.

10. **(C)** It is most likely that the apartments at which the interviewer had difficulty finding someone home were apartments with fewer students living in them. Replacing these with other randomly picked apartments most likely replaces smaller-occupancy apartments with larger-occupancy ones.

11. **(E)** While the procedure does use some element of chance, all possible groups of size 50 do not have the same chance of being picked, and so the result is not a simple random sample. There is a very real chance of selection bias. For example, a number of relatives with the same name and similar long-distance calling patterns might be selected. The typical methodology of a systematic sample involves picking every *n*th member from the list, where *n* is roughly the population size divided by the desired sample size.

12. **(A)** The natural variation in samples is called sampling error. Embarrassing questions and resulting untruthful answers are an example of response bias. Inaccuracies and mistakes due to human error are one of the real concerns of researchers.

13. **(C)** Surveying people coming out of any church results in a very unrepresentative sample of the adult population, especially given the question under consideration. Using chance and obtaining a high response rate will not change the selection bias and make this into a well-designed survey.

Free-Response

1. (a) Both studies were observational because no treatments were applied.

   (b) Typical cell phone use today, especially among younger people, is well over half an hour, so half an hour does not seem to be a reasonable split between moderate and heavy use.

   (c) This absolutely affects conclusions in that both studies look for relationships with brain cancer. While voice conversation involves holding the phone against one's head, text messaging does not.

   (d) The Denmark study looks at how many years individuals used their cell phones, but not at the extent of daily use, while the WHO study does consider daily usage.

2. There are many possible examples, such as Are you in favor of protecting the habitat of the spotted owl, which is almost extinct and desperately in need of help from an environmentally conscious government? and Are you in favor of protecting the habitat of the spotted owl no matter how much unemployment and resulting poverty this causes among hard-working loggers?

3. (a) To be a simple random sample, every possible group of size 25 has to be equally likely to be selected, and this is not true here. For example, if there are 40 students who always rush to be first in line, this procedure will allow for only 2 of them to be in the sample. Or if each homeroom of size 20 arrives as a unit, this procedure will allow for only 1 person from each homeroom to be in the sample.

   (b) A simple random sample of the students can be obtained by numbering them from 001 to 500 and then picking three digits at a time from a random number table, ignoring numbers over 500 and ignoring repeats, until a group of 25 numbers is obtained. The students corresponding to these 25 numbers will be a simple random sample.

4. The direct telephone and mailing options will both suffer from undercoverage bias. For example, especially affected by the legislation under discussion are the homeless, and they do not have telephones or mailing addresses. The pollster interviews will result in a convenience sample, which can be highly unrepresentative of the population. In this case, there might be a real question concerning which members of her constituency spend any time in the downtown area where her office is located. The radio appeal will lead to a voluntary response sample, which typically gives too much emphasis to persons with strong opinions.

5. In numbering the people 0 through 9, each digit stands for whose coat someone receives. Pick the digits, omitting repeats, until a group of ten different digits is obtained. Check for a match (1 appearing in the first position corresponding to person 1, or 2 appearing in the next position corresponding to person 2, and so on, up to 0 appearing in the last position corresponding to person 10).

6. (a) To obtain an SRS, you might use a random number table and note the first two different numbers between 1 and 5 that appear. Or you could use a calculator to generate numbers between 1 and 5, again noting the first two different numbers that result.

   (b) Time and cost considerations would be the benefit of substitution. However, substitution rather than returning to the same home later could lead to selection bias because certain types of people are not and will not be home at 9 a.m. With substitution the sample would no longer be a simple random sample.

   (c) Corner lot homes like homes 1 and 5 might have different residents (perhaps with higher income levels) than other homes.

7. (a) Method A is an example of *cluster sampling,* where the population is divided into heterogeneous groups called *clusters* and individuals from a random sample of the clusters are surveyed. It is often more practical to simply survey individuals from a random sample of clusters (in this case, a random sample of city blocks) than to try to randomly sample a whole population (in this case the entire city population).

   (b) Method B is an example of *stratified sampling,* where the population is divided into homogeneous groups called *strata* and random individuals from each stratum are chosen. Stratified samples can often give useful information about each stratum (in this case, about each of the five neighborhoods) in addition to information about the whole population (the city population).

Planning and Conducting Experiments

TOPIC 8

- Experiments versus Observational Studies
- Confounding, Control Groups, Placebo Effects, Blinding
- Treatments, Experimental Units, Randomization
- Replication, Blocking, Generalizability of Results

There are several primary principles dealing with the proper planning and conducting of experiments. First, possible confounding variables must be controlled for, usually through the use of comparison. Second, chance should be used in assigning which subjects are to be placed in which groups for which treatment. Third, natural variation in outcomes can be lessened by using more subjects.

EXPERIMENTS VERSUS OBSERVATIONAL STUDIES VERSUS SURVEYS

In an experiment we impose some change or treatment and measure the result or response. In an observational study we simply observe and measure something that has taken place or is taking place, while trying not to cause any changes by our presence. A sample survey is an observational study in which we draw conclusions about an entire population by considering an appropriately chosen sample to look at. An experiment often suggests a causal relationship, while an observational study may show only the existence of associations.

EXAMPLE 8.1

A study is to be designed to determine whether daily calcium supplements benefit women by increasing bone mass. How can an observational study be performed? An experiment? Which is more appropriate here?

Answer: An observational study might interview and run tests on women seen purchasing calcium supplements in a pharmacy. Or perhaps all patients hospitalized during a particular time period could be interviewed with regard to taking calcium and then their bone mass measured. The bone mass measurements of those taking calcium supplements could then be compared to that of those not taking supplements.

(continued)

An experiment could be performed by selecting some number of subjects, using chance to pick half to receive calcium supplements while the other half receives similar-looking placebos, and noting the difference in bone mass before and after treatment for each group.

The experimental approach is more appropriate here. With the observational study there could be many explanations for any bone mass difference noted between patients who take calcium and those who don't. For example, women who have voluntarily been taking calcium supplements might be precisely those who take better care of themselves in general and thus have higher bone mass for other reasons. The experiment tries to control for lurking variables by randomly giving half the subjects calcium.

EXAMPLE 8.2

A study is to be designed to examine the life expectancies of tall people versus those of short people. Which is more appropriate, an observational study or an experiment?

Answer: An observational study, examining medical records of heights and ages at time of death, seems straightforward. An experiment where subjects are randomly chosen to be made short or tall, followed by recording age at death, would be groundbreaking (and, of course, nonsensical).

EXAMPLE 8.3

A study is to be designed to examine the GPAs of students who take marijuana regularly and those who don't. Which is more appropriate, an observational study or an experiment?

Answer: As much as some researchers might want to randomly require half the subjects to take an illegal drug, this would be unethical. The proper procedure here is an observational study, having students anonymously fill out questionnaires asking about marijuana usage and GPA.

Experiments involve *explanatory variables,* called *factors,* that are believed to have an effect on *response variables.* A group is treated with some level of the explanatory variable, and the outcome on the response variable is measured.

EXAMPLE 8.4

To test the value of help sessions outside the classroom, students could be divided into three groups, with one group receiving 4 hours of help sessions per week outside the classroom, a second group receiving 2 hours of help sessions outside the classroom, and a third group receiving no help outside the classroom. What are the explanatory and response variables and what are the levels?

Answer: The explanatory variable, help sessions outside the classroom, is being given at three levels: 4 hours weekly, 2 hours weekly, and 0 hours weekly. The response variable is not specified but might be a final exam score or performance on a particular test.

The different factor-level combinations are called *treatments.* In Example 8.4, there are three treatments (corresponding to the three levels of the one factor). Suppose the students were further randomly divided into a morning class and an afternoon class. There would then be two factors, one with three levels and one with two levels, and a total of six treatments (AM class with 4 hours help, AM class with 2 hours help, AM class with 0 hours help, PM class with 4 hours help, PM class with 2 hours help, and PM class with 0 hours help).

CONFOUNDING, CONTROL GROUPS, PLACEBO EFFECTS, AND BLINDING

When there is uncertainty with regard to which variable is causing an effect, we say the variables are *confounded.* For example, suppose two fertilizers require different amounts of watering. In an experiment it might be difficult to determine if the difference in fertilizers or the difference in watering is the real cause of observed differences in plant growth. Sometimes we can control for confounding. For example, we can have many test plots using one or the other of the fertilizers, with equal numbers of sunny and shady plots for each fertilizer, so that fertilizer and sun are not confounded.

A *lurking variable* is a variable that drives two other variables, creating the mistaken impression that the two other variables are related by cause and effect. For example, elementary school students with larger shoe sizes appear to have higher reading levels. However, there is a lurking variable, age, which drives both the other variables. That is, older students tend to wear larger shoes than younger students, and older students also tend to have higher reading levels. Wearing larger shoes will not improve reading skills! Instead of *lurking variable,* another standard terminology is simply to say there is a *common response;* that is, changes in both shoe size and reading level are caused by changes in age.

In an experiment there is a group that receives the treatment, and there is a *control group* that doesn't. The experiment compares the responses in the treatment group to the responses in the control group. Randomly putting subjects into treatment and control groups can help reduce the problems posed by confounding and lurking variables. Thus these problems are easier to control for when doing experiments than when doing observational studies.

It is a fact that many people respond to any kind of perceived treatment. This is called the *placebo effect.* For example, when given a sugar pill after surgery but told that it is a strong pain reliever, many patients feel immediate relief from their pain. In many studies, subjects appear to consciously or subconsciously want to help the researcher prove a point. Thus when responses are noticed in any experiment, there is concern whether real physical responses are being caused by the psychological placebo effect. *Blinding* occurs when the subjects or the response evaluators don't know which subjects are receiving different treatments such as placebos.

TIP

Blinding and placebos in experiments are important but are not always feasible. You can still have "experiments" without these.

EXAMPLE 8.5

A study is intended to test the effects of vitamin E and beta carotene on heart attack rates. How should it be set up?

Answer: Using randomization, the subjects should be split into four groups: those who will be given just vitamin E, just beta carotene, both vitamin E and beta carotene, and neither vitamin E nor beta carotene. For example, as each subject joins the test, the next digit in a random number table can be read off, ignoring 0 and 5–9, and with 1, 2, 3, and 4 designating which group the subject is placed in. Or if the total number of subjects is known and available, for example, 800, then each can be assigned a number and three digits at a time be read off the random number table. With repeats and numbers over 800 thrown away, the first 200 numbers picked represent one group, the next 200 another, and so on. More meaningful results will be obtained if the study is double-blind, that is, if not only are the subjects unaware of what kind of tablets they are taking but so are the doctors evaluating whether or not they have heart problems. Many diagnoses are not clear-cut, and doctors can be influenced if they know exactly which potential preventive their patients are taking.

TREATMENTS, EXPERIMENTAL UNITS, AND RANDOMIZATION

An experiment is performed on objects called *experimental units*, and if the units are people, they are called *subjects*. The experimental units or subjects are typically divided into two groups. One group receives a *treatment* and is called the *treatment group*. A comparison is made between the response noted in the treatment group and the response noted in the *control group*, the group that receives no treatment.

To help minimize the effect of lurking variables, and of confounding, it is important to use *randomization*, that is, to use chance in deciding which subjects go into which group. It is not sufficient to try to systematically match characteristics between the two groups. It seems reasonable, for example, to hand-sort subjects so that both the treatment group and the control group have the same number of women, the same number of Catholics, the same number of Hispanics, the same number of short people, and so on, but this method does not work well. There are always other variables that one might not think of considering until after the results of the experiment start coming in. The best method to use is randomization employing a computer, a hat with names in it, or a random number table.

Note that *randomization* usually refers to how given subjects are assigned to treatments, not to how a group of subjects are chosen from an entire population. The object of an experiment is to see if different treatments lead to different responses, and so we randomly assign subjects to treatments to balance unknown sources of variability. Random assignment to treatments is critical, especially if the subjects are not randomly selected, as is the case in medical/drug experiments. Generalizing the findings of the study is a separate question, one that depends on how the initial group of subjects was assembled.

COMPLETELY RANDOMIZED DESIGN FOR TWO TREATMENTS

Comparing two treatments using randomization is often the design of choice. To help minimize hidden bias, it is best if subjects do not know which treatment they are receiving. This is called *single-blinding*. Another precaution is the use of *double-blinding*, in which neither the subjects nor those evaluating their responses know who is receiving which treatment.

EXAMPLE 8.6

There is a pressure point on the wrist that some doctors believe can be used to help control the nausea experienced following certain medical procedures. The idea is to place a band containing a small marble firmly on a patient's wrist so that the marble is located directly over the pressure point. Describe how an experiment might be run on 50 postoperative patients.

Answer: Assign each patient a number from 01 to 50. From a random number table read off two digits at a time, throwing away repeats, 00, and numbers over 50, until 25 numbers have been selected. Put wristbands with marbles over the pressure point on the patients with these assigned numbers. Put wristbands with marbles on the remaining patients also, but *not* over the pressure point. Have a researcher check by telephone with all 50 patients at designated time intervals to determine the degree of nausea being experienced. Neither the patients nor the researcher on the telephone should know which patients have the marbles over the correct pressure point.

EXAMPLE 8.7

A chemical fertilizer company wishes to test whether using their product results in superior vegetables. After dividing a large field into small plots, how might the experiment proceed?

Answer: If the company has one recommended fertilizer application level, half the plots can be randomly selected (assigning the plots numbers and using a random number table) to receive the prescribed dosage of fertilizer. This random selection of plots is to ensure that neither fertilized plants nor unfertilized plants are inadvertently given land with better rainfall, sunshine, soil type, and so on. To avoid possible bias on the part of employees who will weed and water the plants, they should not know which plots have received the fertilizer. It might be necessary to have containers, one for each plot, of a similar-looking, similar-smelling substance, half of which contain the fertilizer while the rest contain a chemically inactive material. Finally, if the vegetables are to be judged by quantity and size, the measurements will be less subject to bias. However, if they are to be judged qualitatively, for example, by taste, the judges should not know which vegetables were treated with the fertilizer and which were not.

If the researchers also wish to consider level, that is, the amount of fertilizer, randomization should be used for more groupings. For example, if there are 60 plots on which to test four levels of fertilizer, the first 12 different two-digit numbers in the range 01–60 appearing on a random number table might receive one level, the next 12 new two-digit numbers a second level, and so on, with the last 12 plots receiving the "placebo" treatment.

RANDOMIZED PAIRED COMPARISON DESIGN

Two treatments can be compared based on the responses of paired subjects, one of whom receives one treatment while the other receives the second treatment. Often the paired subjects are really single subjects who are given both treatments, one at a time.

EXAMPLE 8.8

The famous Pepsi-Coke tests had subjects compare the taste of samples of each drink. How could such a paired comparison test be set up?

Answer: It is crucial that such a test be blind, that is, that the subjects not know which cup contains which drink. Furthermore, to help avoid hidden bias, which drink the subjects taste first should be decided by chance. For example, as each subject arrives, the researcher could read off the next digit from a random number table, with the subject receiving Pepsi or Coke first depending on whether the digit is odd or even.

Note: Even though the subjects are being given a drink, and there is some randomization going on, some statisticians consider this to be a sample survey aimed at estimating a population proportion rather than a true experiment.

EXAMPLE 8.9

Does seeing pictures of accidents caused by drunk drivers influence one's opinion on penalties for drunk drivers? How could a comparison test be designed?

Answer: The subjects could be asked questions about drunk driving penalties before and then again after seeing the pictures, and any change in answers noted. This would be a poor design because there is no control group, there is no use of randomization, and subjects might well change their answers because they realize that that is what is expected of them after seeing the pictures.

A better design is to use randomization to split the subjects into two groups, half of whom simply answer the questions while the other half first see the pictures and then answer the questions.

Another possibility is to use a group of twins as subjects. One of each set of twins is randomly picked (e.g., based on choosing an odd or even digit from a random number table) to answer the questions without seeing the pictures, while the other first sees the pictures and then answers the questions. The answers could be compared from each set of twins. This is a paired comparison test that might help minimize lurking variables due to family environment, heredity, and so on.

REPLICATION, BLOCKING, AND GENERALIZABILITY OF RESULTS

When differences are observed in a comparison test, the researcher must decide whether these differences are *statistically significant* or whether they can be explained by natural variation. One important consideration is the size of the sample—the larger the sample, the more significant the observation. This is the principle of *replication*; that is, the treatment should be repeated on a sufficient number of subjects so that real response differences are more apparent.

Just as stratification in sampling design first divides the population into representative groups called strata, *blocking* in experiment design first divides the subjects into representative groups called blocks. One can think of blocking as running a separate experiment on each block. This technique helps control certain lurking variables by bringing them directly into the picture and helps make conclusions more specific. The paired comparison design is a special case of blocking in which each pair (or each subject if the subjects serve as their own controls) can be considered a block.

TIP

Use proper terminology! For example, the language of experiments is different from the language of observational studies—you shouldn't mix up *blocking* and *stratification.*

EXAMPLE 8.10

There is a rising trend for star college athletes to turn professional without finishing their degrees. A study is performed to assess whether reading an article about professional salaries has an impact on such decisions. Randomization can be used to split the subjects into two groups, and those in one group given the article before answering questions. How can a block design be incorporated into the design of this experiment?

Answer: The subjects can be split into two blocks, underclass and upperclass, before using randomization to assign some to read the article before questioning. With this design, the impact of the salary article on freshmen and sophomores can be distinguished from the impact on juniors and seniors.

Similarly, blocking can be used to separately analyze men and women, those with high GPAs and those with low GPAs, those in different sports, those with different majors, and so on.

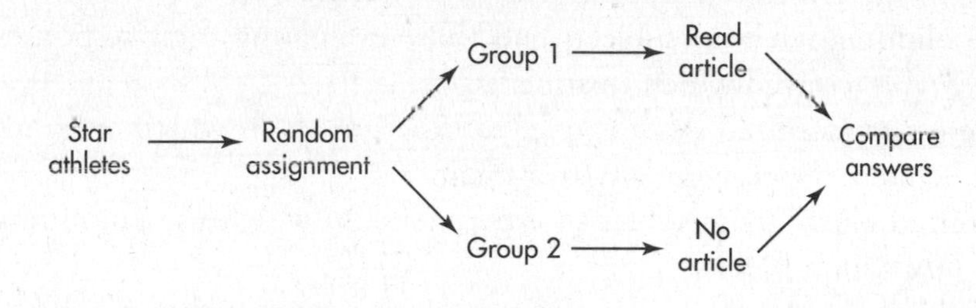

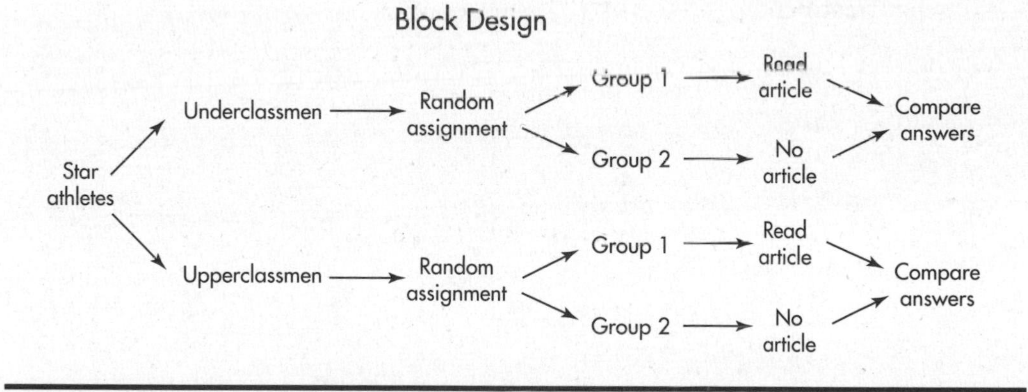

A major goal of experiments is to be able to *generalize* the results to broader populations. Often an experiment must be repeated in a variety of settings. For example, it is hard to generalize from the effect a television commercial has on students at a private midwestern high school to the effect the same commercial has on retired senior citizens in Florida. Generally, comparison and randomization are important, blinding is sometimes critical, and taking care to avoid hidden bias as much as possible is always indicative of a well-designed experiment. However, knowledge of the subject so that realistic situations can be created in testing should also be emphasized. Testing and experimenting on people does not put them in natural states, and this situation can lead to artificial responses.

Summary

- Experiments involve applying a treatment to one or more groups and observing the responses.
- Experiments often have a treatment group and a control group.
- Blocking is the process of dividing the subjects into representative groups to bring certain differences into the picture (for example, blocking by gender, age, or race).
- Randomization of who receives what treatments is extremely important in handling unknown and uncontrollable differences.
- Variables are said to be confounded when there is uncertainty as to which variable is causing an effect.
- The placebo effect refers to the fact that many people respond to any kind of perceived treatment.
- Blinding refers to subjects not knowing which treatment they are receiving.
- Double-blinding refers to subjects and those evaluating their responses not knowing who received which treatments.
- Completely randomized designs refer to experiments in which everyone has an equal chance of receiving any treatment.
- Randomized block designs refer to experiments in which the randomization occurs only within blocks.
- Randomized paired comparison designs refer to experiments in which subjects are paired and randomization is used to decide who in each pair receives what treatment.

Questions on Topic Eight: Planning and Conducting Experiments

Multiple-Choice Questions

Directions: The questions or incomplete statements that follow are each followed by five suggested answers or completions. Choose the response that best answers the question or completes the statement.

1. A study is made to determine whether studying Latin helps students achieve higher scores on the verbal section of the SAT exam. In comparing records of 200 students, half of whom have taken at least 1 year of Latin, it is noted that the average SAT verbal score is higher for those 100 students who have taken Latin than for those who have not. Based on this study, guidance counselors begin to recommend Latin for students who want to do well on the SAT exam. Which of the following is *incorrect*?

   (A) While this study indicates a relation, it does not prove causation.
   (B) There could well be a confounding variable responsible for the seeming relationship.
   (C) Self-selection here makes drawing the counselors' conclusion difficult.
   (D) A more meaningful study would be to compare an SRS from each of the two groups of 100 students.
   (E) This is an observational study, not an experiment.

2. In a 1927–32 Western Electric Company study on the effect of lighting on worker productivity, productivity increased with each increase in lighting but then also increased with every decrease in lighting. If it is assumed that the workers knew a study was in progress, this is an example of

   (A) the effect of a treatment unit.
   (B) the placebo effect.
   (C) the control group effect.
   (D) sampling error.
   (E) voluntary response bias.

3. When the estrogen-blocking drug tamoxifen was first introduced to treat breast cancer, there was concern that it would cause osteoporosis as a side effect. To test this concern, cancer subjects were randomly selected and given tamoxifen, and their bone density was measured before and after treatment. Which of the following is a true statement?

   (A) This study was an observational study.
   (B) This study was a sample survey of randomly selected cancer patients.
   (C) This study was an experiment in which the subjects were used as their own controls.
   (D) With the given procedure, there cannot be a placebo effect.
   (E) Causation cannot be concluded without knowing the survival rates.

4. In designing an experiment, blocking is used

   (A) to reduce bias.
   (B) to reduce variation.
   (C) as a substitute for a control group.
   (D) as a first step in randomization.
   (E) to control the level of the experiment.

5. Which of the following is *incorrect*?

   (A) Blocking is to experiment design as stratification is to sampling design.
   (B) By controlling certain variables, blocking can make conclusions more specific.
   (C) The paired comparison design is a special case of blocking.
   (D) Blocking results in increased accuracy because the blocks have smaller size than the original group.
   (E) In a randomized block design, the randomization occurs within the blocks.

6. Consider the following studies being run by three different nursing home establishments.

   I. One nursing home has pets brought in for an hour every day to see if patient morale is improved.
   II. One nursing home allows hourly visits every day by kindergarten children to see if patient morale is improved.
   III. One nursing home administers antidepressants to all patients to see if patient morale is improved.

   Which of the following is true?

   (A) None of these studies uses randomization.
   (B) None of these studies uses control groups.
   (C) None of these studies uses blinding.
   (D) Important information can be obtained from all these studies, but none will be able to establish causal relationships.
   (E) All of the above

7. A consumer product agency tests miles per gallon for a sample of automobiles using each of four different octanes of gasoline. Which of the following is true?

   (A) There are four explanatory variables and one response variable.
   (B) There is one explanatory variable with four levels of response.
   (C) Miles per gallon is the only explanatory variable, but there are four response variables corresponding to the different octanes.
   (D) There are four levels of a single explanatory variable.
   (E) Each explanatory level has an associated level of response.

8. Is hot oatmeal with fruit or a Western omelet with home fries a more satisfying breakfast? Fifty volunteers are randomly split into two groups. One group is fed oatmeal with fruit, while the other is fed Western omelets with home fries. Each volunteer then rates his/her breakfast on a one to ten scale for satisfaction. If the Western omelet with home fries receives a substantially higher average score, what is a reasonable conclusion?

   (A) In general, people find Western omelets with home fries more satisfying for breakfast than hot oatmeal with fruit.
   (B) There is no reasonable conclusion because the subjects were volunteering rather than being randomly selected from the general population.
   (C) There is no reasonable conclusion because of the small size of the sample.
   (D) There is no reasonable conclusion because blinding was not used.
   (E) There is no reasonable conclusion because there are too many possible confounding variables such as age, race, and ethnic background of the individual volunteers and season when the study was performed.

9. Which of the following is a true statement?

   (A) In well-designed observational studies, responses are systematically influenced during the collection of data.
   (B) In well-designed experiments, the treatments result in responses that are as similar as possible.
   (C) A well-designed experiment always has a single treatment but may test that treatment at different levels.
   (D) Causation and association are unrelated concepts.
   (E) In well-designed, well-conducted experiments, strong association implies cause and effect.

10. Which of the following is *not* important in the design of experiments?

   (A) Control of confounding variables
   (B) Randomization in assigning subjects to different treatments
   (C) Replication of the experiment using sufficient numbers of subjects
   (D) Care in observing without imposing change
   (E) Isolating variability due to differences between blocks

11. Which of the following is a true statement about the design of matched-pair experiments?

    (A) Each subject might receive both treatments.
    (B) Each pair of subjects receives the identical treatment, and differences in their responses are noted.
    (C) Blocking is one form of matched-pair design.
    (D) Stratification into two equal sized strata is an example of matched pairs.
    (E) Randomization is unnecessary in true matched pair designs.

12. Do teenagers prefer sports drinks colored blue or green? Two different colorings, which have no effect on taste, are used on the identical drink to result in a blue and a green beverage; volunteer teenagers are randomly assigned to drink one or the other colored beverage; and the volunteers then rate the beverage on a one to ten scale. Because of concern that sports interest may affect the outcome, the volunteers are first blocked by whether or not they play on a high school team. Is blinding possible in this experiment?

    (A) No, because the volunteers know whether they are drinking a blue or green drink.
    (B) No, because the volunteers know whether or not they play on a high school team.
    (C) Yes, by having the experimenter in a separate room randomly pick one of two containers and remotely have a drink poured from that container.
    (D) Yes, by having the statistician analyzing the results not know which volunteer sampled which drink.
    (E) Yes, by having the volunteers drink out of solid colored thermoses, so that they don't know the color of the drink they are tasting.

13. Some researchers believe that too much iron in the blood can raise the level of cholesterol. The iron level in the blood can be lowered by making periodic blood donations. A study is performed by randomly selecting half of a group of volunteers to give periodic blood donations while the rest do not. Is this an experiment or an observational study?

    (A) An experiment with a single factor
    (B) An experiment with control group and blinding
    (C) An experiment with blocking
    (D) An observational study with comparison and randomization
    (E) An observational study with little if any bias

Free-Response Questions

Directions: You must show all work and indicate the methods you use. You will be graded on the correctness of your methods and on the accuracy of your final answers.

Eleven Open-Ended Questions

1. The belief that sugar causes hyperactivity is the most popular example of how people believe that food influences behavior.
   (a) Many parents, witnessing the aftermath of cake and ice cream at birthday parties, attest to the relationship between sugar and hyperactivity. Are these observational studies or experiments? Explain.
   (b) Name a confounding variable to the above and explain how it is confounded with sugar.
   (c) Design a study to allow a parent to determine whether sugar causes hyperactivity in his/her child and explain why double blinding is so important here.

2. Suppose a new drug is developed that appears in laboratory settings to completely prevent people who test positive for human immunodeficiency virus (HIV) from ever developing full-blown acquired immunodeficiency syndrome (AIDS). Putting all ethical considerations aside, design an experiment to test the drug. What ethical considerations might arise during the testing that would force an early end to the experiment?

3. A new weight-loss supplement is to be tested at three different levels (once, twice, and three times a day). Design an experiment, including a control group and including blocking for gender, for 80 overweight volunteers, half of whom are men. Explain carefully how you will use randomization.

4. Two studies are run to measure the health benefits of long-time use of daily high doses of vitamin C. Researchers in the first study send a questionnaire to all 50,000 subscribers to a health magazine, asking whether they have taken large doses of vitamin C for at least a 2-year period and what they perceive to be the health benefits, if any. The response rate is 80%. The 10,000 people who did not respond to the first mailing receive follow-up telephone calls, and eventually responses are registered from 98% of the magazine subscribers. Researchers in a second study take a group of 200 volunteers and randomly select 100 to receive high doses of vitamin C while the others receive a similar-looking, similar-tasting placebo. The volunteers are not told whether they are receiving the vitamin, but their doctors know and are asked to note health changes during a 2-year period. Comment on the designs of the two studies, remarking on their good points and on possible sources of error.

5. Explain how you would design an experiment to evaluate whether subliminal advertising (flashing "BUY POPCORN" on the screen for a fraction of a second) results in more popcorn being sold in a movie theater. Show how you will incorporate comparison, randomization, and blinding.

6. Throughout history millions of people have used garlic to obtain a variety of perceived health benefits. A vitamin production company decides to run a scientific test to assess the value of garlic in promoting a general sense of well-being. They randomly pick 250 of their employees, and once a day for 2 months the employees fill out questionnaires about their sense of well-being that day. For the next 2 months the employees take garlic capsules daily and again fill out the same questionnaires. Finally, for 2 concluding months the employees stop taking the pills and continue to fill out the daily questionnaires. Comment on the design of this experiment.

7. A new pain control procedure has been developed in which the patient uses a small battery pack to vary the intensity and duration of electric signals to electrodes surgically embedded in the afflicted area. Putting all ethical considerations aside, design an experiment to test the procedure. What ethical considerations might arise during the testing that would force an early end to the experiment?

8. A new vegetable fertilizer is to be tested at two different levels (regular concentration and double concentration). Design an experiment, including a control, for 30 test plots, half of which are in shade. Explain carefully how you will use randomization.

9. Two studies are run to measure the extent to which taking zinc lozenges helps to shorten the duration of the common cold. Researchers in the first study send questionnaires to all 5000 employees of a major teaching hospital asking whether they have taken zinc lozenges to fight the common cold and what they perceive to be the benefits, if any. The response rate is 90%. The 500 people who did not respond to the first mailing receive follow-up telephone calls, and eventually responses are obtained from over 99% of the hospital employees. Researchers in the second study take a group of 100 volunteers and randomly select 50 to receive zinc lozenges while the others receive a similar-looking, similar-tasting placebo. The volunteers are not told whether they are taking the zinc lozenges, but their doctors know and are asked to accurately measure the duration of common cold symptoms experienced by the volunteers. Comment on the designs of the two studies, remarking on their good points and on possible sources of error.

10. Explain how you would design an experiment to evaluate whether praying for a hospitalized heart attack patient leads to a speedier recovery. Show how you would incorporate comparison, randomization, and blinding.

11. The computer science department plans to offer three introductory-level CS courses: one using Pascal, one using C++, and one using Java.
    (a) The department chairperson plans to give all students the same general programming exam at the end of the year and to compare the relative effectiveness of using each of the programming languages by comparing the mean grades of the students from each course. What is wrong, if anything, with the chairperson's plan?
    (b) The chairperson also wishes to determine whether math majors or science majors do better in the courses. Suppose he calculates that the average grade of science majors was higher than the average grade of math majors in each of the courses. Does it follow that the average grade of all the science majors taking the three courses must be higher than the average grade of all the math majors? Explain.
    (c) Suppose 300 students wish to take introductory programming. How would you randomly assign 100 students to each of the three courses?
    (d) How would you randomly assign students to the three courses if you wanted the assignment to be independent from student to student with each student in turn having a one-third probability of taking each of the three classes.
    (e) Name a lurking variable that all the above methods miss.

An Investigative Task

A high school offers two precalculus courses, one that uses a traditional lecture and drill method, and a second that divides students into small groups to work on open-ended problems. To compare the effectiveness of the two methods, the administration proposes to compare average SAT math scores for the students in the two courses.

(a) What is wrong with the administration's proposal?

(b) Suppose a group of 50 students are willing to take either course. Explain how you would use a random number table to set up an experiment comparing the effectiveness of the two courses.

(c) Apply your setup procedure to the given random number table:

84177 06757 17613 15582 51506 81435 41050 92031 06449
05059 59884 31180 53115 84469 94868 57967 05811 84514
75011 13006 63395 55041 15866 06589 13119 71020 85940
91932 06488 74987 54355 52704 90359 02649 47496 71567
94268 08844 26294 64759 08989 57024 97284 00637 89283
03514 59195 07635 03309 72605 29357 23737 67881 03668
33876 35841 52869 23114 15864 38942

(d) Discuss any lurking variables that your setup doesn't consider.

Answer Key

1. **D**	4. **B**	7. **D**	10. **D**	13. **A**
2. **B**	5. **D**	8. **A**	11. **A**	
3. **C**	6. **E**	9. **E**	12. **A**	

Answers Explained

Multiple-Choice

1. **(D)** It may well be that very bright students are the same ones who both take Latin and do well on the SAT verbal exam. If students could be randomly assigned to take or not take Latin, the results would be more meaningful. Of course, ethical considerations might make it impossible to isolate the confounding variable in this way. Only using a sample from the observations gives less information.

2. **(B)** The desire of the workers for the study to be successful led to a placebo effect.

3. **(C)** In experiments on people, the subjects can be used as their own controls, with responses noted before and after the treatment. However, with such designs there is always the danger of a placebo effect. Thus the design of choice would involve a separate control group to be used for comparison.

4. **(B)** Blocking divides the subjects into groups, such as men and women, or political affiliations, and thus reduces variation.

5. **(D)** Blocking in experiment design first divides the subjects into representative groups called blocks, just as stratification in sampling design first divides the population into representative groups called strata. This procedure can control certain variables by bringing them directly into the picture, and thus conclusions are more specific. The paired comparison design is a special case of blocking in which each pair can be considered a block. Unnecessary blocking detracts from accuracy because of smaller sample sizes.

6. **(E)** None of the studies has any controls, such as randomization, control groups, or blinding, and so while they may give valuable information, they cannot establish cause and effect.

7. **(D)** Octane is the only explanatory variable, and it is being tested at four levels. Miles per gallon is the single response variable.

8. **(A)** There is nothing wrong with using volunteers—what is important is to randomly assign the volunteers into the two treatment groups. There is no way to use blinding in this study—the subjects will clearly know which breakfast

they are eating. The main idea behind randomly assigning subjects to the different treatments is to control for various possible confounding variables—it is reasonable to assume that people of various ages, races, ethnic backgrounds, etc., are assigned to receive each of the treatments.

9. **(E)** In good observational studies, the responses are not influenced during the collecting of data. In good experiments, treatments are compared as to differences in responses. In an experiment, there can be many treatments, each at a different level. Well-designed experiments can show cause and effect.

10. **(D)** Control, randomization, and replication are all important aspects of well-designed experiments. Care in observing without imposing change refers to observational studies, not experiments.

11. **(A)** Each subject might receive both treatments, as, for example, in the Pepsi-Coke taste comparison study. The point is to give each subject in a matched pair a different treatment and note any difference in responses. Matched-pair experiments are a particular example of blocking, not vice versa. Stratification refers to a sampling method, not to experimental design. Randomization is used to decide which of a pair gets which treatment or which treatment is given first if one subject is to receive both.

12. **(A)** Blinding does have to do with whether or not the subjects know which treatment (color in this experiment) they are receiving. However, drinking out of solid colored thermoses makes no sense since the beverages are identical except for color and the point of the experiment is the teenager's reaction to color. Blinding has nothing to do with blocking (team participation in this experiment).

13. **(A)** This study is an experiment because a treatment (periodic removal of a pint of blood) is imposed. There is no blinding because the subjects clearly know whether or not they are giving blood. There is no blocking because the subjects are not divided into blocks before random assignment to treatments. For example, blocking would have been used if the subjects had been separated by gender or age before random assignment to give or not give blood donations. There is a single factor—giving or not giving blood.

Free-Response

1. (a) These are observational studies as there is no randomization of treatments to subjects.

   (b) The excitement of a birthday party is a confounding variable. Without conducting a proper experiment, there is no way of telling whether observed hyperactivity is caused by sugar or by the excitement of a party or by some other variable.

(c) The parent should randomly give the child sugar or sugar-free sweets at parties and observe the child's behavior. It is important that the parent not know which the child is receiving (double blinding), because the parent might perceive a difference in behavior which is not really there if he/she knows whether or not the child is being given a sugary food.

2. Ask doctors, hospitals, or blood testing laboratories to make known that you are looking for HIV-positive volunteers. As the volunteers arrive, use a random number table to give each one the drug or a placebo (e.g., if the next digit in the table is odd, the volunteer gets the drug, while if the next digit is even, the volunteer gets a placebo). Use double-blinding; that is, both the volunteers and their doctors should not know if they are receiving the drug or the placebo. Ethical considerations will arise, for example, if the drug is very successful. If volunteers on the placebo are steadily developing full-blown AIDS while no one on the drug is, then ethically the test should be stopped and everyone put on the drug. Or if most of the volunteers on the drug are dying from an unexpected fatal side effect, the test should be stopped and everyone taken off the drug.

> **TIP**
>
> Simply saying to "randomly assign" subjects to treatment groups is usually an incomplete response. You need to explain *how* to make the assignments—for example, by using a random number table or through generating random numbers on a calculator.

3. To achieve blocking by gender, first separate the men and women. Label the 40 men 01 through 40. Use a random number table to pick two digits at a time, ignoring 00 and numbers greater than 40, and ignoring repeats, until a group of ten such numbers is obtained. These men will receive the supplement at the once-a-day level. Follow along in the table, continuing to ignore repeats, until another group of ten is selected. These men will receive the supplement at the twice-a-day level. Again ignore repeats until a third group of ten is selected to receive the supplement at the three-times-a-day level, while the remaining men will be a control group and not receive the supplement. Now repeat the entire procedure, starting by labeling the women 01 through 40. A decision should be made whether or not to use a placebo and have all participants take "something" three times a day. Weigh all 80 overweight volunteers before and after a predetermined length of time. Calculate the change in weight for each individual. Calculate the average change in weight among the ten people in each of the eight groups. Compare the four averages from each block (men and women) to determine the effect, if any, of different levels of the supplement for men and for women.

4. The first study, an observational study, does not suffer from nonresponse bias, as do most mailed questionnaires, because it involved follow-up telephone calls and achieved a high response rate. However, this study suffers terribly from selection bias because people who subscribe to a health magazine are not representative of the general population. One would expect most of them to strongly believe that vitamins improve their health. The second study, a controlled experiment, used comparison between a treatment group and a control group, used randomization in selecting who went into each group, and used blinding to control for a placebo effect on the part of the volunteers. However, it did not use double-blinding; that is, the doctors knew whether their patients

were receiving the vitamin, and this could have introduced hidden bias when they made judgments regarding their patients' health.

5. Every day for some specified period of time, look at the next digit on a random number table. If it is odd, flash the subliminal message all day on the screen, while if it is even, don't flash the message that day (randomization). Don't let the customers know what is happening (blinding) and don't let the clerks selling the popcorn know what is happening (double-blinding). Compare the quantity of popcorn bought by the treatment group, that is, by the people who receive the subliminal message, to the quantity bought by the control group, the people who don't receive the message (comparison).

6. Any conclusions would probably be meaningless. There is a substantial danger of the placebo effect here; that is, real physical responses could be caused by the psychological effect of knowing the intent of the research. The experiment would be considerably strengthened by using a control group taking a look-alike capsule. Any conclusions are further suspect because of the choice of subjects. Rather than making a random selection from the intended population, the company is using a sample from its own employees, a sample almost guaranteed to have concerns, interests, and backgrounds that will confound the responses or limit their generalizability.

7. Ask doctors and hospitals to make known that you are looking for volunteers from among intractable pain sufferers. As the volunteers arrive, use a random number table to decide which will have the electrodes properly embedded in their pain centers and which will have the electrodes harmlessly embedded in wrong positions. For example, if the next digit in the table is odd, the volunteer receives the proper embedding, while if the next digit is even, the volunteer does not. Use double-blinding; that is, both the volunteers and their doctors should not know if the volunteers are receiving the proper embedding. Ethical considerations will arise, for example, if the procedure is very successful. If volunteers with the wrong embeddings are in constant pain, while everyone with proper embedding is pain-free, then ethically the test should be stopped and everyone given the proper embedding. Or if most of the volunteers with proper embedding develop an unexpected side effect of the pain spreading to several nearby sites, then the test should be stopped and the procedure discontinued for everyone.

8. To achieve blocking by sunlight, first separate the sunlit and shaded plots. Label the 15 sunlit plots 01 through 15. Using a random number table, pick two digits at a time, ignoring 00 and numbers above 15 and ignoring repeats, until a group of five such numbers is obtained. These sunlit plots will receive the fertilizer at regular concentration. Continue in the table, ignoring repeats, until another group of five is selected. These sunlit plots will receive the fertilizer at double concentration, while the remaining sunlit plots will be a control group receiving no fertilizer. Now repeat the procedure, this time labeling the

shaded plots 01 through 15. Assuming size is the pertinent outcome, weigh all vegetables at the end of the season, compare the average weights among the three sunlit groups, and compare the average weights among the three shaded groups to determine the effect of the fertilizer, if any, at different levels on sunlit plots and separately on shaded plots.

9. The first study, an observational study, does not suffer from nonresponse bias, as do most studies involving mailed questionnaires, because the researchers made follow-up telephone calls and achieved a very high response rate. However, the first study suffers terribly from selection bias. People who work at a teaching hospital are not representative of the general population. One would expect many of them to have heard about how zinc coats the throat to hinder the propagation of viruses. The second study, a controlled experiment, used comparison between a treatment group and a control group, used randomization in selecting who went into each group, and used blinding to control for a placebo effect on the part of the volunteers. However, they did not use double-blinding; that is, the doctors knew whether their patients were receiving the zinc lozenges, and this could introduce hidden bias as the doctors make judgments about their patients' health.

10. For each new heart attack patient entering the hospital, look at the next digit from a random number table. If it is odd, give the name to a group of people who will pray for the patient throughout his or her hospitalization, while if it is even, don't ask the group to pray (randomization). Don't let the patients know what is happening (blinding) and don't let the doctors know what is happening (double-blinding). Compare the lengths of hospitalization of patients who receive prayers with those of control group patients who don't receive prayers (comparison).

11. (a) Allowing the students to self-select which class to take leads to a lurking variable that could be significant. For example, perhaps the brighter students all want to learn a certain one of the three languages.

    (b) It is possible for the average score of all science majors to be lower than the average for all math majors even though the science majors averaged higher in each class. For example, suppose that the students taking Java scored much higher than the students in the other two classes. Furthermore, only one science major took Java, and she scored tops in the class. Then the overall average of the math majors could well be higher. This is an example of Simpson's paradox, in which a comparison can be reversed when more than one group is combined to form a single group.

    (c) Number the students 001 through 300. Read off three digits at a time from a random number table, noting all triplets between 001 and 300 and ignoring repeats, until 100 such numbers have been selected. Keep reading off three digits, ignoring repeats, until 100 new numbers between 001 and 300 are selected. These get C++, while the remaining 100 get Java. Even quicker would be to use a calculator to generate random digits between 001 and 300.

(d) Go through the list of students, flipping a die for each. If a 1 or a 2 shows, the student takes Pascal, if a 3 or a 4 shows, C++, and if a 5 or a 6 shows, Java.

(e) A possible lurking variable are the teachers. For example, perhaps the better teachers teach Java.

Investigative Task

(a) There is no reason to believe that there was anything random about which students took which course. Perhaps all the weaker students self-selected or were advised to choose the traditional course.

(b) The students could be labeled 01 through 50. Pairs of digits could then be read off a random number table, ignoring numbers over 50 and ignoring duplicates, until a set of 25 numbers is obtained. The students corresponding to these numbers could be enrolled in the traditional course, and the remaining students in the other.

(c) Applying the above procedure results in {17, 31, 14, 35, 41, 05, 09, 20, 06, 44, 50, 43, 11, 18, 45, 01, 13, 33, 04, 19, 02, 08, 40, 49, 03}. Enroll the students with these numbers in the traditional course.

(d) Which teachers teach which courses is not considered. Perhaps the more interesting, exciting teachers teach the new version. Even though a control group is selected, there is no blinding, and so students in the new version might work harder because they realize they are part of an experiment.

THEME THREE: PROBABILITY

Probability as Relative Frequency

- Law of Large Numbers
- Addition Rule
- Multiplication Rule
- Conditional Probabilities
- Independence
- Multistage Probability Calculations
- Discrete Random Variables and Probability Distributions
- Simulation of Probability Distributions
- Mean and Standard Deviation of a Random Variable

In the world around us, unlikely events sometimes take place. At other times, events that seem inevitable do not occur. Because of the myriad and minute origins of various happenings, it is often impracticable, or simply impossible, to predict exact outcomes. However, while we may not be able to foretell a specific result, we can sometimes assign what is called a *probability* to indicate the likelihood that a particular event will occur.

THE LAW OF LARGE NUMBERS

The *relative frequency* of an event is the proportion of times the event happened, that is, the number of times the event happened divided by the total number of trials.

EXAMPLE 9.1

If there were 12 cloudy days during a 30-day period, the relative frequency of cloudy days was $\frac{12}{30} = .4$.

Relative frequencies may change every time an experiment is performed. However, when an experiment is performed a large number of times, the relative frequency of an event tends to become closer to what is called the probability of the event. (In some cases, *probability* can be defined as long-term relative frequency.)

EXAMPLE 9.2

A nurse notes that on one day 19 out of 50 new patients had a fever, while on another day 28 out of 56 had a fever. The two relative frequencies $\frac{19}{50} = .38$ and $\frac{28}{56} = .50$, are different. What would be the outcome of summing up a very long series of daily measurements? Although individual daily measurements are uncertain, the long-term proportion of patients arriving with a fever will "settle down" to a fixed number, called the probability of this event. Whatever the probability of a patient arriving with a fever is—for example, perhaps it is .412—summing up the outcomes from an indefinitely long series of days should arrive at this idealized result.

EXAMPLE 9.3

Suppose the numbers in the following random number table correspond to people arriving for work at a large factory. Let 0, 1, and 2 be smokers and 3–9 be nonsmokers. After each ten arrivals, calculate the total relative frequency of smokers and graph the results.

84177	06757	17613	15582	51506	81435	41050	92031	06449	05059
59884	31180	53115	84469	94868	57967	05811	84514	75011	13006
63395	55041	15866	06589	13119	71020	85940	91932	06488	74987
54355	52704	90359	02649	47496	71567	94268	08844	26294	64759
08989	57024	97284	00637	89283	03514	59195	07635	03309	72605
29357	23737	67881	03668	33876	35841	52869	23114	15864	38942

Answer: The cumulative relative frequencies are $\frac{2}{10} = .2$, $\frac{6}{20} = .3$, $\frac{9}{30} = .3$, $\frac{15}{40} = .375$, $\frac{18}{50} = .36$, $\frac{21}{60} = .35$, $\frac{23}{70} = .329$, $\frac{23}{80} = .288$, $\frac{27}{90} = .3$, $\frac{33}{100} = .33$, and so on. Graphing these and further values gives

Note how the cumulative relative frequency at first fluctuates considerably but then seems to level off. The concept that the relative frequency tends closer and closer to a certain number (the probability) as an experiment is repeated more and more times is called the *law of large numbers.*

The *probability* of a particular outcome of an experiment is a mathematical statement about the likelihood of that event occurring. Probabilities are always between 0 and 1, with a probability close to 0 meaning that an event is unlikely to occur and a probability close to 1 meaning that the event is likely to occur. The sum of the probabilities of all the separate outcomes of an experiment is always 1.

ADDITION RULE, MULTIPLICATION RULE, CONDITIONAL PROBABILITIES, AND INDEPENDENCE

Complementary Events

The probability that an event will not occur, that is, the probability of its complement, is equal to 1 minus the probability that the event will occur:

$$P(A^C) = 1 - P(A)$$

EXAMPLE 9.4

If the probability that a company will win a contract is .3, what is the probability that it will not win the contract?

Answer: 1 – .3 = .7

TIP

Don't add probabilities unless the events are mutually exclusive.

Addition Principle

If two events are mutually exclusive, that is, they cannot occur simultaneously, the probability that at least one event will occur is equal to the sum of the respective probabilities of the two events:

$$\text{If } P(A \cap B) = 0 \quad \text{then} \quad P(A \cup B) = P(A) + P(B)$$

where $A \cap B$, (read "A intersect B") means that both A and B occur, while $A \cup B$ (read "A union B") means that either A or B or both occur.

EXAMPLE 9.5

If the probabilities that Jane, Tom, and Mary will be chosen chairperson of the board are .5, .3, and .2, respectively, the probability that the chairperson will be either Jane or Mary is .5 + .2 = .7.

When two events are not mutually exclusive, the sum of their probabilities counts their shared occurrence twice. This leads to the following rule.

General Addition Rule

For any pair of events A and B,

$$P(A \cup B) = P(A) + P(B) - P(A \cap B)$$

EXAMPLE 9.6

Suppose the probability that a construction company will be awarded a certain contract is .25, the probability that it will be awarded a second contract is .21, and the probability that it will get both contracts is .13. What is the probability that the company will win at least one of the two contracts?

Answer: .25 + .21 – .13 = .33

TIP

Don't multiply probabilities unless the events are independent.

Multiplication Rule

If the chance that one event will happen is not influenced by whether or not a second event happens, the probability that *both* events will occur is the product of their separate probabilities.

EXAMPLE 9.7

TIP

Don't confuse independence with mutally exclusive.

The probability that a student will receive a state grant is $\frac{1}{3}$, while the probability that she will be awarded a federal grant is $\frac{1}{2}$. If whether or not she receives one grant is not influenced by whether or not she receives the other, what is the probability of her receiving both grants?

Answer: $\frac{1}{3} \times \frac{1}{2} = \frac{1}{6}$

The multiplication rule can be extended to more than two events; that is, given a sequence of *independent* events, the probability that *all* will happen is equal to the product of their individual probabilities.

EXAMPLE 9.8

Suppose a reputed psychic in an extrasensory perception (ESP) experiment has called heads or tails correctly on ten successive tosses of a coin, What is the probability that guessing would have yielded this perfect score?

Answer: $\left(\frac{1}{2}\right)\left(\frac{1}{2}\right)\cdots\left(\frac{1}{2}\right) = \left(\frac{1}{2}\right)^{10} = \frac{1}{1024}$

Conditional Probability and Independence

To best understand independence as mentioned above, we first examine the idea of *conditional probability*, that is, the probability of an event given that another event has occurred.

EXAMPLE 9.9

The table below gives the results of a survey of the drinking and smoking habits of 1200 college students. Rows and columns have also been summed.

	Drink beer	Don't drink	
Smoke	315	165	480
Don't smoke	585	135	720
	900	300	1200

What is the probability that someone in this group smokes?

Answer: $P(\text{smokes}) = \frac{480}{1200} = .4$

What is the probability a student smokes given that she is a beer drinker?

(continued)

Answer: We now narrow our attention to the 900 beer drinkers, and thus we calculate $\frac{315}{900} = .35$. We use the following notation: P(smoke|drink) (read "the probability that people smoke given that they drink").

What is the probability that someone drinks?
Answer: $P(\text{drink}) = \frac{900}{1200} = .75$

What is the probability that students drink given that they smoke?
Answer: $P(\text{drink}|\text{smoke}) = \frac{315}{480} = .65625$

A formula for conditional probability is given by

$$P(A|B) = \frac{P(A \cap B)}{P(B)}$$

EXAMPLE 9.10

If 90% of the households in a certain region have answering machines and 50% have both answering machines and call waiting, what is the probability that a household chosen at random and found to have an answering machine also has call waiting?
Answer:

$$P(\text{call waiting}|\text{answering machine})$$
$$= \frac{P(\text{call waiting} \cap \text{answering machine})}{P(\text{answering machine})} = \frac{.5}{.9} = \frac{5}{9}$$

EXAMPLE 9.11

A psychologist interested in right-handedness versus left-handedness and in IQ scores collected the following data from a random sample of 2000 high school students.

	Right-handed	Left-handed	
High IQ	190	10	200
Normal IQ	1710	90	1800
	1900	100	2000

What is the probability that a student from this group has a high IQ?
Answer: $P(\text{highIQ}) = \frac{200}{2000} = .1$

What is the probability that a student has a high IQ given that she is left-handed?
Answer: Shrinking the population of interest to the 100 left-handed students gives $P(\text{highIQ}|\text{left-handed}) = \frac{10}{100} = .1$

Note that the probability of having a high IQ is not affected by the fact that the student is left-handed. We say that the two events, high IQ and left-handed, are *independent*.

That is, A and B are independent if $P(A|B) = P(A)$, and earlier it was noted that in this case we also have

$$P(A \cap B) = P(A)P(B)$$

Whether events are mutually exclusive or are independent are two very different properties! One refers to events being disjoint, the other to an event having no effect on whether or not the other occurs. Note that mutually exclusive (disjoint) events are *not* independent (except in the special case that one of the events has probability 0). That is, mutually exclusive gives that $P(A \cap B) = 0$, while independence gives that $P(A \cap B) = P(A)P(B)$ (and the only way these are ever simultaneously true is in the very special case when $P(A) = 0$ or $P(B) = 0$).

MULTISTAGE PROBABILITY CALCULATIONS

EXAMPLE 9.12

TIP

Tree diagrams can be very useful in working with conditional probabilities.

A videocassette recorder (VCR) manufacturer receives 70% of his parts from factory F1 and the rest from factory F2. Suppose that 3% of the output from F1 are defective, while only 2% of the output from F2 are defective. What is the probability a received part is defective?

Answer: In such problems it is helpful to draw a tree diagram like the following.

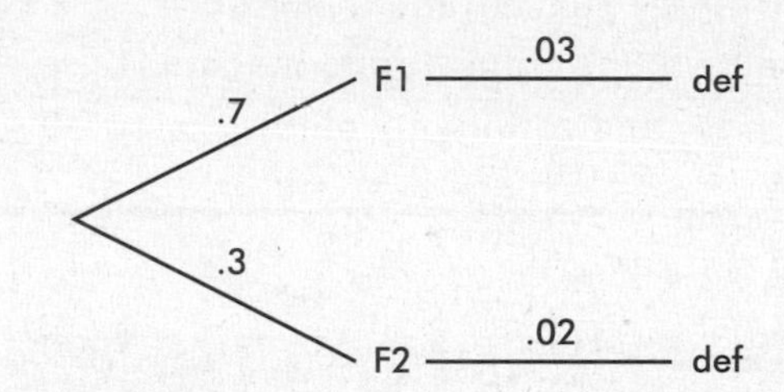

We then have

$$P(\text{F1} \cap \text{def}) = P(\text{F1})P(\text{def}|\text{F1}) = (.7)(.03) = .021$$
$$P(\text{F2} \cap \text{def}) = P(\text{F2})P(\text{def}|\text{F2}) = (.3)(.02) = .006$$

At this stage a Venn diagram is helpful in finishing the problem:

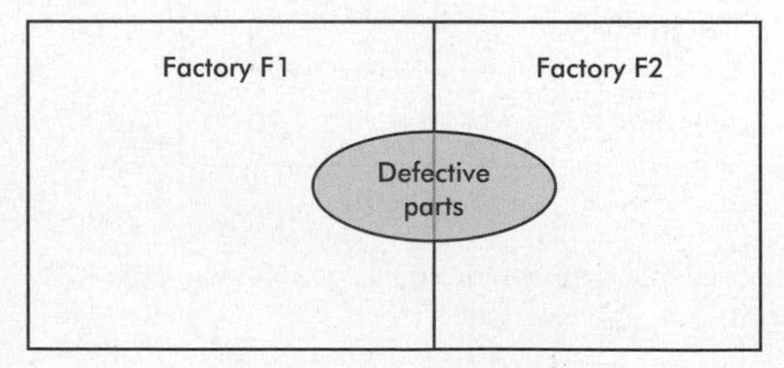

$$P(\text{def}) = P(\text{F1} \cap \text{def}) + P(\text{F2} \cap \text{def}) = .021 + .006 = .027$$

We can take the above analysis one stage further and answer such questions as: If a randomly chosen part is defective, what is the probability it came from factory F1? From factory F2?

$$P(\text{F1}|\text{def}) = \frac{P(\text{F1} \cap \text{def})}{P(\text{def})} = \frac{.021}{.027} \approx .778$$

$$P(\text{F2}|\text{def}) = \frac{P(\text{F2} \cap \text{def})}{P(\text{def})} = \frac{.006}{.027} \approx .222$$

EXAMPLE 9.13

Suppose that three branches of a local bank average, respectively, 120, 180, and 100 clients per day. Suppose further that the probabilities that a client will transact business involving more than $100 during a visit are, respectively, .5, .6, and .7. A client is chosen at random. What is the probability that the client will transact business involving over $100? What is the probability that the client went to the first branch given that she transacted business involving over $100?

Answer: $P(B1) = \frac{120}{400} = .3$, $P(B2) = \frac{180}{400} = .45$, and $P(B3) = \frac{100}{400} = .25$. Then we have

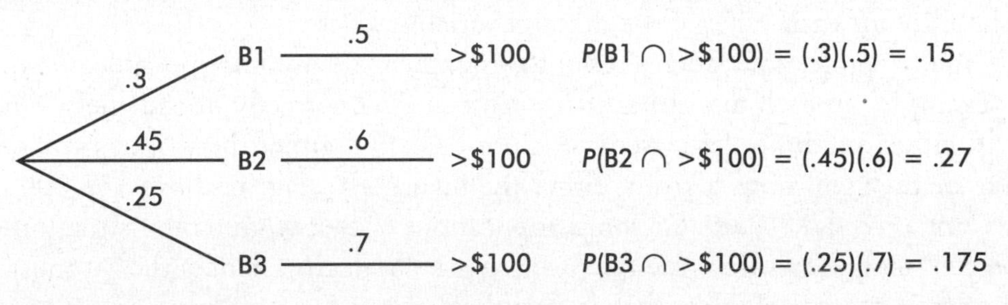

Now $P(> \$100) = .15 + .27 + .175 = .595$, and $P(B1 \mid > \$100) = \frac{.15}{.595} \approx .252$.

TIP

Naked or *bald answers* will receive little or *no* credit. You must show where answers come from.

DISCRETE RANDOM VARIABLES AND THEIR PROBABILITY DISTRIBUTIONS

Often each outcome of an experiment has not only an associated probability but also an associated *real number.* For example, the probability may be $\frac{1}{2}$ that there are five defective batteries, the probability may be .01 that a company receives seven contracts, or the probability may be .95 that three people recover from a disease. If X represents the different numbers associated with the potential outcomes of some chance situation, we call X a *random variable.*

EXAMPLE 9.14

A prison official knows that $\frac{1}{2}$ of the inmates he admits stay only 1 day, $\frac{1}{4}$ stay 2 days, $\frac{1}{5}$ stay 3 days, and $\frac{1}{20}$ stay 4 days before they are either released or are sent on to the county jail. If X represents the number of days, then X is a random variable that takes the value 1 with probability $\frac{1}{2}$, the value 2 with probability $\frac{1}{4}$, the value 3 with probability $\frac{1}{5}$, and the value 4 with probability $\frac{1}{20}$.

The random variable in Example 9.14 is called *discrete* because it can assume only a countable number of values. The random variable in Example 9.15, however, is said to be *continuous* because it can assume values associated with a whole line interval.

EXAMPLE 9.15

Let X be a random variable whose values correspond to the speeds at which a jet plane can fly. The jet may be traveling at 623.478 . . . miles per hour or at any other speed in some whole interval. We might ask what the probability is that the plane is flying at between 300 and 400 miles per hour.

A *probability distribution* for a discrete variable is a listing or formula giving the probability for each value of the random variable.

In many applications, such as coin tossing, there are only *two* possible outcomes. For example, on each toss either the psychic guesses correctly or she guesses incorrectly, either a radio is defective or it is not defective, either the workers will go on strike or they will not walk out, either the manager's salary is above $50,000 or it does not exceed $50,000. In some applications such two-outcome situations are repeated many times. For example, we can ask about the chances that at most one out of four tires is defective, that at least three out of five unions will vote to go on strike, or that exactly two out of three executives have salaries above $50,000. For applications in which a two-outcome situation is repeated a certain number of times and the probability of each of the two outcomes remains the same for each repetition, the resulting calculations involve what are known as *binomial probabilities.*

EXAMPLE 9.16

Suppose the probability that a lightbulb is defective is .1. What is the probability that four lightbulbs are all defective?

Answer: Because of independence (i.e., whether one lightbulb is defective is not influenced by whether any other lightbulb is defective), we can multiply individual probabilities of being defective to find the probability that all the bulbs are defective:

$$(.1)(.1)(.1)(.1) = (.1)^4 = .0001$$

EXAMPLE 9.17

Again suppose the probability that a lightbulb is defective is .1. What is the probability that exactly two out of three lightbulbs are defective?

Answer: We subdivide the problem as follows: The probability that the first two bulbs are defective and the third is good is $(.1)(.1)(.9) = .009$. (Note that if the probability of being defective is .1, the probability of being good is .9.) The probability that the first bulb is good and the other two are defective is $(.9)(.1)(.1) = .009$. Finally, the probability that the second bulb is good and the other two are defective is $(.1)(.9)(.1) = .009$. Summing, we find that the probability that exactly two out of three bulbs are defective is $.009 + .009 + .009 = .027$.

EXAMPLE 9.18

If the probability that a lightbulb is defective is .1, what is the probability that exactly three out of eight lightbulbs are defective?

Answer: Again we can subdivide the problem. For example, the probability that the first, third, and seventh bulbs are defective and the rest are good is

$$(.1)(.9)(.1)(.9)(.9)(.9)(.1)(.9) = (.1)^3(.9)^5 = .00059049$$

The probability that the second, third, and fifth bulbs are defective and the rest are good is

$$(.9)(.1)(.1)(.9)(.1)(.9)(.9)(.9) = (.1)^3(.9)^5 = .00059049$$

As can be seen, the probability of any particular arrangement of three defective and five good bulbs is $(.1)^3(.9)^5 = .00059049$. How many such arrangements are there? In other words, in how many ways can we pick three of eight positions for the defective bulbs (the remaining five positions are for good bulbs)? The answer is given by *combinations*:

$$\binom{8}{3} = \frac{8!}{5!3!} = \frac{8 \times 7 \times 6}{3 \times 2} = 56$$

Each of these 56 arrangements has a probability of .00059049. Thus, the probability that exactly three out of eight lightbulbs are defective is $56 \times .00059049 = .03306744$:

$$\binom{8}{3}(.1)^3(.9)^5 = \frac{8!}{5!3!}(.1)^3(.9)^5 = .03306744$$

[On the TI-84 one can calculate binompdf(8, .1, 3) = .03306744.]

Note the application `CATALOG HELP` on the TI-84. For example, go to `2nd DISTR` and scroll down to `binompdf(` . Instead of `ENTER`, click on the + button. You are then reminded of the order in which the variables must be entered: first the number of trials, then the probability of success p, and finally, optionally, the number of successes x.

EXAMPLE 9.19

Suppose 30% of the employees in a large factory are smokers. What is the probability that there will be exactly two smokers in a randomly chosen five-person work group?

Answer: We reason as follows. The probability that a person smokes is 30% = .3, and so the probability that he or she does not smoke is 1 – .3 = .7. The probability of a particular arrangement of two smokers and three nonsmokers is $(.3)^2(.7)^3 = .03087$. The number of such arrangements is

$$\binom{5}{2} = \frac{5!}{2!3!} = 10$$

Each such arrangement has probability .03087, and so the final answer is $10 \times .03087 = .3087$.

$$\binom{5}{2}(.3)^2(.7)^3 = \frac{5!}{2!3!}(.3)^2(.7)^3 = .3087$$

[Or binompdf(5, .3, 2) = .3087.]

We can state the general principle as follows.

Binomial Formula

Suppose an experiment has two possible outcomes, called *success* and *failure*, with the probability of success equal to p and the probability of failure equal to q (of course, $p + q = 1$ and $q = 1 - p$). Suppose further that the experiment is repeated n times and that the outcome at any particular time has no influence over the outcome at any other time. Then the probability of exactly k successes (and thus $n–k$ failures) is

$$\binom{n}{k} p^k q^{n-k} = \frac{n!}{k!(n-k)!} p^k q^{n-k}$$

EXAMPLE 9.20

A manager notes that there is a .125 probability that any employee will arrive late for work. What is the probability that exactly one person in a six-person department will arrive late?

Answer: If the probability of being late is .125, the probability of being on time is 1 – .125 = .875. If one person out of six is late, 6 – 1 = 5 will be on time. Thus the desired probability is

$$\binom{6}{1}(.125)^1(.875)^5 = 6(.125)(.875)^5 = .385$$

[Or going to 2nd DISTR on the TI-84, binompdf(6, .125, 1) ≈ .385.]

Many, perhaps most, applications of probability involve such phrases as *at least, at most, less than,* and *more than.* In these cases, solutions involve summing two or more cases. For such calculations, the TI-84 `binomcdf` is very useful. `binomcdf(n,p,x)` gives the probability of x or fewer successes in a binomial distribution with number of trials n and probability of success p.

EXAMPLE 9.21

A manufacturer has the following quality control check at the end of a production line: If at least eight of ten randomly picked articles meet all specifications, the whole shipment is approved. If, in reality, 85% of a particular shipment meet all specifications, what is the probability that the shipment will make it through the control check?

Answer: The probability of an item meeting specifications is .85, and so the probability of it not meeting specifications must be .15. We want to determine the probability that at least eight out of ten articles will meet specifications, that is, the probability that exactly eight or exactly nine or exactly ten articles will meet specifications. We sum the three binomial probabilities:

Exactly 8 of 10 meet specifications	Exactly 9 of 10 meet specifications	Exactly 10 of 10 meet specifications
$\binom{10}{8}(.85)^8(.15)^2$	$+\binom{10}{9}(.85)^9(.15)^1$	$+\binom{10}{10}(.85)^{10}(.15)^0$

$$= \frac{10!}{8!2!}(.85)^8(.15)^2 + 10(.85)^9(.15) + (.85)^{10} \approx .820$$

[On the TI-84 one can calculate 1 – binomcdf(10, .85, 7) ≈ .820 or binomcdf(10, .15, 2) ≈ .820.]

EXAMPLE 9.22

For the problem in Example 9.21, what is the probability that a shipment in which only 70% of the articles meet specifications will make it through the control check?

Answer:

$$\binom{10}{8}(.7)^8(.3)^2+\binom{10}{9}(.7)^9(.3)^1+\binom{10}{10}(.7)^{10}(.3)^0$$
$$=45(.7)^8(.3)^2+10(.7)^9(.3)+(.7)^{10}$$
$$\approx .383$$

[Or binomcdf(10, .3, 2) ≈ .383.]

In some situations it is easier to calculate the probability of the complementary event and subtract this value from 1.

EXAMPLE 9.23

Joe DiMaggio had a career batting average of .325. What was the probability that he would get at least one hit in five official times at bat?

Answer: We could sum the probabilities of exactly one hit, two hits, three hits, four hits, and five hits. However, the complement of "at least one hit" is "zero hits." The probability of no hit is

$$\binom{5}{0}(.325)^0(.675)^5=(.675)^5\approx .140$$

and thus the probability of at least one hit in five times at bat is $1 - .140 = .860$.

[Or binomcdf(5, .675, 4) ≈ .860.]

EXAMPLE 9.24

A grocery store manager notes that 35% of customers who buy a particular product make use of a store coupon to receive a discount. If seven people purchase the product, what is the probability that fewer than four will use a coupon?

Answer: In this situation, "fewer than four" means zero, one, two, or three.

$$\binom{7}{0}(.35)^0(.65)^7+\binom{7}{1}(.35)^1(.65)^6+\binom{7}{2}(.35)^2(.65)^5+\binom{7}{3}(.35)^3(.65)^4$$
$$=(.65)^7+7(.35)(.65)^6+21(.35)^2(.65)^5+35(.35)^3(.65)^4$$
$$\approx .800$$

[Or binomcdf(7, .35, 3) ≈ .800.]

Sometimes we are asked to calculate the probability of each of the possible outcomes (the results should sum to 1).

EXAMPLE 9.25

If the probability that a male birth will occur is .51, what is the probability that a five-child family will have all boys? Exactly four boys? Exactly three boys? Exactly two boys? Exactly one boy? All girls?

Answer:

$$P(5\text{ boys}) = \binom{5}{5}(.51)^5(.49)^0 = (.51)^5 = .0345$$

$$P(4\text{ boys}) = \binom{5}{4}(.51)^4(.49)^1 = 5(.51)^4(.49) = .1657$$

$$P(3\text{ boys}) = \binom{5}{3}(.51)^3(.49)^2 = 10(.51)^3(.49)^2 = .3185$$

$$P(2\text{ boys}) = \binom{5}{2}(.51)^2(.49)^3 = 10(.51)^2(.49)^3 = .3060$$

$$P(1\text{ boys}) = \binom{5}{1}(.51)^1(.49)^4 = 5(.51)(.49)^4 = .1470$$

$$P(0\text{ boys}) = \binom{5}{0}(.51)^0(.49)^5 = (.49)^5 = \underline{.0283}$$

$$1.0000$$

[Or binompdf(5, .49) = {.0345 .1657 .3185 .3060 .1470 .0283}.]

A list such as the one in Example 9.25 shows the entire *probability distribution*, which in this case refers to a listing of all outcomes and their probabilities.

Geometric Probabilities

Suppose an experiment has two possible outcomes, called *success* and *failure*, with the probability of success equal to p and the probability of failure equal to $q = 1 - p$, and the trials are independent. Then the probability that the first success is on trial number $X = k$ is

$$q^{k-1}p$$

EXAMPLE 9.26

Suppose only 12% of men in ancient Greece were honest. What is the probability that the first honest man Diogenes encounters will be the third man he meets?

Answer: $(.88)^2(.12) = .092928$ [or geometpdf (.12, 3) = .092928]

What is the probability that the first honest man he encounters will be no later than the fourth man he meets?

Answer: $(.12) + (.88)(.12) + (.88)^2(.12) + (.88)^3(.12) = .40030464$
[or geometcdf (.12, 4) = .40030464]

SIMULATION OF PROBABILITY DISTRIBUTIONS, INCLUDING BINOMIAL AND GEOMETRIC

Instead of algebraic calculations, sometimes we can use simulation to answer probability questions.

EXAMPLE 9.27

A study reported by Liu and Schutz states that in hockey the better team has a 65% chance of scoring the first goal in overtime. If a team is considered the better team in all five of its overtime games, is it reasonable to believe that it will score the first goal in at least four of these games? Answer the question using simulation.

Answer: Since the team will score first 65% of the time, let the digits 01–65 represent this outcome. To simulate five games select ten digits from the random number table and look at them two at a time. Among the five pairs, note how many times 01–65 appear. (Repeats are OK.) Underlining these pairs gives

8417706757 1761315582 5150681435 4105092031 0644905059

5988431180 5311584469 9486857967 0581184514 7501113006

6339555041 1586606589 1311971020 8594091932 0648874987

5435552704 9035902649 4749671567 9426808844 2629464759

0898957024 9728400637 8928303514 5919507635 0330972605

2935723737 6788103668 3387635841 5286923114 1586438942

For example, in the first group of five pairs, two (17 and 57) out of five are between 01 and 65. In the second group, four (17, 61, 31, and 55) out of five are in the range. Tabulating the frequencies of each result gives

Number of First Goals	Frequency
0	1
1	0
2	4
3	8
4	13
5	4

In this simulation we see that the better team will score the first goal in at least four of the five overtime games 17 out of 30 times. Thus the answer to the original question is that it is not unreasonable to believe that the better team will win at least four of the five games.

It is easy to confuse *number of samples* with *sample size.* Note that above we have simulated 30 samples of five games each.

TIP

Be able to describe simulations so that others can repeat your procedure.

In Example 9.27 simulation is used with regard to a binomial distribution, while in the example below it is used with regard to a geometric distribution. A *geometric distribution* shows the number of trials needed until a success is achieved.

EXAMPLE 9.28

Suppose the probability that a company will land a contract is .3. Use simulation to find the probability that the company will first land a contract on the first bid, on the second, on the third, and so on.

(continued)

Answer: Let 0, 1, and 2 represent landing a contract, while 3–9 represent not getting the contract. Read the random number table, one digit at a time, noting how many digits must be read until a 0, 1, or 2 is encountered. Using the table in Example 9.27 results in

841(first success in third try)
770(first success in third try)
67571(first success in fifth try)
761(first success in third try)
31(first success in second try).

Continuing in this fashion and tabulating the results, we have

Number of the Try at Which the First Success Occurred	Frequency
1	21
2	19
3	16
4	7
5	9
Over 5	12

Thus we estimate the probabilities as follows:

Number of the Try at Which the First Success Occurred	Estimated Probability
1	$\frac{21}{84} = .25$
2	$\frac{19}{84} = .23$
3	$\frac{16}{84} = .19$
4	$\frac{7}{84} = .08$
5	$\frac{9}{84} = .11$
Over 5	$\frac{12}{84} = .14$

The actual probabilities are .3, $(.7)(.3) = .21$, $(.7)^2(.3) = .147$, $(.7)^3(.3) = .1029$, $(.7)^4(.3) = .07203$, and $1 - (.3 + .21 + .147 + .1029 + .07203) = .16807$.

[On the TI-84 one can calculate the geometric probabilities geometpdf(.3, 1) = .3, . . . , geometpdf(.3, 5) = .07203, and 1 – geometcdf(.3, 5) = .16807.]

One example of simulating a nonbinomial, nongeometric distribution is the following.

EXAMPLE 9.29

If a two-person committee is chosen at random from a group consisting of three supervisors and six employees, what is the probability that the committee will have exactly one supervisor and one employee? Use simulation to estimate the answer.

Answer: Let the digits 1, 2, and 3 represent the supervisors, and the digits 4, 5, 6, 7, 8, and 9 represent the employees. Our procedure is to pick digits one at a time from a random number table, ignoring repeats and ignoring 0's, until two different digits have been chosen. Reading from the random number table in Example 9.27 results in the committees (8, 4) with two employees, (1, 7) with one of each, (7, 6) with two employees, (7, 5) with two employees, (7, 1) with one of each, (7, 6) with two employees, (1, 3) with two supervisors, and so on. Tabulating from the table gives

Two supervisors:	11
One of each:	53
Two employees:	66.

Thus an estimate of the probability that the committee will consist of one supervisor and one employee is $\frac{53}{130} = .41$.

[The actual probability is $\dfrac{\binom{3}{1}\binom{6}{1}}{\binom{9}{2}} = \dfrac{18}{36} = .5$.]

In performing a simulation, you must:

1. Set up a correspondence between outcomes and random numbers.
2. Give a procedure for choosing the random numbers (for example, pick three digits at a time from a designated row in a random number table).
3. Give a stopping rule.
4. Note what is to be counted (what is the purpose of the simulation), and give the count if requested.

MEAN (EXPECTED VALUE) AND STANDARD DEVIATION OF A RANDOM VARIABLE, INCLUDING BINOMIAL

A bettor placing a chip on a number in roulette has a small chance of winning a lot and a large chance of losing a little. To determine whether the bet is "fair," we must be able to calculate the *expected value* of the game to the bettor. An insurance company in any given year pays out a large sum of money to each of a relatively small number of people, while collecting a small sum of money from each of many other individuals. To determine policy premiums an actuary must calculate the expected values relating to deaths in various age groups of policyholders.

EXAMPLE 9.30

Concessionaires know that attendance at a football stadium will be 60,000 on a clear day, 45,000 if there is light snow, and 15,000 if there is heavy snow. Furthermore, the probability of clear skies, light snow, or heavy snow on any particular day is $\frac{1}{2}$, $\frac{1}{3}$, and $\frac{1}{6}$, respectively. (Here we have a random variable X that takes the values 60,000, 45,000, and 15,000.)

(continued)

Outcome:	Clear skies	Light snow	Heavy snow
Probability:	$\frac{1}{2}$,	$\frac{1}{3}$,	$\frac{1}{6}$,
Random variable:	60,000	45,000	15,000

What average attendance should be expected for the season?

Answer: We reason as follows. Suppose on, say, 12 game days we have clear skies $\frac{1}{2}$ of the time (6 days), a light snow $\frac{1}{3}$ of the time (4 days), and a heavy snow $\frac{1}{6}$ of the time (2 days). Then the average attendance will be

$$\frac{(6\times 60{,}000)+(4\times 45{,}000)+(2\times 15{,}000)}{12}=47{,}500$$

Note that we could have divided the 12 into each of the three terms in the numerator to obtain

$$\left(\frac{6}{12}\times 60{,}000\right)+\left(\frac{4}{12}\times 45{,}000\right)+\left(\frac{2}{12}\times 15{,}000\right)$$

or, equivalently,

$$\left(\frac{1}{2}\times 60{,}000\right)+\left(\frac{1}{3}\times 45{,}000\right)+\left(\frac{1}{6}\times 15{,}000\right)=47{,}500$$

Actually, there was no need to consider 12 days. We could have simply multiplied probabilities by corresponding attendances and summed the resulting products.

The final calculation in Example 9.30 motivates our definition of expected value. The *expected value* (or *average* or *mean*) of a random variable X is the sum of the products obtained by multiplying each value x_i by the corresponding probability p_i:

$$E(X)=\mu_x=\sum x_i p_i$$

EXAMPLE 9.31

In a lottery, 10,000 tickets are sold at $1 each with a prize of $7500 for one winner. What is the average result for each bettor?

Answer: The actual winning payoff is $7499 because the winner paid $1 for a ticket, so we have

Outcome:	Win	Lose
Probability:	$\frac{1}{10{,}000}$	$\frac{9{,}999}{10{,}000}$
Random variable:	7499	−1

$$\text{Expected Value}=7499\left(\frac{1}{10{,}000}\right)+(-1)\left(\frac{9{,}999}{10{,}000}\right)=-0.25$$

Thus the *average* result for each person betting on the lottery is a $0.25 loss. Alternatively, we can say that the expected payoff to the lottery system is $0.25 for each ticket sold.

EXAMPLE 9.32

A manager must choose among three options. Option A has a 10% chance of resulting in a \$250,000 gain but otherwise will result in a \$10,000 loss. Option B has a 50% chance of gaining \$40,000 and a 50% chance of losing \$2000. Finally, option C has a 5% chance of gaining \$800,000 but otherwise will result in a loss of \$20,000. Which option should the manager choose?

Answer: Calculate the expected values of the three options:

	Option A		Option B		Option C	
Outcome	Gain	Loss	Gain	Loss	Gain	Loss
Probability	.10	.90	.50	.50	.05	.95
Random variable	250,000	−10,000	40,000	−2,000	800,000	−20,000

$$E(A) = .10(250,000) + .90(-10,000) = \$16,000$$

$$E(B) = .50(40,000) + .50(-2000) = \$19,000$$

$$E(C) = .05(800,000) + .95(-20,000) = \$21,000$$

The manager should choose option C! However, although option C has the greatest mean (expected value), the manager may well wish to consider the relative riskiness of the various options. If, for example, a \$5000 loss would be disastrous for the company, the manager might well decide to choose option B with its maximum possible loss of \$2000. (It should be intuitively clear that some concept of variance would be helpful here in measuring the risk; later in this topic the variance of a random variable will be defined.)

Suppose we have a *binomial random variable,* that is, a random variable whose values are the numbers of "successes" in some binomial probability distribution.

EXAMPLE 9.33

Of the automobiles produced at a particular plant, 40% had a certain defect. Suppose a company purchases five of these cars. What is the expected value for the number of cars with defects?

Answer: We might guess that the average or mean or expected value is 40% of 5 = $.4 \times 5 = 2$, but let's calculate from the definition. Letting X represent the number of cars with the defect, we have

$$P(0) = \binom{5}{0}(.4)^0(.6)^5 = (.6)^5 = .07776$$

$$P(1) = \binom{5}{1}(.4)^1(.6)^4 = 5(.4)(.6)^4 = .25920$$

$$P(2) = \binom{5}{2}(.4)^2(.6)^3 = 10(.4)^2(.6)^3 = .34560$$

$$P(3) = \binom{5}{3}(.4)^3(.6)^2 = 10(.4)^3(.6)^2 = .23040$$

$$P(4) = \binom{5}{4}(.4)^4(.6)^1 = 5(.4)^4(.6) = .07680$$

$$P(5) = \binom{5}{5}(.4)^5(.6)^0 = (.4)^5 = .01024$$

(continued)

Outcome (no. of cars):	0	1	2	3	4	5
Probability:	.07776	.25920	.34560	.23040	.07680	.01024
Random variable:	0	1	2	3	4	5

$$\begin{aligned} E(X) &= 0(.07776) + 1(.25920) + 2(.34560) + 3(.23040) \\ &\quad + 4(.07680) + 5(.01024) \\ &= 2 \end{aligned}$$

Thus the answer turns out to be the same as would be obtained by simply multiplying the probability of a "success" by the number of cases.

The following is true: If we have a binomial probability situation with the probability of a success equal to p and the number of trials equal to n, the expected value or mean number of successes for the n trials is np.

EXAMPLE 9.34

An insurance salesperson is able to sell policies to 15% of the people she contacts. Suppose she contacts 120 people during a 2-week period. What is the expected value for the number of policies she sells?

Answer: We have a binomial probability with the probability of a success .15 and the number of trials 120, and so the mean or expected value for the number of successes is $120 \times .15 = 18$.

We have seen that the mean of a random variable is $\Sigma x_i p_i$. However, not only is the mean important, but also we would like to measure the variability for the values taken on by a random variable. Since we are dealing with chance events, the proper tool is variance (along with standard deviation). *Variance* was defined in Topic Two to be the mean average of the squared deviations $(x - \mu)^2$. If we regard the $(x - \mu)^2$ terms as the values of some random variable (whose probability is the same as the probability of x), the mean of this new random variable is $\Sigma(x_i - \mu_x)^2 p_i$, which is precisely how we define the variance σ^2 of a random variable:

$$var(X) = \sigma_x^2 = \sum (x_i - \mu_x)^2 p_i$$

As before, the standard deviation σ is the square root of the variance.

EXAMPLE 9.35

A highway engineer knows that his crew can lay 5 miles of highway on a clear day, 2 miles on a rainy day, and only 1 mile on a snowy day. Suppose the probabilities are as follows:

Outcome:	Clear	Rain	Snow
Probability:	.6	.3	.1
Random variable (miles of highway):	5	2	1

(continued)

What are the mean (expected value) and the variance?

Answer:

$$\mu_x = \sum x_i p_i = 5(.6) + 2(.3) + 1(.1) = 3.7$$

$$\sigma_x^2 = \sum (x_i - \mu_x)^2 p_i = (5 - 3.7)^2(.6) + (2 - 3.7)^2(.3) + (1 - 3.7)^2(.1)$$
$$= 2.61$$

EXAMPLE 9.36

Look again at Example 9.33. We calculated the mean to be 2. What is the variance?

Answer: $\sigma^2 = (0-2)^2(.07776) + (1-2)^2(.2592) + (2-2)^2(.3456) + (3-2)^2(.2304) + (4-2)^2(.0768) + (5-2)^2(.01024) = 1.2$

Could we have calculated the above result more easily? In this case, we have a binomial random variable and the variance can be simply calculated as $npq = np(1-p)$.

EXAMPLE 9.37

How can we use this method to calculate the variance in Example 9.33 more simply?

Answer: $np(1-p) = 5(.4)(.6) = 1.2$

Thus, for a random variable X,

Mean or expected value: $\mu_x = \Sigma x_i p_i$

Variance: $\sigma_x^2 = \Sigma(x_i - \mu_x)^2 p_i$

Standard deviation: $\sigma_x = \sqrt{\sum (x_i - \mu_x)^2 p_i}.$

In the case of a binomial probability distribution with probability of a success equal to p and number of trials equal to n, if we let X be the number of successes in the n trials, the above equations become

Mean or expected value: $\mu_x = np$

Variance: $\sigma_x^2 = npq = np(1-p)$

Standard deviation: $\sigma_x = \sqrt{npq} = \sqrt{np(1-p)}.$

EXAMPLE 9.38

Sixty percent of all buyers of new cars choose automatic transmissions. For a group of five buyers of new cars, calculate the mean and standard deviation for the number of buyers choosing automatic transmissions.

Answer:

$$\mu_x = np = 5(.6) = 3.0$$

$$\sigma_x = \sqrt{np(1-p)} = \sqrt{5(.6)(.4)} \approx 1.1$$

Summary

- The Law of Large Numbers states that in the long run, a cumulative relative frequency tends closer and closer to what is called the probability of an event.
- The probability of the complement of an event is equal to 1 minus the probability of the event.
- Two events are mutually exclusive if they cannot occur simultaneously.
- If two events are mutually exclusive, the probability that at least one will occur is the sum of their respective probabilities.
- More generally, $P(A \cup B) = P(A) + P(B) - P(A \cap B)$.
- The conditional probability of A given B is given by: $P(A|B) = \frac{P(A \cap B)}{P(B)}$.
- Two events A and B are independent if $P(A|B) = P(A)$; that is, the probability of one event occurring does not influence whether or not the other occurs.
- If two events A and B are independent, then $P(A \cap B) = P(A) \times P(B)$.
- A random variable takes various numeric values, each with a given probability.
- The mean or expected value of a random variable is calculated by $\mu = \Sigma\, xP(x)$, while the variance is $\sigma^2 = \Sigma(x - \mu)^2\, P(x)$.
- The binomial distribution arises when there are two possible outcomes, the probability of success is constant, and the trials are independent.
- In a binomial distribution, the mean number of successes is np and the standard deviation is $\sqrt{np(1-p)}$.

Questions on Topic Nine: Probability as Relative Frequency

Multiple-Choice Questions

Directions: The questions or incomplete statements that follow are each followed by five suggested answers or completions. Choose response that best answers the question or completes the statement.

1. A student reasons that either he will or will not receive a 5 on the AP Statistics exam, and therefore the probability of receiving a 5 is 0.5. Why is this incorrect reasoning?

   (A) The events are mutually exclusive.
   (B) The events are independent.
   (C) The events are not independent.
   (D) The events are complements.
   (E) The events are not equally probable.

2. In the November 27, 1994, issue of *Parade* magazine, the "Ask Marilyn" section contained this question: "Suppose a person was having two surgeries performed at the same time. If the chances of success for surgery A are 85%, and the chances of success for surgery B are 90%, what are the chances that both would fail?" What do you think of Marilyn's solution: (.15)(.10) = .015 or 1.5%?

   (A) Her solution is mathematically correct but not explained very well.
   (B) Her solution is both mathematically correct and intuitively obvious.
   (C) Her use of complementary events is incorrect.
   (D) Her use of the general addition formula is incorrect.
   (E) She assumed independence of events, which is most likely wrong.

3. A weighted die comes up spots with the following probabilities:

Spots	1	2	3	4	5	6
Probability	.10	.15	.20	.25	.20	.10

   If two of these dice are thrown, what is the probability the sum is 10?

   (A) $(.25)(.10) + (.20)^2$
   (B) $2(.25)(.10) + (.20)^2$
   (C) $2(.25)(.10) + 2(.20)^2$
   (D) $(.10)(.20)(.10) + (.20)^2$
   (E) $(.10)(.20)(.10) + (.15)(.25)^2 + (.15)(.20)(.20) + (.25)(.10)$

4. According to a CBS/*New York Times* poll taken in 1992, 15% of the public have responded to a telephone call-in poll. In a random group of five people, what is the probability that exactly two have responded to a call-in poll?

   (A) .138
   (B) .165
   (C) .300
   (D) .835
   (E) .973

5. In a 1974 "Dear Abby" letter a woman lamented that she had just given birth to her eighth child, and all were girls! Her doctor had assured her that the chance of the eighth child being a girl was only 1 in 100. What was the real probability that the eighth child would be a girl?

   (A) .01
   (B) .5
   (C) $(.5)^7$
   (D) $(.5)^8$
   (E) $\frac{(.5)^7 + (.5)^8}{2}$

6. The yearly mortality rate for American men from prostate cancer has been constant for decades at about 25 of every 100,000 men. (This rate has not changed in spite of new diagnostic techniques and new treatments.) In a group of 100 American men, what is the probability that at least 1 will die from prostate cancer in a given year?

   (A) .00025
   (B) .0247
   (C) .025
   (D) .9753
   (E) .99975

7. Alan Dershowitz, one of O. J. Simpson's lawyers, has stated that only 1 out of every 1000 abusive relationships ends in murder each year. If he is correct, and if there are approximately 1.5 million abusive relationships in the United States, what is the expected value for the number of people who are killed each year by an abusive partner?

   (A) 1
   (B) 500
   (C) 1000
   (D) 1500
   (E) None of the above

8. For an advertising promotion, an auto dealer hands out 1000 lottery tickets with a prize of a new car worth $25,000. For someone with a single ticket, what is the standard deviation for the amount won?

   (A) $7.07
   (B) $25.00
   (C) $49.95
   (D) $790.17
   (E) $624,375

9. Suppose that among the 6000 students at a high school, 1500 are taking honors courses and 1800 prefer watching basketball to watching football. If taking honors courses and preferring basketball are independent, how many students are both taking honors courses and prefer basketball to football?

   (A) 300
   (B) 330
   (C) 450
   (D) 825
   (E) There is insufficient information to answer this question.

10. An inspection procedure at a manufacturing plant involves picking three items at random and then accepting the whole lot if at least two of the three items are in perfect condition. If in reality 90% of the whole lot are perfect, what is the probability that the lot will be accepted?

    (A) .600
    (B) .667
    (C) .729
    (D) .810
    (E) .972

11. Suppose that, in a certain part of the world, in any 50-year period the probability of a major plague is .39, the probability of a major famine is .52, and the probability of both a plague and a famine is .15. What is the probability of a famine given that there is a plague?

    (A) .240
    (B) .288
    (C) .370
    (D) .385
    (E) .760

12. Suppose that, for any given year, the probabilities that the stock market declines, that women's hemlines are lower, and that both events occur are, respectively, .4, .35, and .3. Are the two events independent?

    (A) Yes, because $(.4)(.35) \neq .3$.
    (B) No, because $(.4)(.35) \neq .3$.
    (C) Yes, because $.4 > .35 > .3$.
    (D) No, because $.5(.3 + .4) = .35$.
    (E) There is insufficient information to answer this question.

13. If $P(A) = .2$ and $P(B) = .1$, what is $P(A \cup B)$ if A and B are independent?
    (A) .02
    (B) .28
    (C) .30
    (D) .32
    (E) There is insufficient information to answer this question.

14. The following data are from *The Commissioner's Standard Ordinary Table of Mortality:*

Age	Number Surviving
0	10,000,000
20	9,664,994
40	9,241,359
70	5,592,012

What is the probability that a 20-year-old will survive to be 70?
    (A) .407
    (B) .421
    (C) .559
    (D) .579
    (E) .966

15. At a warehouse sale 100 customers are invited to choose one of 100 identical boxes. Five boxes contain \$700 color television sets, 25 boxes contain \$540 camcorders, and the remaining boxes contain \$260 cameras. What should a customer be willing to pay to participate in the sale?
    (A) \$260
    (B) \$352
    (C) \$500
    (D) \$540
    (E) \$699

16. There are two games involving flipping a coin. In the first game you win a prize if you can throw between 40% and 60% heads. In the second game you win if you can throw more than 75% heads. For each game would you rather flip the coin 50 times or 500 times?
    (A) 50 times for each game
    (B) 500 times for each game
    (C) 50 times for the first game, and 500 for the second
    (D) 500 times for the first game, and 50 for the second
    (E) The outcomes of the games do not depend on the number of flips.

17. Two cards are picked, without replacement, from a standard deck. Which of the following pairs are independent events?

    (A) Getting a heart on the first pick; getting a diamond on the second.
    (B) Getting a red card on the first pick; getting a black card on the second.
    (C) Getting an ace on the first pick; getting an ace on the second.
    (D) Getting two aces; getting two kings.
    (E) Getting two aces; getting a black card on the second pick.

18. Two cards are picked, with replacement, from a standard deck. Which of the following pairs are independent events?

    (A) Getting two aces; getting two kings.
    (B) Getting an ace on the first pick; getting a king on the second.
    (C) Getting at least one ace; getting at least one king.
    (D) Getting exactly one ace; getting exactly one king.
    (E) Getting no aces; getting no kings.

19. The average annual incomes of high school and college graduates in a midwestern town are $21,000 and $35,000, respectively. If a company hires only personnel with at least a high school diploma and 20% of its employees have been through college, what is the mean income of the company employees?

    (A) $23,800
    (B) $27,110
    (C) $28,000
    (D) $32,200
    (E) $56,000

20. An insurance company charges $800 annually for car insurance. The policy specifies that the company will pay $1000 for a minor accident and $5000 for a major accident. If the probability of a motorist having a minor accident during the year is .2, and of having a major accident, .05, how much can the insurance company expect to make on a policy?

    (A) $200
    (B) $250
    (C) $300
    (D) $350
    (E) $450

21. You can choose one of three boxes. Box A has four $5 bills and a single $100 bill, box B has 400 $5 bills and 100 $100 bills, and box C has 24 $1 bills. You can have all of box C or blindly pick one bill out of either box A or box B. Which offers the greatest expected winning?

    (A) Box A
    (B) Box B
    (C) Box C
    (D) Either A or B, but not C
    (E) All offer the same expected winning.

22. Given that 52% of the U.S. population are female and 15% are older than age 65, can we conclude that (.52)(.15) = 7.8% are women older than age 65?

    (A) Yes, by the multiplication rule.
    (B) Yes, by conditional probabilities.
    (C) Yes, by the law of large numbers.
    (D) No, because the events are not independent.
    (E) No, because the events are not mutually exclusive.

Questions 23–27 refer to the following study: One thousand students at a city high school were classified both according to GPA and whether or not they consistently skipped classes.

	GPA			
	<2.0	2.0–3.0	>3.0	
Many skipped classes	80	25	5	110
Few skipped classes	175	450	265	890
	255	475	270	

23. What is the probability that a student has a GPA between 2.0 and 3.0?

    (A) .025
    (B) .227
    (C) .450
    (D) .475
    (E) .506

24. What is the probability that a student has a GPA under 2.0 and has skipped many classes?

    (A) .080
    (B) .281
    (C) .285
    (D) .314
    (E) .727

25. What is the probability that a student has a GPA under 2.0 or has skipped many classes?

    (A) .080
    (B) .281
    (C) .285
    (D) .314
    (E) .727

26. What is the probability that a student has a GPA under 2.0 given that he has skipped many classes?

    (A) .080
    (B) .281
    (C) .285
    (D) .314
    (E) .727

27. Are "GPA between 2.0 and 3.0" and "skipped few classes" independent?

    (A) No, because .475 ≠ .506.
    (B) No, because .475 ≠ .890.
    (C) No, because .450 ≠ .475.
    (D) Yes, because of conditional probabilities.
    (E) Yes, because of the product rule.

28. Mathematically speaking, casinos and life insurance companies make a profit because of

    (A) their understanding of sampling error and sources of bias.
    (B) their use of well-designed, well-conducted surveys and experiments.
    (C) their use of simulation of probability distributions.
    (D) the central limit theorem.
    (E) the law of large numbers.

29. Consider the following table of ages of U.S. senators:

Age (yr):	<40	40–49	50–59	60–69	70–79	>79
Number of senators:	5	30	36	22	5	2

    What is the probability that a senator is under 70 years old given that he or she is at least 50 years old?

    (A) .580
    (B) .624
    (C) .643
    (D) .892
    (E) .969

Questions 30–33 refer to the following study: Five hundred people used a home test for HIV, and then all underwent more conclusive hospital testing. The accuracy of the home test was evidenced in the following table.

	HIV	Healthy	
Positive test	35	25	60
Negative test	5	435	440
	40	460	

30. What is the *predictive value* of the test? That is, what is the probability that a person has HIV and tests positive?

(A) .070
(B) .130
(C) .538
(D) .583
(E) .875

31. What is the *false-positive* rate? That is, what is the probability of testing positive given that the person does not have HIV?

(A) .054
(B) .050
(C) .130
(D) .417
(E) .875

32. What is the *sensitivity* of the test? That is, what is the probability of testing positive given that the person has HIV?

(A) .070
(B) .130
(C) .538
(D) .583
(E) .875

33. What is the *specificity* of the test? That is, what is the probability of testing negative given that the person does not have HIV?

(A) .125
(B) .583
(C) .870
(D) .950
(E) .946

34. Suppose that 2% of a clinic's patients are known to have cancer. A blood test is developed that is positive in 98% of patients with cancer but is also positive in 3% of patients who do not have cancer. If a person who is chosen at random from the clinic's patients is given the test and it comes out positive, what is the probability that the person actually has cancer?

    (A) .02
    (B) .4
    (C) .5
    (D) .6
    (E) .98

35. Suppose the probability that you will receive an A in AP Statistics is .35, the probability that you will receive A's in both AP Statistics and AP Biology is .19, and the probability that you will receive an A in AP Statistics but not in AP Biology is .17. Which of the following is a proper conclusion?

    (A) The probability that you will receive an A in AP Biology is .36.
    (B) The probability that you didn't take AP Biology is .01.
    (C) The probability that you will receive an A in AP Biology but not in AP Statistics is .18.
    (D) The given probabilities are impossible.
    (E) None of the above

36. Suppose you are one of 7.5 million people who send in their name for a drawing with 1 top prize of $1 million, 5 second-place prizes of $10,000, and 20 third-place prizes of $100. Is it worth the $0.44 postage it cost you to send in your name?

    (A) Yes, because $\frac{1{,}000{,}000}{0.44} = 2{,}272{,}727$, which is less than 7,500,000.
    (B) No, because your expected winnings are only $0.14.
    (C) Yes, because $\frac{7{,}500{,}000}{(1+5+20)} = 288{,}462$.
    (D) No, because 1,052,000 < 7,500,000.
    (E) Yes, because $\frac{1{,}052{,}000}{26} = 40{,}462$.

37. You have a choice of three investments, the first of which gives you a 10% chance of making $1 million, otherwise you lose $25,000; the second of which gives you a 50% chance of making $500,000, otherwise you lose $345,000; and the third of which gives you an 80% chance of making $50,000, otherwise you make only $1,000. Assuming you will go bankrupt if you don't show a profit, which option should you choose for the best chance of avoiding bankruptcy?

    (A) First choice
    (B) Second choice
    (C) Third choice
    (D) Either the first or the second choice
    (E) All the choices give an equal chance of avoiding bankruptcy.

38. Can the function $f(x) = \frac{x+6}{24}$, for $x = 1, 2,$ and 3, be the probability distribution for some random variable?

    (A) Yes.
    (B) No, because probabilities cannot be negative.
    (C) No, because probabilities cannot be greater than 1.
    (D) No, because the probabilities do not sum to 1.
    (E) Not enough information is given to answer the question.

39. Sixty-five percent of all divorce cases cite incompatibility as the underlying reason. If four couples file for a divorce, what is the probability that exactly two will state incompatibility as the reason?

    (A) .104
    (B) .207
    (C) .254
    (D) .311
    (E) .423

40. Suppose we have a random variable X where the probability associated with the value $\binom{10}{k}(.37)^k(.63)^{10-k}$ for $k = 0, \ldots, 10$. What is the mean of X?

    (A) 0.37
    (B) 0.63
    (C) 3.7
    (D) 6.3
    (E) None of the above

41. A computer technician notes that 40% of computers fail because of the hard drive, 25% because of the monitor, 20% because of a disk drive, and 15% because of the microprocessor. If the problem is not in the monitor, what is the probability that it is in the hard drive?

    (A) .150
    (B) .400
    (C) .417
    (D) .533
    (E) .650

42. Suppose you toss a coin ten times and it comes up heads every time. Which of the following is a true statement?

    (A) By the law of large numbers, the next toss is more likely to be tails than another heads.
    (B) By the properties of conditional probability, the next toss is more likely to be heads given that ten tosses in a row have been heads.
    (C) Coins actually do have memories, and thus what comes up on the next toss is influenced by the past tosses.
    (D) The law of large numbers tells how many tosses will be necessary before the percentages of heads and tails are again in balance.
    (E) The probability that the next toss will again be heads is .5.

43. A city water supply system involves three pumps, the failure of any one of which crashes the system. The probabilities of failure for each pump in a given year are .025, .034, and .02, respectively. Assuming the pumps operate independently of each other, what is the probability that the system does crash during the year?

    (A) Less than .05
    (B) .077
    (C) .079
    (D) .081
    (E) .923

44. Which of the following are true statements?

    I. The histogram of a binomial distribution with $p = .5$ is always symmetric no matter what n, the number of trials, is.
    II. The histogram of a binomial distribution with $p = .9$ is skewed to the right.
    III. The histogram of a binomial distribution with $p = .9$ is almost symmetric if n is very large.

    (A) I and II
    (B) I and III
    (C) II and III
    (D) I, II, and III
    (E) None of the above gives the complete set of true responses.

45. Suppose that 60% of students who take the AP Statistics exam score 4 or 5, 25% score 3, and the rest score 1 or 2. Suppose further that 95% of those scoring 4 or 5 receive college credit, 50% of those scoring 3 receive such credit, and 4% of those scoring 1 or 2 receive credit. If a student who is chosen at random from among those taking the exam receives college credit, what is the probability that she received a 3 on the exam?

    (A) .125
    (B) .178
    (C) .701
    (D) .813
    (E) .822

46. Given the probabilities $P(A) = .4$ and $P(A \cup B) = .6$, what is the probability $P(B)$ if A and B are mutually exclusive? If A and B are independent?

    (A) .2, .4
    (B) .2, .33
    (C) .33, .2
    (D) .6, .33
    (E) .6, .4

47. There are 20 students in an AP Statistics class. Suppose all are unprepared and randomly guess on each of ten multiple-choice questions, each with a choice of five possible answers. Which of the following is one run of a simulation to estimate the probability at least one student guesses correctly on over half the questions?

    (A) Assume "1" is the correct answer and "2, 3, 4, 5" are incorrect answers. Randomly pick 200 numbers between 1 and 5. Arrange the numbers into 20 groups of 10 each, each group corresponding to one of the students. For each student, count how many correct ("1") answers there are. Record whether at least one student has six or more correct answers.
    (B) Assume "2" is the correct answer and "1, 3, 4, 5" are incorrect answers. Assign the numbers 1–20 to the students. Randomly choose 200 numbers between 1 and 20. For each number chosen, randomly pick a number between 1 and 5. Record how many correct ("2") answers there are, divide by 20, and record whether this quotient is at least six.
    (C) Assume "3" is the correct answer and "1, 2, 4, 5" are incorrect answers. Assign the numbers 1–10 to the questions. Randomly choose 200 numbers between 1 and 10. For each number chosen, randomly pick a number between 1 and 5. Record how many correct ("3") answers there are, divide by 20, and record whether this quotient is at least six.
    (D) Assume "4" is the correct answer and "1, 2, 3, 5" are incorrect answers. Randomly choose 200 numbers between 1 and 5. Record how many correct ("1") answers there are, divide by 20, and record whether this quotient is at least six.
    (E) Assume "5" is the correct answer and "1, 2, 3, 4" are incorrect answers. Randomly pick 20 numbers between 1 and 5. Do this 10 times, and for each group of 20 note how many correct ("5") answers there are. Record whether at least one group has over half correct answers.

Free-Response Questions

Directions: You must show all work and indicate the methods you use. You will be graded on the correctness of your methods and on the accuracy of your final answers.

Seven Open-Ended Questions

1. A manufacturer is considering two options for a quality control check at the end of a production line:

   Option A: If at least five of six randomly picked articles meet all specifications, the day's production is approved.

   Option B: If at least ten of twelve randomly picked articles meet all specifications, the day's production is approved.

   If you are a buyer wanting the most assurance that a day's production will not be accepted if there were only 75% of the articles meeting all specifications, which option would you request the manufacturer to use? Give statistical justification.

2. A sample of applicants for a management position yields the following numbers with regard to age and experience:

	Years of experience		
	0–5	6–10	>10
Less than 50 years old	80	125	20
More than 50 years old	10	75	50

   (a) What is the probability that an applicant is less than 50 years old? Has more than 10 years' experience? Is more than 50 years old and has five or fewer years' experience?

   (b) What is the probability that an applicant is less than 50 years old given that she has between 6 and 10 years' experience?

   (c) Are the two events "less than 50 years old" and "more than 10 years' experience" independent events? How about the two events "more than 50 years old" and "between 6 and 10 years' experience"? Explain.

3. Suppose USAir accounts for 20% of all U.S. domestic flights. As of mid-1994, USAir was involved in four of the previous seven major disasters. "That's enough to begin getting suspicious but not enough to hang them," said Dr. Brad Efton in *The New York Times* (September 11, 1994, Sec. 4, p. 4). Using a binomial distribution, comment on Dr. Efton's remark.

4. Assume there is no overlap between the 56% of the population who wear glasses and the 4% who wear contacts. If 55% of those who wear glasses are women and 63% of those who wear contacts are women, what is the probability that the next person you encounter on the street will be a woman with glasses? A woman with contacts? A man with glasses? A man with contacts? A person not wearing glasses or contacts? Explain your reasoning.

5. Explain what is wrong with each of the following statements:

   (a) The probability that a student will score high on the AP Statistics exam is .43, while the probability that she will not score high is .47.

   (b) The probability that a student plays tennis is .18, while the probability that he plays basketball is six times as great.

   (c) The probability that a student enjoys her English class is .64, while the probability that she enjoys both her English and her social studies classes is .71.

   (d) The probability that a student will be accepted by his first choice for college is .38, while the probability that he will be accepted by his first or second choice is .32.

   (e) The probability that a student fails AP Statistics and will still be accepted by an Ivy League school is –.17.

6. Player A rolls a die with 7 on four sides and 11 on two sides. Player B flips a coin with 6 on one side and 10 on the other. Assume a fair die and a fair coin.

   (a) Suppose the winner is the player with the higher number showing. Explain who you would rather be. If player A receives $0.25 for every win, what should player B receive for every win to make the game fair?

   (b) Suppose player A receives $0.70 from player B when the 7 shows and $1.10 when the 11 shows, while player B receives $0.60 from player A when the 6 shows and $1.00 when the 10 shows. Explain whom you would rather be.

7. Could hands-free, automatic faucets actually be housing more bacteria than the old-fashioned, manual kind? The concern is that decreased water flow may increase the chance that bacteria grows, because the automatic faucets are not being thoroughly flushed through. It is known that 15% of water cultures from older faucets in hospital patient care areas test positive for Legionella bacteria. A recent study at Johns Hopkins Hospital found Legionella bacteria growing in 10 of cultured water samples from 20 electronic faucets.

   (a) If the probability of Legionella bacteria growing in a faucet is .15, what is the probability that in a sample of 20 faucets, 10 or more have the bacteria growing?

   (b) Does the Johns Hopkins study provide sufficient evidence that the probability of Legionella bacteria growing in electronic faucets is greater than 15%? Explain.

An Investigative Task

If a car is stopped for speeding, the probability that the driver has illegal drugs hidden in the car is .14. A police officer manning a speed trap plans to thoroughly search cars he stops for speeding until he finds one with illegal drugs.

(a) Use simulation to estimate the probability that illegal drugs will turn up before he stops the tenth car. Explain your work. Use the following random number table to run as many trials as possible.

8417706757 1761315582 5150681435 4105092031 0644905059
5988431180 5311584469 9486857967 0581184514 7501113006
6339555041 1586606589 1311971020 8594091932 0648874987
5435552704 9035902649 4749671567 9426808844 2629464759
0898957024 9728400637 8928303514 5919507635 0330972605
2935723737 6788103668 3387635841 5286923114 1586438942

(b) The results of a 100-trial simulation are shown below. Each trial ends when a car with illegal drugs is found.

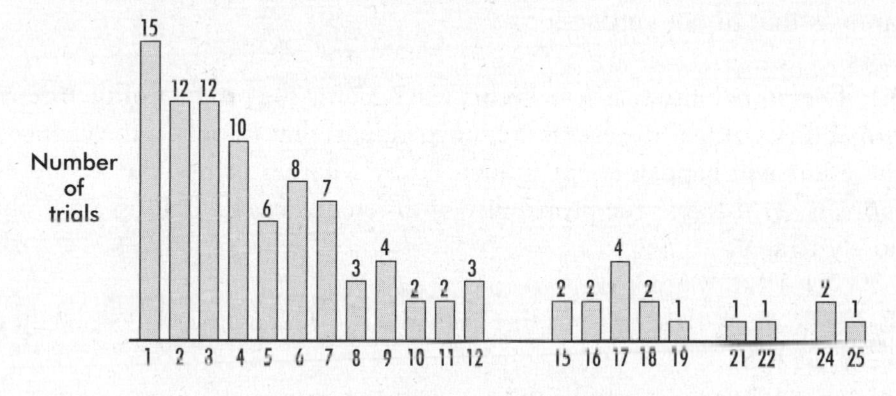

Use the above 100-trial results to estimate the mean number of cars searched before finding one with illegal drugs.

(c) Use probability to calculate the probability that illegal drugs will be found before he stops the tenth car. Show your work.

Answer Key

1. **E**	11. **D**	21. **E**	31. **A**	41. **D**
2. **E**	12. **B**	22. **D**	32. **E**	42. **E**
3. **B**	13. **B**	23. **D**	33. **E**	43. **B**
4. **A**	14. **D**	24. **A**	34. **B**	44. **B**
5. **B**	15. **B**	25. **C**	35. **D**	45. **B**
6. **B**	16. **D**	26. **E**	36. **B**	46. **B**
7. **D**	17. **E**	27. **A**	37. **C**	47. **A**
8. **D**	18. **B**	28. **E**	38. **A**	
9. **C**	19. **A**	29. **D**	39. **D**	
10. **E**	20. **D**	30. **A**	40. **C**	

Answers Explained

Multiple-Choice

1. **(E)** There is no reason to assume that the probability of getting a 5 is the same as that of not getting a 5.

2. **(E)** The probability that *both* events will occur is the product of their separate probabilities only if the events are independent, that is, only if the chance that one event will happen is not influenced by whether or not the second event happens. In this case the probability of different surgeries failing are probably closely related.

3. **(B)** P(1st die is 4)P(2nd die is 6) + P(1st die is 6)P(2nd die is 4) + P(1st die is 5) P(2nd die is 5)

4. **(A)** $10(.15)^2(.85)^3 = .138$ or binompdf(5, .15, 2) = .138

5. **(B)** The probability of the next child being a girl is independent of the gender of the previous children.

6. **(B)** $1 - (.99975)^{100} = .0247$

7. **(D)** $.001(1{,}500{,}000) = 1500$

8. **(D)** $\bar{x} = E(X) = \sum xP(x) = 0(.999) + 25{,}000(.001) = 25$

   $$\sigma = \sqrt{\sum (x - \bar{x})^2 P(x)} = \sqrt{(0 - 25)^2(.999) + (25{,}000 - 25)^2(.001)} = 790.17$$

   [Or on the TI-84, put returns and probabilities into two lists and run `1-Var Stats L1,L2`]

9. **(C)** $\frac{1500}{6000} = .25$ are honors students, and $\frac{1800}{6000} = .3$ prefer basketball. Because of independence, their intersection is $(.25)(.3) = .075$ of the students, and $.075(6000) = 450$.

10. **(E)** $(.9)^3 + 3(.9)^2(.1) = .972$

11. **(D)** $P(\text{famine} \mid \text{plague}) = \dfrac{P(\text{famine} \cap \text{plague})}{P(\text{plague})} = \dfrac{.15}{.39} = .385$

12. **(B)** If E and F are independent, then $P(E \cap F) = P(E)P(F)$; however, in this problem, $(.4)(.35) \neq .3$.

13. **(B)** Because A and B are independent, we have $P(A \cap B) = P(A)P(B)$, and thus $P(A \cup B) = .2 + .1 - (.2)(.1) = .28$.

14. **(D)** $\dfrac{5{,}592{,}012}{9{,}664{,}994} = .579$

15. **(B)** $E(X) = \mu_x = \Sigma x_i p_i = 700(.05) + 540(.25) + 260(.7) = 352$

16. **(D)** The probability of throwing heads is .5. By the law of large numbers, the more times you flip the coin, the more the relative frequency tends to become closer to this probability. With fewer tosses there is a greater chance of wide swings in the relative frequency.

17. **(E)** P(diamond on 2nd pick) = 1/4; however,
 P(diamond on 2nd pick | heart on first pick) = 13/51
 P(black on 2nd pick) = 1/2; however, P(black on 2nd pick | red on 1st pick) = 26/51
 P(ace on 2nd pick) = 1/13; however, P(ace on 2nd pick | ace on 1st pick) = 3/51
 P(two kings) = (1/13)(3/51); however, P(two kings | two aces) = 0
 P(black on 2nd pick) = 1/2 and P(black on 2nd pick | two aces) = 6/12

18. **(B)** P(two kings) = (1/13)(1/13); however, P(two kings | two aces) = 0
 P(king on 2nd pick) = 1/13, and P(king on 2nd pick | ace on 1st pick) = 1/13
 P(at least one king) = $1 - (12/13)^2 = 25/169$; however,
 P(at least one king | at least one ace) =
 [(1/13)(1/13) + (1/13)(1/13)]/[25/169] = 2/25
 P(exactly one king) = (1/13)(12/13) + (12/13)(1/13) = 24/169; however,
 P(exactly one king | exactly one ace) = (2/169)/(24/169) = 1/12
 P(no kings) = $(12/13)^2$; however, P(no kings | no aces) = $(11/12)^2$

19. **(A)** $E(X) = \mu_x = \Sigma x_i p_i = 21{,}000(.8) + 35{,}000(.2) = 23{,}800$

20. **(D)** 1000(.2) + 5000(.05) = 450, and 800 – 450 = 350.

21. **(E)** $5\left(\frac{4}{5}\right) + 100\left(\frac{1}{5}\right) = 24$, and $5\left(\frac{400}{500}\right) + 100\left(\frac{100}{500}\right) = 24$

22. **(D)** $P(E \cap F) = P(E)P(F)$ only if the events are independent. In this case, women live longer than men and so the events are not independent.

For questions 23–27, we first sum the rows and columns:

	GPA <2.0	GPA 2.0–3.0	GPA >3.0	
Many skipped classes	80	25	5	110
Few skipped classes	175	450	265	890
	255	475	270	1000

23. **(D)** $\frac{25+450}{1000} = .475$

24. **(A)** $\frac{80}{1000} = .080$ (probability of an intersection)

25. **(C)** $\frac{80+25+5+175}{1000} = .285$ (probability of a union)

26. **(E)** $\frac{80}{80+25+5} = .727$ (conditional probability)

27. **(A)** $P(2.0-3.0\text{ GPA}) = \frac{475}{1000} = .475$; however, $P(2.0-3.0\text{ GPA}|\text{few skips}) = \frac{450}{890} = .506$. If independent, these would have been equal.

28. **(E)** While the outcome of any single play on a roulette wheel or the age at death of any particular person is uncertain, the law of large numbers gives that the relative frequencies of specific outcomes in the long run tend to become closer to numbers called probabilities.

29. **(D)** $\frac{36+22}{36+22+5+2} = .892$

30. **(A)** $\frac{35}{500} = .070$

31. **(A)** $\frac{25}{460} = .054$

32. **(E)** $\frac{35}{40} = .875$

33. **(E)** $\frac{435}{460} = .946$

34. **(B)**

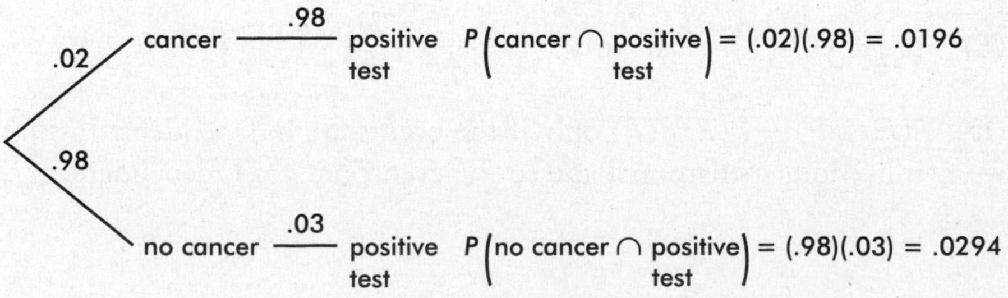

.02 → cancer —.98→ positive test $P(\text{cancer} \cap \text{positive test}) = (.02)(.98) = .0196$

.98 → no cancer —.03→ positive test $P(\text{no cancer} \cap \text{positive test}) = (.98)(.03) = .0294$

$$\begin{aligned} P(\text{positive test}) &= P(\text{cancer} \cap \text{positive test}) \\ &\quad + P(\text{no cancer} \cap \text{positive test}) \\ &= .0196 + .0294 \\ &= .0490 \end{aligned}$$

$$\begin{aligned} P(\text{cancer} \mid \text{positive test}) &= \frac{P(\text{cancer} \cap \text{positive test})}{P(\text{positive test})} \\ &= \frac{.0196}{.0490} \\ &= .4 \end{aligned}$$

35. **(D)** The probability that you will receive an A in AP Statistics but not in AP Biology must be $.35 - .19 = .16$, not .17.

36. **(B)** Your expected winnings are only

$$\frac{1}{7{,}500{,}000}(1{,}000{,}000) + \frac{5}{7{,}500{,}000}(10{,}000) + \frac{20}{7{,}500{,}000}(100) = 0.14$$

37. **(C)** Even though the first two choices have a higher expected value than the third choice, the third choice gives a 100% chance of avoiding bankruptcy.

38. **(A)** The probabilities $\frac{7}{24}$, $\frac{8}{24}$, and $\frac{9}{24}$ are all nonnegative, and they sum to 1.

39. **(D)** $6(.65)^2(.35)^2 = .311$ or binompdf(4, .65, 2) = .311

40. **(C)** This is a binomial with $n = 10$ and $p = .37$, and so the mean is $np = 10(.37) = 3.7$.

41. **(D)** $\frac{.40}{.40+.20+.15} = .533$

42. **(E)** Coins have no memory, and so the probability that the next toss will be heads is .5 and the probability that it will be tails is .5. The law of large numbers says that as the number of tosses becomes larger, the percentage of heads tends to become closer to .5.

43. **(B)** The probabilites of each pump not failing are $1 - .025 = .975$, $1 - .034 = .966$, and $1 - .02 = .98$, respectively. The probability of none failing is $(.975)(.966)(.98) = .923$, and so the probability of at least one failing is $1 - .923 = .077$.

44. **(B)** If $p > .5$, the more likely numbers of successes are to the right and the lower numbers of successes have small probabilities, and so the histogram is skewed to the left. No matter what p is, if n is sufficiently large, the histogram will look almost symmetric.

45. **(B)**

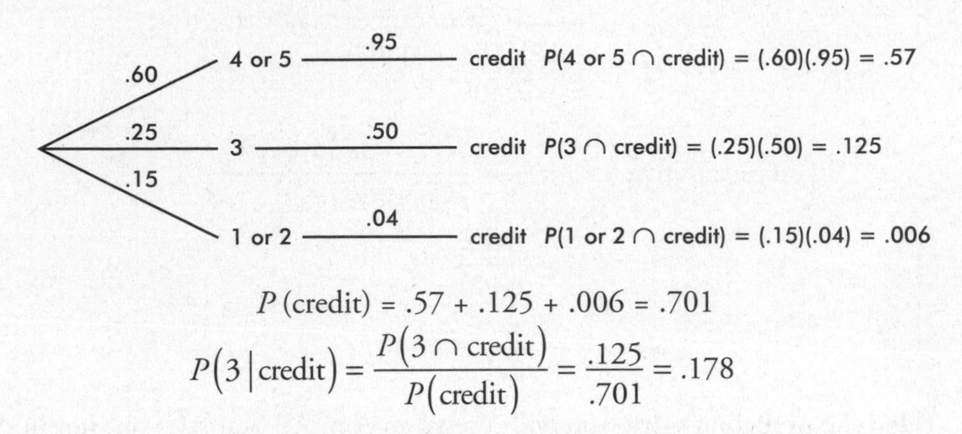

$$P(\text{credit}) = .57 + .125 + .006 = .701$$

$$P(3 \mid \text{credit}) = \frac{P(3 \cap \text{credit})}{P(\text{credit})} = \frac{.125}{.701} = .178$$

46. **(B)** If A and B are mutually exclusive, $P(A \cap B) = 0$. Thus $.6 = .4 + P(B) - 0$, and so $P(B) = .2$. If A and B are independent, then $P(A \cap B) = P(A)P(B)$. Thus $.6 = .4 + P(B) - .4P(B)$, and so $P(B) = \frac{1}{3}$.

47. **(A)** For each of the 20 students, we must generate an answer to each of the 10 questions, and record if any student has at least 6 out of 10 correct answers.

Free-Response

1. With Option A, if in reality only 75% of the articles meet all specifications, the probability of rejecting the day's production is:

$$P(\text{rejection}) = 1 - P(\text{acceptance}) = 1 - \left[\binom{6}{5}(.75)^5(.25) + (.75)^6\right] = .466$$

[On the TI-84, `binomcdf(6,.75,4) = .466`]

With Option B, if in reality only 75% of the articles meet all specifications, the probability of rejecting the day's production is:

$$P(\text{rejection}) = 1 - P(\text{acceptance}) =$$

$$1 - \left[\binom{12}{10}(.75)^{10}(.25)^2 + \binom{12}{11}(.75)^{11}(.25) + (.75)^{12}\right] = .609$$

[On the TI-84, `binomcdf(12,.75,9) = .609`]

For the greatest probability of rejecting the day's production if only 75% of the articles meet all specifications, the buyer should request the manufacturer to use Option B with a probability of rejection of .609 as opposed to Option A with a probability of rejection of only .466.

2. It's easiest to first sum the rows and columns:

	Years of experience 0–5	6–10	>10	
Less than 50 years old	80	125	20	225
More than 50 years old	10	75	50	135
	90	200	70	360

(a) $P(\text{age} < 50) = \frac{225}{360} = .625$

$P(\text{experience} > 10) = \frac{70}{360} = .194$

$P(\text{age} > 50 \cap \text{experience } 6\text{–}10) = \frac{75}{360} = .208$

(b) $P(\text{age} < 50 \mid \text{experience } 6\text{–}10) = \frac{125}{200} = .625$

(c) $P(\text{age} < 50) = \frac{225}{360} = .625$; however, $P(\text{age} < 50 \mid \text{experience} > 10) = \frac{20}{70}$ $= .286$ and so they are not independent.

$P(\text{age} > 50) = \frac{135}{360} = .375$ and also $P(\text{age} > 50 \mid \text{experience } 6\text{–}10) = \frac{75}{200}$ $= .375$ and so these two are independent.

3. If USAir accounted for 20% of the major disasters, the chance that it would be involved in at least four of seven such disasters is

$$\binom{7}{4}(.2)^4(.8)^3 + \binom{7}{5}(.2)^5(.8)^2 + \binom{7}{6}(.2)^6(.8)^1 + (.2)^7$$
$$= .029 + .004 + .000 + 000$$
$$= .033$$

[Or binomcdf(7, .8, 3) = .033.]

Mathematically, if USAir accounted for only 20% of the major disasters, there is only a .033 chance of it being involved in four of seven such disasters. This seems more than enough evidence to be suspicious!

4.

$$P(E|F) = \frac{P(E \cap F)}{P(F)} \quad \text{or} \quad P(E \cap F) = P(E|F)P(F),$$

and so

$$P(\text{woman} \cap \text{glasses}) = (.55)(.56) = .308$$

$$P(\text{woman} \cap \text{contacts}) = (.63)(.04) = .0252$$

If 55% of those who wear glasses are women, 45% of those who wear glasses must be men, and if 63% of those who wear contacts are women, 37% of those who wear contacts must be men. Thus we have

$$P(\text{man} \cap \text{glasses}) = (.45)(.56) = .252$$

$$P(\text{man} \cap \text{contacts}) = (.37)(.04) = .0148$$

The probability that you will encounter a person not wearing glasses or contacts is 1 – (.308 + .0252 + .252 + .0148) = .4.

5. (a) The probability of the complement is 1 minus the probability of the event, but $1 - .43 \neq .47$.

   (b) Probabilities are never greater than 1, but 6(.18) = 1.08.

   (c) The probability of an intersection cannot be greater than the probability of one of the separate events.

   (d) The probability of a union cannot be less than the probability of one of the separate events.

   (e) Probabilities are never negative.

6. (a) Player B wins only if both a 10 shows on the coin (a probability of $\frac{1}{2}$) and a 7 shows on the die (a probability of $\frac{2}{3}$). These events will both happen $\left(\frac{1}{2}\right)\left(\frac{2}{3}\right) = \frac{1}{3}$ of the time, and so player A wins $\frac{2}{3}$ of the time or twice as often as player B. Thus, to make this a fair game, player B should receive \$0.50 each time he wins.

   (b) Player A's expected payoff is

$$\sum xP(x) = (0.70)\left(\frac{2}{3}\right) + (1.10)\left(\frac{1}{3}\right) \approx \$0.83$$

   while player B's expected payoff is

$$\sum xP(x) = (0.60)\left(\frac{1}{2}\right) + (1.00)\left(\frac{1}{2}\right) \approx \$0.80$$

TIP

When using a formula, write down the formula and then substitute the values.

7. (a) We have a binomial with $n = 20$ and $p = .15$, so

$$P(X \geq 10) = \binom{20}{10}(.15)^{10}(.85)^{10} + \binom{20}{11}(.15)^{11}(.85)^{9} + \cdots + (.15) = .00025$$

   [The answer comes quickly from a calculator calculation such as 1-binomcdf(20,.15,9) on the TI-84.]

(b) If the probability of Legionella bacteria growing in an electric faucet is .15, then the probability of a result as extreme or more extreme than what was obtained in the Johns Hopkins study is only .00025. With such a low probability, there is strong evidence to conclude that the probability of Legionella bacteria growing in electronic faucets is greater than .15. That is, there is strong evidence that automatic faucets actually house more bacteria than the old-fashioned, manual kind!

Investigative Task

(a) Let the numbers 01 through 14 represent finding an illegal drug, while 15 through 99 and 00 represent no drugs. Read off pairs of digits from the random number table until a number representing illegal drugs is found or until nine clean cars are allowed to pass. Note whether a car with illegal drugs is found before nine clean cars pass. Repeat this procedure. Underlining numbers representing the presence of illegal drugs gives

8417706757 1761315582 5150681435 4105092031 0644905059 5988431180
5311584469 9486857967 0581184514 7501113006 6339555041 1586606589
1311971020 8594091932 0648874987 5435552704 9035902649 4749671567
9426808844 2629464759 0898957024 9728400637 8928303514 5919507635
0330972605 2935723737 6788103668 3387635841 5286923114 1586438942

The first nine cars are clean. Then 14 is found after four clean cars, then 05 after two clean cars, then 09 before any clean cars, then 06 after two clean cars, then 11 after seven free cars, and so on. Tabulating gives

First nine cars clean: 5

Illegal drugs found before tenth car: 24

The probability that illegal drugs will be found before the tenth car is estimated to be $\frac{24}{5+24} = .83$.

(b) $\Sigma xP(x) = 1(.15) + 2(.12) + \cdots + 25(.01) = 6.81$

(c) The probability that the first stopped car will have illegal drugs is .14. The probability that the first will be clean but that the second will have drugs is (.86)(.14). The probability that two clean cars will be followed by one with drugs is $(.86)^2(.14)$. The probability that three clean ones will be followed by one with drugs is $(.86)^3(.14)$, and so on. The probability of drugs being found before the tenth car is

$$.14 + (.86)(.14) + (.86)^2(.14) + \cdots + (.86)^8(.14) = .7427$$

Or more simply we could solve by subtracting the complementary probability from 1, that is, $1 - (.86)^9 = .7427$.

Combining Independent Random Variables

TOPIC **10**

- Independence Versus Dependence
- Mean and Standard Deviation for Sums and Differences of Independent Random Variables

In many experiments a pair of numbers is associated with each outcome. This situation leads to a study of pairs of random variables. When the random variables are independent, there is an easy calculation for finding both the mean and standard deviation of the sum (and difference) of the two random variables.

INDEPENDENCE VERSUS DEPENDENCE

EXAMPLE 10.1

An automobile salesperson sells three models of vans with selling prices of \$20,000, \$25,000, and \$30,000, respectively. For each sale, the salesperson receives a bonus of either \$500 or \$750. The probabilities of the various outcomes are given by the following table:

		Bonus	
		\$500	\$750
Amount of sale	\$20,000	.30	.05
	\$25,000	.20	.20
	\$30,000	.05	.20

If X is the "amount of sale" random variable, what is the probability distribution of X?
Answer: Summing each row gives

x(\$):	20,000	25,000	30,000
$P(x)$:	.35	.40	.25

If Y is the "bonus" random variable, what is the probability distribution of Y?
Answer: Summing each column gives

y(\$):	500	750
$P(y)$:	.55	.45

(continued)

What is the probability of a $500 bonus given that the sale was $30,000?
Answer:

$$P(\$500\text{bonus} \mid \$30{,}000\text{sale}) = \frac{P(\$500\text{bonus} \cap \$30{,}000\text{sale})}{P(\$30{,}000\text{sale})}$$
$$= \frac{.05}{.25} = .2$$

However, $P(\$500 \text{ bonus}) = .55$, and so $P(\$500 \text{ bonus} \mid \$30{,}000 \text{ sale}) \neq P(\$500 \text{ bonus})$. We say that the amount of sale random variable and the bonus random variable are not independent.

We say that the two random variables X and Y are independent if $P(x|y) = P(x)$ for all values of x and y. Equivalently, X and Y are independent if $P(x \cap y) = P(x)P(y)$ for all values of x and y. [This expression is also written $P(X = x, Y = y) = P(X = x)P(Y = y)$.]

EXAMPLE 10.2

In a study on vitamin C and the common cold, people were randomly assigned to take either no vitamin C or 500 milligrams daily, and it was noted whether or not they came down with a cold. Letting X, the "vitamin C" random variable, take the values 0 and 500, and Y, the "colds" random variable, take the values 0 (no colds) and 1 (at least one cold), the following table gives the observed probabilities:

		Colds (Y)	
		0	1
Vitamin C (X)	0	.10	.30
	500	.15	.45

Are the vitamin C random variable and the colds random variable independent?
Answer: Summing rows and columns gives

$$P(X = 0) = .10 + .30 = .40$$
$$P(X = 500) = .15 + .45 = .60$$
$$P(Y = 0) = .10 + .15 = .25$$
$$P(Y = 1) = .30 + .45 = .75$$

Now checking the conditional probabilities,

$$P(X = 0 \mid Y = 0) = \frac{.10}{.25} = .4 = P(X = 0)$$

$$P(X = 0 \mid Y = 1) = \frac{.30}{.75} = .4 = P(X = 0)$$

$$P(X = 500 \mid Y = 0) = \frac{.15}{.25} = .6 = P(X = 500)$$

$$P(X = 500 \mid Y = 1) = \frac{.45}{.75} = .6 = P(X = 500)$$

Thus we conclude that X and Y are independent.

Note that an alternative check for independence is the following: $P(X = 0, Y = 0) = .10$ and $P(X = 0)P(Y = 0) = (.40)(.25) = .10$. Similarly, $P(X = 0, Y = 1) = P(X = 0)P(Y = 1) = .30$, $P(X = 500, Y = 0) = P(X = 500)P(Y = 0) = .15$, and $P(X = 500, Y = 1) = P(X = 500)P(Y = 1) = .45$.

MEAN AND STANDARD DEVIATION FOR SUMS AND DIFFERENCES OF INDEPENDENT RANDOM VARIABLES

One way of comparing two populations is to analyze their set of differences.

EXAMPLE 10.3

Suppose set $X = \{2, 9, 11, 22\}$ and set $Y = \{5, 7, 15\}$. Note that the mean of set X is $\mu_x = \frac{2+9+11+22}{4} = 11$ and the mean of set Y is $\mu_y = \frac{5+7+15}{3} = 9$. Form the set Z of differences by subtracting each element of Y from each element of X:

$$\begin{aligned} Z = \{&2-5, 2-7, 2-15, 9-5, 9-7, 9-15, 11-5, 11-7, \\ &11-15, 22-5, 22-7, 22-15\} \\ = \{&-3, -5, -13, 4, 2, -6, 6, 4, -4, 17, 15, 7\} \end{aligned}$$

What is the mean of Z?

Answer:

$$\mu = \frac{-3-5-13+4+2-6+6+4-4+17+15+7}{12} = \frac{24}{12} = 2$$

Note that $\mu_z = \mu_x - \mu_y$.

In general, the mean of a set of differences is equal to the difference of the means of the two original sets. Even more generally, if a sum is formed by picking one element from each of several sets, the mean of such sums is simply the sum of the means of the various sets.

How is the variance of Z related to the variances of the original sets?

Answer:

$$\mu_x = 11 \quad \sigma_x^2 = \frac{(2-11)^2 + (9-11)^2 + (11-11)^2 + (22-11)^2}{4} = \frac{206}{4} = 51.5$$

$$\mu_y = 9 \quad \sigma_y^2 = \frac{(5-9)^2 + (7-9)^2 + (15-9)^2}{3} = \frac{56}{3} = 18.67$$

$$\mu_z = 2 \quad \sigma_z^2 = \frac{(-3-2)^2 + (-5-2)^2 + (-13-2)^2 + \cdots + (7-2)^2}{12} = \frac{842}{12} = 70.17$$

Note that in the above example $\sigma_z^2 = \sigma_x^2 + \sigma_y^2$. This is true for the variance of any set of differences. More generally, if a total is formed by a procedure that adds or subtracts one element from each of several independent sets, the variance of the resulting totals is simply the sum of the variances of the several sets.

Not only can we sum variances as shown above to calculate total variance, but we can also reverse the process and determine how the total variance is split up among its various sources. For example, we can find what portion of the variance in numbers of sales made by a company's sales representatives is due to the individual salesperson, what portion is due to the territory, what portion is due to the particular products sold by each salesperson, and so on.

The above principles extend more broadly to random variables.

EXAMPLE 10.4

Suppose two random variables X and Y have the following joint probability distribution table:

		Y: 5	Y: 8	Y: 9
X	3	.10	.05	.15
	4	.20	.10	.00
	6	.05	.20	.15

What are the expected values (means) of the random variables X, Y, and $X + Y$?

Answer:

$$\begin{aligned} E(X) &= \sum x_i p_i \\ &= 3(.10 + .05 + .15) + 4(.20 + .10 + .00) + 6(.05 + .20 + .15) \\ &= 4.50 \end{aligned}$$

$$\begin{aligned} E(Y) &= \sum y_i p_i \\ &= 5(.10 + .20 + .05) + 8(.05 + .10 + .20) + 9(.15 + .00 + .15) \\ &= 7.25 \end{aligned}$$

$$\begin{aligned} E(X + Y) &= (3+5)(.10) + (3+8)(.05) + (3+9)(.15) + (4+5)(.20) \\ &\quad + (4+8)(.10) + (4+9)(.00) + (6+5)(.05) + (6+8)(.20) \\ &\quad + (6+9)(.15) \\ &= 11.75 \end{aligned}$$

Note that $E(X + Y) = E(X) + E(Y)$. This is true in general for expected values of sums of random variables. It is also true that $E(X - Y) = E(X) - E(Y)$.

What are the variances of the random variables X, Y, and $X + Y$?

Answer:

$$\begin{aligned} \text{var}(X) &= \sum (x_i - \mu_x)^2 p_i \\ &= (3 - 4.50)^2(.30) + (4 - 4.50)^2(.30) + (6 - 4.50)^2(.40) \\ &= 1.65 \end{aligned}$$

$$\begin{aligned} \text{var}(Y) &= (5 - 7.25)^2(.35) + (8 - 7.25)^2(.35) + (9 - 7.25)^2(.30) \\ &= 2.8875 \end{aligned}$$

$$\begin{aligned} \text{var}(X + Y) &= (8 - 11.75)^2(.10) + (11 - 11.75)^2(.05) + (12 - 11.75)^2(.15) \\ &\quad + (9 - 11.75)^2(.20) + (12 - 11.75)^2(.10) + (13 - 11.75)^2(.00) \\ &\quad + (11 - 11.75)^2(.05) + (14 - 11.75)^2(.20) + (15 - 11.75)^2(.15) \\ &= 5.5875 \end{aligned}$$

Note that $\text{var}(X + Y) \neq \text{var}(X) + \text{var}(Y)$.

Do the variances ever sum like the means do? Consider the following example of independent random variables X and Y.

EXAMPLE 10.5

Suppose that two random variables X and Y have the following joint probability distribution table:

		Y: 2	Y: 5	Y: 10	
X	5	.07	.05	.08	.20
X	7	.28	.20	.32	.80
		.35	.25	.40	

If one computes all the conditional probabilities, it can be shown that $P(x|y) = P(x)$ for all values of x and y, and thus X and Y are independent. What are the variances of the random variables X, Y, and $X + Y$?

Answer: First we calculate

$$E(X) = 5(.20) + 7(.80) = 6.60$$

$$E(Y) = 2(.35) + 5(.25) + 10(.40) = 5.95$$

$$E(X+Y) = E(X) + E(Y) = 12.55$$

Then we calculate the variances

$$\text{var}(X) = (5 - 6.60)^2(.20) + (7 - 6.60)^2(.80) = 0.64$$

$$\begin{aligned}\text{var}(Y) &= (2 - 5.95)^2(.35) + (5 - 5.95)^2(.25)\\ &\quad + (10 - 5.95)^2(.40)\\ &= 12.2475\end{aligned}$$

$$\begin{aligned}\text{var}(X+Y) &= (7 - 12.55)^2(.07) + (10 - 12.55)^2(.05)\\ &\quad + (15 - 12.55)^2(.08) + (9 - 12.55)^2(.28)\\ &\quad + (12 - 12.55)^2(.20) + (17 - 12.55)^2(.32)\\ &= 12.8875\end{aligned}$$

Note that 0.64 + 12.2475 = 12.8875; that is, var($X + Y$) = var(X) + var(Y).

TIP

Variances of independent random variables add, but standard deviations don't!

More generally, *if two random variables are independent,* the variance of the sum (or the difference) of the two random variables is equal to the *sum* of the two individual variances.

EXAMPLE 10.6

Suppose the mean SAT verbal score is 425 with standard deviation 100, while the mean SAT math score is 475 with standard deviation 100. What can be said about the mean and standard deviation of the combined math and verbal scores?

Answer: The mean is 425 + 475 = 900. Can we add the two variances to obtain $100^2 + 100^2 = 20{,}000$ and then take the square root to obtain 141.4? No, because we don't have independence—students with high math scores tend to have high verbal scores, and those with low math scores tend to have low verbal scores. The standard deviation of the combined scores cannot be calculated from the given information.

For random variables X and Y the generalizations for transforming X and for combining X and Y are sometimes written as

Transforming X: $E(X \pm a) = E(X) \pm a$ $\quad$ $\text{var}(X \pm a) = \text{var}(X)$

$E(bX) = bE(X)$ $\quad$ $\text{var}(bX) = b^2\,\text{var}(X)$

Combining X and Y: $E(X \pm Y) = E(X) \pm E(Y)$

$\text{var}(X \pm Y) = \text{var}(X) + \text{var}(Y)$ if X and Y are independent

Does it make sense that variances *add* for both sums and differences of independent random variables? *Ranges* may illustrate this point more clearly: suppose prices for primary residences in a community go from $210,000 to $315,000, while prices for summer cottages go from $50,000 to $120,000. Someone wants to purchase a primary residence and a summer cottage. The minimum *total* spent will be $260,000, the maximum possible is $435,000, and thus the range for the total is $435,000 – $260,000 = $175,000. The minimum *difference* spent between the two homes will be $90,000, the maximum difference will be $265,000, and thus the range for the difference is $265,000 – $90,000 = $175,000, the same as the range for the total. (Variances and ranges are both measures of spread and behave the same way with regard to the above rule.)

Summary

- Two random variables X and Y are independent if $P(x|y) = P(x)$ for all values of x and y.
- When combining two random variables, $E(X + Y) = E(X) + E(Y)$ and $E(X - Y) = E(X) - E(Y)$.
- When combining two independent random variables, the variances always add: $\text{var}(X \pm Y) = \text{var}(X) + \text{var}(Y)$.

Questions on Topic Ten: Combining Independent Random Variables

Multiple-Choice Questions

Directions: The questions or incomplete statements that follow are each followed by five suggested answers or completions. Choose the response that best answers the question or completes the statement.

Questions 1–4 refer to the following: Students are classified by television usage (unending, average, and infrequent) and how often they exercise (regular, occasional, and never), resulting in the following joint probability table.

	Exercise		
TV usage	Regular (Y = 1)	Occasional (Y = 2)	Never (Y = 3)
Unending (X = 1)	.05	.05	.10
Average (X = 2)	.20	.15	.10
Infrequent (X = 3)	.15	.15	.05

1. What is the probability distribution for X?

   (A) $P(X = 1) = .05$, $P(X = 2) = .20$, $P(X = 3) = .15$
   (B) $P(X = 1) = .10$, $P(X = 2) = .35$, $P(X = 3) = .30$
   (C) $P(X = 1) = .20$, $P(X = 2) = .45$, $P(X = 3) = .35$
   (D) $P(X = 1) = .40$, $P(X = 2) = .35$, $P(X = 3) = .25$
   (E) It cannot be determined from the given information.

2. What is the probability $P(X = 2, Y = 3)$, that is, the probability that a student has average television usage but never exercises?

   (A) .10
   (B) .1125
   (C) .22
   (D) .40
   (E) .60

3. What is the probability $P(X = 2 | Y = 3)$, that is, the probability that a student has average television usage given that he never exercises?

   (A) .10
   (B) .17
   (C) .22
   (D) .40
   (E) .60

4. Are X and Y independent?

   (A) Yes, because the conditional probabilities $P(X = x \mid Y = y)$ equal the corresponding unconditional probabilities $P(X = x)$.
   (B) Yes, because the joint probabilities are equal to the product of the respective probabilities.
   (C) Yes, because of either of the above answers.
   (D) No.
   (E) The answer cannot be determined from the given information.

5. Following are parts of the probability distributions for the random variables X and Y.

x	$P(x)$
1	?
2	.2
3	.3
4	?

y	$P(y)$
1	.4
2	?
3	.1

   If X and Y are independent and the joint probability $P(X = 1, Y = 2) = .1$, what is $P(X = 4)$?

   (A) .1
   (B) .2
   (C) .3
   (D) .4
   (E) .5

6. Suppose X and Y are random variables with $E(X) = 25$, $\text{var}(X) = 3$, $E(Y) = 30$, and $\text{var}(Y) = 4$. What are the expected value and variance of the random variable $X + Y$?

   (A) $E(X + Y) = 55$, $\text{var}(X + Y) = 3.5$
   (B) $E(X + Y) = 55$, $\text{var}(X + Y) = 5$
   (C) $E(X + Y) = 55$, $\text{var}(X + Y) = 7$
   (D) $E(X + Y) = 27.5$, $\text{var}(X + Y) = 7$
   (E) There is insufficient information to answer this question.

7. Suppose X and Y are random variables with $\mu_x = 10$, $\sigma_x = 3$, $\mu_y = 15$, and $\sigma_y = 4$. Given that X and Y are independent, what are the mean and standard deviation of the random variable $X + Y$?

   (A) $\mu_{x+y} = 25$, $\sigma_{x+y} = 3.5$
   (B) $\mu_{x+y} = 25$, $\sigma_{x+y} = 5$
   (C) $\mu_{x+y} = 25$, $\sigma_{x+y} = 7$
   (D) $\mu_{x+y} = 12.5$, $\sigma_{x+y} = 7$
   (E) There is insufficient information to answer this question.

8. Suppose X and Y are random variables with $E(X) = 500$, $\text{var}(X) = 50$, $E(Y) = 400$, and $\text{var}(Y) = 30$. Given that X and Y are independent, what are the expected value and variance of the random variable $X - Y$?

   (A) $E(X - Y) = 100$, $\text{var}(X - Y) = 20$
   (B) $E(X - Y) = 100$, $\text{var}(X - Y) = 80$
   (C) $E(X - Y) = 900$, $\text{var}(X - Y) = 20$
   (D) $E(X - Y) = 900$, $\text{var}(X - Y) = 80$
   (E) There is insufficient information to answer this question.

9. Suppose the average height of policemen is 71 inches with a standard deviation of 4 inches, while the average for policewomen is 66 inches with a standard deviation of 3 inches. If a committee looks at all ways of pairing up one male with one female officer, what will be the mean and standard deviation for the difference in heights for the set of possible partners?

   (A) Mean of 5 inches with a standard deviation of 1 inch
   (B) Mean of 5 inches with a standard deviation of 3.5 inches
   (C) Mean of 5 inches with a standard deviation of 5 inches
   (D) Mean of 68.5 inches with a standard deviation of 1 inch
   (E) Mean of 68.5 inches with a standard deviation of 3.5 inches

10. Consider the set of scores of all students taking the AP Statistics exam. What is true about the variance of this set?

    (A) Given a boxplot of this distribution, if the whisker lengths are equal, the variance will equal the interquartile range.
    (B) If the distribution of scores is bell-shaped, the variance will be between −3 and +3.
    (C) If there are a few high scores with the bulk of the scores less than the mean, the variance will be skewed to the right.
    (D) If the distribution is symmetric, the variance will equal the standard deviation.
    (E) The variance is the sum of variances coming from a variety of factors such as study time, course grade, IQ, and so on.

Answer Key

1. **C** 4. **D** 7. **B** 10. **E**
2. **A** 5. **C** 8. **B**
3. **D** 6. **E** 9. **C**

Answers Explained

Multiple-Choice

1. **(C)** Summing each row gives $P(X = 1) = .20$, $P(X = 2) = .45$, and $P(X = 3) = .35$.

2. **(A)** This question asks for the probability of an intersection, and the answer is found in the second row, third column.

3. **(D)** $P(X = 2|Y = 3) = \frac{P(X=2, Y=3)}{P(Y=3)} = \frac{.10}{.10+.10+.05} = .40$

4. **(D)** For example, $P(X = 2|Y = 3) = .40$, but $P(X = 2) = .45$. For independence this has to hold for all values of x and y.

5. **(C)** $P(Y = 2) = 1 - (.4 + .1) = .5$. By independence, $P(X = 1, Y = 2) = P(X = 1)P(Y = 2)$, and so $.1 = P(X = 1)(.5)$ and $P(X = 1) = .2$. Then $P(X = 4) = 1 - (.2 + .2 + .3) = .3$.

6. **(E)** Without independence we cannot determine var$(X + Y)$ from the information given.

7. **(B)** The means and the variances add. Thus the new variance is $3^2 + 4^2 = 25$, and the new standard deviation is 5.

8. **(B)** If two random variables are independent, the mean of the difference of the two random variables is equal to the difference of the two individual means; however, the variance of the difference of the two random variables is equal to the sum of the two individual variances.

9. **(C)** The mean of a set of differences is the difference of the means of the two sets, while the variance is the sum of the two variances. Thus $71 - 66 = 5$, while $4^2 + 3^2 = 25$ and $\sqrt{25} = 5$.

10. **(E)** The total variance can be split up among its various sources. Also, the alternative choices are all nonsense.

The Normal Distribution

TOPIC 11

- Properties
- Using Tables
- A Model for Measurement
- Commonly Used Probabilities and z-scores
- Finding Means and Standard Deviations
- Normal Approximation to the Binomial
- Checking Normality

Some of the most useful probability distributions are symmetric and bell-shaped:

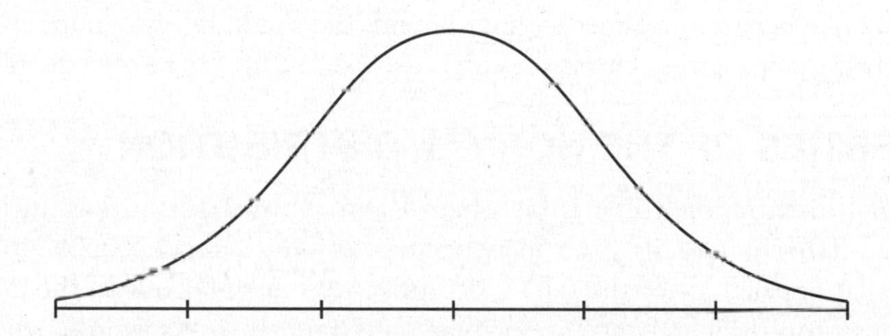

In this topic we study one such distribution called the *normal distribution* (not *all* symmetric, unimodal distributions are normal). The normal distribution is valuable in describing various natural phenomena, especially those involving growth or decay. However, the real importance of the normal distribution in statistics is that it can be used to describe the results of sampling procedures.

The normal distribution can be viewed as a limiting case of the binomial. More specifically, we start with any fixed probability of success p and consider what happens as n becomes arbitrarily large. To obtain a visual representation of the limit, we draw histograms using an increasingly large n for each fixed p.

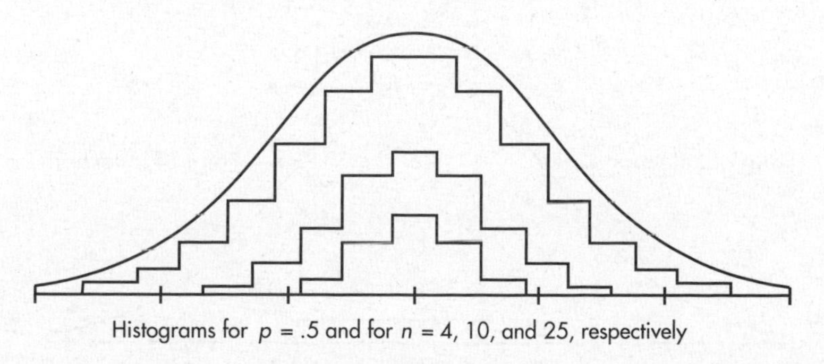

Histograms for $p = .5$ and for $n = 4$, 10, and 25, respectively

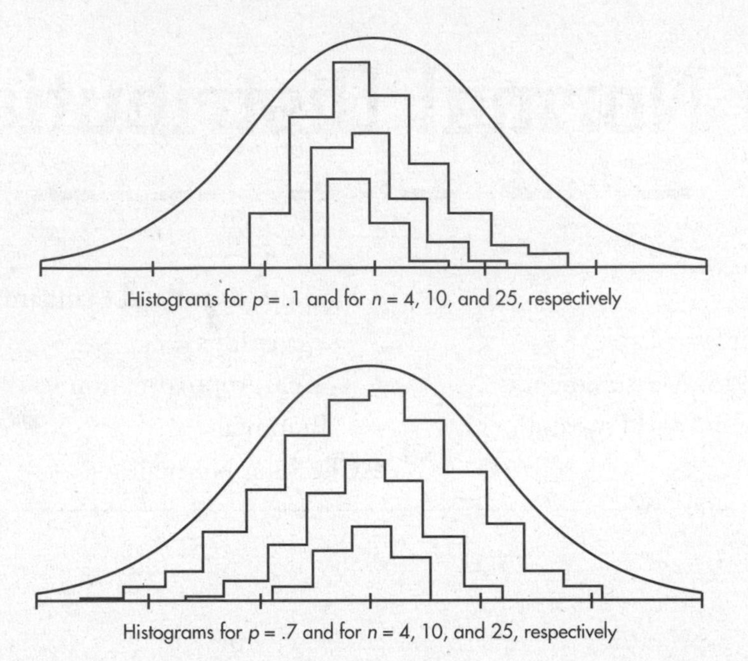

Histograms for $p = .1$ and for $n = 4, 10$, and 25, respectively

Histograms for $p = .7$ and for $n = 4, 10$, and 25, respectively

For greater clarity, different scales have been used for the histograms associated with different values of n. From the diagrams above it is reasonable to accept that as n becomes larger without bound, the resulting histograms approach a smooth bell-shaped curve. This is the curve associated with the normal distribution.

PROPERTIES OF THE NORMAL DISTRIBUTION

The normal distribution curve is bell-shaped and symmetric and has an infinite base. Long, flat-looking tails cover many values but only a small proportion of the area. The flat appearance of the tails is deceptive. Actually, far out in the tails, the curve drops proportionately at an ever-increasing rate. In other words, when two intervals of equal length are compared, the one closer to the center may experience a greater numerical drop, but the one further out in the tail experiences a greater drop when measured as a proportion of the height at the beginning of the interval.

The mean here is the same as the median and is located at the center. We want a unit of measurement that applies equally well to any normal distribution, and we choose a unit that arises naturally out of the curve's shape. There is a point on each side where the slope is steepest. These two points are called *points of inflection*, and the distance from the mean to either point is precisely equal to one standard deviation. Thus, it is convenient to measure distances under the normal curve in terms of z-scores (recall from Topic 2 that z-scores are fractions or multiples of standard deviations from the mean).

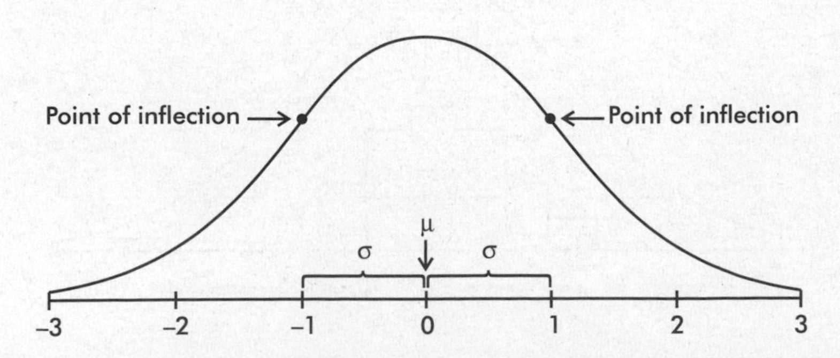

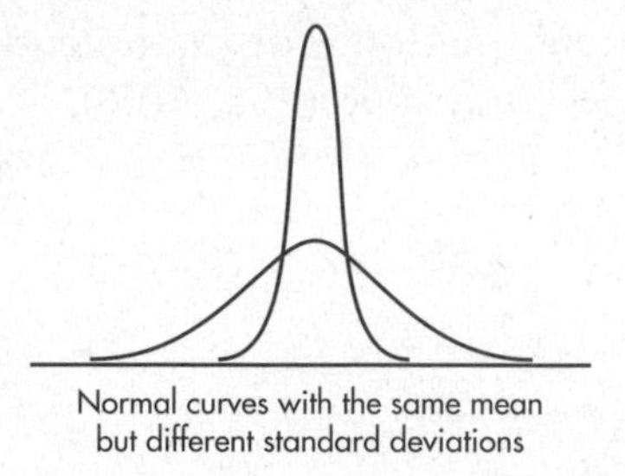
Normal curves with the same mean but different standard deviations

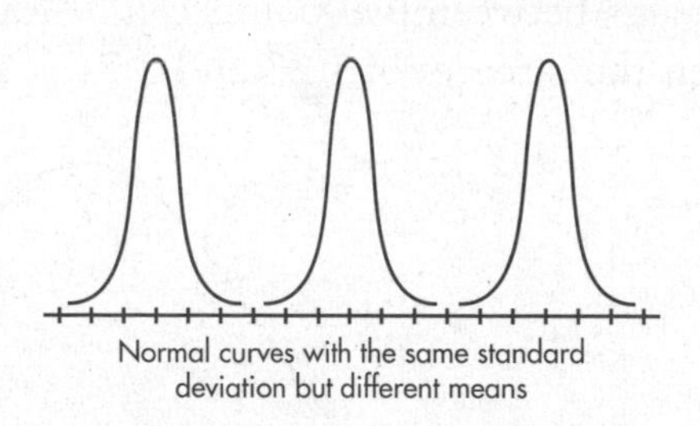
Normal curves with the same standard deviation but different means

Mathematically, the formula for the normal curve turns out to be $y = e^{-z^2/2}$, where y is the relative height above z-score. (By *relative height* we mean proportion of the height above the mean.) However, our interest is not so much in relative heights under the normal curve as it is in *proportionate areas.* The probability associated with any interval under the normal curve is equal to the proportionate area found under the curve and above the interval.

A useful property of the normal curve is that approximately 68% of the area (and thus 68% of the observations) falls within one standard deviation of the mean, approximately 95% of the area falls within two standard deviations of the mean, and approximately 99.7% of the area falls within three standard deviations of the mean.

TIP

The notation $N(\mu,\sigma)$ stands for a normal model with mean μ and standard deviation σ.

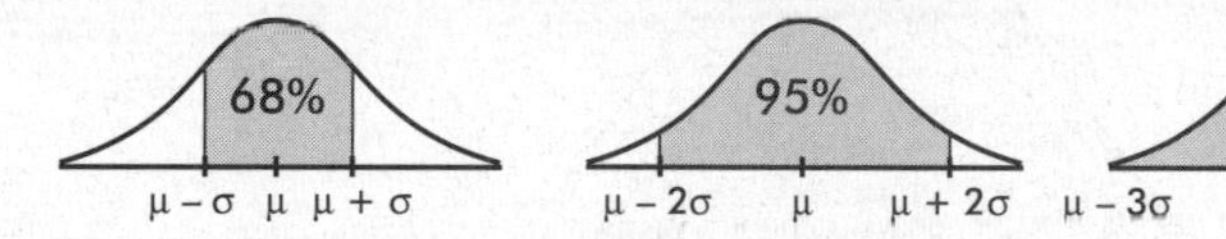

USING TABLES OF THE NORMAL DISTRIBUTION

Table A in the Appendix shows areas under the normal curve. Specifically, this table shows the area to the left of a given point. It shows, for example, that to the left of a z-score of 1.2 there is .8849 of the area:

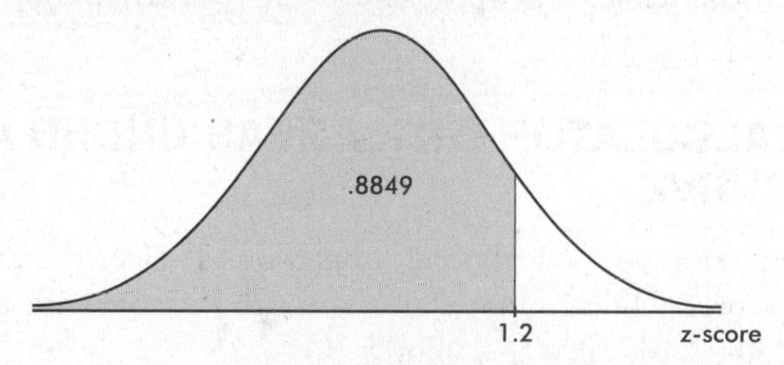

Because the total area is 1, the area to the right of a point can be calculated by calculating 1 minus the area to the left of the point. For example, the area to the right of a z-score of – 2.13 is 1 – .0166 = .9834. (Alternatively, by symmetry, we could have said that the area to the right of –2.13 is the same as that to the left of +2.13 and then read the answer directly from Table A.)

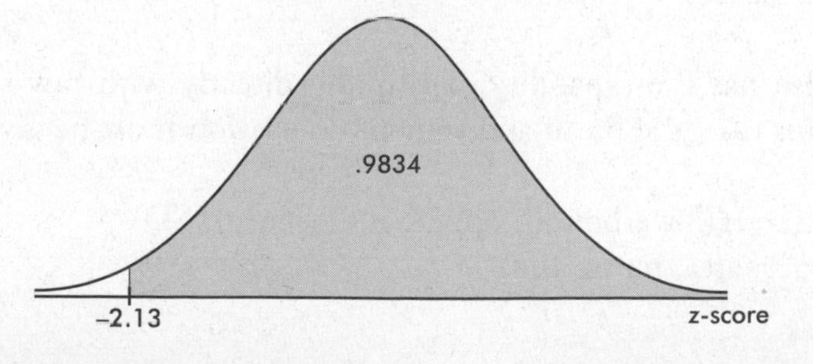

The area between two points can be found by subtraction. For example, the area between the *z*-scores of 1.23 and 2.71 is equal to .9966 – .8907 = .1059.

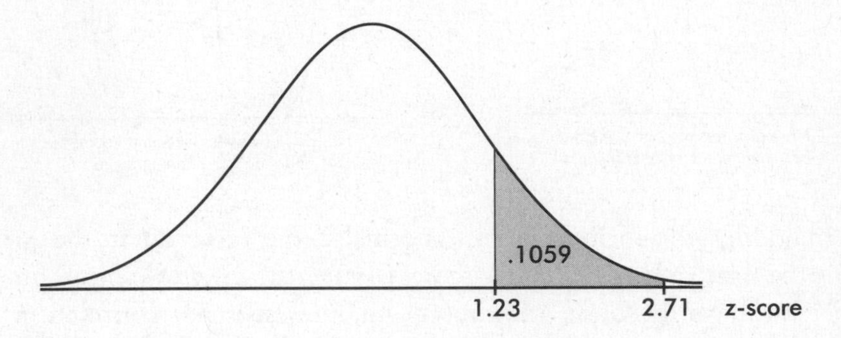

Given a probability, we can use Table A in reverse to find the appropriate *z*-score. For example, to find the *z*-score that has an area of .8982 to the left of it, we search for the probability closest to .8982 in the bulk of the table and read off the corresponding *z*-score. In this case we find .8980 with the *z*-score 1.27.

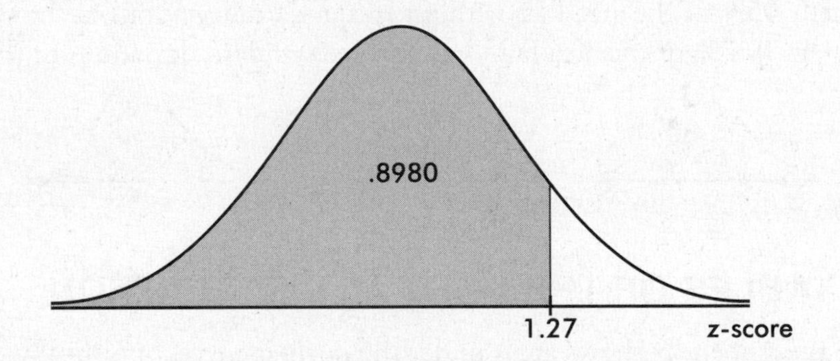

To find the *z*-score with an area of .72 to the right of it, we look for 1– .72 = .28 in the bulk of the table. The probability .2810 corresponds to the *z*-score –0.58.

USING A CALCULATOR WITH AREAS UNDER A NORMAL CURVE

On the TI-84, `normalcdf`(lowerbound, upperbound) gives the area (probability) between two *z*-scores, while `invNorm`(area) gives the *z*-score with the given area to its left. So we could have calculated above:

```
normalcdf(–100, 1.2) = .8849
normalcdf(–2.13, 100) = .9834
normalcdf(1.23, 2.71) = .1060
invNorm(.8982) = 1.271
invNorm(.28) = –0.5828
```

The TI-84 also has the capability of working directly with raw scores instead of *z*-scores. In this case, the mean and standard deviation must be given:

```
normalcdf(lowerbound, upperbound, mean, SD)
invNorm(area, mean, SD)
```

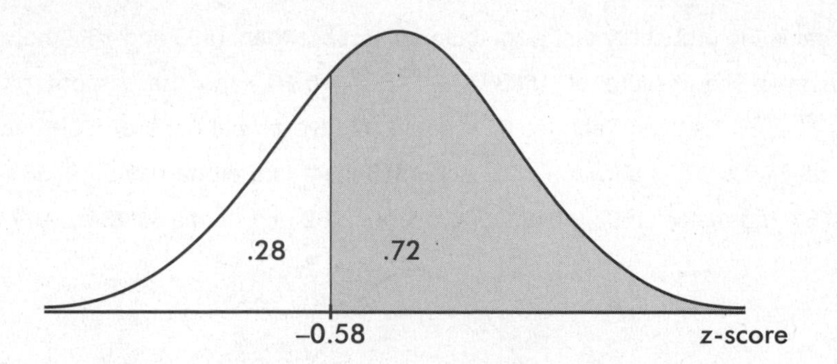

THE NORMAL DISTRIBUTION AS A MODEL FOR MEASUREMENT

In solving a problem using the normal distribution as a model, drawing a picture of a normal curve with horizontal lines showing raw scores and *z*-scores is usually helpful.

> **TIP**
> Don't use the normal model if the distribution isn't symmetric and unimodal.

EXAMPLE 11.1

The life expectancy of a particular brand of lightbulb is normally distributed with a mean of 1500 hours and a standard deviation of 75 hours.

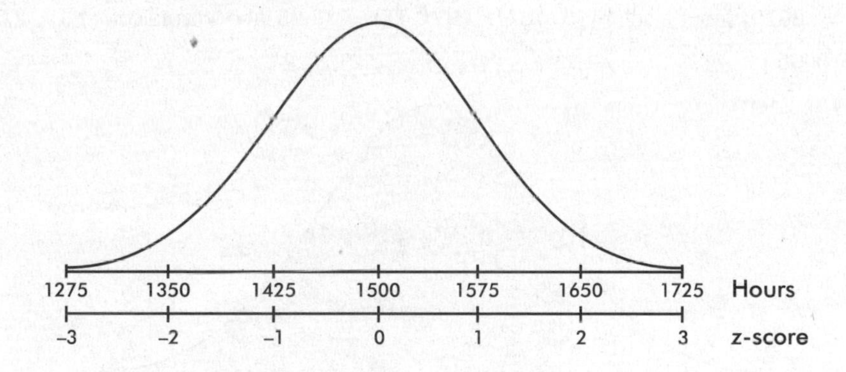

> **TIP**
> Draw a picture!

a. What is the probability that a lightbulb will last less than 1410 hours?
Answer: The *z*-score of 1410 is $\frac{1410-1500}{75} = -1.2$, and from Table A the probability to the left of −1.2 is .1151. [On the TI-84, normalcdf(0, 1410, 1500, 75) = .1151 and normalcdf(−10, −1.2) = .1151.]

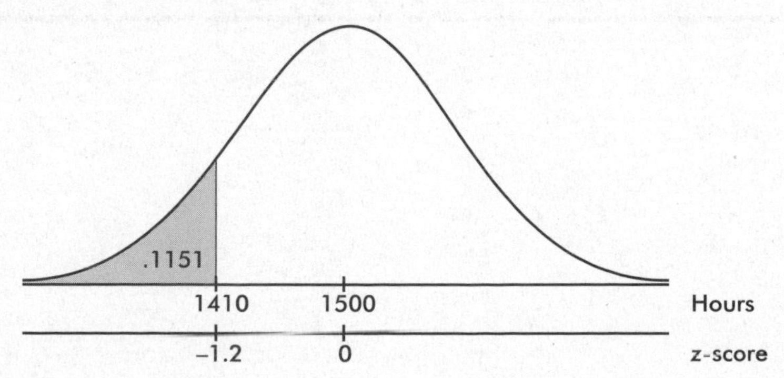

(continued)

b. What is the probability that a lightbulb will last between 1563 and 1648 hours?

Answer: The *z*-score of 1563 is $\frac{1563-1500}{75} = 0.84$, and the *z*-score of 1648 is $\frac{1648-1500}{75} = 1.97$. In Table A, 0.84 and 1.97 give probabilities of .7995 and .9756, respectively. Thus between 1563 and 1648 there is a probability of .9756 – .7995 = .1761. [normalcdf(1563, 1648, 1500, 75) = .1762 and normalcdf(.84, 1.97) = .1760.]

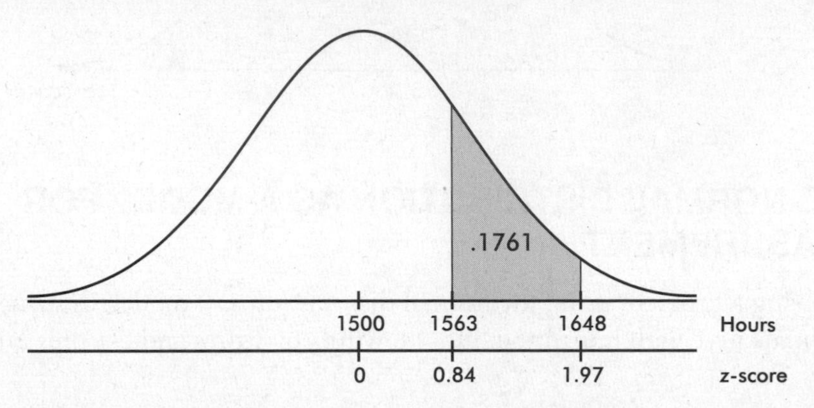

c. What is the probability that a lightbulb will last between 1416 and 1677 hours?

Answer: The *z*-score of 1416 is $\frac{1416-1500}{75} = -1.12$, and the *z*-score of 1677 is $\frac{1677-1500}{75} = 2.36$. In Table A, –1.12 gives a probability of .1314, and 2.36 gives a probability of .9909. Thus between 1416 and 1677 there is a probability of .9909 – .1314 = .8595. [normalcdf(1416, 1677, 1500, 75) = .8595 and normalcdf(–1.12, 2.36) = .8595.]

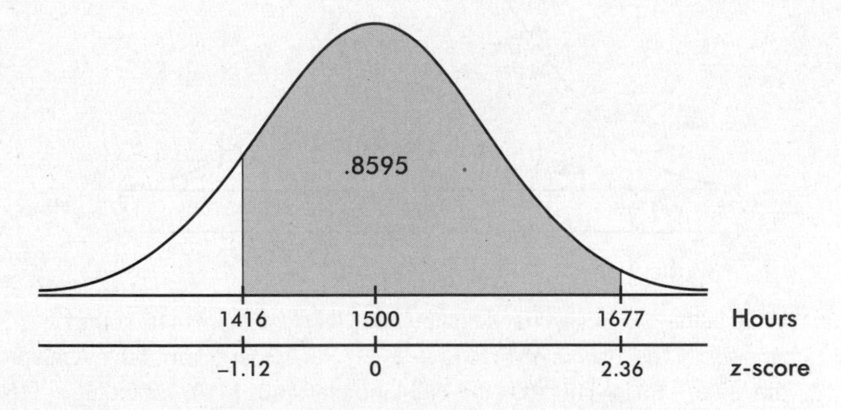

EXAMPLE 11.2

A packing machine is set to fill a cardboard box with a mean of 16.1 ounces of cereal. Suppose the amounts per box form a normal distribution with a standard deviation equal to 0.04 ounce.

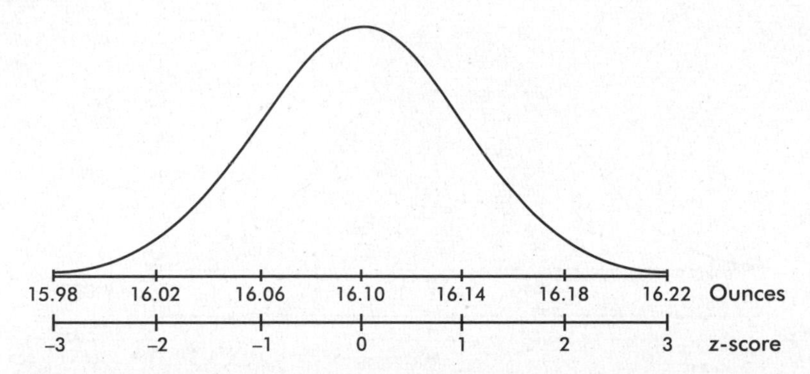

a. What percentage of the boxes will end up with at least 1 pound of cereal?
Answer: The *z*-score of 16 is $\frac{16-16.1}{0.04}=-2.5$, and –2.5 in Table A gives a probability of .0062. Thus the probability of having more than 1 pound in a box is 1 – .0062 = .9938 or 99.38%. [normalcdf(16, 1000, 16.10, .04) = .9938 and normalcdf(–2.5, 10) = .9938.]

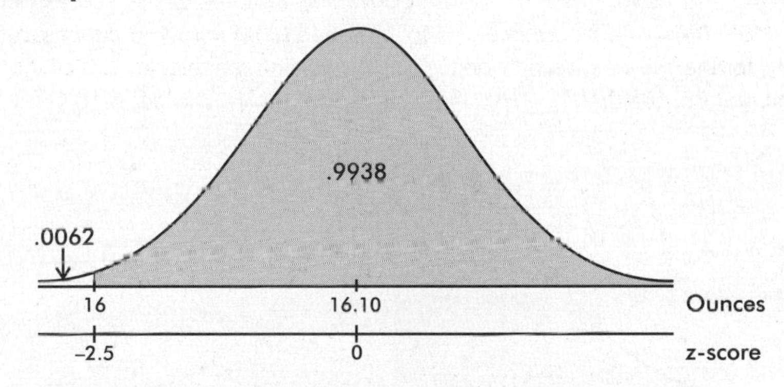

b. Ten percent of the boxes will contain less than what number of ounces?
Answer: In Table A, we note that .1 area (actually .1003) is found to the left of a –1.28 *z*-score. [On the TI-84, invNorm(.1) = –1.2816.] Converting the *z*-score of –1.28 into a raw score yields $-1.28=\frac{x-16.1}{0.04}$ or 16.1 – 1.28(0.04) = 16.049 ounces. [Or directly on the TI-84, invNorm(.1, 16.10, .04) = 16.049.]

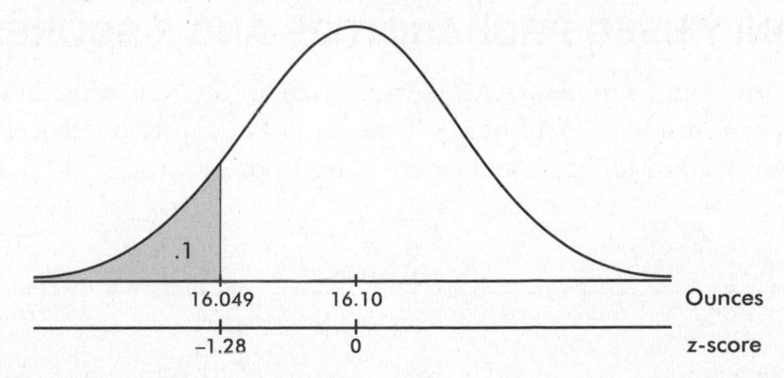

(continued)

c. Eighty percent of the boxes will contain more than what number of ounces?
Answer: In Table A, we note that 1 – .8 = .2 area (actually .2005) is found to the left of a *z*-score of –0.84, and so to the right of a –0.84 *z*-score must be 80% of the area. [Or invNorm(.2) = .8416.] Converting the *z*-score of –0.84 into a raw score yields 16.1 – 0.84(0.04) = 16.066 ounces. [Or directly on the TI-84, invNorm(.2, 16.10, .04) = 16.066.]

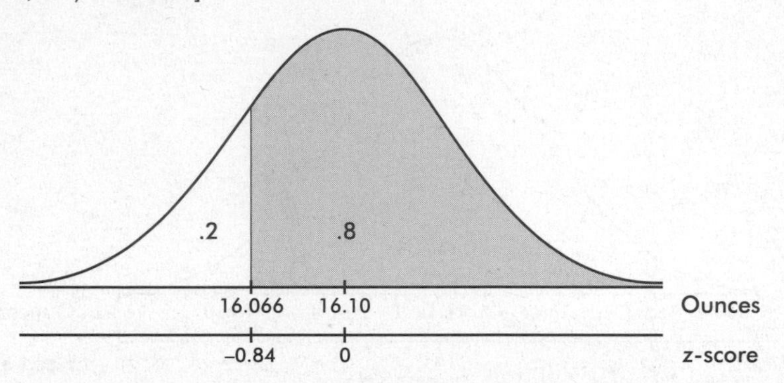

d. The middle 90% of the boxes will be between what two weights?
Answer: Ninety percent in the middle leaves five percent in each tail. In Table A, we note that .0505 area is found to the left of a *z*-score of –1.64, while 0.495 area is found to the left of a *z*-score of –1.65, and so we use the *z*-score of –1.645. Then 90% of the area is between *z*-scores of –1.645 and 1.645. [Or invNorm(.05) = –1.6449 and invNorm(.95) = 1.6449.] Converting *z*-scores to raw scores yields 16.1 – 1.645(0.04) = 16.034 ounces and 16.1 + 1.645(0.04) = 16.166 ounces, respectively, for the two weights between which we will find the middle 90% of the boxes. [invNorm(.05, 16.10, .04) = 16.034 and invNorm(.95, 16.10, .04) = 16.166.]

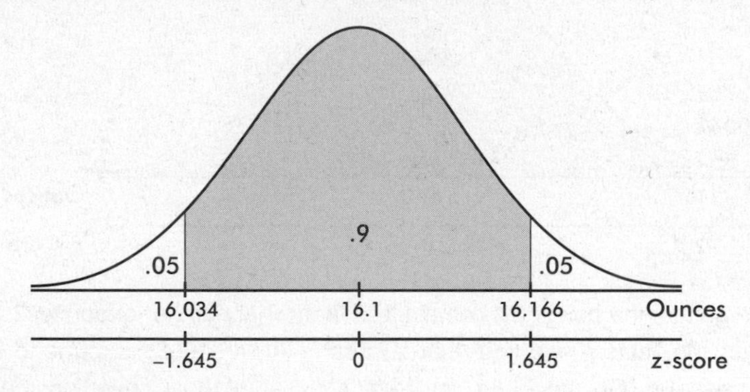

COMMONLY USED PROBABILITIES AND *Z*-SCORES

As can be seen from Example 11.2*d*, there is often an interest in the limits enclosing some specified middle percentage of the data. For future reference, the limits most frequently asked for are noted below in terms of *z*-scores.

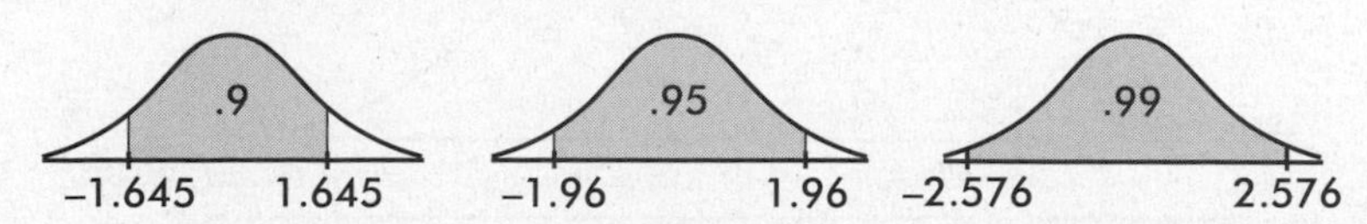

Ninety percent of the values are between *z*-scores of –1.645 and +1.645, 95% of the values are between *z*-scores of –1.96 and +1.96, and 99% of the values are between *z*-scores of –2.576 and +2.576.

Sometimes the interest is in values with particular percentile rankings. For example,

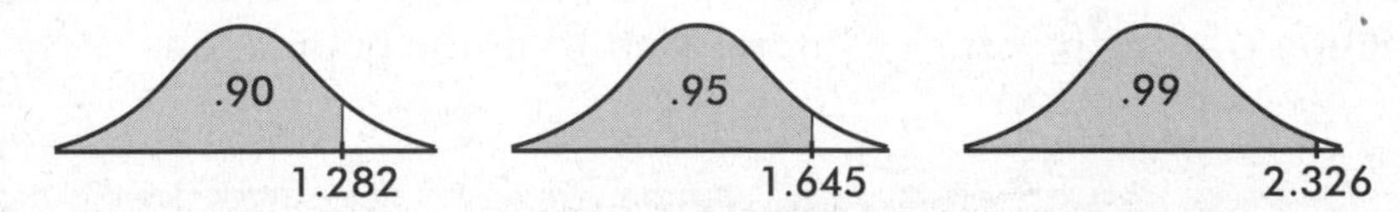

Ninety percent of the values are below a *z*-score of 1.282, 95% of the values are below a *z*-score of 1.645, and 99% of the values are below a *z*-score of 2.326.

There are corresponding conclusions for negative *z*-scores:

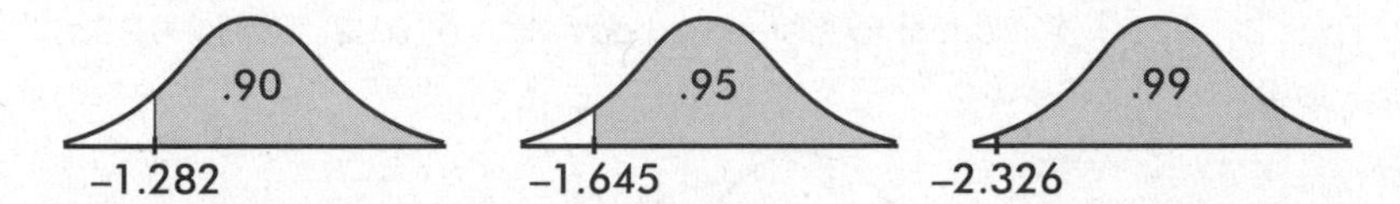

Ninety percent of the values are above a *z*-score of –1.282, 95% of the values are above a *z*-score of –1.645, and 99% of the values are above a *z*-score of –2.326.

It is also useful to note the percentages corresponding to values falling between integer *z*-scores. For example,

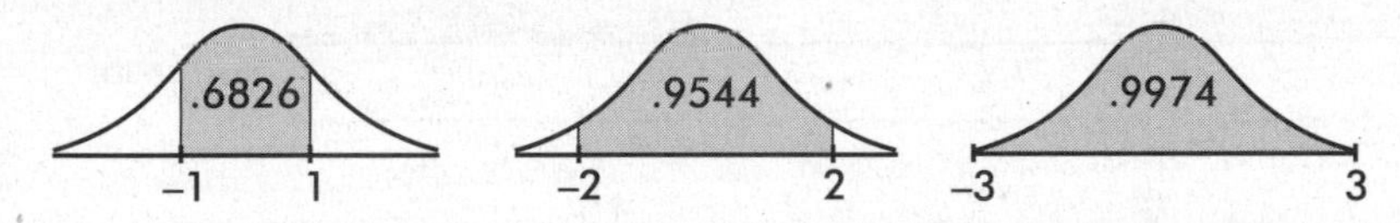

Note that 68.26% of the values are between *z*-scores of –1 and +1, 95.44% of the values are between *z*-scores of –2 and +2, and 99.74% of the values are between *z*-scores of –3 and +3.

EXAMPLE 11.3

Suppose that the average height of adult males in a particular locality is 70 inches with a standard deviation of 2.5 inches.

a. If the distribution is normal, the middle 95% of males are between what two heights?
Answer: As noted above, the critical *z*-scores in this case are ±1.96, and so the two limiting heights are 1.96 standard deviations from the mean. Therefore, 70 ± 1.96(2.5) = 70 ± 4.9, or from 65.1 to 74.9 inches.

b. Ninety percent of the heights are below what value?
Answer: The critical *z*-score is 1.282, and so the value in question is 70 + 1.282(2.5) = 73.205 inches.

c. Ninety-nine percent of the heights are above what value?
Answer: The critical *z*-score is –2.326, and so the value in question is 70 – 2.326(2.5) = 64.185 inches.

d. What percentage of the heights are between *z*-scores of ±1? Of ±2? Of ±3?
Answer: 68.26%, 95.44%, and 99.74%, respectively.

FINDING MEANS AND STANDARD DEVIATIONS

If we know that a distribution is normal, we can calculate the mean μ and the standard deviation σ using percentage information from the population.

EXAMPLE 11.4

Given a normal distribution with a mean of 25, what is the standard deviation if 18% of the values are above 29?

Answer: Looking for a .82 probability in Table A, we note that the corresponding *z*-score is 0.92. Thus 29 − 25 = 4 is equal to a standard deviation of 0.92, that is, $0.92\sigma = 4$, and $\sigma = \frac{4}{0.92} = 4.35$.

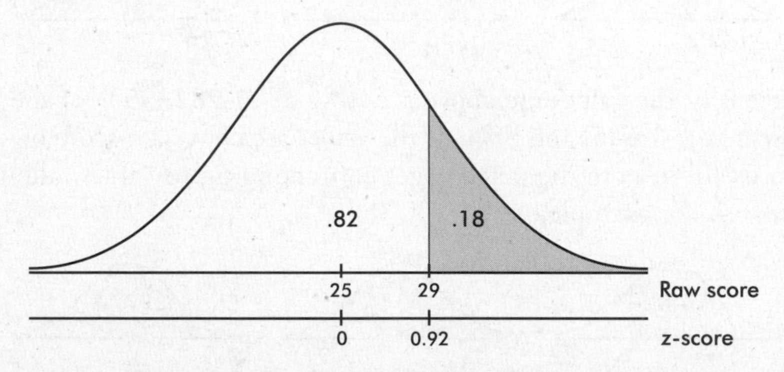

EXAMPLE 11.5

Given a normal distribution with a standard deviation of 10, what is the mean if 21% of the values are below 50?

Answer: Looking for a .21 probability in Table A leads us to a *z*-score of −0.81. Thus 50 is −0.81 standard deviation from the mean, and so $\mu = 50 + 0.81(10) = 58.1$.

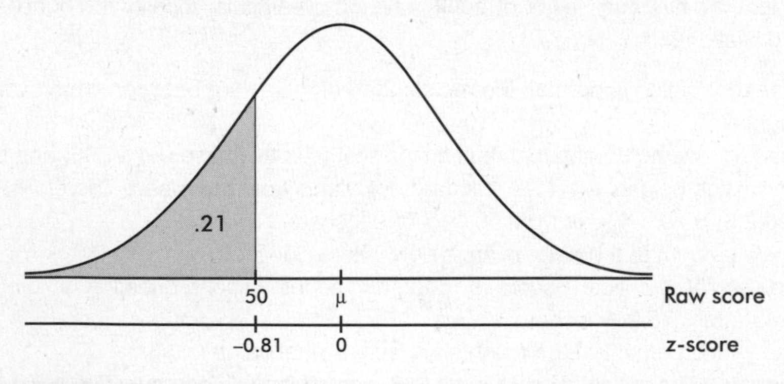

EXAMPLE 11.6

Given a normal distribution with 80% of the values above 125 and 90% of the values above 110, what are the mean and standard deviation?

Answer: Table A gives critical *z*-scores of –0.84 and –1.28. Thus we have $\frac{125-\mu}{\sigma} = -0.84$ and $\frac{110-\mu}{\sigma} = -1.28$. Solving the system $\{125 - \mu = -0.84\sigma,\ 110 - \mu = -1.28\sigma\}$ simultaneously gives $\mu = 153.64$ and $\sigma = 34.09$.

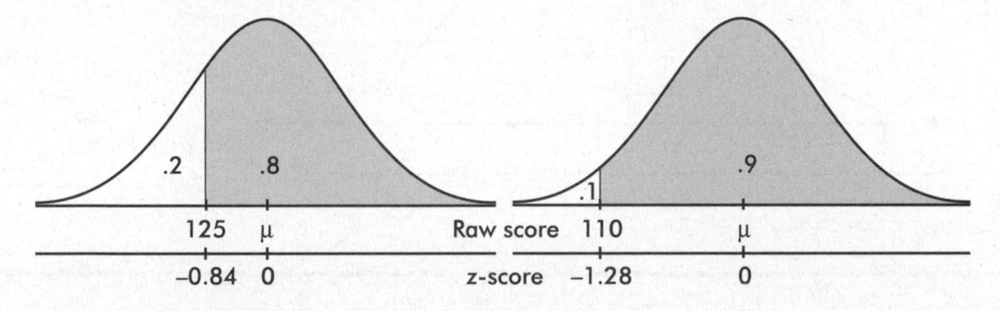

NORMAL APPROXIMATION TO THE BINOMIAL

Many practical applications of the binomial involve examples in which *n* is large. However, for large *n*, binomial probabilities can be quite tedious to calculate. Since the normal can be viewed as a limiting case of the binomial, it is natural to use the normal to approximate the binomial in appropriate situations.

TIP

When *n* is small, you cannot use the normal model to approximate the binomial model.

The binomial takes values only at integers, while the normal is continuous with probabilities corresponding to areas over intervals. Therefore, we establish a technique for converting from one distribution to the other. For approximation purposes we do as follows. Each binomial probability corresponds to the normal probability over a unit interval centered at the desired value. Thus, for example, to approximate the binomial probability of eight successes we determine the normal probability of being between 7.5 and 8.5.

EXAMPLE 11.7

Suppose that 15% of the cars coming out of an assembly plant have some defect. In a delivery of 40 cars what is the probability that exactly 5 cars have defects?

Answer: The actual answer is $\left(\frac{40!}{35!5!}\right)(.15)^5(.85)^{35}$, but clearly this involves a nontrivial calculation. [If one has a calculator such as the TI-84, then binompdf(40, .15, 5) = .1692.] To approximate the answer using the normal, we first calculate the mean μ and the standard deviation σ as follows:

$$\mu = np = 40(.15) = 6$$
$$\sigma = \sqrt{np(1-p)} = \sqrt{40(.15)(.85)} = 2.258$$

(continued)

We then calculate the appropriate *z*-scores: $\frac{4.5-6}{2.258} = -0.66$, and $\frac{5.5-6}{2.258} = -0.22$. Looking up the corresponding probabilities in Table A, we obtain a final answer of .4129 – .2546 = .1583. (The actual answer is .1692.)

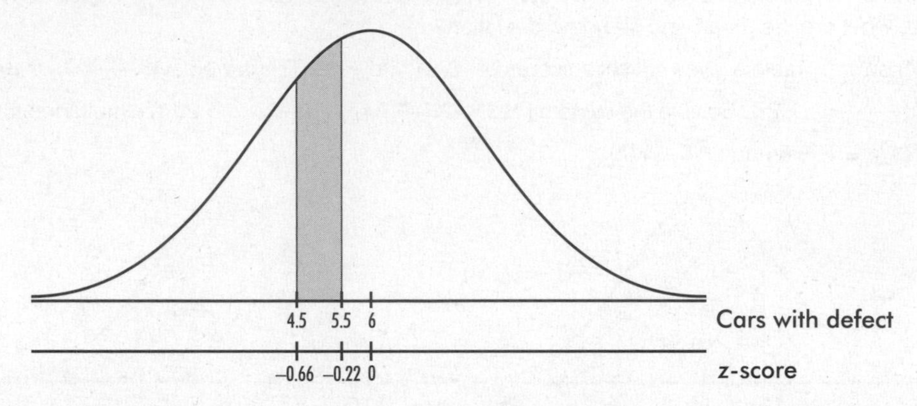

Even more useful are approximations relating to probabilities over intervals.

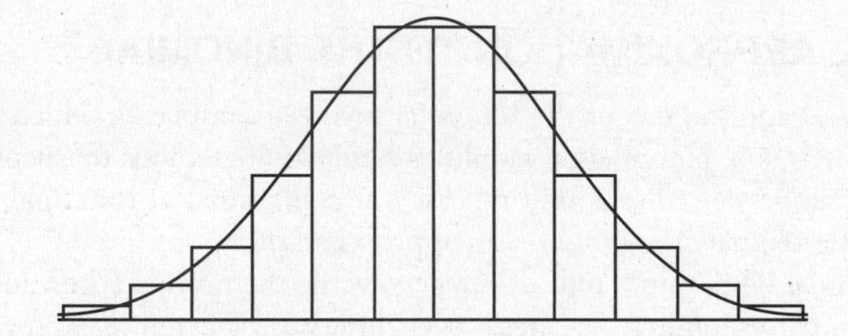

EXAMPLE 11.8

If 60% of the population support massive federal budget cuts, what is the probability that in a survey of 250 people at most 155 people support such cuts?

Answer: The actual answer is the sum of 156 binomial expressions:

$$(.4)^{250} + \cdots + \frac{250!}{155!95!}(.6)^{155}(.4)^{95}$$

However, a good approximation can be obtained quickly and easily by using the normal. We calculate μ and σ:

$$\mu = np = 250(.6) = 150$$

$$\sigma = \sqrt{np(1-p)} = \sqrt{250(.6)(.4)} = 7.746$$

The binomial of at most 155 successes corresponds to the normal probability of ≤155.5. The *z*-score of 155.5 is $\frac{155.5-150}{7.746} = 0.71$. Using Table A leads us to a final answer of .7611. [Or binomcdf(250, .6, 155) = .7605.]

(continued)

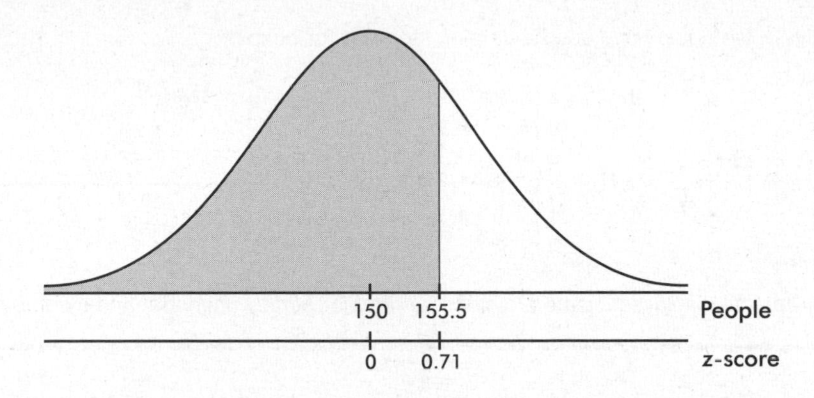

Is the normal a good approximation? The answer, of course, depends on the error tolerances in particular situations. A general rule of thumb is that the normal is a good approximation to the binomial whenever both np and $n(1 - p)$ are greater than 10. (Some authors use 5 instead of 10 here.)

CHECKING NORMALITY

In the examples in this Topic we have assumed the population has a normal distribution. When you collect your own data, before you can apply the procedures developed above, you must decide whether it is reasonable to assume the data come from a normal population. In later Topics we will have to check for normality before applying certain inference procedures.

The initial check should be to draw a picture. Dotplots, stemplots, boxplots, and histograms are all useful graphical displays to show that data is unimodal and roughly symmetric.

EXAMPLE 11.9

The ages at inauguration of U.S. presidents from Washington to G. W. Bush were: {57, 61, 57, 57, 58, 57, 61, 54, 68, 51, 49, 64, 50, 48, 65, 52, 56, 46, 54, 49, 51, 47, 55, 55, 54, 42, 51, 56, 55, 51, 54, 51, 60, 61, 43, 55, 56, 61, 52, 69, 64, 46, 54}. Can we conclude that the distribution is roughly normal?

Answer: Entering the 43 data points in a graphing calculator gives the following histogram:

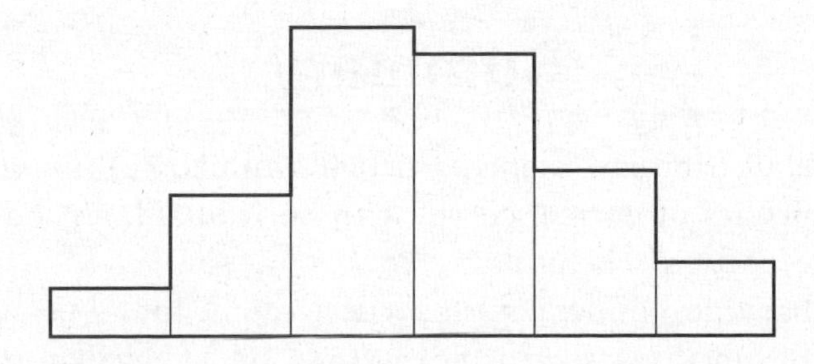

(continued)

Alternatively, we could have used a dotplot, stemplot, or boxplot:

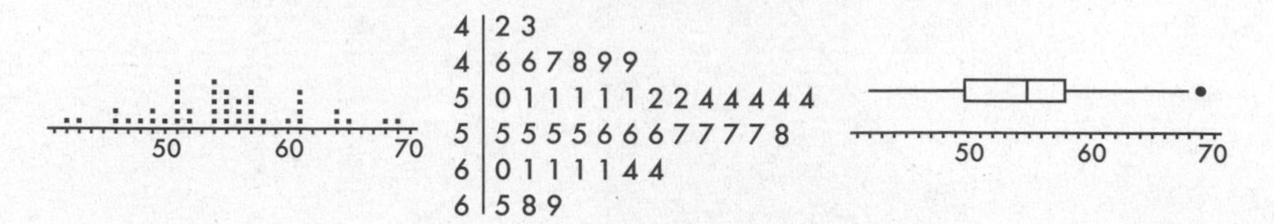

All the graphical displays indicate a distribution that is *roughly* unimodal and symmetric.

A more specialized graphical display to check normality is the *normal probability plot.* When the data distribution is roughly normal, the plot is roughly a diagonal straight line. While this plot more clearly shows deviations from normality, it is not as easy to understand as a histogram. The normal probability plot is difficult to calculate by hand; however, technology such as the TI-84 readily plots the graph. (On the TI-84, in STATPLOT the sixth choice in TYPE gives the normal probability plot.)

The normal probability plot for the data in Example 11.9 is:

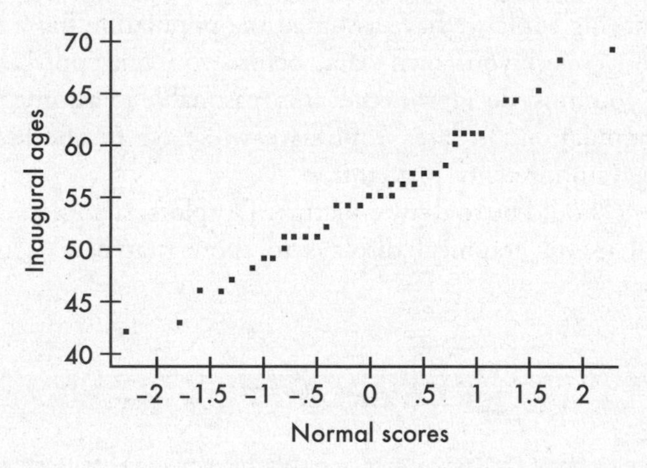

A graph this close to a diagonal straight line indicates that the data have a distribution very close to normal.

Note that alternatively, one could have plotted the data (in this case, inaugural ages) on the *x*-axis and the normal scores on the *y*-axis.

Summary

- The normal distribution is one particular symmetric, unimodal distribution.
- Drawing pictures of normal curves with horizontal lines showing raw scores and *z*-scores is usually helpful.
- Areas (probabilities) under a normal curve can be found using the given table or calculator functions, such as `normalcdf` on TI calculators.
- Critical values corresponding to given probabilities can be found using the given table or calculator functions, such as `invNorm` on TI calculators.
- The normal probability plot is very useful in gauging whether a given distribution of data is approximately normal: If the plot is nearly straight, the data is nearly normal.

Questions on Topic Eleven: The Normal Distribution

Multiple-Choice Questions

Directions: The questions or incomplete statements that follow are each followed by five suggested answers or completions. Choose the response that best answers the question or completes the statement.

1. Which of the following is a true statement?

   (A) The area under a normal curve is always equal to 1, no matter what the mean and standard deviation are.
   (B) All bell-shaped curves are normal distributions for some choice of μ and σ.
   (C) The smaller the standard deviation of a normal curve, the lower and more spread out the graph.
   (D) Depending upon the value of the standard deviation, normal curves with different means may be centered around the same number.
   (E) Depending upon the value of the standard deviation, the mean and median of a particular normal distribution may be different.

2. Which of the following is a true statement?

   (A) The area under the standard normal curve between 0 and 2 is twice the area between 0 and 1.
   (B) The area under the standard normal curve between 0 and 2 is half the area between −2 and 2.
   (C) For the standard normal curve, the interquartile range is approximately 3.
   (D) For the standard normal curve, the range is 6.
   (E) For the standard normal curve, the area to the left of 0.1 is the same as the area to the right of 0.9.

3. Populations P1 and P2 are normally distributed and have identical means. However, the standard deviation of P1 is twice the standard deviation of P2. What can be said about the percentage of observations falling within two standard deviations of the mean for each population?

   (A) The percentage for P1 is twice the percentage for P2.
   (B) The percentage for P1 is greater, but not twice as great, as the percentage for P2.
   (C) The percentage for P2 is twice the percentage for P1.
   (D) The percentage for P2 is greater, but not twice as great, as the percentage for P1.
   (E) The percentages are identical.

4. Consider the following two normal curves:

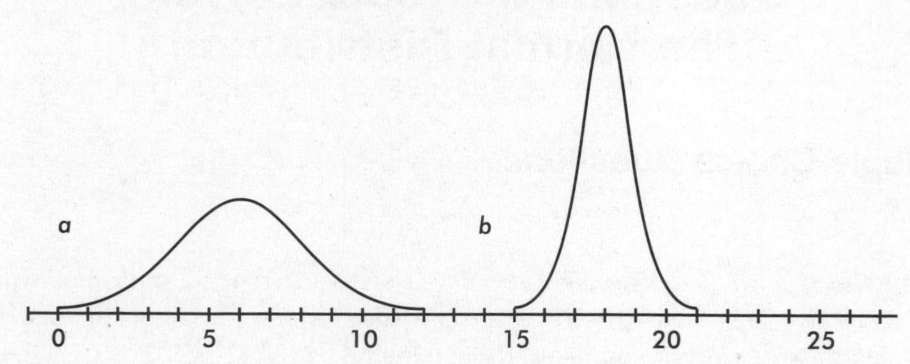

Which has the larger mean and which has the larger standard deviation?

(A) Larger mean, *a*; larger standard deviation, *a*
(B) Larger mean, *a*; larger standard deviation, *b*
(C) Larger mean, *b*; larger standard deviation, *a*
(D) Larger mean, *b*; larger standard deviation, *b*
(E) Larger mean, *b*; same standard deviation

In Questions 5–15, assume the given distributions are normal.

5. A trucking firm determines that its fleet of trucks averages a mean of 12.4 miles per gallon with a standard deviation of 1.2 miles per gallon on cross-country hauls. What is the probability that one of the trucks averages fewer than 10 miles per gallon?

(A) .0082
(B) .0228
(C) .4772
(D) .5228
(E) .9772

6. A factory dumps an average of 2.43 tons of pollutants into a river every week. If the standard deviation is 0.88 tons, what is the probability that in a week more than 3 tons are dumped?

(A) .2578
(B) .2843
(C) .6500
(D) .7157
(E) .7422

7. An electronic product takes an average of 3.4 hours to move through an assembly line. If the standard deviation is 0.5 hour, what is the probability that an item will take between 3 and 4 hours?

(A) .2119
(B) .2295
(C) .3270
(D) .3811
(E) .6730

8. The mean score on a college placement exam is 500 with a standard deviation of 100. Ninety-five percent of the test takers score above what?

(A) 260
(B) 336
(C) 405
(D) 414
(E) 664

9. The average noise level in a restaurant is 30 decibels with a standard deviation of 4 decibels. Ninety-nine percent of the time it is below what value?

(A) 20.7
(B) 32.0
(C) 33.4
(D) 37.8
(E) 39.3

10. The mean income per household in a certain state is $9500 with a standard deviation of $1750. The middle 95% of incomes are between what two values?

(A) $5422 and $13,578
(B) $6070 and $12,930
(C) $6621 and $12,379
(D) $7260 and $11,740
(E) $8049 and $10,951

11. One company produces movie trailers with mean 150 seconds and standard deviation 40 seconds, while a second company produces trailers with mean 120 seconds and standard deviation 30 seconds. What is the probability that two randomly selected trailers, one produced by each company, will combine to less than three minutes?

(A) .000
(B) .036
(C) .099
(D) .180
(E) .405

12. Jay Olshansky from the University of Chicago was quoted in *Chance News* as arguing that for the average life expectancy to reach 100, 18% of people would have to live to 120. What standard deviation is he assuming for this statement to make sense?

(A) 21.7
(B) 24.4
(C) 25.2
(D) 35.0
(E) 111.1

13. Cucumbers grown on a certain farm have weights with a standard deviation of 2 ounces. What is the mean weight if 85% of the cucumbers weigh less than 16 ounces?

    (A) 13.92
    (B) 14.30
    (C) 14.40
    (D) 14.88
    (E) 15.70

14. If 75% of all families spend more than $75 weekly for food, while 15% spend more than $150, what is the mean weekly expenditure and what is the standard deviation?

    (A) $\mu = 83.33, \sigma = 12.44$
    (B) $\mu = 56.26, \sigma = 11.85$
    (C) $\mu = 118.52, \sigma = 56.26$
    (D) $\mu = 104.39, \sigma = 43.86$
    (E) $\mu = 139.45, \sigma = 83.33$

15. A coffee machine can be adjusted to deliver any fixed number of ounces of coffee. If the machine has a standard deviation in delivery equal to 0.4 ounce, what should be the mean setting so that an 8-ounce cup will overflow only 0.5% of the time?

    (A) 6.97 ounces
    (B) 7.22 ounces
    (C) 7.34 ounces
    (D) 7.80 ounces
    (E) 9.03 ounces

16. Assume that a baseball team has an average pitcher, that is, one whose probability of winning any decision is .5. If this pitcher has 30 decisions in a season, what is the probability that he will win at least 20 games?

    (A) .0505
    (B) .2514
    (C) .2743
    (D) .3333
    (E) .4300

17. Given that 10% of the nails made using a certain manufacturing process have a length less than 2.48 inches, while 5% have a length greater than 2.54 inches, what are the mean and standard deviation of the lengths of the nails? Assume that the lengths have a normal distribution.

    (A) $\mu = 2.506, \sigma = 0.0205$
    (B) $\mu = 2.506, \sigma = 0.0410$
    (C) $\mu = 2.516, \sigma = 0.0825$
    (D) $\mu = 2.516, \sigma = 0.1653$
    (E) The mean and standard deviation cannot be computed from the information given.

Free-Response Questions

Directions: You must show all work and indicate the methods you use. You will be graded on the correctness of your methods and on the accuracy of your final answers.

Two Open-Ended Questions

1. The time it takes Steve to walk to school follows a normal distribution with mean 30 minutes and standard deviation 5 minutes, while the time it takes Jan to walk to school follows a normal distribution with mean 25 minutes and standard deviation 4 minutes. Assume their walking times are independent of each other.
   (a) If they leave at the same time, what is the probability that Steve arrives before Jan?
   (b) How much earlier than Jan should Steve leave so that he has a 90% chance of arriving before Jan?
2. (a) Two components are in series so the failure of either will cause the system to fail. The time to failure for a new component is normally distributed with a mean of 3000 hours and a standard deviation of 400 hours. One of the components has already run for 2500 hours, while the other has run for 2800 hours. Assuming component failures are independent, what is the probability the system survives for 10 more days (240 hours)?
   (b) If the components are put in parallel, the system will fail only if both components fail. If the two components in (a) are put in parallel, what is the probability that the system survives for 10 more days?

Two Investigative Tasks

1. When Michael Jordan came up to try for a sixth free throw after having made five straight free throws, the announcer commented that the law of averages would be working against Jordan.

   (a) What did the announcer mean? Was this a correct interpretation of probability? Explain.

   (b) Jordan makes 90% of his free throws. What is the probability that he will make six straight free throws? That he will make five straight free throws and then miss the next? That he will make the next free throw given that he has made the last five in a row?

   (c) How could you set up a simulation using a random number table to analyze the situation the announcer was commenting on?

   (d) If Jordan shoots six times from the free throw line, what are the mean and standard deviation for the number of shots he is expected to make?

   (e) If Jordan makes only 35 out of 40 free throw tries during the playoffs, is this sufficient evidence that the probability of his making a free throw is really below .90? Explain.

2. Many people have lactose intolerance leading to cramps and diarrhea when they eat dairy products. Substantial relief can be obtained by taking dietary supplements of lactase, an enzyme. One product consists of caplets containing a mean of 9000 FCC lactase units with a standard deviation of 590 units and a normal distribution of lactase units. A person with severe lactose intolerance may need to take two such tablets whenever they eat dairy products. Tablets with under 8500 FCC lactase units can produce noticeably less relief.

   (a) Determine the probability that a caplet has less than 8500 FCC lactase units. Round off to the nearest tenth.

   (b) Using the random number table below, run five simulations for the number of tablets a lab technician samples before finding two with less than 8500 units each.

   84177 06757 17613 15582 51506 81435 41050 92031 06449 05059
   59884 31180 53115 84469 94868 57967 05811 84514 75011 13006
   63395 55041 15866 06589 13119 71020 85940 91932 06488 74987
   54355 52704 90359 02649 47496 71567 94268 08844 26294 64759
   08989 57024 97284 00637 89283 03514 59195 07635 03309 72605
   29357 23737 67881 03668 33876 35841 52869 23114 15864 38942

(c) The results of two 100-trial simulations, one looking for two tablets each less than 8500 units and one looking for two tablets each greater than 9000 units, are shown below. Which distribution is which? Explain your answer.

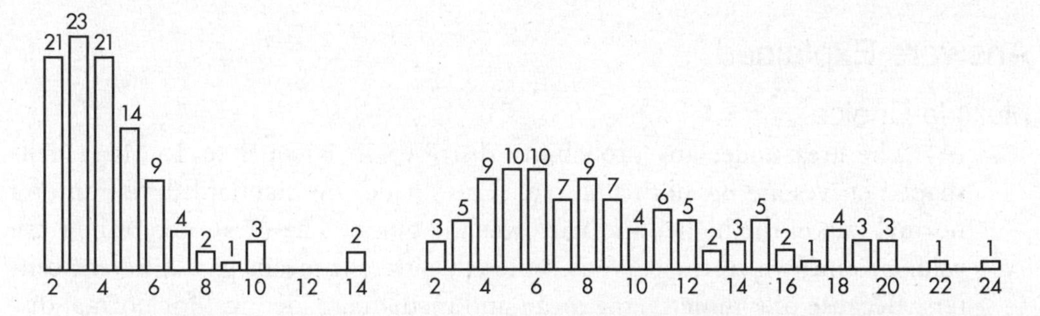

(d) Using the correct barplot above, estimate the expected number of caplets a laboratory will sample before finding two with less than 8500 units each.

Answer Key

1. **A**	5. **B**	9. **E**	13. **A**	17. **A**
2. **B**	6. **A**	10. **B**	14. **D**	
3. **E**	7. **E**	11. **B**	15. **A**	
4. **C**	8. **B**	12. **A**	16. **A**	

Answers Explained

Multiple-Choice

1. **(A)** The area under any probability distribution is equal to 1. Many bell-shaped curves are *not* normal curves. The smaller the standard deviation of a normal curve, the higher and narrower the graph. The mean determines the value around which the curve is centered; different means give different centers. Because of symmetry, the mean and median are identical for normal distributions.

2. **(B)** Statement (B) is true by symmetry of the normal curve, however .4772 is not twice .3413, 0.67 – (–0.67) is not 3, the range is not finite, and the $P(z < 0.1)$ is more than .5 while $P(z > 0.9)$ is less than .5.

3. **(E)** All normal distributions have about 95% of their observations within two standard deviations of the mean.

4. **(C)** Curve *a* has mean 6 and standard deviation 2, while curve *b* has mean 18 and standard deviation 1.

5. **(B)** The *z*-score of 10 is $\frac{10-12.4}{1.2} = -2$. From Table A, to the left of –2 is an area of .0228. [normalcdf(0, 10, 12.4, 1.2) = .0228.]

6. **(A)** The *z*-score of 3 is $\frac{3-2.43}{0.88} = 0.65$ From Table A, to the left of 0.65 is an area of .7422, and so to the right must be 1 – .7422 = .2578. [normalcdf(3, 1000, 2.43, .88) = .2589.]

7. **(E)** The *z*-scores of 3 and 4 are $\frac{3-3.4}{0.5} = -0.8$ and $\frac{4-3.4}{0.5} = 1.2$, respectively. From Table A, to the left of –0.8 is an area of .2119, and to the left of 1.2 is an area of .8849. Thus between 3 and 4 is an area of .8849 – .2119 = .6730. [normalcdf(3, 4, 3.4, .5) = .6731.]

8. **(B)** If 95% of the area is to the right of a score, 5% is to the left. Looking for .05 in the body of Table A, we note the *z*-score of –1.645. Converting this to a raw score gives 500 – 1.645(100) = 336. [invNorm(.05, 500, 100) = 336.]

9. **(E)** The critical *z*-score associated with 99% to the left is 2.326, and 30 + 2.326(4) = 39.3. [invNorm(.99, 30, 4) = 39.3.]

10. **(B)** The critical z-scores associated with the middle 95% are ±1.96, and 9500 ± 1.96(1750) = 6070 and 12,930. [invNorm(.025, 9500, 1750) = 6070 and invNorm(.975, 9500, 1750) = 12,930.]

11. **(B)** $\mu_{X+Y} = 150 + 120 = 270$ and $\sigma_{X+Y} = \sqrt{40^2 + 30^2} = 50$. Then

$$P\left(z < \frac{180-270}{50}\right) = .036$$

12. **(A)** The critical z-score associated with 18% to the right or 82% to the left is 0.92. Then $100 + 0.92\sigma = 120$ gives $\sigma = 21.7$.

13. **(A)** The critical z-score associated with 85% to the left is 1.04. Then $\mu + 1.04(2) = 16$ gives $\mu = 13.92$.

14. **(D)** The critical z-scores associated with 75% to the right (25% to the left) and with 15% to the right (85% to the left) are –0.67 and 1.04, respectively. Then $\{\mu - 0.67\sigma = 75, \mu + 1.04\sigma = 150\}$ gives $\mu = 104.39$ and $\sigma = 43.86$.

15. **(A)** The critical z-score associated with .5% to the right (99.5% to the left) is 2.576. Then $c + 2.576(0.4) = 8$ gives $c = 6.97$.

16. **(A)** Using the normal as an approximation to the binomial, we have $\mu = 30(.5) = 15$, $\sigma^2 = 30(.5)(.5) = 7.5$, $\sigma = 2.739$, $\frac{19.5-15}{2.739} = 1.64$, and $1 - .9495 = .0505$. [Or binomcdf(30, .5, 10) = .0494.]

17. **(A)** The critical z-scores for 10% to the left and 5% to the right are –1.282 and 1.645, respectively. Then $\{\mu - 1.282\sigma = 2.48, \mu + 1.645\sigma = 2.54\}$ gives $\mu = 2.506$ and $\sigma = 0.0205$.

Free-Response

1. (a) Let the random variable D be the difference in walking times (Steve–Jan). Then $\mu_D = 30 - 25 = 5$ and $\sigma_D = \sqrt{5^2 + 4^2} = 6.403$. For Steve to arrive before Jan, the difference in walking times must be < 0.

$$P(D < 0) = P\left(z < \frac{0-5}{6.403}\right) = .217$$

(b) If m is the number of minutes he should leave early, then the new mean of the differences is $5 - m$, and we want $P\left(z < \frac{0-(5-m)}{6.403}\right) = .9$, which gives $\frac{-5+m}{6.403}$ and $m = 13.2$ minutes.

2. (a) $$P(2740|2500) = \frac{P\left(z > \frac{2740-3000}{400}\right)}{P\left(z > \frac{2500-3000}{400}\right)} = \frac{0.742154}{0.894350} = 0.829825 \text{ and}$$

$$P(3040|2800) = \frac{P\left(z > \frac{3040-3000}{400}\right)}{P\left(z > \frac{2800-3000}{400}\right)} = \frac{0.460172}{0.691462} = 0.665506.$$

Given independence, the probability that both components last 240 more hours is (0.829825)(0.665506) = 0.552.

(b) The probability that both fail is (1 – 0.829825)(1 – 0.665506) = 0.056923.
The probability that at least one doesn't fail is $1 - 0.056923 \approx 0.943$.

Investigative Tasks

1. (a) The announcer meant that to maintain his tabulated free throw percentage, Jordan would have to soon miss one. This is not a correct use of probability. If Jordan makes a certain percentage of free throws, that probability applies to each throw irrespective of the previous throws. By the law of large numbers, in the long run the relative frequency tends toward the correct probability, but no conclusion is possible about any given outcome.

(b) P(six in a row) = $(.9)^6$ = .531
P(five, then a miss) = $(.9)^5(.1)$ = .059
P(next|previous five) = P(next) = .9

(c) Let the digits 1 through 9 stand for making a free throw, while 0 stands for a miss. Look at blocks of five random numbers. If all stand for "makes" (no 0), look at the very next digit to see if it is a 0 or not. Keep a tally of how many times five makes in a row are followed by a make and how many times five makes in a row are followed by a miss.

(d) This is a binomial distribution with n = 6 and p = .9, and so the mean is np = 6(9) = 5.4 while the standard deviation is $\sqrt{np(1-p)}$ $\sqrt{6(.9)(.1)} = .735$.

(e) Since nq = 40(1 – .90) = 4, the normal approximation to the binomial is not recommended. However, a direct binomial calculation yields

$$\begin{aligned} P(\leq 35) &= 1 - P(\geq 36) \\ &= 1 - \left[\binom{40}{36}(.9)^{36}(.1)^4 + \binom{40}{37}(.9)^{37}(.1)^3 \right. \\ &\quad \left. + \binom{40}{38}(.9)^{38}(.1)^2 + \binom{40}{39}(.9)^{39}(.1) + (.9)^{40}\right] \\ &\approx .37 \end{aligned}$$

[Or simply use binomcdf(40, .9, 35) on the TI-84.] With such a high probability, there is no evidence to conclude that Jordan's average has dropped!

2. (a) With $z = \frac{8500-9000}{590} = -0.85$, Table A gives a probability of $.1977 \approx .2$.

   (b) For example, letting 0 and 1 represent caplets containing less than 8500 units and 2 through 9 represent caplets containing more than 8500 units, we can read off digits until we find two containing less than 8500 caplets. Five simulations would yield

   841770 67571761 31558251 50681 435410

   giving 6, 8, 8, 5, and 6 tablets sampled before finding two with less than 8500 units apiece.

   (c) The probability of a caplet having more than 9000 units is .5, and thus one would expect to find two caplets with more than 9000 units much quicker than finding two caplets with less than 8500 units. Thus the first histogram results from looking for two caplets with more than 9000 units and the second histogram results from looking for two caplets with less than 8500 units.

   (d) An estimate for the expected value is obtained by

$$\sum xP(x) = 2(.03) + 3(.05) + 4(.09) + 5(.10) + \cdots + 24(.01) = 9.44.$$

Sampling Distributions

TOPIC 12

- Sample Proportion
- Sample Mean
- Central Limit Theorem
- Difference Between Two Independent Sample Proportions
- Difference Between Two Independent Sample Means
- The *t*-distribution
- The Chi-Square Distribution
- The Standard Error

TIP

A sampling distribution is not the same thing as the distribution of a sample.

The *population* is the complete set of items of interest. A *sample* is a part of a population used to represent the population. The population mean μ and population standard deviation σ are examples of *population parameters.* The sample mean $\bar{x}$ and the sample standard deviation s are examples of *statistics.* Statistics are used to make inferences about population parameters. While a population parameter is a fixed quantity, statistics vary depending on the particular sample chosen. The probability distribution showing how a statistic varies is called a *sampling distribution.* The sampling distribution is *unbiased* if its mean is equal to the associated population parameter.

We want to know some important truth about a population, but in practical terms this truth is unknowable. What's the average adult human body weight? What proportion of people have high cholesterol? What we can do is carefully collect data from as large and representative a group of individuals as possible and then use this information to estimate the value of the population parameter. How close are we to the truth? We know that different samples would give different estimates, and so sampling error is unavoidable. What is wonderful, and what we will learn in this Topic, is that we can quantify this sampling error! We can make statements like "the average weight must almost surely be within 5 pounds of 176 pounds" or "the proportion of people with high cholesterol must almost surely be within ±3% of 37%."

SAMPLING DISTRIBUTION OF A SAMPLE PROPORTION

Whereas the mean is basically a quantitative measurement, the proportion represents essentially a qualitative approach. The interest is simply in the presence or absence of some attribute. We count the number of yes responses and form a proportion. For example, what proportion of drivers wear seat belts? What proportion of SCUD missiles can be intercepted? What proportion of new stereo sets have a certain defect?

This separation of the population into "haves" and "have-nots" suggests that we can make use of our earlier work on binomial distributions. We also keep in mind that, when n (trials, or in this case sample size) is large enough, the binomial can be approximated by the normal.

In this topic we are interested in estimating a population proportion p by considering a single sample proportion $\hat{p}$. This sample proportion is just one of a whole universe of sample proportions, and to judge its significance we must know how sample proportions vary. Consider the set of proportions from all possible samples of a specified size n. It seems reasonable that these proportions will cluster around the population proportion (the sample proportion is an unbiased statistic of the population proportion) and that the larger the chosen sample size, the tighter the clustering.

How do we calculate the mean and standard deviation of the set of population proportions? Suppose the sample size is n and the actual population proportion is p. From our work on binomial distributions, we remember that the mean and standard deviation for the number of successes in a given sample are pn and $\sqrt{np(1-p)}$, respectively, and for large n the complete distribution begins to look "normal."

Here, however, we are interested in the proportion rather than in the number of successes. From Topic Two we remember that when we multiply or divide every element by a constant, we multiply or divide both the mean and the standard deviation by the same constant. In this case, to change number of successes to proportion of successes, we divide by n:

$$\mu_{\hat{p}} = \frac{pn}{n} = p \quad \text{and} \quad \sigma_{\hat{p}} = \frac{\sqrt{np(1-p)}}{n} = \sqrt{\frac{p(1-p)}{n}}$$

Furthermore, if each element in an approximately normal distribution is divided by the same constant, it is reasonable that the result will still be an approximately normal distribution.

Thus the principle forming the basis of the following discussion is

> Start with a population with a given proportion p. Take all samples of size n. Compute the proportion in each of these samples. Then
>
> 1. the set of all sample proportions is approximately normally distributed (often stated: the distribution of sample proportions is approximately normal).
> 2. the mean $\mu_{\hat{p}}$ of the set of sample proportions equals p, the population proportion.
> 3. the standard deviation $\sigma_{\hat{p}}$ of the set of sample proportions is approximately equal to $\sqrt{\frac{p(1-p)}{n}}$.
>
> Alternatively, we say that the sampling distribution of $\hat{p}$ is approximately normal with mean p and standard deviation $\sqrt{\frac{p(1-p)}{n}}$.

Since we are using the normal approximation to the binomial, both np and $n(1 - p)$ should be at least 10. Furthermore, in making calculations and drawing conclusions from a specific sample, it is important that the sample be a *simple random sample.*

Finally, because sampling is usually done without replacement, the sample cannot be too large; the sample size n should be no larger than 10% of the population. (We're actually worried about *independence*, but randomly selecting a relatively small sample allows us to assume independence. Of course, it's always better to have larger samples—it's just that if the sample is large relative to the population, then the proper inference techniques are different from those taught in introductory statistics classes.)

EXAMPLE 12.1

Suppose that 70% of all dialysis patients will survive for at least 5 years. In a simple random sample (SRS) of 100 new dialysis patients, what is the probability that the proportion surviving for at least 5 years will exceed 80%?

Answer: Both $np = (100)(.7) = 70 > 10$ and $n(1 - p) = (100)(.3) = 30 > 10$, and our sample is clearly less than 10% of all dialysis patients. So the set of sample proportions is approximately normally distributed with mean .70 and standard deviation

$$\sigma_{\hat{p}} = \sqrt{\frac{(.7)(.3)}{100}} = .0458$$

With a z-score of $\frac{.80-.70}{.0458} = 2.18$, the probability that the sample proportion will exceed 80% is $1 - .9854 = .0146$. [normalcdf(.8, 1, .7, .0458) = .0145.]

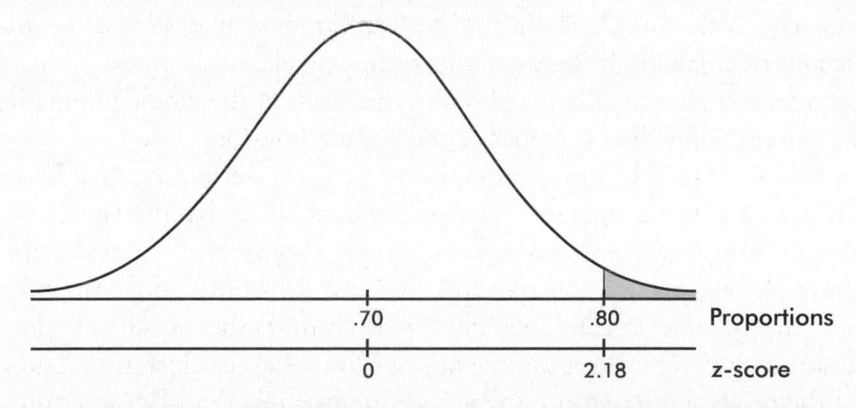

EXAMPLE 12.2

It is estimated that 48% of all motorists use their seat belts. If a police officer observes 400 cars go by in an hour, what is the probability that the proportion of drivers wearing seat belts is between 45% and 55%?

(continued)

Answer: Both $np = (400)(.48) = 192 > 10$ and $n(1 - p) = (400)(.52) = 208 > 10$, and our sample is clearly less than 10% of all motorists. So the set of sample proportions is approximately normally distributed with mean .48 and standard deviation

$$\sigma_{\hat{p}} = \sqrt{\frac{(.48)(.52)}{400}} = .0250$$

The *z*-scores of .45 and .55 are $\frac{.45-.48}{.0250} = -1.2$ and $\frac{.55-.48}{.0250} = 2.8$, respectively. From Table A, the area between .45 and .55 is $.9974 - .1151 = .8823$. Thus there is a .8823 probability that between 45% and 55% of the drivers are wearing seat belts. [normalcdf(.45, .55, .48, .0250) = .8824.]

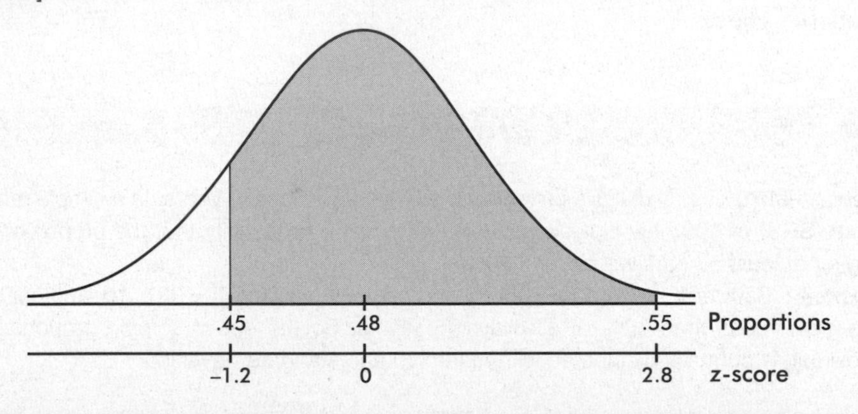

SAMPLING DISTRIBUTION OF A SAMPLE MEAN

Suppose we are interested in estimating the mean μ of a population. For our estimate we could simply randomly pick a single element of the population, but then we would have little confidence in our answer. Suppose instead that we pick 100 elements and calculate their average. It is intuitively clear that the resulting sample mean has a greater chance of being closer to the mean of the whole population than the value for any individual member of the population does.

When we pick a sample and measure its mean $\bar{x}$, we are finding exactly one sample mean out of a whole universe of sample means. To judge the significance of a single sample mean, we must know how sample means vary. Consider the set of means from all possible samples of a specified size. It is both apparent and reasonable that the sample means are clustered around the mean of the whole population; furthermore, these sample means have a tighter clustering than the elements of the original population. In fact, we might guess that the larger the chosen sample size, the tighter the clustering.

How do we calculate the standard deviation $\sigma_{\bar{x}}$ of the set of sample means? Suppose the variance of the population is σ^2 and we are interested in samples of size n. Sample means are obtained by first summing together n elements and then dividing by n. A set of sums has a variance equal to the sum of the variances associated with the original sets. In our case, $\sigma^2_{\text{sums}} = \sigma^2 + \cdots + \sigma^2 = n\sigma^2$. When each element of a set is divided by some constant, the new variance is the old one divided by the square of the constant. Since the sample means are obtained by dividing the sums by n, the variance of the sample means is obtained by dividing the variance of the sums by n^2. Thus if $\sigma_{\bar{x}}$ symbolizes the standard deviation of the sample means, we find that

$$\sigma_{\bar{x}}^2 = \frac{\sigma_{\text{sums}}^2}{n^2} = \frac{n\sigma^2}{n^2} = \frac{\sigma^2}{n}$$

In terms of standard deviations, we have $\sigma_{\bar{x}} = \frac{\sigma}{\sqrt{n}}$.

We have shown the following:

> Start with a population with a given mean μ and standard deviation σ. Compute the mean of all samples of size n. Then the mean of the set of sample means will equal μ, the mean of the population, and the standard deviation $\sigma_{\bar{x}}$ of the set of sample means will be approximately equal to $\frac{\sigma}{\sqrt{n}}$, that is, the standard deviation of the whole population divided by the square root of the sample size.

Note that the variance of the set of sample means varies directly as the variance of the original population and inversely as the size of the samples, while the standard deviation of the set of sample means varies directly as the standard deviation of the original population and inversely as the square root of the size of the samples.

EXAMPLE 12.3

Suppose that tomatoes weigh an average of 10 ounces with a standard deviation of 3 ounces and a store sells boxes containing 12 tomatoes each. If customers determine the average weight of the tomatoes in each box they buy, what will be the mean and standard deviation of these averages?

Answer: We have samples of size 12. The mean of these sample means will equal the population mean, 10 ounces. The standard deviation of these sample means will equal $\frac{3}{\sqrt{12}} = 0.866$ ounces.

Note that while giving the mean and standard deviation of the set of sample means, we did not describe the shape of the distribution. If we are also given that the original population is normal, then we can conclude that the set of sample means has a normal distribution.

EXAMPLE 12.4

Suppose that the distribution for total amounts spent by students vacationing for a week in Florida is normally distributed with a mean of $650 and a standard deviation of $120. What is the probability that an SRS of 10 students will spend an average of between $600 and $700?

Answer: The mean and standard deviation of the set of all sample means of size 10 are:

$$\mu_{\bar{x}} = 650 \text{ and } \sigma_{\bar{x}} = \frac{120}{\sqrt{10}} = 37.95$$

(continued)

The z-scores of 600 and 700 are $\frac{600-650}{37.95} = -1.32$ and $\frac{700-650}{37.95} = 1.32$, respectively. Using Table A, we find the desired probability is .9066 – .0934 = .8132. [normalcdf(600, 700, 650, 37.95) = .8123.]

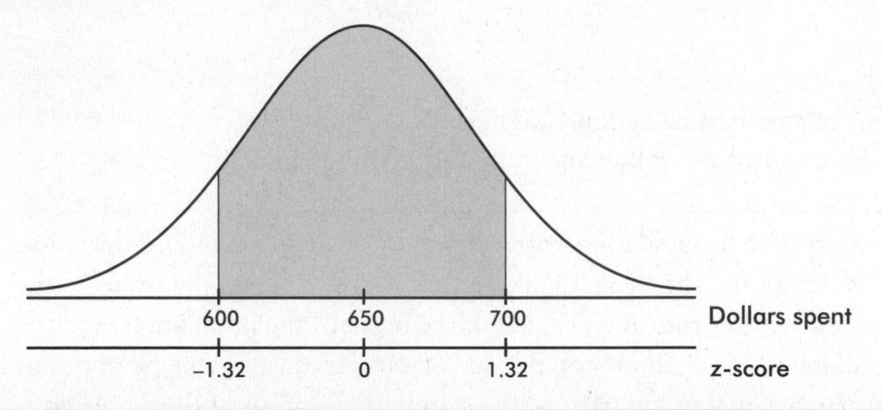

CENTRAL LIMIT THEOREM

We assumed above that the original population had a normal distribution. Unfortunately, few populations are normal, let alone exactly normal. However, it can be shown mathematically that no matter how the original population is distributed, if n is large enough, then the set of sample means is approximately normally distributed. For example, there is no reason to suppose that the amounts of money that different people spend in grocery stores are normally distributed. However, if each day we survey 30 people leaving a store and determine the average grocery bill, these daily averages will have a nearly normal distribution.

The following principle forms the basis of much of what we discuss in this topic and in those following. It is a simplified statement of the *central limit theorem* of statistics.

> Start with a population with a given mean μ, a standard deviation σ, and any shape distribution whatsoever. Pick n sufficiently large (at least 30) and take all samples of size n. Compute the mean of each of these samples. Then
>
> 1. the set of all sample means is approximately normally distributed (often stated: the distribution of sample means is approximately normal).
> 2. the mean of the set of sample means equals μ, the mean of the population.
> 3. the standard deviation $\sigma_{\bar{x}}$ of the set of sample means is approximately equal to $\frac{\sigma}{\sqrt{n}}$, that is, equal to the standard deviation of the whole population divided by the square root of the sample size.
>
> Alternatively, we say that the sampling distribution of $\bar{x}$ is approximately normal with mean μ and standard deviation $\frac{\sigma}{\sqrt{n}}$.

While we mention $n \geq 30$ as a rough rule of thumb, $n \geq 40$ is often used, and n should be chosen even larger if more accuracy is required or if the original population is far from normal. As with proportions, we have the assumptions of a simple random sample and of sample size n no larger than 10% of the population.

EXAMPLE 12.5

Suppose that the average outstanding credit card balance for young couples is $650 with a standard deviation of $420. In an SRS of 100 couples, what is the probability that the mean outstanding credit card balance exceeds $700?

Answer: The sample size is over 30, we have an SRS, and our sample is less than 10% of all couples with outstanding balances, and so by the central limit theorem the set of sample means is approximately normally distributed with mean 650 and standard deviation $\frac{420}{\sqrt{100}} = 42$. With a *z*-score of $\frac{700-650}{42} = 1.19$, the probability that the sample mean exceeds 700 is 1 −.8830 = .1170. [normalcdf(700, 10000, 650, 42) = .1169.]

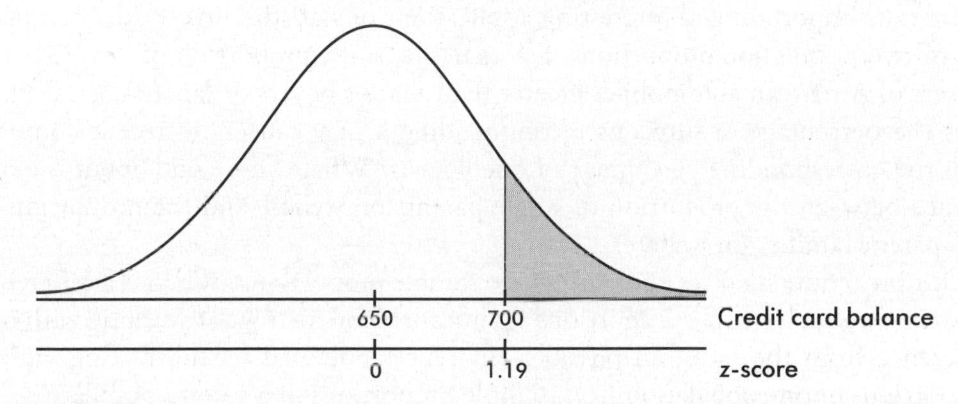

EXAMPLE 12.6

The strength of paper coming from a manufacturing plant is known to be 25 pounds per square inch with a standard deviation of 2.3. In a simple random sample of 40 pieces of paper, what is the probability that the mean strength is between 24.5 and 25.5 pounds per square inch?

Answer: We have a large ($n = 40$) SRS that is still smaller than 10% of all papers coming from the plant. $\mu_{\bar{x}} = 25$ and $\sigma_{\bar{x}} = \frac{2.3}{\sqrt{40}} = 0.364$. The *z*-scores of 24.5 and 25.5 are $\frac{24.5-25}{0.364} = -1.37$ and $\frac{25.5-25}{0.364} = 1.37$, respectively. The probability that the mean strength in the sample is between 24.5 and 25.5 pounds per square inch is .9147 – .0853 = .8294. [normalcdf(24.5, 25.5, 25, .364) = .8304.]

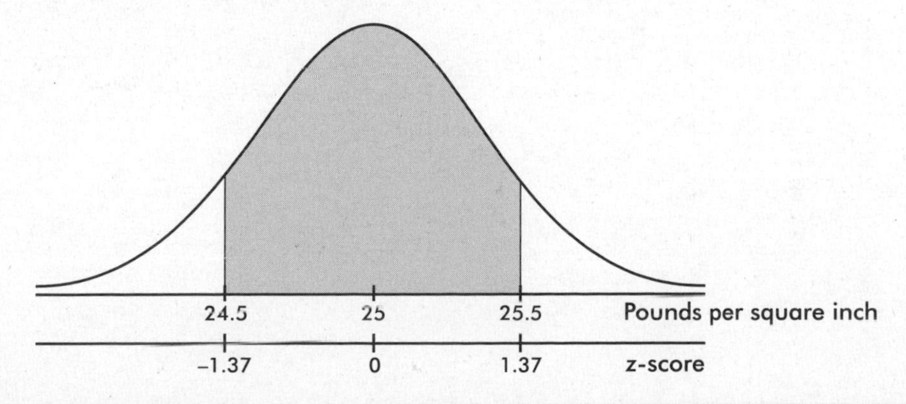

Don't be confused by the several different distributions being discussed! First, there's the distribution of the original population, which may be uniform, bell-shaped, strongly skewed – anything at all. Second, there's the distribution of the data in the sample, and the larger the sample size, the more this will look like the population distribution. Third, there's the distribution of the means of many samples of a given size, and the amazing fact is that this *sampling distribution* can be described by a normal model, regardless of the shape of the original population.

SAMPLING DISTRIBUTION OF A DIFFERENCE BETWEEN TWO INDEPENDENT SAMPLE PROPORTIONS

Numerous important and interesting applications of statistics involve the comparison of two population proportions. For example, is the proportion of satisfied purchasers of American automobiles greater than that of buyers of Japanese cars? How does the percentage of surgeons recommending a new cancer treatment compare with the corresponding percentage of oncologists? What can be said about the difference between the proportion of single parents on welfare and the proportion of two-parent families on welfare?

Our procedure involves comparing two sample proportions. When is a difference between two such sample proportions significant? Note that we are dealing with one difference from the set of all possible differences obtained by subtracting sample proportions of one population from sample proportions of a second population. To judge the significance of one particular difference, we must first determine how the differences vary among themselves. Remember that the variance of a set of differences is equal to the sum of the variances of the individual sets; that is,

$$\sigma_d^2 = \sigma_1^2 + \sigma_2^2$$

Now if

$$\sigma_1 = \sqrt{\frac{p_1(1-p_1)}{n_1}} \quad \text{and} \quad \sigma_2 = \sqrt{\frac{p_2(1-p_2)}{n_2}}$$

then

$$\sigma_d^2 = \frac{p_1(1-p_1)}{n_1} + \frac{p_2(1-p_2)}{n_2} \quad \text{and} \quad \sigma_d = \sqrt{\frac{p_1(1-p_1)}{n_1} + \frac{p_2(1-p_2)}{n_2}}$$

Then we have the following about the sampling distribution of $\hat{p}_1 - \hat{p}_2$:

> Start with two populations with given proportions p_1 and p_2. Take all samples of sizes n_1 and n_2, respectively. Compute the difference $\hat{p}_1 - \hat{p}_2$ of the two proportions in each pair of samples. Then
>
> 1. the set of all differences of sample proportions is approximately normally distributed (alternatively stated: the distribution of differences of sample proportions is approximately normal).
> 2. the mean of the set of differences of sample proportions equals $p_1 - p_2$, the difference of population proportions.
> 3. the standard deviation σ_d of the set of differences of sample proportions is approximately equal to
>
> $$\sqrt{\frac{p_1(1-p_1)}{n_1} + \frac{p_2(1-p_2)}{n_2}}$$

Since we are using the normal approximation to the binomial, n_1p_1, $n_1(1 - p_1)$, n_2p_2, and $n_2(1 - p_2)$ should all be at least 10. Furthermore, in making calculations and drawing conclusions from specific samples, it is important both that the samples be *simple random samples* and that they be taken *independently* of each other. Finally, each sample cannot be too large; the sample sizes should be no larger than 10% of the populations.

EXAMPLE 12.7

A promoter knows that 23% of males enjoy watching boxing matches; however, only 12% of females enjoy watching this sport. In an SRS of 100 men and an independent SRS of 125 women, what is the probability that the difference in the percentages of men and women who enjoy watching boxing is more than 10%?

Answer: We note that $n_1p_1 = 23$, $n_1(1 - p_1) = 77$, $n_2p_2 = 15$, and $n_2(1 - p_2) = 110$ are all >10; we have independent SRSs, and the samples are less than 10% of their respective populations. For the sampling distribution of $\hat{p}_1 - \hat{p}_2$, the mean is .23 – .12 = .11 and the standard deviation is

$$\sqrt{\frac{(.23)(.77)}{100} + \frac{(.12)(.88)}{125}} = .0511$$

The *z*-score of .10 is $\frac{.10-.11}{.0511} = -0.20$. Using Table A, we find 1 – .4207 = .5793 for the probability that the difference is more than 10%. [normalcdf(.10, 1, .11, .0511) = .5776.]

(continued)

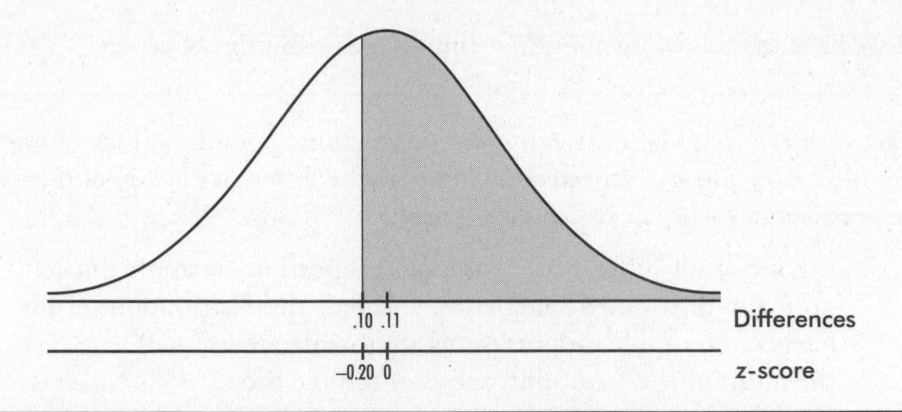

EXAMPLE 12.8

In urban America 43% of married couples own their own homes while only 19% of single people own their own homes. In an SRS of 200 married couples and an independent SRS of 180 single people, the probability is .90 that the difference in percentages of married couples and single people who are homeowners is greater than what percentage?

Answer: We note that $n_1p_1 = 86$, $n_1(1 - p_1) = 114$, $n_2p_2 = 34.2$, and $n_2(1 - p_2) = 145.8$ are all >10; we have independent SRSs, and the samples are less than 10% of their respective populations. For the sampling distribution of $\hat{p}_1 - \hat{p}_2$, the mean is .43 − .19 = .24 and the standard deviation is

$$\sqrt{\frac{(.43)(.57)}{200} + \frac{(.19)(.81)}{180}} = .0456$$

The .90 area to the right corresponds to a *z*-score of −1.282 and thus a difference of .24 − 1.282(.0456) = .182. Thus there is a .90 probability that the difference in percentages is greater than 18.2%. [invNorm(.10, .24, .0456) = .1816.]

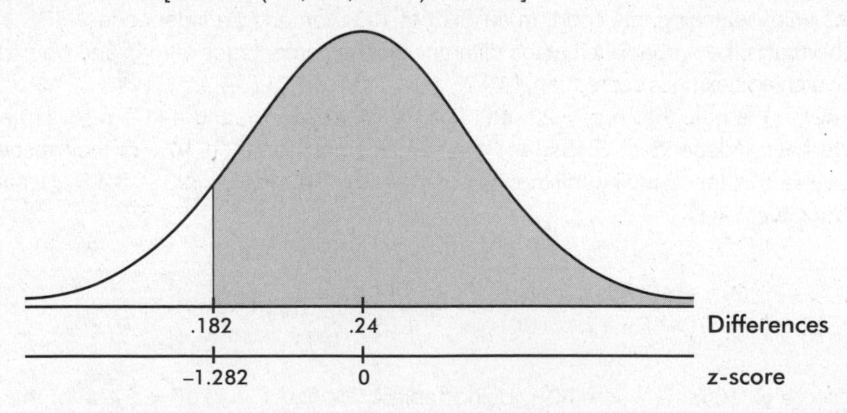

SAMPLING DISTRIBUTION OF A DIFFERENCE BETWEEN TWO INDEPENDENT SAMPLE MEANS

Many real-life applications of statistics involve comparisons of two population means. For example, is the average weight of laboratory rabbits receiving a special diet greater than that of rabbits on a standard diet? Which of two accounting firms pays a higher mean starting salary? Is the life expectancy of a coal miner less than that of a school teacher?

First we consider how to compare the means of samples, one from each population. When is a difference between two such sample means significant? The answer is more apparent when we realize that what we are looking at is one difference from a set of differences. That is, there is the set of all possible differences obtained by subtracting sample means from one set from sample means from a second set. To judge the significance of one particular difference we must first determine how the differences vary among themselves. The necessary key is the fact that the variance of a set of differences is equal to the sum of the variances of the individual sets. Thus,

$$\sigma^2_{\bar{x}_1-\bar{x}_2} = \sigma^2_{\bar{x}_1} + \sigma^2_{\bar{x}_2}$$

Now if

$$\sigma_{\bar{x}_1} = \frac{\sigma_1}{\sqrt{n_1}} \quad \text{and} \quad \sigma_{\bar{x}_2} = \frac{\sigma_2}{\sqrt{n_2}}$$

then

$$\sigma^2_{\bar{x}_1-\bar{x}_2} = \frac{\sigma_1^2}{n_1} + \frac{\sigma_2^2}{n_2} \quad \text{and} \quad \sigma_{\bar{x}_1-\bar{x}_2} = \sqrt{\frac{\sigma_1^2}{n_1} + \frac{\sigma_2^2}{n_2}}$$

Then we have the following about the sampling distribution of $\bar{x}_1 - \bar{x}_2$:

> Start with two normal populations with means μ_1 and μ_2 and standard deviations σ_1 and σ_2. Take all samples of sizes n_1 and n_2, respectively. Compute the difference $\bar{x}_1 - \bar{x}_2$ of the two means in each pair of these samples. Then
>
> 1. the set of all differences of sample means is approximately normally distributed (alternatively stated: the distribution of differences of sample means is approximately normal).
> 2. the mean of the set of differences of sample means equals $\mu_1 - \mu_2$, the difference of population means.
> 3. the standard deviation $\sigma_{\bar{x}_1-\bar{x}_2}$ of the set of differences of sample means is approximately equal to $\sqrt{\frac{\sigma_1^2}{n_1} + \frac{\sigma_2^2}{n_2}}$.

The more either population varies from normal, the greater should be the corresponding sample size. We also have the assumptions of independent simple random samples, and of sample sizes no larger than 10% of the populations.

EXAMPLE 12.9

When fertilizer A is used, the vegetable yield is 4.5 tons per acre with a standard deviation of 0.7 tons, while the yield when fertilizer B is used is 4.3 tons per acre with a standard deviation of 0.4 tons. In 45 sample plots using fertilizer A and 50 plots using fertilizer B, what is the probability that the difference in average yields will be negative, that is, that the average yield for the plots using fertilizer A will be less than the average yield for the plots using fertilizer B?

Answer: We have large ($n_1 = 45$ and $n_2 = 50$) SRSs that are still smaller than 10% of all possible plots. The mean of the differences is 4.5 – 4.3 = 0.2, while the standard deviation is $\sqrt{\frac{(0.7)^2}{45}+\frac{(0.4)^2}{50}} = 0.119$. The *z*-score of a difference of 0 is $\frac{0-0.2}{0.119} = -1.68$. From Table A, the area to the left of –1.68 is .0465. Thus there is a 4.65% chance that the average yield for the sample plots using fertilizer A will be less than the average yield for the sample plots using fertilizer B. [normalcdf(–1000, 0, .2, .119) = .0464.]

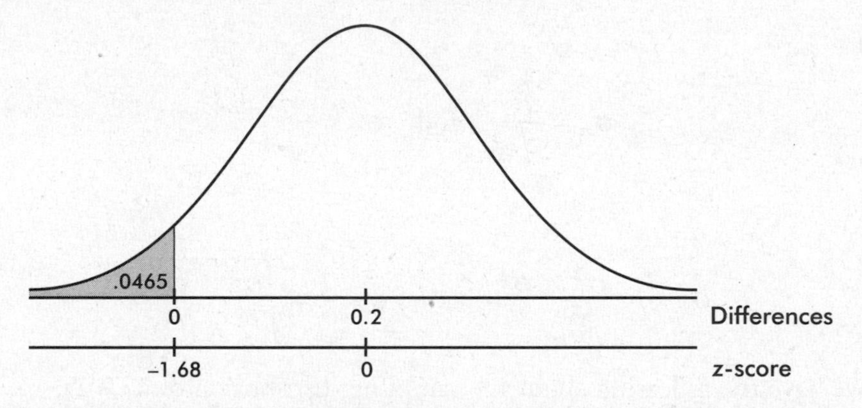

EXAMPLE 12.10

The average number of missed school days for students going to public schools is 8.5 with a standard deviation of 4.1, while students going to private schools miss an average of 5.3 with a standard deviation of 2.9. In an SRS of 200 public school students and an SRS of 150 private school students, with a probability of .95 the difference in average missed days (public average minus private average) is above what number?

Answer: We have large ($n_1 = 200$ and $n_2 = 150$) SRSs that are still smaller than 10% of all public and private school students. The mean of the differences is 8.5 – 5.3 = 3.2, while the standard deviation is $\sqrt{\frac{(4.1)^2}{200}+\frac{(2.9)^2}{150}} = 0.374$. An area of .95 to the right corresponds to a *z*-score of –1.645. The difference in days is 3.2 – 1.645(0.374) = 2.6. Thus there is a .95 probability that the difference in average missed days between public and private school samples is over 2.6 days. [invNorm(.05, 3.2, .374) = 2.585.]

(continued)

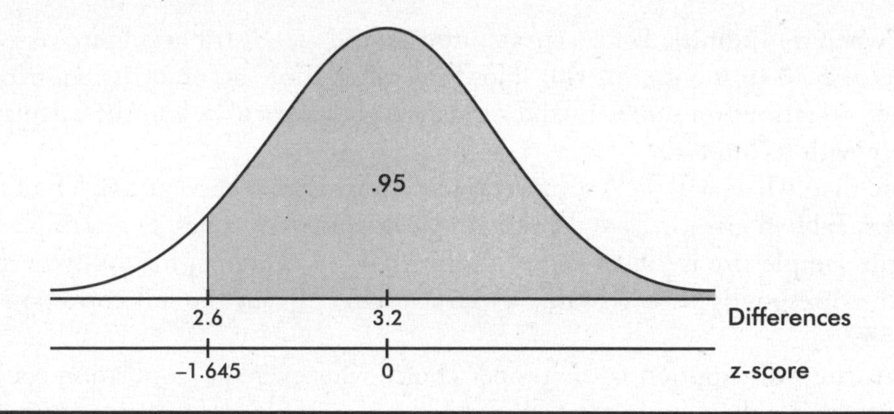

THE *T*-DISTRIBUTION

When the population standard deviation σ is unknown, we use the sample standard deviation s as an estimate for σ. But then $\frac{\bar{x} - \mu}{s/\sqrt{n}}$ does not follow a normal distribution. If, however, **the original population is normally distributed**, there is a distribution that can be used when working with the $s/\sqrt{n}$ ratios. This *Student t-distribution* was introduced in 1908 by W. S. Gosset, a British mathematician employed by the Guiness Breweries. (When we are working with small samples from a population that is *not* nearly normal, we must use very different "nonparametric" techniques not discussed in this review book.)

Thus, for a sample from a normally distributed population, we work with the variable

$$t = \frac{\bar{x} - \mu}{s/\sqrt{n}}$$

with a resulting *t*-distribution that is bell-shaped and symmetric, but lower at the mean, higher at the tails, and so more spread out than the normal distribution.

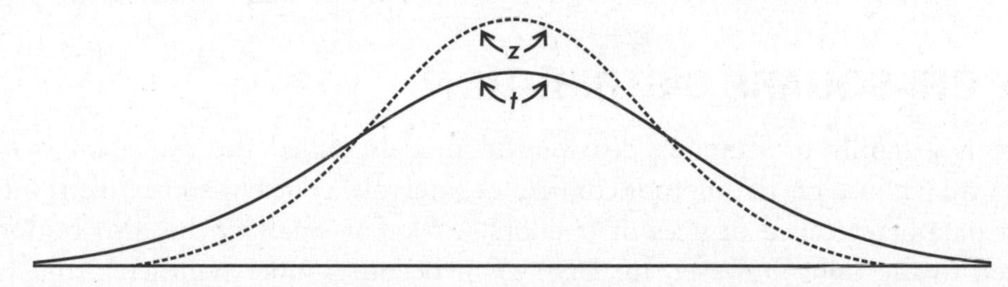

Like the binomial distribution, the *t*-distribution is different for different values of *n*. In the tables these distinct *t*-distributions are associated with the values for degrees of freedom (*df*). For this discussion the *df* value is equal to the sample size minus 1. The smaller the *df* value, the larger the dispersion in the distribution. The larger the *df* value, that is, the larger the sample size, the closer the distribution to the normal distribution.

Since there is a separate *t*-distribution for each *df* value, fairly complete tables would involve many pages; therefore, in Table B of the Appendix we list areas and *t*-values for only the more commonly used percentages or probabilities. The last row of Table B is the normal distribution, which is a special case of the *t*-distribution taken when n is infinite. For practical purposes, the two distributions are very close for any $n \geq 30$ (some use $n \geq 40$). However, when more accuracy is required, the Student t-distribution can be used for much larger values of n, and the calculations are easy with technology.

Note that, whereas Table A gives areas under the normal curve to the left of given z-values, Table B gives areas to the right of given positive t-values. For example, suppose the sample size is 20, and so $df = 20 - 1 = 19$. Then a probability of .05 in the tail corresponds with a t-value of 1.729, while .01 in the tail corresponds to $t = 2.539$.

Thus the t-distribution is the proper choice whenever the population standard deviation σ is unknown. In the real world σ is almost always unknown, and so we should almost always use the t-distribution. In the past this was difficult because extensive tables were necessary for the various t-distributions. However, with calculators such as the TI-84, this problem has diminished. It is no longer necessary to assume that the t-distribution is close enough to the z-distribution whenever n is greater than the arbitrary number 30.

The issue of sample size is refined even further by some statisticians:

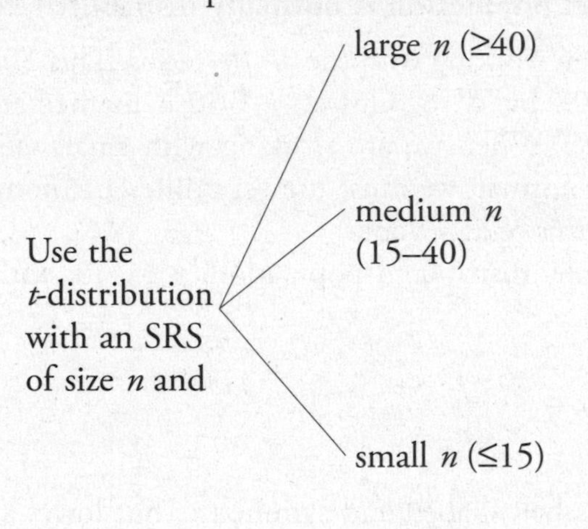

large n (≥ 40): Unnecessary to make any assumptions about parent population.

medium n (15–40): Sample should show no extreme values and little, if any, skewness; or assume parent population is normal.

small n (≤ 15): Sample should show no outliers and no skewness; or assume parent population is normal.

THE CHI-SQUARE DISTRIBUTION

There is a family of sampling distribution models, called the *chi-square models*, which we will use for testing (not confidence intervals). The chi-square distribution has a parameter called degrees of freedom (df). For small df the distribution is skewed to the right; however, for large df it becomes more symmetric and bell-shaped (as does the t-distribution). For one or two degrees of freedom the peak occurs at 0, while for three or more degrees of freedom the peak is at $df - 2$.

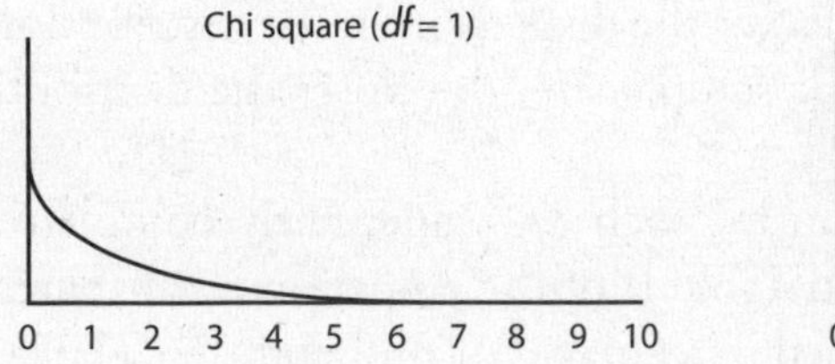

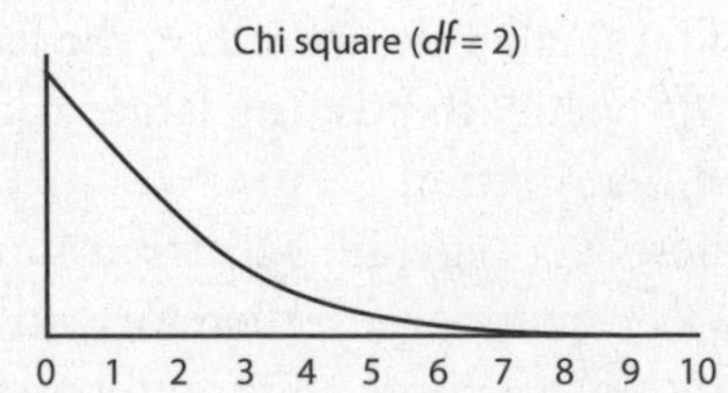

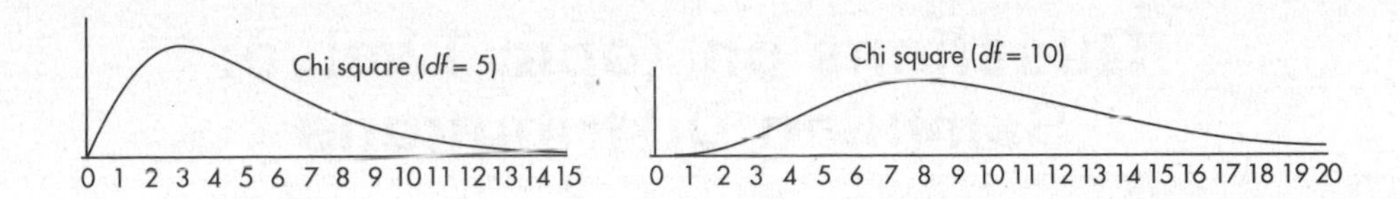

The chi-square distribution, like the *t*-distribution, has a separate curve for each *df* value; therefore, in Table C of the Appendix we list critical chi-square values for only the more commonly used tail probabilities. The chi-square distribution has numerous applications, two of the best known of which are goodness of fit of an observed distribution to a theoretical one, and independence of two criteria of classification of qualitative data. While we will be using chi-square for categorical data, the chi-square distribution as seen above is a *continuous* distribution, and applying it to counting data is just an approximation.

THE STANDARD ERROR

With proportions and means we typically do not know population parameters. So in calculating standard deviations of the sampling models, we actually estimate using sample statistics. In this case, we use the term *standard error*. That is, for proportions,

$$\sigma_{\hat{p}} = \sqrt{\frac{pq}{n}}, \text{ and we have } SE(p) = \sqrt{\frac{\hat{p}\hat{q}}{n}}$$

Similarly, for means,

$$\sigma_{\bar{x}} = \frac{\sigma}{\sqrt{n}}, \text{ and we have } SE(\bar{x}) = \frac{s}{\sqrt{n}}.$$

Summary

- Provided that the sample size n is large enough, the sampling distribution of sample proportions is approximately normal with mean p and standard deviation $\sqrt{\frac{p(1-p)}{n}}$ (where p is the population proportion).
- Provided that the sample size n is large enough, the sampling distribution of sample means is approximately normal with mean μ and standard deviation $\frac{\sigma}{\sqrt{n}}$ (where μ and σ are the population mean and standard deviation).
- When the population standard deviation σ is unknown (which is almost always the case), and provided that the original population is normally distributed or the sample size n is large enough, the sampling distribution of sample means follows a t-distribution with mean μ and standard deviation $\frac{s}{\sqrt{n}}$ (where μ is the population mean and s is the sample standard deviation).

Questions on Topic Twelve: Sampling Distributions

Multiple-Choice Questions

Directions: The questions or incomplete statements that follow are each followed by five suggested answers or completions. Choose the response that best answers the question or completes the statement.

1. Which of the following is a true statement?

   (A) The larger the sample, the larger the spread in the sampling distribution.
   (B) Bias has to do with the spread of a sampling distribution.
   (C) Provided that the population size is significantly greater than the sample size, the spread of the sampling distribution does not depend on the population size.
   (D) Sample parameters are used to make inferences about population statistics.
   (E) Statistics from smaller samples have less variability.

2. Which of the following is an *incorrect* statement?

   (A) The sampling distribution of $\bar{x}$ has mean equal to the population mean μ even if the population is not normally distributed.
   (B) The sampling distribution of $\bar{x}$ has standard deviation $\frac{\sigma}{\sqrt{n}}$ even if the population is not normally distributed.
   (C) The sampling distribution of $\bar{x}$ is normal if the population has a normal distribution.
   (D) When n is large, the sampling distribution of $\bar{x}$ is approximately normal even if the population is not normally distributed.
   (E) The larger the value of the sample size n, the closer the standard deviation of the sampling distribution of $\bar{x}$ is to the standard deviation of the population.

3. Which of the following is a true statement?

   (A) The sampling distribution of $\hat{p}$ has a mean equal to the population proportion p.
   (B) The sampling distribution of $\hat{p}$ has a standard deviation equal to $\sqrt{np(1-p)}$.
   (C) The sampling distribution of $\hat{p}$ has a standard deviation which becomes larger as the sample size becomes larger.
   (D) The sampling distribution of $\hat{p}$ is considered close to normal provided that $n \geq 30$.
   (E) The sampling distribution of $\hat{p}$ is always close to normal.

4. In a school of 2500 students, the students in an AP Statistics class are planning a random survey of 100 students to estimate the proportion who would rather drop lacrosse rather than band during this time of severe budget cuts. Their teacher suggests instead to survey 200 students in order to

   (A) reduce bias.
   (B) reduce variability.
   (C) increase bias.
   (D) increase variability.
   (E) make possible stratification between lacrosse and band.

5. The ages of people who died last year in the United States is skewed left. What happens to the sampling distribution of sample means as the sample size goes from $n = 50$ to $n = 200$?

   (A) The mean gets closer to the population mean, the standard deviation stays the same, and the shape becomes more skewed left.
   (B) The mean gets closer to the population mean, the standard deviation becomes smaller, and the shape becomes more skewed left.
   (C) The mean gets closer to the population mean, the standard deviation stays the same, and the shape becomes closer to normal.
   (D) The mean gets closer to the population mean, the standard deviation becomes smaller, and the shape becomes closer to normal.
   (E) The mean stays the same, the standard deviation becomes smaller, and the shape becomes closer to normal.

6. Which of the following is the best reason that the sample maximum is not used as an estimator for the population maximum?

   (A) The sample maximum is biased.
   (B) The sampling distribution of the sample maximum is not binomial.
   (C) The sampling distribution of the sample maximum is not normal.
   (D) The sampling distribution of the sample maximum has too large a standard deviation.
   (E) The sample mean plus three sample standard deviations gives a much superior estimate for the population maximum.

7. Which of the following are unbiased estimators for the corresponding population parameters?

   I. Sample means
   II. Sample proportions
   III. Difference of sample means
   IV. Difference of sample proportions

   (A) None are unbiased.
   (B) I and II
   (C) I and III
   (D) III and IV
   (E) All are unbiased.

8. Suppose that 35% of all business executives are willing to switch companies if offered a higher salary. If a headhunter randomly contacts an SRS of 100 executives, what is the probability that over 40% will be willing to switch companies if offered a higher salary?

   (A) .1469
   (B) .1977
   (C) .4207
   (D) .8023
   (E) .8531

9. Given that 58% of all gold dealers believe next year will be a good one to speculate in South African gold coins, in a simple random sample of 150 dealers, what is the probability that between 55% and 60% believe that it will be a good year to speculate?

   (A) .0500
   (B) .1192
   (C) .3099
   (D) .4619
   (E) .9215

10. The average outstanding bill for delinquent customer accounts for a national department store chain is $187.50 with a standard deviation of $54.50. In a simple random sample of 50 delinquent accounts, what is the probability that the mean outstanding bill is over $200?

    (A) .0526
    (B) .0667
    (C) .4090
    (D) .5910
    (E) .9474

11. The average number of daily emergency room admissions at a hospital is 85 with a standard deviation of 37. In a simple random sample of 30 days, what is the probability that the mean number of daily emergency admissions is between 75 and 95?

    (A) .1388
    (B) .2128
    (C) .8612
    (D) .8990
    (E) .9970

12. Two companies offer classes designed to improve students' SAT scores. Suppose that 83% of students enrolling in the first program improve their scores, while 74% of those signing up for the second program also raise their scores. In an SRS of 60 students taking the first program and an independent SRS of 50 taking the second, what is the probability that the difference between the percentages of students improving their scores (first program minus second) is more than 15%?

    (A) .0281
    (B) .2236
    (C) .2764
    (D) .3632
    (E) .7764

13. Pepper plants watered lightly every day for a month show an average growth of 27 centimeters with a standard deviation of 8.3 centimeters, while pepper plants watered heavily once a week for a month show an average growth of 29 centimeters with a standard deviation of 7.9 centimeters. In a simple random sample of 60 plants, half of which are given each of the watering treatments, what is the probability that the difference in average growth between the two halves is between –3 and +3 centimeters?

    (A) .3156
    (B) .3240
    (C) .4639
    (D) .6760
    (E) .6844

14. Which of the following statements are true?

    I. Like the normal, *t*-distributions are always symmetric.
    II. Like the normal, *t*-distributions are always mound-shaped.
    III. The *t*-distributions have less spread than the normal, that is, they have less probability in the tails and more in the center than the normal.

    (A) II only
    (B) I and II
    (C) I and III
    (D) II and III
    (E) I, II, and III

15. Which of the following statements about *t*-distributions are true?

    I. The greater the number of degrees of freedom, the narrower the tails.
    II. The smaller the number of degrees of freedom, the closer the curve is to the normal curve.
    III. Thirty degrees of freedom gives the normal curve.

    (A) I only
    (B) I and II
    (C) I and III
    (D) II and III
    (E) I, II, and III

Free-Response Questions

Directions: You must show all work and indicate the methods you use. You will be graded on the correctness of your methods and on the accuracy of your final answers.

Four Open-Ended Questions

1. Car insurance policies for teenagers are typically higher than insurance rates for adult drivers because teenagers, with their lesser driving experience, are considered to be a "high risk" population. The average yearly cost of teenage auto insurance is $2171 with a standard deviation of $612. An SRS is taken of 100 teenage drivers.
   (a) Explain why there is insufficient information to determine the probability a randomly chosen teenage driver pays over $2300 a year for auto insurance.
   (b) What are the mean and standard deviation for the sampling distribution for $\bar{x}$, the mean amount paid for insurance.
   (c) What is the probability that the average amount paid in the sample is over $2300?
2. A health foods magazine conducts a large random sample of honey prices with the resulting data showing a roughly normal distribution with mean $1.87/lb and standard deviation $0.193/lb.
   (a) What is the interquartile range (IQR) of the distribution of honey prices?
   (b) What is the probability that a majority in a random sample of three honey prices are over $2.00/lb?
   (c) What is the probability that the mean price in a random sample of three honey prices is over $2.00/lb?
3. Suppose that the heights of college basketball players are normally distributed with a mean of 74 inches and a standard deviation of 4 inches.
   (a) What percentage of players are over 7 feet?
   (b) What is the probability that at least one of ten randomly selected players is over 7 feet?
   (c) What is the probability that the mean height in an SRS of size 10 is over 6 feet?
   (d) If an outlier is defined to be any value more than 1.5 interquartile ranges above the third quartile or below the first quartile, what percentage of heights of players are outliers?

4. The mathematics department at a state university notes that the SAT math scores of high school seniors applying for admission into their program are normally distributed with a mean of 610 and standard deviation of 50.

   (a) What is the probability that a randomly chosen applicant to the department has an SAT math score above 700?

   (b) What are the shape, mean, and standard deviation of the sampling distribution of the mean of a sample of 40 randomly selected applicants?

   (c) What is the probability that the mean SAT math score in an SRS of 40 applicants is above 625?

   (d) Would your answers to (a), (b), or (c) be affected if the original population of SAT math scores were highly skewed instead of normal? Explain.

An Investigative Task

Consider a population of size $N = 5$ consisting of the elements 3, 6, 8, 10, and 18.

(a) Determine the population mean μ and the population standard deviation σ.

(b) List all possible samples of size $n = 2$ [there are $C(5, 2) = 10$ of them] and determine the mean $\bar{x}$ of each.

(c) Show that the mean of the set of ten sample means is equal to the population mean.

(d) Show that the standard deviation of the set of ten sample means is

$$\sigma_{\bar{x}} = \frac{\sigma}{\sqrt{n}}\sqrt{\frac{N-n}{N-1}}$$

(e) More generally, what can be said about $\frac{\sigma}{\sqrt{n}}\sqrt{\frac{N-n}{N-1}}$ if the population size N is very large?

Answer Key

1. **C**	4. **B**	7. **E**	10. **A**	13. **D**
2. **E**	5. **E**	8. **A**	11. **C**	14. **B**
3. **A**	6. **A**	9. **D**	12. **B**	15. **A**

Answers Explained

Multiple-Choice

1. **(C)** The larger the sample, the smaller the spread in the sampling distribution. Bias has to do with the center, not the spread, of a sampling distribution. Sample statistics are used to make inferences about population proportions. Statistics from smaller samples have more variability.

2. **(E)** It is always true that the sampling distribution of $\bar{x}$ has mean μ and standard deviation $\frac{\sigma}{\sqrt{n}}$. In addition, the sampling distribution will be normal if the population is normal and will be approximately normal if n is large even if the population is not normal. $\frac{\sigma}{\sqrt{n}}$ and σ are not equal unless $n = 1$.

3. **(A)** The sampling distribution of $\hat{p}$ has a standard deviation $\sqrt{\frac{p(1-p)}{n}}$ which is smaller with larger n. While the sampling distribution of $\hat{p}$ is never exactly normal, it is considered close to normal *provided that* both np and $n(1-p)$ are large enough (greater than 10 is a standard guide).

4. **(B)** Sample proportions are an unbiased estimator for the population proportion, and larger sample sizes lead to reduced variability.

5. **(E)** The sampling distribution of $\bar{x}$ has mean μ, standard deviation $\frac{\sigma}{\sqrt{n}}$, and shape, which becomes closer to normal with larger n.

6. **(A)** The maximum of a sample is never larger than the maximum of the population, so the mean of the sample maximums will not be equal to the population maximum. (A sampling distribution is unbiased if its mean is equal to the population parameter.)

7. **(E)** All are unbiased estimators for the corresponding population parameters; that is, the means of their sampling distributions are equal to the population parameters.

8. **(A)** Both $np = (100)(.35) = 35 > 10$ and $n(1 - p) = (100)(.65) = 65 > 10$. The set of sample proportions is approximately normally distributed with mean .35 and standard deviation $\sigma_{\hat{p}} = \sqrt{\frac{(.35)(.65)}{100}} = .0477$. With a *z*-score of $\frac{.40-.35}{.0477} = 1.05$, the probability that the sample proportion exceeds 40% is $1 - .8531 = .1469$. [normalcdf(.40, 1, .35, .0477) = .1473.]

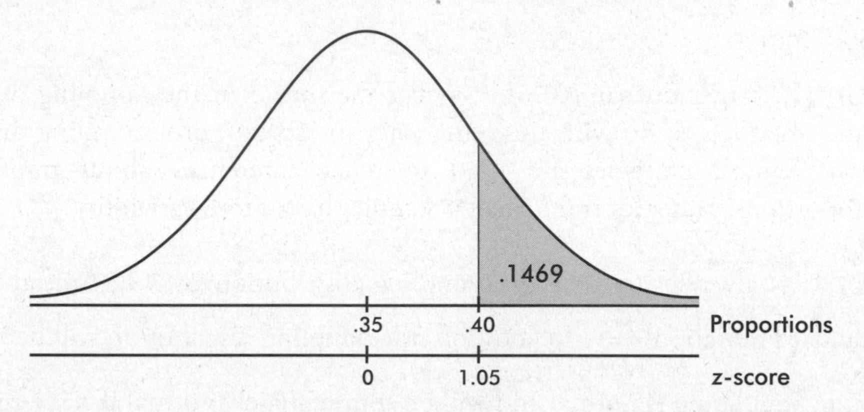

9. **(D)** Both $np = (150)(.58) = 87 > 10$ and $n(1 - p) = (150)(.42) = 63 > 10$. The set of sample proportions is approximately normally distributed with mean .58 and standard deviation $\sigma_{\hat{p}} = \sqrt{\frac{(.58)(.42)}{150}} = .0403$. With *z*-scores of $\frac{.55-.58}{.0403} = -0.74$ and $\frac{.60-.58}{.0403} = 0.50$, the probability the sample proportion is between .55 and .60 is $.6915 - .2296 = .4619$. [normalcdf(.55, .60, .58, .0403) = .4618.]

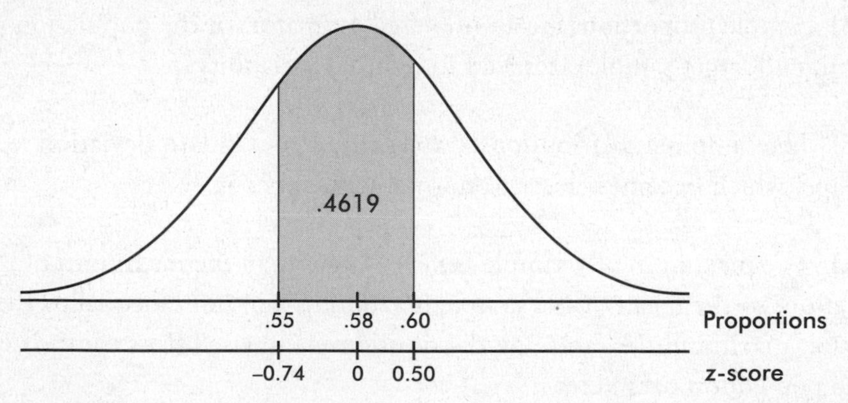

10. **(A)** The set of sample means is approximately normally distributed with $\mu_{\bar{x}} = 187.50$ and $\sigma_{\bar{x}} = \frac{54.50}{\sqrt{50}} = 7.707$. The *z*-score of 200 is $\frac{200 - 187.50}{7.707} = 1.62$, and the probability that the mean outstanding bill is over $200 is 1 – .9474 = .0526. [normalcdf(200, 10000, 187.5, 7.707) = .0524.]

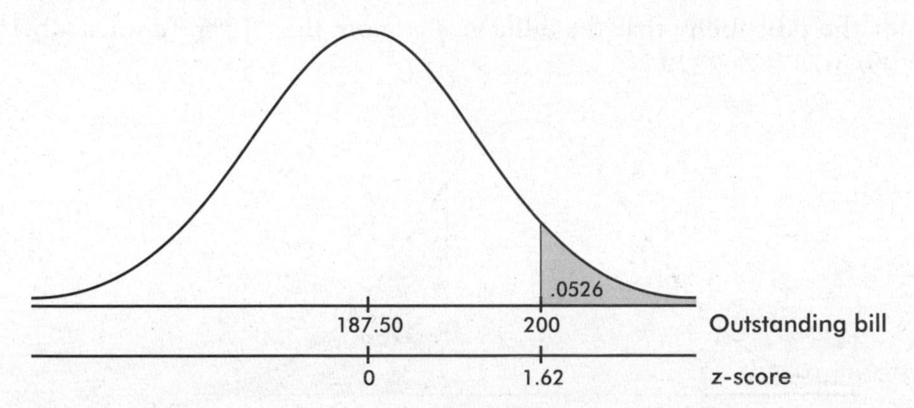

11. **(C)** The set of sample means is approximately normally distributed with $\mu_{\bar{x}} = 85$ and $\sigma_{\bar{x}} = \frac{37}{\sqrt{30}} = 6.755$. The *z*-scores of 75 and 95 are $\frac{75-85}{6.755} = -1.48$ and $\frac{95-85}{6.755} = 1.48$, respectively. The probability that the mean admissions in the sample are between 75 and 95 is .9306 – .0694 = .8612. [normalcdf(75, 95, 85, 6.755) = .8612.]

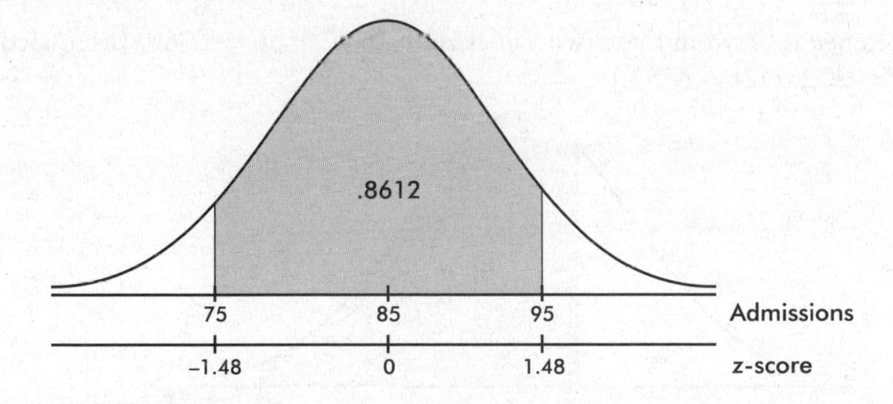

12. **(B)** We note that $n_1p_1 = 49.8$, $n_1(1 - p_1) = 10.2$, $n_2p_2 = 37$, and $n_2(1 - p_2) = 13$ are all >10. For the sampling distribution of $\hat{p}_1 - \hat{p}_2$, the mean is .83 – .74 = .09 and the standard deviation is $\sqrt{\frac{(.83)(.17)}{60} + \frac{(.74)(.26)}{50}} = .0787$. The *z*-score of .15 is $\frac{.15-.09}{.0787} = 0.76$. Using Table A, we find 1 – .7764 = .2236 for the probability that the difference is more than 15%. [normalcdf(.15, 1, .09, .0787) = .2229.]

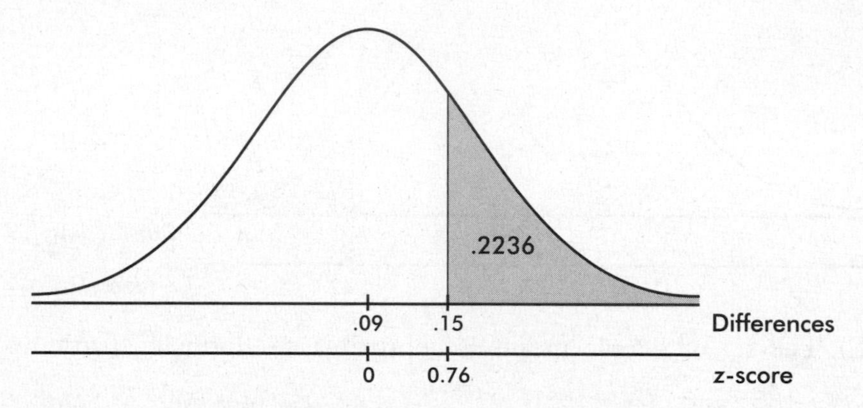

13. **(D)** The mean of the differences equals 27 – 29 = –2, while the standard deviation is $\sqrt{\frac{(8.3)^2}{30} + \frac{(7.9)^2}{30}} = 2.092$. The *z*-scores of +3 and –3 are $\frac{3-(-2)}{2.092} = 2.39$ and $\frac{-3-(-2)}{2.092} = -0.48$, respectively. The probability that the difference is between these two values is .9916 – .3156 = .6760. [normalcdf(–3, 3, –2, 2.092) = .6753.]

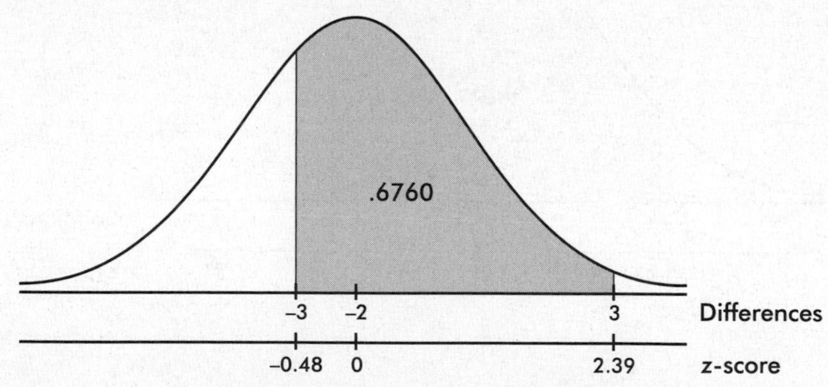

14. **(B)** The *t*-distributions are symmetric and mound-shaped, and they have more, not less, spread than the normal distribution.

15. **(A)** The larger the number of degrees of freedom, the closer the curve to the normal curve. While around the 30 level is often considered a reasonable approximation to the normal curve, it is not the normal curve.

Free-Response

1. (a) We do not know the shape of the distribution of the amount individual teenage drivers pay for insurance, so there is no way of calculating the probability a randomly chosen teenage driver pays over $3000 a year for auto insurance.

   (b) $\mu_{\bar{x}} = \mu = 2171$ and $\sigma_{\bar{x}} = \frac{\sigma}{\sqrt{n}} = \frac{612}{\sqrt{100}} = 61.2$

   (c) With this large a sample size (100 > 30), the Central Limit Theorem tells us that the sampling distribution of sample means is approximately normal. We calculate

   $P(\bar{x} > 2300) = P(z > \frac{2300 - 2171}{61.2}) = P(z > 2.1078) = .0175$

   [On the TI-84, `normalcdf(2.1078,1000)` = .0175]

2. (a) The *z*-scores corresponding to cumulative probabilities of 25% and 75% are ±0.674. Thus, Q_1 = 1.87 – 0.674(0.193) = 1.740, Q_3 = 1.87 + 0.674(0.193) = 2.000, and IQR = $Q_3 - Q_1$ = 2.000 – 1.740 = $0.260/lb.

   (b) From part (a) we have that the probability a honey price is below $2.00/lb is .75, and thus the probability of being above $2.00/lb is 1 – .75 = .25. Now we have a binomial with $n = 3$ and $p = .25$. *P*(majority > $2.00/lb) = *P*(2 or 3 are > $2.00/lb) = $3(.25)^2(.75) + (.25)^3 = 0.15625$.

   (c) The distribution of $\bar{x}$ is normal with a mean of 1.87 and a standard deviation of $\frac{0.193}{\sqrt{3}} = 0.1114$. Thus, $P(\bar{x} > 2.00) = P(z > \frac{2.00 - 1.87}{0.1114}) =$

   $P(z > 1.167) = .01216$

   [On the TI-84, `normalcdf(1.167,1000)` = .1216]

3. (a) The *z*-score of 84 is $\frac{84-74}{4} = 2.5$, which gives a probability of .0062 or 0.62%. [normalcdf(84, 1000, 74, 4) = .0062.]

   (b) $1 - (.9938)^{10} = .06$

   (c) Checking conditions: the original population is given to be normal, we have an SRS, and our sample is less than 10% of all college basketball players.

   The *z*-score of 72 is $\frac{72-74}{\frac{4}{\sqrt{10}}} = -1.58$ with a resulting probability of .9429. [normalcdf(72, 1000, 74, 1.265) = .9431.]

   (d) We have Q_1 = 74 – 0.674(4) = 71.3, Q_3 = 74 + 0.674(4) = 76.7, and 1.5(IQR) = 1.5($Q_3 - Q_1$) = 8.1. Then 71.3 – 8.1 = 63.2 has a *z*-score of –2.7, while 76.7 + 8.1 = 84.8 has a *z*-score of 2.7. The corresponding probabilities give .0035 + .0035 = .007 or 0.7%. The problem could also have been worked simply in terms of *z*-scores: 1.5(0.674 + 0.674) = 2.02 and 0.674 + 2.02 ≈ 2.7, and so on.

4. (a)

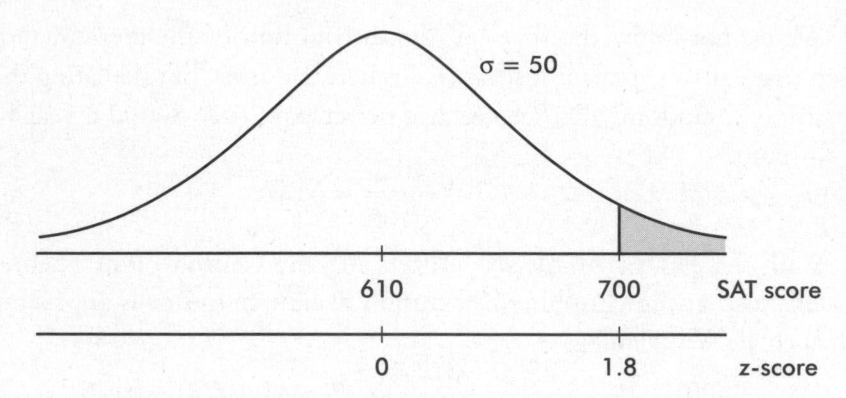

The critical z-score is $\frac{700-610}{50} = 1.8$, and so from Table A the asked-for probability is 1 – .9641 = .0359. [normalcdf(700, 10000, 610, 50) = .0359.]

(b) It is roughly normal with a mean equal to the population mean 610 and a standard deviation equal to the population standard deviation divided by the square root of the sample size, that is, $\frac{50}{\sqrt{40}} = 7.91$.

(c) Checking conditions: the original population is given to be normal, we have an SRS, and our sample comes from less than 10% of all applicants.

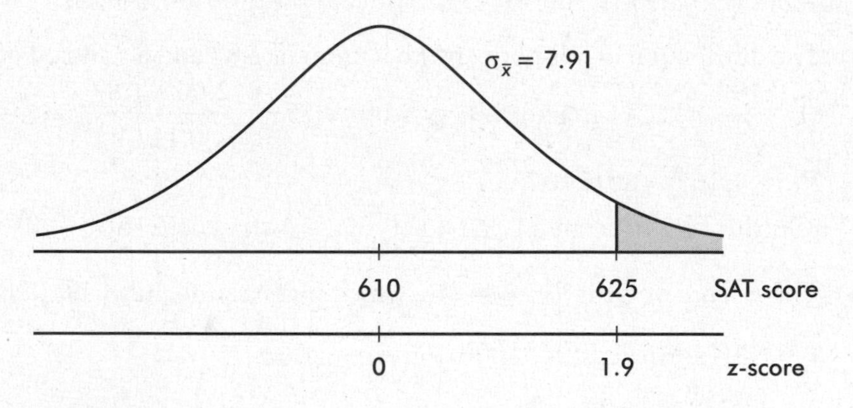

The critical z-score is $\frac{625-610}{7.91} = 1.90$, and from Table A the asked-for probability is 1 – .9713 = .0287. [normalcdf(625, 10000, 610, 7.91) = .0290.]

(d) The answer to part *a* would be affected because it assumes a normal population. The other answers would not be affected because for large enough *n*, the central limit theorem gives that the sampling distribution will be roughly normal regardless of the distribution of the original population.

Investigative Task

(a) Using a calculator, find $\mu = 9$ and $\sigma = 5.0596$.

(b)

Sample	Mean
{3, 6}	4.5
{3, 8}	5.5
{3, 10}	6.5
{3, 18}	10.5
{6, 8}	7
{6, 10}	8
{6, 18}	12
{8, 10}	9
{8, 18}	13
{10, 18}	14

(c) $\frac{(4.5+5.5+6.5+10.5+7+8+12+9+13+14)}{10} = 9.$

(d) Using a calculator, find $\sigma_{\bar{x}} = 3.0984$ and then note that

$$\frac{5.0596}{\sqrt{2}}\sqrt{\frac{5-2}{5-1}} = 3.0984$$

(e) If N is very large, $\frac{N-n}{N-1}$ is approximately equal to 1 and so the expression simplifies to $\frac{\sigma}{\sqrt{n}}$.

THEME FOUR: STATISTICAL INFERENCE

Confidence Intervals

TOPIC **13**

- Definition
- For a Proportion
- For a Difference of Two Proportions
- For a Mean
- For a Difference Between Two Means
- For the Slope of a Least Squares Regression Line

Using a measurement from a sample, we are never able to say *exactly* what a population proportion or mean is; rather we always say we have a certain *confidence* that the population proportion or mean lies in a particular *interval.* The particular interval is centered around a sample proportion or mean (or other statistic) and can be expressed as the sample estimate plus or minus an associated *margin of error.*

THE MEANING OF A CONFIDENCE INTERVAL

Using what we know about sampling distributions, we are able to establish a certain confidence that a sample proportion or mean lies within a specified interval around the population proportion or mean. However, we then have the same confidence that the population proportion or mean lies within a specified interval around the sample proportion or mean (e.g., the distance from Missoula to Whitefish is the same as the distance from Whitefish to Missoula).

Typically we consider 90%, 95%, and 99% confidence interval estimates, but any percentage is possible. The percentage is the percentage of samples that would pinpoint the unknown p or μ within plus or minus a certain margin or error. We do *not* say there is a .90, .95, or .99 probability that p or μ is within a certain margin of error of a given sample proportion or mean. For a given sample proportion or mean, p or μ either is or isn't within the specified interval, and so the probability is either 1 or 0.

As will be seen, there are two aspects to this concept. First, there is the *confidence interval,* usually expressed in the form:

$$\text{estimate} \pm \text{margin of error}$$

Second, there is the *success rate for the method,* that is, the proportion of times repeated applications of this method would capture the true population parameter.

CONFIDENCE INTERVAL FOR A PROPORTION

We are interested in estimating a population proportion p by considering a single sample proportion $\hat{p}$. This sample proportion is just one of a whole universe of sample proportions, and from Topic 12 we remember the following:

1. The set of all sample proportions is approximately normally distributed.
2. The mean $\mu_{\hat{p}}$ of the set of sample proportions equals p, the population proportion.
3. The standard deviation $\sigma_{\hat{p}}$ of the set of sample proportions is approximately equal to $\sqrt{\frac{p(1-p)}{n}}$.

In finding confidence interval estimates of the population proportion p, how do we find $\sqrt{\frac{p(1-p)}{n}}$ since p is unknown? The reasonable procedure is to use the sample proportion $\hat{p}$:

$$\sigma_{\hat{p}} \approx \sqrt{\frac{\hat{p}(1-\hat{p})}{n}}$$

When the standard deviation is estimated in this way (using the sample), we use the term *standard error.* That is,

$$\text{SE}_{\hat{p}} = \sqrt{\frac{\hat{p}(1-\hat{p})}{n}}$$

Remember that we are really using a normal approximation to the binomial, so $n\hat{p}$ and $n(1-\hat{p})$ should both be at least 10. Furthermore, in making calculations and drawing conclusions from a specific sample, it is important that the sample be a *simple random sample.* Finally, the population should be large, typically checked by the assumption that the sample is less than 10% of the population. (If the population is small and the sample exceeds 10% of the population, then models other than the normal are more appropriate.)

TIP

Verifying assumptions and conditions means more than simply listing them with little check marks. You must show work or give some reason to confirm verification.

EXAMPLE 13.1

If 64% of an SRS of 550 people leaving a shopping mall claim to have spent over $25, determine a 99% confidence interval estimate for the proportion of shopping mall customers who spend over $25.

Answer: We check that $n\hat{p} = 550(.64) = 352 > 10$ and $n(1-\hat{p}) = 550(.36) = 198 > 10$, we are given that the sample is an SRS, and it is reasonable to assume that 550 is less than 10% of all mall shoppers. Since $\hat{p} = .64$, the standard deviation of the set of sample proportions is

$$\sigma_{\hat{p}} \approx \sqrt{\frac{(.64)(.36)}{550}} = .0205$$

(continued)

From Topic 11 we know that 99% of the sample proportions should be within 2.576 standard deviations of the population proportion. Equivalently, we are 99% certain that the population proportion is within 2.576 standard deviations of any sample proportion.[1] Thus the 99% confidence interval estimate for the population proportion is .64 ± 2.576(.0205) = .64 ± .053. We say that the *margin of error* is ±.053. We are 99% certain that the proportion of shoppers spending over $25 is between .587 and .693.

We can also say, using the definition of confidence *level*, that if the interviewing procedure were repeated many times, about 99% of the resulting confidence intervals would contain the true proportion (thus we're 99% confident that the method worked for the interval we got).

EXAMPLE 13.2

In a simple random sample of machine parts, 18 out of 225 were found to have been damaged in shipment. Establish a 95% confidence interval estimate for the proportion of machine parts that are damaged in shipment.

Answer: We check that $n\hat{p} = 18 > 10$ and $n(1-\hat{p}) = 225 - 18 = 207 > 10$, we are given that the sample is an SRS, and it is reasonable to assume that 225 is less than 10% of all shipped machine parts. The sample proportion is $\hat{p} = \frac{18}{225} = .08$, and the standard deviation of the set of sample proportions is

$$\sigma_{\hat{p}} \approx \sqrt{\frac{(.08)(.92)}{225}} = .0181$$

The 95% confidence interval estimate for the population proportion is .08 ± 1.96(.0181) = .08 ± .035. Thus we are 95% certain that the proportion of machine parts damaged in shipment is between .045 and .115.

[On the TI-84, under STAT and then TESTS, go to 1-PropZInt. With x:18, n:225, and C-Level:.95, Calculate gives (.04455, .11545).]

On the TI-Nspire the result shows as:

zInterval_1Prop 18,225,0.95: *stat.results*		
	"Title"	"1-Prop z Interval"
	"CLower"	0.044552
	"CUpper"	0.115448
	"p̂"	0.08
	"ME"	0.035448
	"n"	225.

Suppose there are 50,000 parts in the entire shipment. We can translate from proportions to actual numbers:

$$.045(50{,}000) = 2250 \text{ and } .115(50{,}000) = 5750$$

and so we can be 95% confident that there are between 2250 and 5750 defective parts in the whole shipment.

---

[1] Note that we cannot say there is a .99 probability that the population proportion is within 2.576 standard deviations of a given sample proportion. For a given sample proportion, the population proportion either is or isn't within the specified interval, and so the probability is either 1 or 0.

EXAMPLE 13.3

A telephone survey of 1000 adults was taken shortly after the United States began bombing Iraq.

a. If 832 voiced their support for this action, with what confidence can it be asserted that 83.2% ± 3% of the adult U.S. population supported the decision to go to war?
Answer: We check that $n\hat{p} = 832 > 10$ and $n(1 - \hat{p}) = 1000 - 832 = 168 > 10$, we assume an SRS, and clearly 1000 is <10% of the adult U.S. population.

$$\hat{p} = \frac{832}{1000} = .832 \text{ and so } \sigma_{\hat{p}} \approx \sqrt{\frac{(.832)(.168)}{1000}} = .0118$$

The relevant z-scores are $\pm\frac{.03}{.0118} = \pm 2.54$. Table A gives probabilities of .0055 and .9945, and so our answer is .9945 – .0055 = .9890. In other words, 83.2% ± 3% is a 98.90% confidence interval estimate for U.S. adult support of the war decision.

b. If the adult U.S. population is 191 million, estimate the actual numerical support.
Answer: Since

$$.802(191{,}000{,}000) \approx 153{,}000{,}000$$

while

$$.862(191{,}000{,}000) \approx 165{,}000{,}000$$

we can be 98.90% confident that between 153 and 165 million adults supported the initial bombing decision.

EXAMPLE 13.4

A U.S. Department of Labor survey of 6230 unemployed adults classified people by marital status, gender, and race. The raw numbers are as follows:

White, 16 yr and older	Married	Widow/Div.	Single
Men	1090	337	1168
Women	952	423	632

Nonwhite, 16 yr and older	Married	Widow/Div.	Single
Men	266	135	503
Women	189	186	349

a. Find a 90% confidence interval estimate for the proportion of unemployed men who are married.
Answer: Totaling the first row across both tables, we find that there are 3499 men in the survey; 1090 + 266 = 1356 of them are married. We check that $n\hat{p} = 1356 > 10$ and $n(1 - \hat{p}) = 3499 - 1356 = 2143 > 10$, we assume an SRS, and clearly the sample is <10% of the total unemployed adult population. Therefore $\hat{p} = \frac{1356}{3499} = .3875$ and

(continued)

$$\sigma_{\hat{p}} \approx \sqrt{\frac{(.3875)(.6125)}{3499}} = .008236$$

Thus the 90% confidence interval estimate is .3875 ± 1.645(.008236) = .3875 ± .0135. [On the TI-84, 1-PropZInt gives (.37399, .40109).]

b. Find a 98% confidence interval estimate of the proportion of unemployed single persons who are women.

Answer: There are 1168 + 632 + 503 + 349 = 2652 singles in the survey. Of these, 632 + 349 = 981 are women [$n\hat{p} = 981 > 10$ and $n(1 - \hat{p}) = 2652 - 981 = 1671 > 10$], so $\hat{p} = \frac{981}{2652} = .3699$ and

$$\sigma_{\hat{p}} \approx \sqrt{\frac{(.3699)(.6301)}{2652}} = .009375$$

The 98% confidence interval estimate is .3699 ± 2.326(.009375) = .3699 ± .0218. [On the TI-84, 1-PropZInt gives (.3481, .39172).]

As we have seen, there are two types of statements that come out of confidence intervals. First, we can interpret the confidence *interval* and say we are 90% confident that between 60% and 66% of all voters favor a bond issue. Second, we can interpret the confidence *level* and say that if this survey were conducted many times, then about 90% of the resulting confidence intervals would contain the true proportion of voters who favor the bond issue.

Incorrect statements include "The percentage of all voters who support the bond issue is between 60% and 66%," "There is a .90 probability that the true percentage of all voters who favor the bond issue is between 60% and 66%," and "If this survey were conducted many times, then about 90% of the sample proportions would be in the interval (.60, .66)."

One important consideration in setting up a survey is the choice of sample size. To obtain a smaller, more precise interval estimate of the population proportion, we must either decrease the degree of confidence or increase the sample size. Similarly, if we want to increase the degree of confidence, we can either accept a wider interval estimate or increase the sample size. Again, while choosing a larger sample size may seem desirable, in the real world this decision involves time and cost considerations.

In setting up a survey to obtain a confidence interval estimate of the population proportion, what should we use for $\sigma_{\hat{p}}$? To answer this question, we first must consider how large $\sqrt{p(1-p)}$ can be. We plot various values of p:

p:	.1	.2	.3	.4	.5	.6	.7	.8	.9
$\sqrt{p(1-p)}$:	.3	.4	.458	.490	.5	.490	.458	.4	.3

which gives .5 as the intuitive answer. Thus $\sqrt{\frac{p(1-p)}{n}}$ is at most $\frac{.5}{\sqrt{n}}$. We make use of this fact to determine sample sizes in problems such as Examples 13.5 and 13.6.

EXAMPLE 13.5

An Environmental Protection Agency (EPA) investigator wants to know the proportion of fish that are inedible because of chemical pollution downstream of an offending factory. If the answer must be within ±.03 at the 96% confidence level, how many fish should be in the sample tested?

Answer: We want $2.05\sigma_{\hat{p}} \le .03$. From the above remark, $\sigma_{\hat{p}}$ is at most $\frac{.5}{\sqrt{n}}$ and so it is sufficient to consider $2.05\left(\frac{.5}{\sqrt{n}}\right) \le .03$. Algebraically, we have $\sqrt{n} \ge \frac{2.05(.5)}{.03} = 34.17$ and $n \ge 1167.4$. Therefore, choosing a sample of 1168 fish gives the inedible proportion to within ±.03 at the 96% level.

Note that the accuracy of the estimate does *not* depend on what fraction of the whole population we have sampled. What is critical is the *absolute size* of the sample. Is some minimal value of n necessary for the procedures we are using to be meaningful? Since we are using the normal approximation to the binomial, both np and $n(1-p)$ should be at least 10 (see Topic 11).

EXAMPLE 13.6

A study is undertaken to determine the proportion of industry executives who believe that workers' pay should be based on individual performance. How many executives should be interviewed if an estimate is desired at the 99% confidence level to within ±.06? To within ±.03? To within ±.02?

Answer: Algebraically, $2.576\left(\frac{.5}{\sqrt{n}}\right) \le .06$ gives $\sqrt{n} \ge \frac{2.576(.5)}{.06} = 21.5$, so $n \ge 462.25$. Similarly, $2.576\left(\frac{.5}{\sqrt{n}}\right) \le .03$ gives $\sqrt{n} \ge \frac{2.576(.5)}{.03} = 42.9$, so $n \ge 1840.4$. Finally, $2.576\left(\frac{.5}{\sqrt{n}}\right) \le .02$ gives $\sqrt{n} \ge \frac{2.576(.5)}{.02} = 64.4$, so $n \ge 4147.4$. Thus 463, 1841, or 4148 executives should be interviewed, depending on the accuracy desired.

Note that to cut the interval estimate in half (from ±.06 to ±.03), we would have to increase the sample size fourfold, and to cut the interval estimate to a third (from ±.06 to ±.02), a ninefold increase in the sample size would be required (answers are not exact because of round-off error.)

More generally, to divide the interval estimate by d without affecting the confidence level, we must increase the sample size by a multiple of d^2.

A formula for the calculations in Examples 13.5 and 13.6 is

$$n = \left[\frac{z}{2(\text{error})}\right]^2$$

(Note: this formula is not given on the AP Exam formula page.)

CONFIDENCE INTERVAL FOR A DIFFERENCE OF TWO PROPORTIONS

From Topic 12, we have the following information about the sampling distribution of $\hat{p}_1 - \hat{p}_2$:

1. The set of all differences of sample proportions is approximately normally distributed.
2. The mean of the set of differences of sample proportions equals $p_1 - p_2$, the difference of population proportions.
3. The standard deviation σ_d of the set of differences of sample proportions is approximately equal to

$$\sqrt{\frac{p_1(1-p_1)}{n_1} + \frac{p_2(1-p_2)}{n_2}}$$

Remember that we are using the normal approximation to the binomial, so $n_1\hat{p}_1$, $n_1(1-\hat{p}_1)$, $n_2\hat{p}_2$, and $n_2(1-\hat{p}_2)$ should all be at least 10. In making calculations and drawing conclusions from specific samples, it is important both that the samples be *simple random samples* and that they be taken *independently* of each other. Finally, the original populations should be large compared to the sample sizes.

EXAMPLE 13.7

Suppose that 84% of an SRS of 125 nurses working 7 a.m. to 3 p.m. shifts in city hospitals express positive job satisfaction, while only 72% of an SRS of 150 nurses on 11 p.m. to 7 a.m. shifts express similar fulfillment. Establish a 90% confidence interval estimate for the difference.

Answer: Note that $n_1\hat{p}_1 = (125)(.84) = 105$, $n_1(1-\hat{p}_1) = (125)(.16) = 20$, $n_2\hat{p}_2 = (150)(.72) = 108$, and $n_2(1-\hat{p}_2) = (150)(.28) = 42$ are all >10, we are given SRSs, and the population of city hospital nurses is assumed to be large.

$$n_1 = 125 \quad n_2 = 150$$
$$\hat{p}_1 = .84 \quad \hat{p}_2 = .72$$
$$\sigma_d \approx \sqrt{\frac{(.84)(.16)}{125} + \frac{(.72)(.28)}{150}} = .0492$$

The observed difference is .84 – .72 = .12, and the critical *z*-scores are ±1.645. The confidence interval estimate is .12 ± 1.645(.0492) = .12 ± .081. We can be 90% certain that the proportion of satisfied nurses on 7 a.m. to 3 p.m. shifts is between .039 and .201 higher than the proportion for nurses on 11 p.m. to 7 a.m. shifts.

EXAMPLE 13.8

A grocery store manager notes that in an SRS of 85 people going through the express checkout line, only 10 paid with checks, whereas, in an SRS of 92 customers passing through the regular line, 37 paid with checks. Find a 95% confidence interval estimate for the difference between the proportion of customers going through the two different lines who use checks.

Answer: Note that $n_1\hat{p}_1 = 10$, $n_1(1-\hat{p}_1) = 75$, $n_2\hat{p}_2 = 37$, and $n_2(1-\hat{p}_2) = 55$ are all at least 10, we are given SRSs, and the total number of customers is large.

$$n_1 = 85 \quad n_2 = 92$$
$$\hat{p}_1 = \frac{10}{85} = .118 \quad \hat{p}_2 = \frac{37}{92} = .402$$
$$\sigma_d \approx \sqrt{\frac{(.118)(.882)}{85} + \frac{(.402)(.598)}{92}} = .0619$$

The observed difference is .118 – .402 = –.284, and the critical *z*-scores are ±1.96. Thus, the confidence interval estimate is –.284 ± 1.96(.0619) = –.284 ± .121. The manager can be 95% sure that the proportion of customers passing through the express line who use checks is between .163 and .405 lower than the proportion going through the regular line who use checks.

With regard to choosing a sample size, $\sqrt{p(1-p)}$ is at most .5. Thus

$$\sqrt{p(1-p)\left(\frac{1}{n_1}+\frac{1}{n_2}\right)} \le (.5)\sqrt{\frac{1}{n_1}+\frac{1}{n_2}}$$

Now, if we simplify by insisting that $n_1 = n_2 = n$, the above statement can be reduced as follows:

$$(.5)\sqrt{\frac{1}{n}+\frac{1}{n}} = (.5)\sqrt{\frac{2}{n}} = \frac{.5\sqrt{2}}{\sqrt{n}}$$

(Note: this formula is not given on the AP Exam formula page.)

EXAMPLE 13.9

A pollster wants to determine the difference between the proportions of high-income voters and low-income voters who support a decrease in the capital gains tax. If the answer must be known to within ±.02 at the 95% confidence level, what size samples should be taken?

Answer: Assuming we will pick the same size samples for the two sample proportions, we have $\sigma_d \le \frac{.5\sqrt{2}}{\sqrt{n}}$ and $1.96\sigma_d \le .02$. Thus $\frac{1.96(.5)\sqrt{2}}{\sqrt{n}} \le .02$. Algebraically we find that $\sqrt{n} \ge \frac{1.96(.5)\sqrt{2}}{.02} = 69.3$. Therefore, $n \ge 69.3^2 = 4802.5$, and the pollster should use 4803 people for each sample.

CONFIDENCE INTERVAL FOR A MEAN

We are interested in estimating a population mean μ by considering a single sample mean $\bar{x}$. This sample mean is just one of a whole universe of sample means, and from Topic 12 we remember that if n is sufficiently large,

1. the set of all sample means is approximately normally distributed.
2. the mean of the set of sample means equals μ, the mean of the population.
3. the standard deviation $\sigma_{\bar{x}}$ of the set of sample means is approximately equal to $\frac{\sigma}{\sqrt{n}}$, that is, equal to the standard deviation of the whole population divided by the square root of the sample size.

Frequently we do not know σ, the population standard deviation. In such cases, we must use s, the *standard deviation of the sample,* as an estimate of σ. In this case $\frac{s}{\sqrt{n}}$ is called the *standard error,* $SE_{\bar{x}}$, and is used as an estimate for $\sigma_{\bar{x}} = \frac{\sigma}{\sqrt{n}}$. (Note that we use t-distributions instead of the standard normal curve whenever σ is unknown, no matter what the sample size, and the population must be assumed approximately normal.)

Remember that in making calculations and drawing conclusions from a specific sample, it is important that the sample be a *simple random sample* and be no more than 10% of the population.

EXAMPLE 13.10

A bottling machine is operating with a standard deviation of 0.12 ounce. Suppose that in an SRS of 36 bottles the machine inserted an average of 16.1 ounces into each bottle.

a. Estimate the mean number of ounces in all the bottles this machine fills. More specifically, give an interval within which we are 95% certain that the mean lies.
Answer: For samples of size 36, the sample means are approximately normally distributed with a standard deviation of $\sigma_{\bar{x}} = \frac{\sigma}{\sqrt{n}} = \frac{0.12}{\sqrt{36}} = 0.02$. From Topic 11 we know that 95% of the sample means should be within 1.96 standard deviations of the population mean. Equivalently, we are 95% certain that the population mean is within 1.96 standard deviations of any sample mean. In our case, 16.1 ± 1.96(0.02) = 16.1 ± 0.0392, and we are 95% sure that the mean number of ounces in all bottles is between 16.0608 and 16.1392. This is called a *95% confidence interval estimate*. [On the TI-84, under STAT and then TESTS, go to ZInterval. With Inpt:Stats, σ:.12, $\bar{x}$:16.1, n:36, and C-Level:.95, Calculate gives (16.061, 16.139).]

b. How about a 99% confidence interval estimate?
Answer: Here, 16.1 ± 2.576(0.02) = 16.1 ± 0.0515, and we are 99% sure that the mean number of ounces in all bottles is between 16.0485 and 16.1515. [On the TI-84, ZInterval gives (16.048, 16.152).] We can also say, using the definition of confidence *level*, that if the sampling procedure were repeated many times, about 99% of the resulting confidence intervals would contain the true population mean (thus we're 99% confident that the method worked for the interval we got).

Note that when we wanted a higher certainty (99% instead of 95%), we had to settle for a larger, less specific interval (±0.0515 instead of ±0.0392)

EXAMPLE 13.11

At a certain plant, batteries are being produced with a life expectancy that has a variance of 5.76 months squared. Suppose the mean life expectancy in an SRS of 64 batteries is 12.35 months.

a. Find a 90% confidence interval estimate of life expectancy for all the batteries produced at this plant.
Answer: The standard deviation of the population is $\sigma = \sqrt{5.76} = 2.4$, and the standard deviation of the sample means is $\sigma_{\bar{x}} = \frac{\sigma}{\sqrt{n}} = \frac{2.4}{\sqrt{64}} = 0.3$. The 90% confidence interval estimate for the population mean is 12.35 ± 1.645(0.3) = 12.35 ± 0.4935. Thus we are 90% certain that the mean life expectancy of the batteries is between 11.8565 and 12.8435 months. [The TI-84 gives (11.857, 12.843).]

b. What would the 90% confidence interval estimate be if the sample mean of 12.35 had come from a sample of 100 batteries?
Answer: The standard deviation of the sample means would then have been $\sigma_{\bar{x}} = \frac{\sigma}{\sqrt{n}} = \frac{2.4}{\sqrt{100}} = 0.24$, and the 90% confidence interval estimate would be 12.35 ± 1.645(0.24) = 12.35 ± 0.3948. [The TI-84 gives (11.955, 12.745).]

Note that when the sample size increased (from 64 to 100), the same sample mean resulted in a narrower, more specific interval (±0.3948 versus ±0.4935).

EXAMPLE 13.12

A new drug results in lowering the heart rate by varying amounts with a standard deviation of 2.49 beats per minute.

a. Find a 95% confidence interval estimate for the mean lowering of the heart rate in all patients if a 50-person SRS averages a drop of 5.32 beats per minute.
Answer: The standard deviation of sample means is $\sigma_{\bar{x}} = \frac{\sigma}{\sqrt{n}} = \frac{2.49}{\sqrt{50}} = 0.352$. We are 95% certain that the mean lowering of the heart rate is in the range 5.32 ± 1.96(0.352) = 5.32 ± 0.69 or between 4.63 and 6.01 heartbeats per minute. [The TI-84 gives (4.6298, 6.0102).]

b. With what certainty can we assert that the new drug lowers the heart rate by a mean of 5.32 ± 0.75 beats per minute?
Answer: Converting ±0.75 to *z*-scores yields $\frac{\pm 0.75}{0.352} = \pm 2.13$. From Table A, these *z*-scores give probabilities of .0166 and .9834, respectively, so our answer is .9834 − .0166 = .9668. In other words, 5.32 ± 0.75 beats per minute is a 96.68% confidence interval estimate of the mean lowering of the heart rate effected by this drug.

Remember, when σ is unknown (which is almost always the case), we use the *t*-distribution instead of the *z*-distribution. Furthermore, we must have that the parent population is normal, or at least nearly normal, which we can roughly check using a dotplot, stemplot, boxplot, histogram, or normal probability plot of the sample data. If we are not given that the parent population is normal, then for small samples ($n \leq 15$) the sample data should be unimodal and symmetric with no outliers and no skewness; for medium samples ($15 < n < 40$) the sample data should be unimodal and reasonably symmetric with no extreme values and little, if any, skewness; while for large samples ($n \geq 40$) the *t*-methods can be used no matter what the sample data show.

EXAMPLE 13.13

When ten cars of a new model were tested for gas mileage, the results showed a mean of 27.2 miles per gallon with a standard deviation of 1.8 miles per gallon. What is a 95% confidence interval estimate for the gas mileage achieved by this model? (Assume that the population of mpg results for all the new model cars is approximately normally distributed.)

Answer: The population standard is unknown, and so we use the *t*-distribution. The standard deviation of the sample means is $\sigma_{\bar{x}} = \frac{1.8}{\sqrt{10}} = 0.569$. With 10 − 1 = 9 degrees of freedom and 2.5% in each tail, the appropriate *t*-scores are ±2.262. Thus we can be 95% certain that the gas mileage of the new model is in the range 27.2 ± 2.262(0.569) = 27.2 ± 1.3 or between 25.9 and 28.5 miles per gallon.

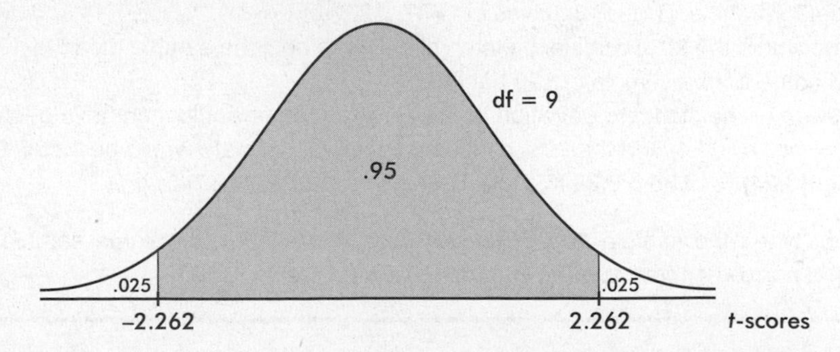

[On the TI-84, go to STAT, then TESTS, then TInterval. Using Stats, put in $\bar{x}$:27.2, Sx:1.8, n:10, and C-level:.95. Then Calculate gives (25.912, 28.488).]

EXAMPLE 13.14

A new process for producing synthetic gems yielded six stones weighing 0.43, 0.52, 0.46, 0.49, 0.60, and 0.56 carats, respectively, in its first run. Find a 90% confidence interval estimate for the mean carat weight from this process. (Assume that the population of carats of all gems produced by this new process is approximately normally distributed.)

Answer: Using the statistical package on your calculator or, much more laboriously, by hand, calculate the sample mean $\bar{x}$ and standard deviation s, and then find $\sigma_{\bar{x}}$ the standard deviation of the set of sample means:

$$\bar{x} = \frac{\Sigma x}{n} = \frac{0.43+0.52+0.46+0.49+0.60+0.56}{6} = \frac{3.06}{6} = 0.51$$

$$\begin{aligned} s &= \sqrt{\frac{\sum(x-\bar{x})^2}{n-1}} \\ &= \sqrt{\frac{(0.08)^2+(0.01)^2+(0.05)^2+(0.02)^2+(0.09)^2+(0.05)^2}{5}} \\ &= 0.0632 \end{aligned}$$

and

$$\sigma_{\bar{x}} \approx \frac{s}{\sqrt{n}} = \frac{0.0632}{\sqrt{6}} = 0.0258$$

With df = 6 − 1 = 5 and 5% in each tail, the *t*-scores are ±2.015. Thus we can be 90% sure that the new process will yield stones weighing 0.51 ± 2.015(0.0258) = 0.51 ± 0.052 or between 0.458 and 0.562 carats.

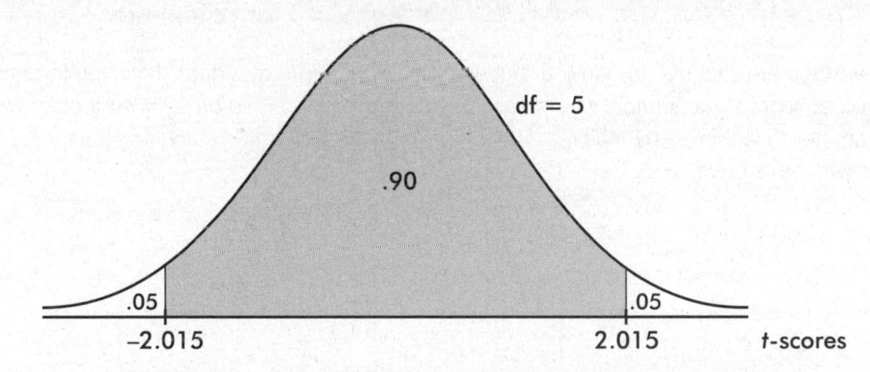

[On the TI-84, put the data in a list and then use Data under TInterval to obtain (.45797, .56203).]

On the TI-Nspire the result shows as:

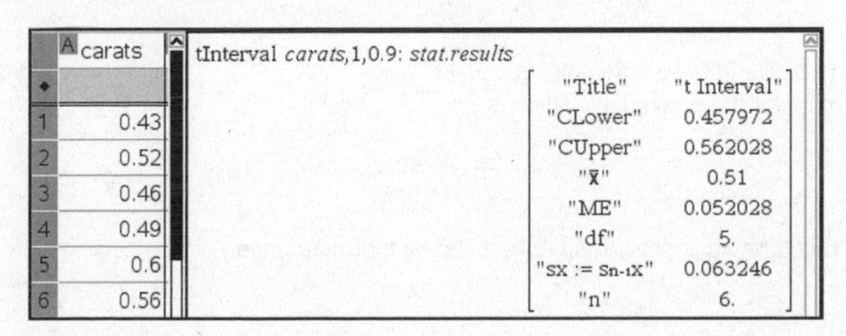

	A carats
1	0.43
2	0.52
3	0.46
4	0.49
5	0.6
6	0.56

tInterval *carats*,1,0.9: *stat.results*

"Title"	"t Interval"
"CLower"	0.457972
"CUpper"	0.562028
"x̄"	0.51
"ME"	0.052028
"df"	5.
"sx := sn-1x"	0.063246
"n"	6.

TIP

Read the question *carefully!* Be sure you understand exactly what you are being asked to do or find or explain.

EXAMPLE 13.15

A survey was conducted involving 250 out of 125,000 families living in a city. The average amount of income tax paid per family in the sample was $3540 with a standard deviation of $1150. Establish a 99% confidence interval estimate for the total taxes paid by all the families in the city. (Assume that the population of all income taxes paid is approximately normally distributed.)

Answer: We are given $\bar{x} = 3540$ and $s = 1150$. We use s as an estimate for σ and calculate the standard deviation of the sample means to be $\sigma_{\bar{x}} \approx \frac{s}{\sqrt{n}} = \frac{1150}{\sqrt{250}} = 72.73$.

Since σ is unknown, we use a *t*-distribution. In this case we obtain $3540 \pm 2.596(72.73) = 3540 \pm 188.8$ and then $125{,}000(3540 \pm 188.8) = 442{,}500{,}000 \pm 23{,}600{,}000$. That is, we are 99% confident that the total tax paid by all families in the city is between $418,900,000 and $466,100,000.

Statistical principles are useful not only in analyzing data but also in setting up experiments, and one important consideration is the choice of a sample size. In making interval estimates of population means, we have seen that each inference must go hand in hand with an associated confidence level statement. Generally, if we want a smaller, more precise interval estimate, we either decrease the degree of confidence or increase the sample size. Similarly, if we want to increase the degree of confidence, we either accept a wider interval estimate or increase the sample size. Thus, choosing a larger sample size always seems desirable; in the real world, however, time and cost considerations are involved.

EXAMPLE 13.16

Ball bearings are manufactured by a process that results in a standard deviation in diameter of 0.025 inch. What sample size should be chosen if we wish to be 99% sure of knowing the diameter to within ±0.01 inch?

Answer: We have

$$\sigma_{\bar{x}} = \frac{\sigma}{\sqrt{n}} = \frac{0.025}{\sqrt{n}} \quad \text{and} \quad 2.576\sigma_{\bar{x}} \le 0.01$$

Thus

$$2.576\left(\frac{0.025}{\sqrt{n}}\right) \le 0.01$$

Algebraically, we find that

$$\sqrt{n} \ge \frac{2.576(0.025)}{0.01} = 6.44$$

so $n \ge 41.5$. We choose a sample size of 42.

A formula for the above calculation is

$$n = \left(\frac{z\sigma}{\text{error}}\right)^2$$

(*Note:* This formula is not given on the AP Exam formula page.)

(continued)

Note that we can obtain a better estimate by recalculating using a *t*-score obtained from $df = n - 1$ where n is the value obtained from using the *z*-score. For example, in the above problem, a 99% confidence interval with $df = 42 - 1 = 41$ gives $t = \pm 2.701$. Then $2.701\left(\frac{0.025}{\sqrt{n}}\right) \le 0.01$ results in $n = 46$.

CONFIDENCE INTERVAL FOR A DIFFERENCE BETWEEN TWO MEANS

We have the following information about the sampling distribution of $\bar{x}_1 - \bar{x}_2$:

1. The set of all differences of sample means is approximately normally distributed.
2. The mean of the set of differences of sample means equals $\mu_1 - \mu_2$, the difference of population means.
3. The standard deviation $\sigma_{\bar{x}_1 - \bar{x}_2}$ of the set of differences of sample means is approximately equal to $\sqrt{\frac{\sigma_1^2}{n_1} + \frac{\sigma_2^2}{n_2}}$.

In making calculations and drawing conclusions from specific samples, it is important both that the samples be *simple random samples* and that they be taken *independently* of each other.

EXAMPLE 13.17

A 30-month study is conducted to determine the difference in the numbers of accidents per month occurring in two departments in an assembly plant. Suppose the first department averages 12.3 accidents per month with a standard deviation of 3.5, while the second averages 7.6 accidents with a standard deviation of 3.4. Determine a 95% confidence interval estimate for the difference in the numbers of accidents per month. (Assume that the two populations are independent and approximately normally distributed.)

Answer:

$$n_1 = 30 \qquad n_2 = 30$$
$$\bar{x}_1 = 12.3 \qquad \bar{x}_2 = 7.6$$
$$s_1 = 3.5 \qquad s_2 = 3.4$$

$$\sigma_{\bar{x}_1 - \bar{x}_2} = \sqrt{\frac{\sigma_1^2}{n_1} + \frac{\sigma_2^2}{n_2}} \approx \sqrt{\frac{s_1^2}{n_1} + \frac{s_2^2}{n_2}} = \sqrt{\frac{(3.5)^2}{30} + \frac{(3.4)^2}{30}} = 0.89$$

The observed difference is $12.3 - 7.6 = 4.7$, and with df $= (n_1 - 1) + (n_2 - 1) = 58$, the critical *t*-scores are ± 2.00. Thus the confidence interval estimate is $4.7 \pm 2.00(0.89) = 4.7 \pm 1.78$. We are 95% confident that the first department has between 2.92 and 6.48 more accidents per month than the second department. (Using the conservative $df = \min(x_1 - 1, x_2 - 1) = 29$ gives $4.7 \pm 2.045(0.89) = 4.7 \pm 1.82$.)

[On the TI-84, 2-SampTInt gives (2.9167, 6.4833).]

EXAMPLE 13.18

A survey is run to determine the difference in the cost of groceries in suburban stores versus inner city stores. A preselected group of items is purchased in a sample of 45 suburban and 35 inner city stores, and the following data are obtained.

Suburban stores	Inner city stores
$n_1 = 45$	$n_2 = 35$
$\bar{x}_1 = \$36.52$	$\bar{x}_2 = \$39.40$
$s_1 = \$1.10$	$s_2 = \$1.23$

Find a 90% confidence interval estimate for the difference in the cost of groceries. (Assume that the two populations are independent and approximately normally distributed.)

Answer:

$$\sigma_{\bar{x}_1-\bar{x}_2} \approx \sqrt{\frac{(1.10)^2}{45}+\frac{(1.23)^2}{35}} = 0.265$$

The observed difference is $36.52 - 39.40 = -2.88$, the critical t-scores are ± 1.664, and the confidence interval estimate is $-2.88 \pm 1.664(0.265) = -2.88 \pm 0.44$. Thus we are 90% certain that the selected group of items costs between \$2.44 and \$3.32 *less* in suburban stores than in inner city stores.

[On the TI-84, 2-SampTInt gives (−3.321, −2.439).]

EXAMPLE 13.19

A hardware store owner wishes to determine the difference between the drying times of two brands of paint. Suppose the standard deviation between cans in each population is 2.5 minutes. How large a sample (same number) of each must the store owner use if he wishes to be 98% sure of knowing the difference to within 1 minute?

Answer:

$$\sigma_{\bar{x}_1-\bar{x}_2} = \sqrt{\frac{(2.5)^2}{n}+\frac{(2.5)^2}{n}} = \frac{3.536}{\sqrt{n}}$$

With a critical z-score of 2.326, we have $2.326\left(\frac{3.536}{\sqrt{n}}\right) \le 1$, so $\sqrt{n} \ge 8.22$ and $n \ge 67.6$. Thus the owner should test samples of 68 paint patches from each brand.

In the case of independent samples, there is also a procedure based on the assumption that both original populations not only are normally distributed but also have equal variances. Then we can get a better estimate of the common population variance by *pooling* the two sample variances. But it's never wrong not to pool.

The analysis and procedure discussed in this section require that the two samples being compared be independent of each other. However, many experiments and tests involve comparing two populations for which the data naturally occur in pairs. In this case of *paired differences*, the proper procedure is to run a one-sample analysis on the single variable consisting of the differences from the paired data.

EXAMPLE 13.20

An SAT preparation class of 30 students produces the following improvement in scores:

Student	First Score	Second Score	Improvement
1	912	1025	113
2	1025	1085	60
3	1295	1350	55
4	1123	1202	79
5	875	982	107
6	890	950	60
7	1002	1089	87
8	998	1159	161
9	1235	1246	11
10	1045	1135	90
11	956	1005	49
12	987	1010	23
13	1028	1015	–13
14	954	1032	78
15	1152	1310	158
16	1215	1302	87
17	948	1010	62
18	1190	1235	45
19	1077	1103	26
20	1223	1200	–23
21	1100	1187	87
22	842	910	68
23	985	1049	64
24	1107	1123	16
25	847	901	54
26	987	1086	99
27	1228	1276	48
28	1005	1029	24
29	1166	1221	55
30	808	874	66

Find a 90% confidence interval estimate of the average improvement in test scores.

Answer: The two sets of test scores (before and after the course) are not independent; they are associated in matched pairs, with one pair of scores for each student. The proper procedure is to run a one-sample analysis on the single variable consisting of the set of improvements. In this case we obtain $\bar{x} = 63.2$, $s = 41.89$, $\sigma_{\bar{x}} \approx \frac{41.89}{\sqrt{30}} = 7.65$, and a 90% confidence interval estimate of $63.2 \pm 1.699(7.648) = 63.2 \pm 13.0$. We can be 90% confident that the average improvement in test scores is between 50.2 and 76.2.

When to do a two-sample analysis versus a matched pair analysis can be confusing. The key idea is whether or not the two groups are independent. The difference has to do with design. What is the average reaction time of people who are sober compared to that of people who have had two beers? If our experimental design calls for randomly assigning half of a group of volunteers to drink two beers and then comparing reaction times with the remaining sober volunteers, then a two-sample analysis is called for. However, if our experimental design calls for testing reaction times of all the volunteers and then giving all the volunteers two beers and retesting, then a matched pair analysis is called for. Matching may come about as above because you have made two measurements on the same person, or measurements might be made on sets of twins, or between salaries of president and provost at a number of universities, etc. The key is whether the two sets of measurements are independent, or related in some way relevant to the question under consideration.

CONFIDENCE INTERVAL FOR THE SLOPE OF A LEAST SQUARES REGRESSION LINE

In Topic 4 we discussed the least squares line

$$\hat{y} = \bar{y} + b_1(x - \bar{x})$$

where b_1 is the slope of the line. This slope is readily found using the statistical software on a calculator. It is an estimate of the slope β of the true regression line. A confidence interval estimate for β can be found using t-scores, where df = $n - 2$, and the appropriate standard deviation is

$$s_{b_1} = \frac{\sqrt{\frac{\Sigma(y_i - \hat{y}_i)^2}{n-2}}}{\sqrt{\Sigma(x_i - \bar{x})^2}}$$

where the sum in the numerator is the sum of the squared residuals (see Topic 4) and the sum in the denominator is the sum of the squared deviations from the mean. The value of s_{b_1} is found at the same time that the statistical software on your calculator finds the value of b_1. Some calculators calculate the numerator $s = \sqrt{\frac{\Sigma(y_i - \hat{y}_i)^2}{n-2}}$, and then the denominator can be calculated by $\sqrt{\Sigma(x_i - \bar{x})^2} = s_x\sqrt{n-1}$.

Assumptions here include: (1) the sample must be randomly selected; (2) the scatterplot should be approximately linear; (3) there should be no apparent pattern in the residuals plot (On the TI-84, after Stat → Calc → LinReg, the list of residuals is stored in RESID under LIST NAMES); and (4) the distribution of the residuals should be approximately normal. (The fourth point can be checked by a histogram, dotplot, stemplot, or a normal probability plot of the residuals.)

EXAMPLE 13.21

A random sample of ten high school students produced the following results for number of hours of television watched per week and GPA.

TV hours	12	21	8	20	16	16	24	0	11	18
GPA	3.1	2.3	3.5	2.5	3.0	2.6	2.1	3.8	2.9	2.6

Determine the least squares line and give a 95% confidence interval estimate for the true slope.

Answer: Checking the assumptions:

We are told that the data come from a *random* sample of students.

Scatterplot is approximately linear.

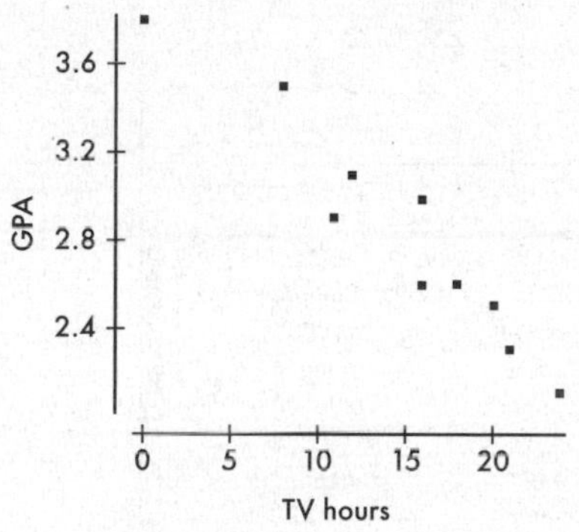

No apparent pattern is evident in residuals plot.

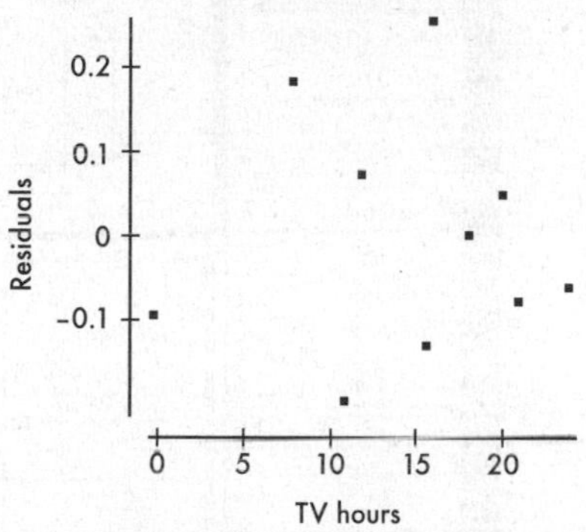

Distribution of residuals is approximately normal.

Checked by histogram.

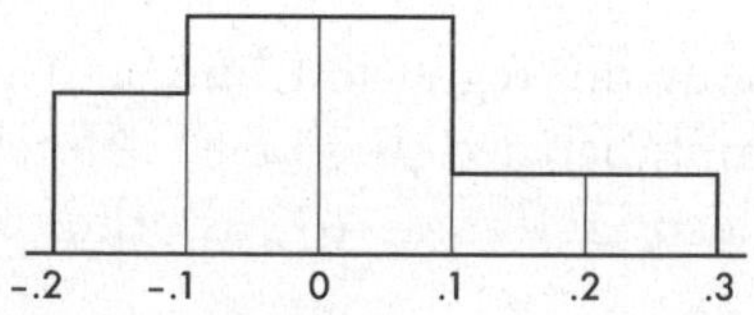

Checked by normal probability plot.

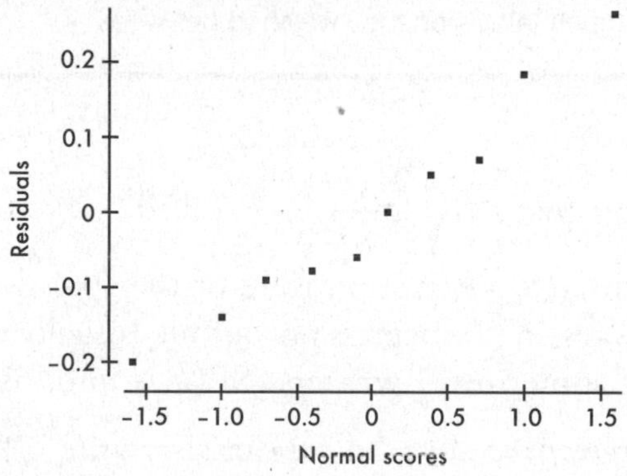

Now using the statistics software on a calculator gives

$$\hat{y} = 3.892 - 0.07202x$$

Using, for example, the TI-84 (LinRegTTest and 1-Var Stats) one also obtains

$$s = 0.1531 \quad \text{and} \quad s_x = 7.074$$

Thus

$$s_{b_1} = \frac{0.1531}{7.074\sqrt{10-1}} = 0.0072$$

(continued)

TIP

If you refer to a graph, whether it is a histogram, boxplot, stemplot, scatterplot, residuals plot, normal probability plot, or some other kind of graph, you should **roughly draw it.** It is not enough to simply say, "I did a normal probability plot of the residuals on my calculator and it looked linear."

[Note: On the TI-84, `LinRegTTest` gives both b_1 and t, and you can also calculate

$$s_{b_1} = \frac{b_1}{t} = \frac{-0.07202}{-9.984} = 0.0072, \text{ which comes from } t = \frac{b_1}{s_{b_1}}.]$$

With ten data points, the degrees of freedom are $df = 10 - 2 = 8$, and the critical t-values (with .025 in each tail) are ± 2.306. The 95% confidence interval of the true slope is

$$b_1 \pm t\, s_{b_1} = -0.07202 \pm 2.306(0.0072) = -0.07202 \pm 0.0166$$

We are thus 95% confident that each additional hour before the television each week is associated with between a 0.055 and a 0.089 drop in GPA. (A TI-84 with `LinRegTInt` will give this result more directly.)

On the TI-Nspire one can show the following:

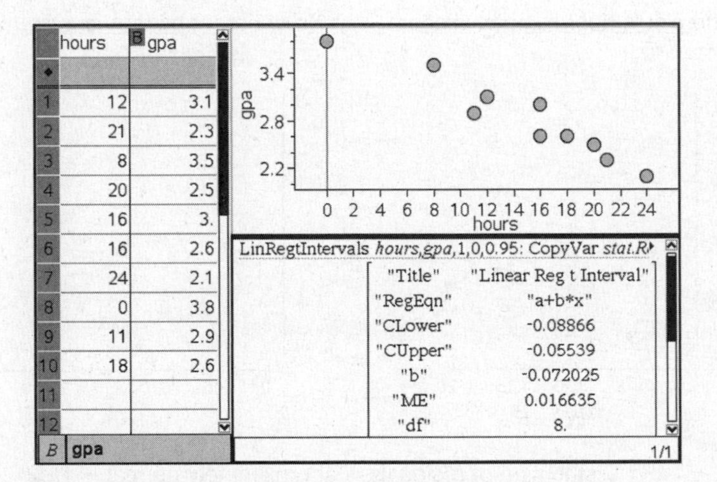

Remember we must be careful about drawing conclusions concerning cause and effect; it is possible that students who have lower GPAs would have these averages no matter how much television they watched per week.

Note that $s = \sqrt{\frac{\sum (y - \hat{y})^2}{n-2}}$, called the *standard deviation of the residuals*, is sometimes written s_e. It is a measure of the spread about the regression line, and is almost always in the regression output (usually simply labeled as s or S). Note that S_{b_1} is sometimes written $\text{SE}(b_1)$ and is also referred to as the *standard error of the slope.* So we can also write $\text{SE}(b_1) = \frac{s_e}{s_x\sqrt{n-1}}$. We see that increasing s_e (that is, increasing spread about the line) *increases* the slope's standard error, while increasing s_x (that is, increasing the range of x-values) or increasing n (the sample size) *decreases* the slope's standard error.

EXAMPLE 13.22

Information concerning SAT verbal scores and SAT math scores was collected from 15 randomly selected subjects. A linear regression performed on the data using a statistical software package produced the following printout:

```
Dependent variable: Math
Variable        Coef          SE Coef         T        Prob
Constant      92.5724          31.75        2.92      0.012
Verbal        0.763604         0.05597      13.6      0.000
S = 16.69    R-Sq = 93.5%    R-Sq(adj) = 93.0%
```

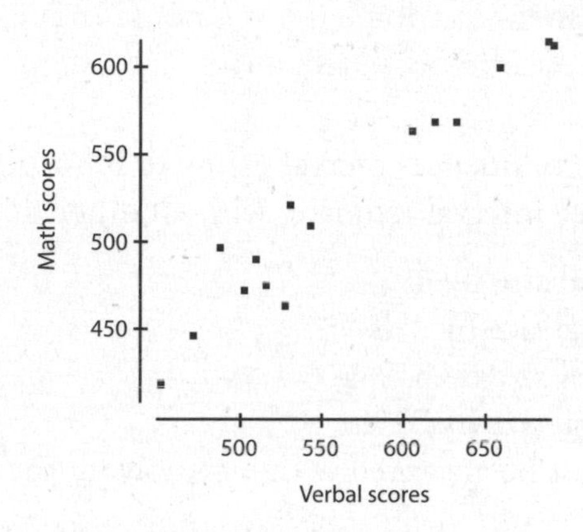

What is the regression equation?

Answer: Assuming that all assumptions for regression are met (we are given that the sample is random, and the scatterplot is approximately linear), the *y*-intercept and slope of the equation are found in the `Coef` column of the above printout

$$\widehat{\text{Math}} = 92.57 + 0.764 \text{ Verbal}$$

What is a 95% confidence interval estimate for the slope of the regression line?

Answer: The standard deviation of the residuals is S = 16.69 and the standard error of the slope is $S_{b_1} = 0.05597$. With 15 data points, df = 15 − 2 = 13, and the critical *t*-values are ± 2.160. The 95% confidence interval of the true slope is:

$$b_1 \pm t\, s_{b_1} = 0.764 \pm 2.160(0.05597) = 0.764 \pm 0.121$$

We are 95% confident that for every 1-point increase in verbal SAT score, the average increase in math SAT score is between .64 and .89

> **TIP**
>
> Practice how to read *generic computer output,* and note that you usually will *not* need to use all the information given.

Summary

- Important assumptions/conditions always must be checked before calculating confidence intervals.
- We are never able to say exactly what a population parameter is; rather, we say that we have a certain confidence that it lies in a certain interval.
- If we want a narrower interval, we must either decrease the confidence level or increase the sample size.
- If we want a higher level of confidence, we must either accept a wider interval or increase the sample size.
- The level of confidence refers to the percentage of samples that produce intervals that capture the true population parameter.

Questions on Topic Thirteen: Confidence Intervals

Multiple-Choice Questions

Directions: The questions or incomplete statements that follow are each followed by five suggested answers or completions. Choose the response that best answers the question or completes the statement.

1. Changing from a 95% confidence interval estimate for a population proportion to a 99% confidence interval estimate, with all other things being equal,

   (A) increases the interval size by 4%.
   (B) decreases the interval size by 4%.
   (C) increases the interval size by 31%.
   (D) decreases the interval size by 31%.
   (E) This question cannot be answered without knowing the sample size.

2. In general, how does doubling the sample size change the confidence interval size?

   (A) Doubles the interval size
   (B) Halves the interval size
   (C) Multiplies the interval size by 1.414
   (D) Divides the interval size by 1.414
   (E) This question cannot be answered without knowing the sample size.

3. A confidence interval estimate is determined from the GPAs of a simple random sample of n students. All other things being equal, which of the following will result in a smaller margin of error?

   (A) A smaller confidence level
   (B) A larger sample standard deviation
   (C) A smaller sample size
   (D) A larger population size
   (E) A smaller sample mean

4. A survey was conducted to determine the percentage of high school students who planned to go to college. The results were stated as 82% with a margin of error of ±5%. What is meant by ±5%?

    (A) Five percent of the population were not surveyed.
    (B) In the sample, the percentage of students who plan to go to college was between 77% and 87%.
    (C) The percentage of the entire population of students who plan to go to college is between 77% and 87%.
    (D) It is unlikely that the given sample proportion result would be obtained unless the true percentage was between 77% and 87%.
    (E) Between 77% and 87% of the population were surveyed.

5. Most recent tests and calculations estimate at the 95% confidence level that the maternal ancestor to all living humans called mitochondrial Eve lived 273,000 ±177,000 years ago. What is meant by "95% confidence" in this context?

    (A) A confidence interval of the true age of mitochondrial Eve has been calculated using z-scores of ±1.96.
    (B) A confidence interval of the true age of mitochondrial Eve has been calculated using t-scores consistent with $df = n - 1$ and tail probabilities of ±.025.
    (C) There is a .95 probability that mitochondrial Eve lived between 96,000 and 450,000 years ago.
    (D) If 20 random samples of data are obtained by this method, and a 95% confidence interval is calculated from each, then the true age of mitochondrial Eve will be in 19 of these intervals.
    (E) 95% of all random samples of data obtained by this method will yield intervals that capture the true age of mitochondrial Eve.

6. One month the actual unemployment rate in France was 13.4%. If during that month you took an SRS of 100 Frenchmen and constructed a confidence interval estimate of the unemployment rate, which of the following would have been true?

    (A) The center of the interval was 13.4.
    (B) The interval contained 13.4.
    (C) A 99% confidence interval estimate contained 13.4.
    (D) The z-score of 13.4 was between ±2.576.
    (E) None of the above are true statements.

7. In a recent Zogby International survey, 11% of 10,000 Americans under 50 said they would be willing to implant a device in their brain to be connected to the Internet if it could be done safely. What is the margin of error at the 99% confidence level?

   (A) $\pm\sqrt{\frac{(1{,}100)(8{,}900)}{10{,}000}}$

   (B) $\pm 1.96 \frac{.5}{\sqrt{10{,}000}}$

   (C) $\pm 2.576 \frac{\sqrt{(.11)(.89)}}{10{,}000}$

   (D) $\pm 1.96 \sqrt{\frac{(.11)(.89)}{10{,}000}}$

   (E) $\pm 2.576 \sqrt{\frac{(.11)(.89)}{10{,}000}}$

8. The margin of error in a confidence interval estimate using *z*-scores covers which of the following?

   (A) Sampling variability
   (B) Errors due to undercoverage and nonresponse in obtaining sample surveys
   (C) Errors due to using sample standard deviations as estimates for population standard deviations
   (D) Type I errors
   (E) Type II errors

9. In an SRS of 50 teenagers, two-thirds said they would rather text a friend than call. What is the 98% confidence interval for the proportion of teens who would rather text than call a friend?

   (A) $\frac{2}{3} \pm 1.96 \sqrt{\frac{(2/3)(1/3)}{50}}$

   (B) $\frac{2}{3} \pm 2.326 \sqrt{\frac{(2/3)(1/3)}{50}}$

   (C) $\frac{2}{3} \pm 2.405 \sqrt{\frac{(2/3)(1/3)}{50}}$

   (D) $\frac{2}{3} \pm 1.96 \frac{\sqrt{(2/3)(1/3)}}{50}$

   (E) $\frac{2}{3} \pm 2.326 \frac{\sqrt{(2/3)(1/3)}}{50}$

10. In a survey funded by Burroughs-Welcome, 750 of 1000 adult Americans said they didn't believe they could come down with a sexually transmitted disease (STD). Construct a 95% confidence interval estimate of the proportion of adult Americans who don't believe they can contract an STD.

    (A) (.728, .772)
    (B) (.723, .777)
    (C) (.718, .782)
    (D) (.713, .787)
    (E) (.665, .835)

11. A 1993 *Los Angeles Times* poll of 1703 adults revealed that only 17% thought the media was doing a "very good" job. With what degree of confidence can the newspaper say that 17% ± 2% of adults believe the media is doing a "very good" job?

    (A) 72.9%
    (B) 90.0%
    (C) 95.0%
    (D) 97.2%
    (E) 98.6%

12. A politician wants to know what percentage of the voters support her position on the issue of forced busing for integration. What size voter sample should be obtained to determine with 90% confidence the support level to within 4%?

    (A) 21
    (B) 25
    (C) 423
    (D) 600
    (E) 1691

13. In a *New York Times* poll measuring a candidate's popularity, the newspaper claimed that in 19 of 20 cases its poll results should be no more than three percentage points off in either direction. What confidence level are the pollsters working with, and what size sample should they have obtained?

    (A) 3%, 20
    (B) 6%, 20
    (C) 6%, 100
    (D) 95%, 33
    (E) 95%, 1068

14. In an SRS of 80 teenagers, the average number of texts handled in a day was 50 with a standard deviation of 15. What is the 96% confidence interval for the average number of texts handled by teens daily?

    (A) $50 \pm 2.054(15)$

    (B) $50 \pm 2.054 \frac{15}{\sqrt{79}}$

    (C) $50 \pm 2.054 \frac{15}{\sqrt{80}}$

    (D) $50 \pm 2.088 \frac{15}{\sqrt{79}}$

    (E) $50 \pm 2.088 \frac{15}{\sqrt{80}}$

15. One gallon of gasoline is put in each of 30 test autos, and the resulting mileage figures are tabulated with $\bar{x} = 28.5$ and $s = 1.2$. Determine a 95% confidence interval estimate of the mean mileage.

    (A) (28.46, 28.54)
    (B) (28.42, 28.58)
    (C) (28.1, 28.9)
    (D) (27.36, 29.64)
    (E) (27.3, 29.7)

16. The number of accidents per day at a large factory is noted for each of 64 days with $\bar{x} = 3.58$ and $s = 1.52$. With what degree of confidence can we assert that the mean number of accidents per day at the factory is between 3.20 and 3.96?

    (A) 48%
    (B) 63%
    (C) 90%
    (D) 95%
    (E) 99%

17. A company owns 335 trucks. For an SRS of 30 of these trucks, the average yearly road tax paid is \$9540 with a standard deviation of \$1205. What is a 99% confidence interval estimate for the total yearly road taxes paid for the 335 trucks?

    (A) \$9540 ± \$103
    (B) \$9540 ± \$567
    (C) \$3,196,000 ± \$606
    (D) \$3,196,000 ± \$35,000
    (E) \$3,196,000 ± \$203,000

18. What sample size should be chosen to find the mean number of absences per month for school children to within ±.2 at a 95% confidence level if it is known that the standard deviation is 1.1?

    (A) 11
    (B) 29
    (C) 82
    (D) 96
    (E) 117

19. Hospital administrators wish to learn the average length of stay of all surgical patients. A statistician determines that, for a 95% confidence level estimate of the average length of stay to within ±0.5 days, 50 surgical patients' records will have to be examined. How many records should be looked at to obtain a 95% confidence level estimate to within ±0.25 days?

    (A) 25
    (B) 50
    (C) 100
    (D) 150
    (E) 200

20. The National Research Council of the Philippines reported that 210 of 361 members in biology are women, but only 34 of 86 members in mathematics are women. Establish a 95% confidence interval estimate of the difference in proportions of women in biology and women in mathematics in the Philippines.

    (A) .187 ± .115
    (B) .187 ± .154
    (C) .395 ± .103
    (D) .543 ± .154
    (E) .582 ± .051

21. In a simple random sample of 300 elderly men, 65% were married, while in an independent simple random sample of 400 elderly women, 48% were married. Determine a 99% confidence interval estimate for the difference between the proportions of elderly men and women who are married.

    (A) $(.65 - .48) \pm 2.326\sqrt{\frac{(.65)(.35)}{300} + \frac{(.48)(.52)}{400}}$

    (B) $(.65 - .48) \pm 2.576\sqrt{\frac{(.65)(.35)}{300} + \frac{(.48)(.52)}{400}}$

    (C) $(.65 - .48) \pm 2.576\left(\frac{(.65)(.35)}{\sqrt{300}} + \frac{(.48)(.52)}{\sqrt{400}}\right)$

    (D) $\left(\frac{.65 + .48}{2}\right) \pm 2.576\sqrt{\frac{(.65)(.35)}{300} + \frac{(.48)(.52)}{400}}$

    (E) $\left(\frac{.65 + .48}{2}\right) \pm 2.807\sqrt{(.565)(.435)\left(\frac{1}{300} + \frac{1}{400}\right)}$

22. A researcher plans to investigate the difference between the proportion of psychiatrists and the proportion of psychologists who believe that most emotional problems have their root causes in childhood. How large a sample should be taken (same number for each group) to be 90% certain of knowing the difference to within ±.03?

    (A) 39
    (B) 376
    (C) 752
    (D) 1504
    (E) 3007

23. In a study aimed at reducing developmental problems in low-birth-weight (under 2500 grams) babies (*Journal of the American Medical Association*, June 13, 1990, page 3040), 347 infants were exposed to a special educational curriculum while 561 did not receive any special help. After 3 years the children exposed to the special curriculum showed a mean IQ of 93.5 with a standard deviation of 19.1; the other children had a mean IQ of 84.5 with a standard deviation of 19.9. Find a 95% confidence interval estimate for the difference in mean IQs of low-birth-weight babies who receive special intervention and those who do not.

    (A) $(93.5 - 84.5) \pm 1.97\sqrt{\frac{(19.1)^2}{347} + \frac{(19.9)^2}{561}}$

    (B) $(93.5 - 84.5) \pm 1.97\left(\frac{19.1}{\sqrt{347}} + \frac{19.9}{\sqrt{561}}\right)$

    (C) $(93.5 - 84.5) \pm 1.65\sqrt{\frac{(19.1)^2}{347} + \frac{(19.9)^2}{561}}$

    (D) $(93.5 - 84.5) \pm 1.65\left(\frac{19.1}{\sqrt{347}} + \frac{19.9}{\sqrt{561}}\right)$

    (E) $(93.5 - 84.5) \pm 1.65\sqrt{\frac{(19.1)^2 + (19.9)^2}{347 + 561}}$

24. Does socioeconomic status relate to age at time of HIV infection? For 274 high-income HIV-positive individuals the average age of infection was 33.0 years with a standard deviation of 6.3, while for 90 low-income individuals the average age was 28.6 years with a standard deviation of 6.3 (*The Lancet*, October 22, 1994, page 1121). Find a 90% confidence interval estimate for the difference in ages of high- and low-income people at the time of HIV infection.

    (A) 4.4 ± 0.963
    (B) 4.4 ± 1.26
    (C) 4.4 ± 2.51
    (D) 30.8 ± 2.51
    (E) 30.8 ± 6.3

25. An engineer wishes to determine the difference in life expectancies of two brands of batteries. Suppose the standard deviation of each brand is 4.5 hours. How large a sample (same number) of each type of battery should be taken if the engineer wishes to be 90% certain of knowing the difference in life expectancies to within 1 hour?

    (A) 10
    (B) 55
    (C) 110
    (D) 156
    (E) 202

26. Two confidence interval estimates from the same sample are (16.4, 29.8) and (14.3, 31.9). What is the sample mean, and if one estimate is at the 95% level while the other is at the 99% level, which is which?

    (A) $\bar{x} = 23.1$; (16.4, 29.8) is the 95% level.
    (B) $\bar{x} = 23.1$; (16.4, 29.8) is the 99% level.
    (C) It is impossible to completely answer this question without knowing the sample size.
    (D) It is impossible to completely answer this question without knowing the sample standard deviation.
    (E) It is impossible to completely answer this question without knowing both the sample size and standard deviation.

27. Two 90% confidence interval estimates are obtained: I (28.5, 34.5) and II (30.3, 38.2).

    a. If the sample sizes are the same, which has the larger standard deviation?
    b. If the sample standard deviations are the same, which has the larger size?

    (A) *a.* I *b.* I
    (B) *a.* I *b.* II
    (C) *a.* II *b.* I
    (D) *a.* II *b.* II
    (E) More information is needed to answer these questions.

28. Suppose (25, 30) is a 90% confidence interval estimate for a population mean μ. Which of the following are true statements?

    (A) There is a .90 probability that $\bar{x}$ is between 25 and 30.
    (B) 90% of the sample values are between 25 and 30.
    (C) There is a .90 probability that μ is between 25 and 30.
    (D) If 100 random samples of the given size are picked and a 90% confidence interval estimate is calculated from each, then μ will be in 90 of the resulting intervals.
    (E) If 90% confidence intervals are calculated from all possible samples of the given size, μ will be in 90% of these intervals.

29. Under what conditions would it be meaningful to construct a confidence interval estimate when the data consist of the entire population?

    (A) If the population size is small ($n < 30$)
    (B) If the population size is large ($n \geq 30$)
    (C) If a higher level of confidence is desired
    (D) If the population is truly random
    (E) Never

30. A social scientist wishes to determine the difference between the percentage of Los Angeles marriages and the percentage of New York marriages that end in divorce in the first year. How large a sample (same for each group) should be taken to estimate the difference to within ±.07 at the 94% confidence level?

    (A) 181
    (B) 361
    (C) 722
    (D) 1083
    (E) 1443

31. What is the critical *t*-value for finding a 90% confidence interval estimate from a sample of 15 observations?

    (A) 1.341
    (B) 1.345
    (C) 1.350
    (D) 1.753
    (E) 1.761

32. Acute renal graft rejection can occur years after the graft. In one study (*The Lancet*, December 24, 1994, page 1737), 21 patients showed such late acute rejection when the ages of their grafts (in years) were 9, 2, 7, 1, 4, 7, 9, 6, 2, 3, 7, 6, 2, 3, 1, 2, 3, 1, 1, 2, and 7, respectively. Establish a 90% confidence interval estimate for the ages of renal grafts that undergo late acute rejection.

    (A) 2.024 ± 0.799
    (B) 2.024 ± 1.725
    (C) 4.048 ± 0.799
    (D) 4.048 ± 1.041
    (E) 4.048 ± 1.725

33. Nine subjects, 87 to 96 years old, were given 8 weeks of progressive resistance weight training (*Journal of the American Medical Association,* June 13, 1990, page 3032). Strength before and after training for each individual was measured as maximum weight (in kilograms) lifted by left knee extension:

Before:	3	3.5	4	6	7	8	8.5	12.5	15
After:	7	17	19	12	19	22	28	20	28

Find a 95% confidence interval estimate for the strength gain.

(A) 11.61 ± 3.03
(B) 11.61 ± 3.69
(C) 11.61 ± 3.76
(D) 19.11 ± 1.25
(E) 19.11 ± 3.69

34. A catch of five fish of a certain species yielded the following ounces of protein per pound of fish: 3.1, 3.5, 3.2, 2.8, and 3.4. What is a 90% confidence interval estimate for ounces of protein per pound of this species of fish?

(A) 3.2 ± 0.202
(B) 3.2 ± 0.247
(C) 3.2 ± 0.261
(D) 4.0 ± 0.202
(E) 4.0 ± 0.247

35. In a random sample of 25 professional baseball players, their salaries (in millions of dollars) and batting averages result in the following regression analysis:

```
The regression equation is    Batting = 0.234 + 0.00805 Salary

Predictor        Coef    SE Coef        T       P
Constant     0.233577   0.005883    39.70   0.000
Salary       0.008051   0.001058     7.61   0.000

S = 0.0169461    R-Sq = 71.6%    R-Sq(adj) = 70.3%
```

Which of the following gives a 98% confidence interval for the slope of the regression line?

(A) 0.008051 ± 2.326(0.001058)
(B) 0.008051 ± 2.326(0.0169461)
(C) $0.008051 \pm 2.485\left(\dfrac{0.001058}{\sqrt{25}}\right)$
(D) $0.008051 \pm 2.492\left(\dfrac{0.0169461}{\sqrt{25}}\right)$
(E) 0.008051 ± 2.500(0.001058)

36. Is there a relationship between mishandled-baggage rates (number of mishandled bags per 1000 passengers) and percentage of on-time arrivals? Regression analysis of ten airlines gives the following TI-Nspire output (baggage rate is independent variable):

Alternate	β&ρ≠0..
RegEqn	a+b*x
t	-2.10163
PVal	0.068763
a	87.2338
b	-2.16661
s	4.33798
SESlope...	1.03092
r^2	0.355714
r	-0.596418

We are 95% confident that if an airline averages one more mishandled bag per 1000 passengers than a second airline, then its percentage of on-time arrivals will average

(A) between 0.15 and 4.19 less than that of the second airline.
(B) between 0.00 and 4.33 less than that of the second airline.
(C) between 4.46 less and 0.13 more than that of the second airline.
(D) between 4.49 less and 0.17 more than that of the second airline.
(E) between 4.54 less and 0.21 more than that of the second airline.

37. Smoking is a known risk factor for cardiovascular and cancer diseases. A recent study (Rusanen et al., *Archives of Internal Medicine*, October 25, 2010) surveyed 21,123 members of a California health care system looking at smoking levels and risk of dementia (as measured by the Cox hazard ratio). Results are summarized in the following graph of 95% confidence intervals.

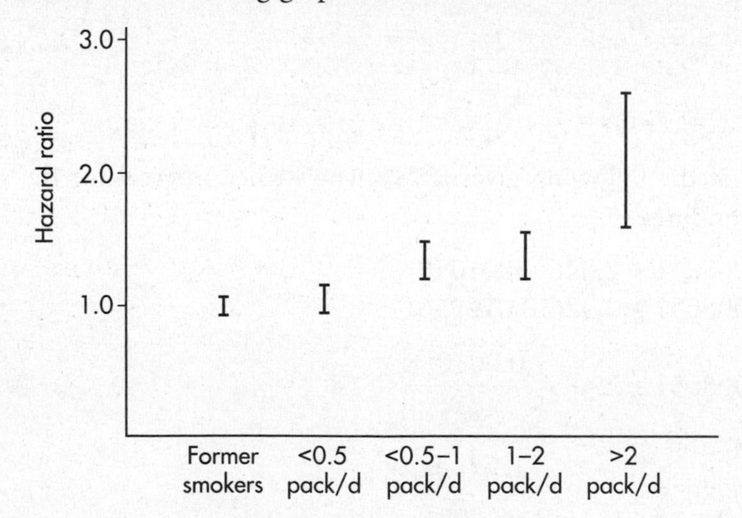

Which of the following is an appropriate conclusion?

(A) Dementia somehow causes an urge to smoke.
(B) Heavier smoking leads to a greater risk of dementia.
(C) Cutting down on smoking will lower one's risk of dementia.
(D) Heavier smoking is associated with more precise confidence intervals of dementia as measured by the hazard ratio.
(E) There is a positive association between smoking and risk of dementia.

Free-Response Questions

Directions: You must show all work and indicate the methods you use. You will be graded on the correctness of your methods and on the accuracy of your final answers.

Eleven Open-Ended Questions

1. An SRS of 1000 voters finds that 57% believe that competence is more important than character in voting for President of the United States.

   (a) Determine a 95% confidence interval estimate for the percentage of voters who believe competence is more important than character.

   (b) If your parents know nothing about statistics, how would you explain to them why you can't simply say that 57% of voters believe that competence is more important.

   (c) Also explain to your parents what is meant by 95% confidence level.

2. During the H1N1 pandemic, one published study concluded that if someone in your family had H1N1, you had a 1 in 8 chance of also coming down with the disease. A state health officer tracks a random sample of new H1N1 cases in her state and notes that 129 out of a potential 876 family members later come down with the disease.

   (a) Calculate a 90% confidence interval for the proportion of family members who come down with H1N1 after an initial family member does in this state.

   (b) Based on this confidence interval, is there evidence that the proportion of family members who come down with H1N1 after an initial family member does in this state is different from the 1 in 8 chance concluded in the published study? Explain.

   (c) Would the conclusion in (b) be any different with a 99% confidence interval? Explain.

3. In a random sample of 500 new births in the United States, 41.2% were to unmarried women, while in a random sample of 400 new births in the United Kingdom, 46.5% were to unmarried women.

   (a) Calculate a 95% confidence interval for the difference in the proportions of new births to unmarried women in the United States and United Kingdom.

   (b) Does the confidence interval support the belief by a UN health care statistician that the proportions of new births to unmarried women is different in the United States and United Kingdom? Explain.

4. An SRS of 40 inner city gas stations shows a mean price for regular unleaded gasoline to be \$3.45 with a standard deviation of \$0.05, while an SRS of 120 suburban stations shows a mean of \$3.38 with a standard deviation of \$0.08.

   (a) Construct 95% confidence interval estimates for the mean price of regular gas in inner city and in suburban stations.

   (b) The confidence interval for the inner city stations is wider than the interval for the suburban stations even though the standard deviation for inner city stations is less than that for suburban stations. Explain why this happened.

   (c) Based on your answer in part (a), are you confident that the mean price of inner city gasoline is less than \$3.50? Explain.

5. In a simple random sample of 30 subway cars during rush hour, the average number of riders per car was 83.5 with a standard deviation of 5.9. Assume the sample data are unimodal and reasonably symmetric with no extreme values and little, if any, skewness.

   (a) Establish a 90% confidence interval estimate for the average number of riders per car during rush hour. Show your work.

   (b) Assuming the same standard deviation of 5.9, how large a sample of cars would be necessary to determine the average number of riders to within ± 1 at the 90% confidence level? Show your work.

6. In a sample of ten basketball players the mean income was \$196,000 with a standard deviation of \$315,000.

   (a) Assuming all necessary assumptions are met, find a 95% confidence interval estimate of the mean salary of basketball players.

   (b) What assumptions are necessary for the above estimate? Do they seem reasonable here?

7. An SRS of ten brands of breakfast cereals is tested for the number of calories per serving. The following data result: 185, 190, 195, 200, 205, 205, 210, 210, 225, 230.

   Establish a 95% confidence interval estimate for the mean number of calories for servings of breakfast cereals. Be sure to check assumptions.

8. Bisphenol A (BPA), a synthetic estrogen found in packaging materials, has been shown to leach into infant formula and beverages. Most recently it has been detected in high concentrations in cash register receipts. Random sample bio-monitoring data to see if retail workers carry higher amounts of BPA in their bodies than non-retail workers is as follows:

		BPA concentration (μg/L)	
	Sample size	Mean	Standard deviation
Non-retail workers	528	2.43	0.45
Retail workers	197	3.28	0.48

   (a) Calculate a 99% confidence interval for the difference in mean BPA body concentrations of non-retail and retail workers.

   (b) Does the confidence interval support the belief that retail workers carry higher amounts of BPA in their bodies than non-retail workers? Explain.

9. A new drug is tested for relief of allergy symptoms. In a double-blind experiment, 12 patients are given varying doses of the drug and report back on the number of hours of relief. The following table summarizes the results:

Dosage (mg)	2	3	4	4	5	6	6	7	8	8	9	10
Duration of relief (hrs)	3	7	6	8	10	8	13	16	15	21	23	24

   (a) Determine the equation of the least squares regression line.

   (b) Construct a 90% confidence interval estimate for the slope of the regression line, and interpret this in context.

10. Information with regard to the assessed values (in $1000) and the selling prices (in $1000) of a random sample of homes sold in an NE market yields the following computer output:

Dependent variable is **Price**

R squared = 89.8% R squared (adjusted) = 89.2%
s = 9.508 with 20 – 2 = 18 degrees of freedom

Source	**SS**	**df**	**MS**	**F-ratio**
Regression	14271.2	1	14271.2	158
Residual	1627.31	18	90.4063	

Variable	**Coeff**	**s.e. of coeff**	**t-ratio**	**prob**
Constant	0.890087	16.16	0.055	0.956
Assessed	1.0292	0.08192	12.6	0.000

   (a) Determine the equation of the least squares regression line.

   (b) Construct a 99% confidence interval estimate for the slope of the regression line, and interpret this in context.

11. In a July 2008 study of 1050 randomly selected smokers, 74% said they would like to give up smoking.
    (a) At the 95% confidence level, what is the margin of error?
    (b) Explain the meaning of "95% confidence interval" in this example.
    (c) Explain the meaning of "95% confidence level" in this example.
    (d) Give an example of possible response bias in this example.
    (e) If we want instead to be 99% confident, would our confidence interval need to be wider or narrower?
    (f) If the sample size were greater, would the margin of error be smaller or greater?

Answer Key

1. **C**	7. **E**	13. **E**	19. **E**	25. **C**	31. **E**	37. **E**
2. **D**	8. **A**	14. **E**	20. **A**	26. **A**	32. **D**	
3. **A**	9. **B**	15. **C**	21. **B**	27. **C**	33. **C**	
4. **D**	10. **B**	16. **D**	22. **D**	28. **E**	34. **C**	
5. **E**	11. **D**	17. **E**	23. **A**	29. **E**	35. **E**	
6. **E**	12. **C**	18. **E**	24. **B**	30. **B**	36. **E**	

Answers Explained

Multiple-Choice

1. **(C)** The critical *z*-scores will go from ±1.96 to ±2.576, resulting in an increase in the interval size: $\frac{2.576}{1.96} = 1.31$ or an increase of 31%.

2. **(D)** Increasing the sample size by a multiple of *d* divides the interval estimate by $\sqrt{d}$.

3. **(A)** The margin of error varies directly with the critical *z*-value and directly with the standard deviation of the sample, but inversely with the square root of the sample size. The value of the sample mean and the population size do not affect the margin of error.

4. **(D)** Although the sample proportion *is* between 77% and 87% (more specifically, it is 82%), this is not the meaning of ±5%. Although the percentage of the entire population is likely to be between 77% and 87%, this is not known for certain.

5. **(E)** The 95% refers to the method: 95% of all intervals obtained by this method will capture the true population parameter. Nothing is certain about any particular set of 20 intervals. For any particular interval, the probability that it captures the true parameter is 1 or 0 depending upon whether the parameter is or isn't in it.

6. **(E)** There is no guarantee that 13.4 is anywhere near the interval, so none of the statements are true.

7. **(E)** The critical *z*-score with .005 in each tail is 2.576 (from last line on Table B or `invNorm(.005)` on the TI-84), and $\sigma_{\hat{p}} = \sqrt{\frac{\hat{p}(1-\hat{p})}{n}} = \sqrt{\frac{(.11)(.89)}{10{,}000}}$.

8. **(A)** The margin of error has to do with measuring chance variation but has nothing to do with faulty survey design. As long as *n* is large, *s* is a reasonable estimate of σ; however, again this is not measured by the margin of error. (With *t*-scores, there is a correction for using *s* as an estimate of σ.)

9. **(B)** The critical *z*-score with .01 in the tails is 2.326 (`invNorm(.01)` on the TI-84), and $\sigma_{\hat{p}} = \sqrt{\frac{\hat{p}(1-\hat{p})}{n}} = \sqrt{\frac{(2/3)(1/3)}{50}}$.

10. **(B)** $\sigma_{\hat{p}} = \sqrt{\frac{(.75)(.25)}{1000}} = .0137$

 $.75 \pm 1.96(.0137) = .75 \pm .027$ or $(.723, .777)$

11. **(D)** $\sigma_{\hat{p}} = \sqrt{\frac{(.17)(.83)}{1703}} = .0091$

 $z(.0091) = .02 \quad z = 2.20, \quad .9861 - .0139 = 97.2\%$

12. **(C)** $1.645\left(\frac{.5}{\sqrt{n}}\right) \le .04, \sqrt{n} \ge 20.563, n \ge 422.8$, so choose $n = 423$.

13. **(E)** $\frac{19}{20} = 95\%; 1.96\left(\frac{.5}{\sqrt{n}}\right) \le .03, \sqrt{n} \ge 32.67, n \ge 1067.1$, and so the pollsters should have obtained a sample size of at least 1068. (They actually interviewed 1148 people.)

14. **(E)** LOL, OMG, the critical *t*-score with .02 in the tails is 2.088 (Table B or `invt(.02,79)` on the TI-84), and $\sigma_{\bar{x}} = \frac{s}{\sqrt{n}} = \frac{15}{\sqrt{80}}$.

15. **(C)** Using *t*-scores: $28.5 \pm 2.045\left(\frac{1.2}{\sqrt{30}}\right) = 28.5 + 0.45$.

16. **(D)** $\sigma_{\bar{x}} = \frac{1.52}{\sqrt{64}} = 0.19, \frac{0.38}{0.19} = 2$, and $.4772 + .4772 = .9544 \approx 95\%$.

17. **(E)** Using *t*-scores:

$$335\left[9540 \pm 2.756\left(\frac{1205}{\sqrt{30}}\right)\right] = 335(9540 \pm 606.3) = \$3{,}196{,}000 \pm \$203{,}000.$$

18. **(E)** $1.96\left(\frac{1.1}{\sqrt{n}}\right) \le 0.2, \sqrt{n} \ge 10.78$, and $n \ge 116.2$; choose $n = 117$.

19. **(E)** To divide the interval estimate by d without affecting the confidence level, multiply the sample size by a multiple of d^2. In this case, $4(50) = 200$.

20. **(A)**

$$n_1 = 361 \qquad n_2 = 86$$
$$\hat{p}_1 = \frac{210}{361} = .582 \quad \hat{p}_2 = \frac{34}{86} = .395$$
$$\sigma_d = \sqrt{\frac{(.582)(.418)}{361} + \frac{(.395)(.605)}{86}} = .0588$$
$$(.582 - .395) \pm 1.96(.0588) = .187 \pm .115$$

21. **(B)**

$$n_1 = 300 \quad n_2 = 400$$
$$\hat{p}_1 = .65 \quad \hat{p}_2 = .48$$
$$\sigma_d = \sqrt{\frac{(.65)(.35)}{300} + \frac{(.48)(.52)}{400}} = .0372$$
$$(.65 - .48) \pm 2.576(.0372) = .17 \pm .096$$

22. **(D)** $1.645\left(\frac{.5\sqrt{2}}{\sqrt{n}}\right) \le .03$, $\sqrt{n} \ge 38.77$, and $n \ge 1503.3$; the researcher should choose a sample size of at least 1504.

23. **(A)** $\sigma_d = \sqrt{\frac{(19.1)^2}{347} + \frac{(19.9)^2}{561}}$ and with df= min(347 – 1, 561 – 1), critical t-scores are ±1.97.

24. **(B)** $\sigma_{\bar{x}_1 - \bar{x}_2} = \sqrt{\frac{(6.3)^2}{274} + \frac{(6.3)^2}{90}} = 0.765$

(33.0 – 28.6) ± 1.645(0.765) = 4.4 ± 1.26

25. **(C)** $\sigma_{\bar{x}_1 - \bar{x}_2} = \sqrt{\frac{(4.5)^2}{n} + \frac{(4.5)^2}{n}} = \frac{6.364}{\sqrt{n}}$

$1.645\left(\frac{6.364}{\sqrt{n}}\right) \le 1$, $\sqrt{n} \ge 10.47$, $n \ge 109.6$; choose n = 110.

26. **(A)** The sample mean is at the center of the confidence interval; the lower confidence level corresponds to the narrower interval.

27. **(C)** Narrower intervals result from smaller standard deviations and from larger sample sizes.

28. **(E)** Only III is true. The 90% refers to the method; 90% of all intervals obtained by this method will capture μ. Nothing is sure about any particular set of 100 intervals. For any particular interval, the probability that it captures μ is either 1 or 0 depending on whether μ is or isn't in it.

29. **(E)** In determining confidence intervals, one uses sample statistics to estimate population parameters. If the data are actually the whole population, making an estimate has no meaning.

30. **(B)** $\frac{1.88(.5)\sqrt{2}}{\sqrt{n}} \le .07$ gives $\sqrt{n} \ge 18.99$ and $n \ge 360.7$.

31. **(E)** With df= 15 – 1 = 14 and .05 in each tail, the critical t-value is 1.761.

32. **(D)** $\bar{x} = 4.048$, $s = 2.765$, $df = 20$, and

$$4.048 \pm 1.725\left(\frac{2.765}{\sqrt{21}}\right) = 4.048 \pm 1.041.$$

33. **(C)** The confidence interval estimate of the set of nine differences is

$$11.61 \pm 2.306\left(\frac{4.891}{\sqrt{9}}\right) = 11.61 \pm 3.76.$$

34. **(C)** $df = 4$, and $3.2 \pm 2.132\left(\frac{0.274}{\sqrt{5}}\right) = 3.2 \pm 0.261.$

35. **(E)** The critical t-values with $df = 25 - 2 = 23$ are ± 2.500. Thus, we have $b_1 \pm t^* \times SE(b_1) = 0.008051 \pm (2.500)(0.001058)$

36. **(E)** The critical t-values with $df = n - 2 = 8$ are ± 2.306. Thus, we have $b \pm t^* \times SE(b) = -2.16661 \pm 2.306(1.03092) = -2.16661 \pm 2.3773$ or $(-4.54, 0.21)$.

37. **(E)** While there is clearly a positive association between smoking levels and hazard ratios, this was an observational study, not an experiment, so cause and effect (as implied in I, II, and III) is not an appropriate conclusion. The margins of error of the confidence intervals become greater (less precision) with heavier smoking.

Free-Response

1. (a) $\hat{p} = .57$ and $\sigma_{\hat{p}} \approx \sqrt{\frac{(.57)(.43)}{1000}} = .0157$. [Note that $n\hat{p} = 1000(.57) = 570$ and $n\hat{q} = 1000(.43) = 430$ are both greater than 10, we are given an SRS, and $1000 < 10\%$ of all voters.] The critical z-scores associated with the 95% level are ± 1.96. Thus the confidence interval estimate is $.57 \pm 1.96(.0157) = .57 \pm .031$, or between 54% and 60%. (We are 95% confident that between 54% and 60% of the voters believe competence is more important than character.)

   (b) Explain to your parents that by using a measurement from a sample we are never able to say *exactly* what a population proportion is; rather we are only able to say we are confident that it is within some range of values, in this case between 54% and 60%.

   (c) In 95% of all possible samples of 1000 voters, the method used gives an estimate that is within three percentage points of the true answer.

2. (a) First, identify the confidence interval and check the conditions:

   This is a one-sample z-interval for the proportion of family members who come down with H1N1 after an initial family member does in this state, that is, $\hat{p} \pm z^* \sqrt{\frac{\hat{p}(1-\hat{p})}{n}}$.

It is given that this is a random sample, and we calculate $n\hat{p} = 129 \geq 10$ and $n(1 - \hat{p}) = 747 \geq 10$.

Second, calculate the interval:

Calculator software (such as `1-PropZInt` on the TI-84) gives (.12757, .16695). [For instructional purposes in this review book, we note that $\hat{p} = \frac{129}{876} = .147$ and $.147 \pm 1.645\sqrt{\frac{(.147)(.853)}{876}} = .147 \pm .020$ or (.127, .167)]

Third, interpret in context:

We are 90% confident that the proportion of family members who come down with H1N1 after an initial family member does in this state is between .127 and .167.

(b) Because 1/8 = .125 is not in the interval of plausible values for the population proportion, there *is* evidence that the proportion of family members who come down with H1N1 after an initial family member does in this state is different from the 1 in 8 chance concluded in the published study.

(c) $.147 \pm 2.576\sqrt{\frac{(.147)(.853)}{876}} = .147 \pm .031$ or (.116, .178) which does include .125, so in this case there is *not* evidence that the proportion of family members who come down with H1N1 after an initial family member does in this state is different from the 1 in 8 chance concluded in the published study.

3. (a) First, identify the confidence interval and check the conditions:
This is a two-sample *z*-interval for $p_{US} - p_{UK}$, the difference in population proportions of new births to unmarried women in the United States and United Kingdom, that is,

$$(\hat{p}_{US} - \hat{p}_{UK}) \pm z^* \sqrt{\frac{\hat{p}_{US}(1-\hat{p}_{US})}{n_{US}} + \frac{\hat{p}_{UK}(1-\hat{p}_{UK})}{n_{UK}}}.$$

It is given that these are random samples, they are clearly independent, and we calculate $n_{US}\hat{p}_{US} = (500)(.412) = 206 \geq 10$,
$n_{US}(1 - \hat{p}_{US}) = (500)(.588) = 294 \geq 10$,
$n_{UK}\hat{p}_{UK} = (400)(.465) = 186 \geq 10$, and
$n_{UK}(1 - \hat{p}_{UK}) = (400)(.535) = 214 \geq 10$.

Second, calculate the interval:

Calculator software (such as `2-PropZInt` on the TI-84) gives (–.1182, .01219). [For instructional purposes in this review book, we note that:

$$(.412 - .465) \pm 1.96\sqrt{\frac{(.412)(.588)}{500} + \frac{(.465)(.535)}{400}} = -.053 \pm .065 \text{ or } (-.118, .012).]$$

Third, interpret in context:

We are 95% confident that the difference in proportions, $p_{US} - p_{UK}$, of new births to unmarried women in the United States and United Kingdom is between –0.118 and 0.012.

(b) Because 0 is in the interval of plausible values for the difference of population proportions, this confidence interval does not support the belief by the UN health care statistician that the proportions of new births to unmarried women is different in the United States and United Kingdom.

4. (a) Checking conditions: We are given random samples and must assume that the parent populations are nearly normal. Then a calculator like the TI-84 gives $\bar{x} \pm t\frac{s}{\sqrt{n}} = 3.45 \pm .016$ for inner city stations and $\bar{x} \pm t\frac{s}{\sqrt{n}} = 3.38 \pm .015$ for suburban stations. That is, we are 95% confident that the mean price for gas in inner city stations is between \$3.434 and \$3.466, and are 95% confident that the mean price for gas in suburban stations is between \$3.366 and \$3.394.

(b) Because the sample size of inner city stations is smaller.

(c) Yes, because the standard deviation of the set of sample means is $\frac{.05}{\sqrt{40}} =$ .0079, and so \$3.50 is more than three standard deviations away from \$3.45.

Thus the probability that the true mean is this far from the sample mean is extremely small.

5. (a) Checking conditions: We are given a simple random sample and must assume that the numbers of riders per car during rush hour are normally distributed. Then $\bar{x} \pm t\frac{s}{\sqrt{n}} = 83.5 \pm 1.699\left(\frac{5.9}{\sqrt{30}}\right) = 83.5 \pm 1.83$. That is, we are 90% confident that the average number of riders per car during rush hour is between 81.67 and 85.33.

(b) $1.645\left(\frac{5.9}{\sqrt{n}}\right) \leq 1$, $\sqrt{n} \geq 9.7055$, and $n \geq 94.2$, so you must choose a random sample of at least 95 subway cars.

6. (a) Calculator software (such as `TIinterval` on the TI-84 gives (–29,337, 421,337), or we are 95% confident that the mean salary of all basketball salaries from the population from which this sample was taken is between –\$29,337 and \$421,337. [For instructional purposes in this review book, we note that with $df = 9$,

$$\bar{x} \pm t\frac{s}{\sqrt{n}} = 196{,}000 \pm 2.262\frac{315{,}000}{\sqrt{10}} = 196{,}000 \pm 225{,}000.]$$

(b) We must assume the sample is an SRS and that basketball salaries are normally distributed. This does *not* seem reasonable—the salaries are probably strongly skewed to the right by a few high ones. With the given mean and standard deviation, and noting that salaries are not negative, clearly we do not have a normal distribution, and furthermore, the sample size, $n = 10$, is too small to invoke the CLT. The calculated confidence interval is not meaningful.

7. This is a one-sample t-interval for the mean number of calories of all breakfast cereals. We are given that we have a random sample, and the nearly normal condition seems reasonable from, for example, either a stemplot or a normal probability plot:

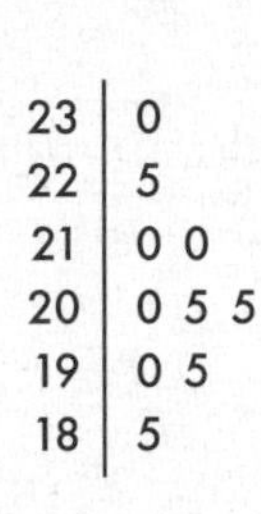

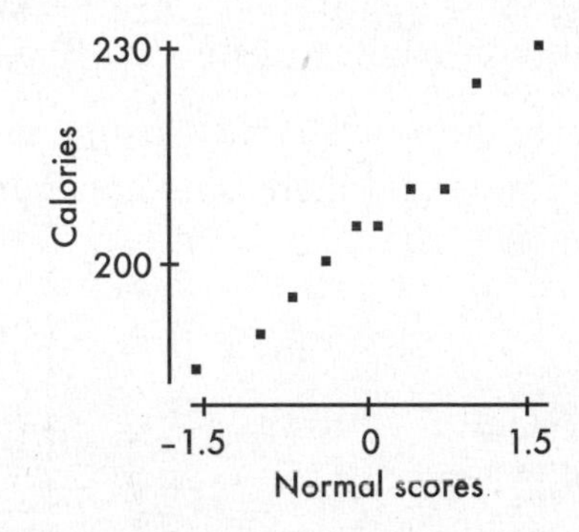

Under these conditions the mean calories can be modeled by a t-distribution with $n - 1 = 10 - 1 = 9$ degrees of freedom.
Calculator software (such as `TIinterval` using `Data` on the TI-84) gives (195.32, 215.68). [For instructional purposes in this review book, we note that a one-sample t-interval for the mean gives:

$$\bar{x} \pm t\frac{s}{\sqrt{n}} = 205.5 \pm 2.262\frac{14.23}{\sqrt{10}} = 205.5 \pm 10.2]$$

We are 95% confident that the true mean number of calories of all breakfast cereals is between 195.3 and 215.7.

8. (a) First, identify the confidence interval and check the conditions:
This is a two-sample t-interval for $\mu_{NRW} - \mu_{RW}$, the difference in population means of BPA body concentrations in non-retail workers and retail workers, that is,

$$(\bar{x}_{NRW} - \bar{x}_{RW}) \pm t^*\sqrt{\frac{(s_{NRW})^2}{n_{NRW}} + \frac{(s_{NR})^2}{n_{RW}}}.$$

It is given that these are random samples, it is reasonable to assume the samples are independent, and both samples sizes (528 and 197) are large enough so that by the CLT, the distributions of sample means are approximately normal and a t-interval may be found.
Second, calculate the interval:

Calculator software (such as `2-SampTInt` on the TI-84) gives (–.9521, –.7479) with $df = 332.3$. [For instructional purposes in this review book, we also write $(2.43 - 3.28) \pm t^*\sqrt{\frac{0.45^2}{528} + \frac{0.48^2}{197}}$.]

Third, interpret in context:

We are 99% confident that the difference in means of BPA body concentrations in non-retail and retail workers is between –0.75 and –0.95 μg/L.

(b) Because 0 is not in the interval of plausible values for the difference of population means, and the entire interval is negative, the interval does support the belief that retail workers carry higher amounts of BPA in their bodies than non-retail workers.

9. We first check conditions: we must assume use of randomization. The scatterplot is roughly linear, and there is no apparent pattern in the residuals plot.

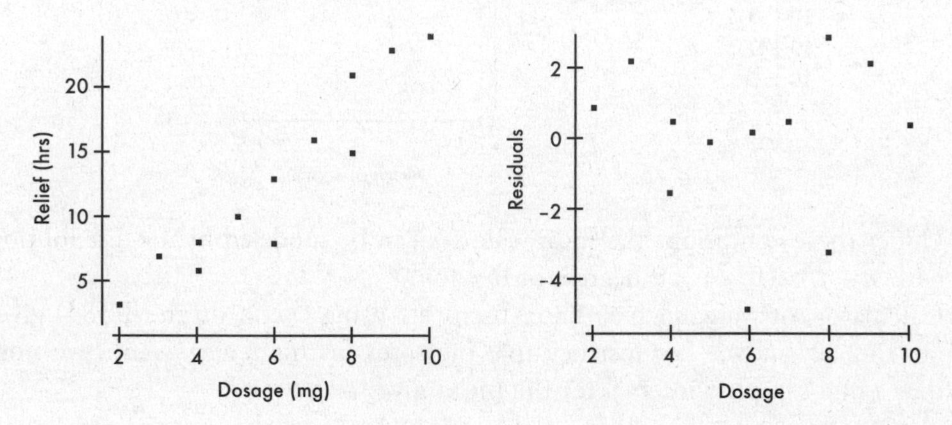

The distribution of the residuals is very roughly normal.

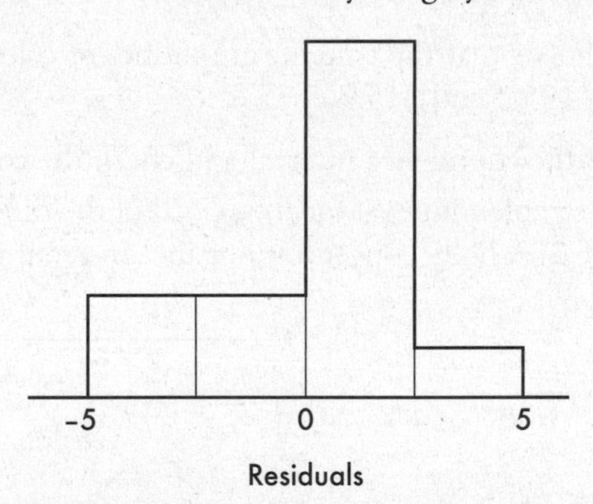

(a) Now using the statistics software on a calculator gives

$$\widehat{Hours} = -3.225 + 2.676\ Dosage$$

(b) Using, for example, the TI-84 (LinRegTTest and 1-Var Stats) one also obtains

$$s = 2.335 \quad \text{and} \quad s_x = 2.486$$

Thus

$$s_{b_1} = \frac{2.335}{2.486\sqrt{12-1}} = 0.2832$$

[Note: On the TI-84, LinRegTTest gives both b_1 and t, and you can also calculate $s_{b_1} = \frac{b_1}{t} = \frac{2.676}{9.450} = 0.2832$, which comes from $t = \frac{b_1}{s_{b_1}}$.]

With 12 data points, the degrees of freedom are $df = 12 - 2 = 10$, and the critical t-values (with .05 in each tail) are ± 1.812. The 99% confidence interval of the true slope is

$$b_1 \pm t\, s_{b_1} = 2.676 \pm 1.812(0.2832) = 2.676 \pm 0.513$$

We are thus 99% confident that each additional gram of the new drug is associated with an average of between 2.2 and 3.2 more hours of allergy relief. (A TI-84+ with `LinRegTInt` will give this result more directly.) We should be careful about extrapolating; note how the scatterplot of Relief versus Dosage seems to be leveling off at the higher end.

10. (a) Assuming that all assumptions for regression are met, the y-intercept and slope of the equation are found in the `Coeff` column of the computer printout.

$$\textit{Selling price} = 0.890 + 1.029\ (\textit{Assessed value})$$

where both the selling price and the assessed value are in $1000.

(b) From the printout, the standard error of the slope is $s_{b_1} = 0.08192$. With $df = 18$ and .005 in each tail, the critical t-values are ± 2.878. The 99% confidence interval of the true slope is:

$$b_1 \pm t\, s_{b_1} = 1.029 \pm 2.878(0.08192) = 1.029 \pm 0.236$$

We are 99% confident that for every $1 increase in assessed value, the average increase in selling price is between $0.79 and $1.27 (or for every $1000 increase in assessed value, the average increase in selling price is between $790 and $1270).

11. (a) $1.96\sqrt{\frac{(.74)(.26)}{1050}} = .0265$. So the margin of error is ±2.65%.

(b) We are 95% confident that between 71.35% and 76.65% of smokers would like to give up smoking.

(c) If this survey were conducted many times, we would expect about 95% of the resulting confidence intervals to contain the true proportion of smokers who would like to give up smoking.

(d) Smoking is becoming more undesirable in society as a whole, so some smokers may untruthfully say they would like to stop.

(e) To be more confident we must accept a *wider* interval.

(f) All other things being equal, the greater the sample size, the *smaller* the margin of error.

Tests of Significance—Proportions and Means

TOPIC 14

- Logic of Significance Testing
- Null and Alternative Hypotheses
- *P*-values
- Type I and Type II Errors
- Concept of Power
- Hypothesis Testing

Closely related to the problem of estimating a population proportion or mean is the problem of testing a hypothesis about a population proportion or mean. For example, a travel agency might determine an interval estimate for the proportion of sunny days in the Virgin Islands or, alternatively, might test a tourist bureau's claim about the proportion of sunny days. A major stockholder in a construction company might ascertain an interval estimate for the proportion of successful contract bids or, alternatively, might test a company spokesperson's claim about the proportion of successful bids. A consumer protection agency might determine an interval estimate for the mean nicotine content of a particular brand of cigarettes or, alternatively, might test a manufacturer's claim about the mean nicotine content of its cigarettes. An agricultural researcher could find an interval estimate for the mean productivity gain caused by a specific fertilizer or, alternatively, might test the developer's claimed mean productivity gain. In each of these cases, the experimenter must decide whether the interest lies in an interval estimate of a population proportion or mean or in a hypothesis test of a claimed proportion or mean.

LOGIC OF SIGNIFICANCE TESTING, NULL AND ALTERNATIVE HYPOTHESIS, *P*-VALUES, ONE- AND TWO-SIDED TESTS, TYPE I AND TYPE II ERRORS, AND THE CONCEPT OF POWER

The general testing procedure is to choose a specific hypothesis to be tested, called the *null hypothesis*, pick an appropriate random sample, and then use measurements from the sample to determine the likelihood of the null hypothesis. If the sample statistic is far enough away from the claimed population parameter, we say that there is sufficient evidence to reject the null hypothesis. We attempt to show that the null hypothesis is unacceptable by showing that it is improbable.

Consider the context of the population proportion. The null hypothesis H_0 is stated in the form of an equality statement about the *population* proportion (for

TIP

Never base your hypotheses on what you find in the sample data.

example, H_0: $p = .37$). There is an *alternative hypothesis,* stated in the form of an inequality (for example, H_a: $p < .37$ or H_a: $p > .37$ or H_a: $p \neq .37$). The strength of the sample statistic $\hat{p}$ can be gauged through its associated *P-value,* which is the probability of obtaining a sample statistic as extreme (or more extreme) as the one obtained if the null hypothesis is assumed to be true. The smaller the *P*-value, the more significant the difference between the null hypothesis and the sample results.

Note that only population parameter symbols appear in H_0 and H_a. Sample statistics like $\bar{x}$ and $\hat{p}$ do NOT appear in H_0 or H_a.

TIP

The *P*-value relates to the probability of the data given that the null hypothesis is true; it is not the probability of the null hypothesis being true.

There are two types of possible errors: the error of mistakenly rejecting a true null hypothesis and the error of mistakenly failing to reject a false null hypothesis. The α-risk, also called the *significance level* of the test, is the probability of committing a *Type I error* and mistakenly rejecting a true null hypothesis. A *Type II error,* a mistaken failure to reject a false null hypothesis, has associated probability β. There is a different value of β for each possible correct value for the population parameter *p*. For each β, $1 - \beta$ is called the power of the test against the associated correct value. The *power* of a hypothesis test is the probability that a Type II error is not committed. That is, given a true alternative, the power is the probability of rejecting the false null hypothesis. Increasing the sample size and increasing the significance level are both ways of increasing the power.

		Population truth	
		H_0 true	H_0 false
Decision based on sample	Reject H_0	Type I error	Correct decision
	Fail to reject H_0	Correct decision	Type II error

A simple illustration of the difference between a Type I and a Type II error is as follows: Suppose the null hypothesis is that all systems are operating satisfactorily with regard to a NASA liftoff. A Type I error would be to delay the liftoff mistakenly thinking that something was malfunctioning when everything was actually OK. A Type II error would be to fail to delay the liftoff mistakenly thinking everything was OK when something was actually malfunctioning. The power is the probability of recognizing a particular malfunction. (Note the complimentary aspect of power, a "good" thing, with Type II error, a "bad" thing.)

TIP

Be able to identify Type I and Type II errors and give possible consequences of each.

Our justice system provides another often quoted illustration. If the null hypothesis is that a person is innocent, then a Type I error results when an innocent person is found guilty, while a Type II error results when a guilty person is not convicted. We try to minimize Type I errors in criminal trials by demanding unanimous jury guilty verdicts. In civil suits, however, many states try to minimize Type II errors by accepting simple majority verdicts.

It should be emphasized that with regard to calculations, questions like "What is the *power* of this test?" and "What is the probability of a *Type II error* in this test?" cannot be answered without reference to a specific alternative hypothesis. Furthermore, AP students are not required to know how to calculate these probabilities. However, they are required to understand the concepts and interactions among the concepts.

HYPOTHESIS TEST FOR A PROPORTION

We assume that we have a simple random sample, that both np_0 and $n(1 - p_0)$ are at least 10, and that the sample size is less than 10% of the population.

EXAMPLE 14.1

A union spokesperson claims that 75% of union members will support a strike if their basic demands are not met. A company negotiator believes the true percentage is lower and runs a hypothesis test at the 10% significance level. What is the conclusion if 87 out of an SRS of 125 union members say they will strike?

Answer: We note that $np_0 = (125)(.75) = 93.75$ and $n(1 - p_0) = (125)(.25) = 31.25$ are both >10 and that we have an SRS, and we assume that 125 is less than 10% of the total union membership.

$$H_0:\ p = .75$$
$$H_a:\ p < .75$$

We use the claimed proportion to calculate the standard deviation of the sample proportions:

$$\sigma_{\hat{p}} = \sqrt{\frac{(.75)(.25)}{125}} = .03873$$

The observed sample proportion is $\hat{p} = \frac{87}{125} = .696$.

To measure the strength of the disagreement between the sample proportion and the claimed proportion, we calculate the *P*-value (also called the *attained significance level*). The z-score for .696 is $\frac{.696-.75}{.03873} = -1.39$, with a resulting *P*-value of .0823. [On the TI-84, normalcdf(−10, −1.39) = .0823 and normalcdf(0, .696, .75, .03873) = .0816.] Since .0823 < .10, there is sufficient evidence to reject H_0 at the 10% significance level. The company negotiator should challenge the union claim at this level. Note, however, that there is not sufficient evidence to reject H_0 at the 5%, or even at the 8%, significance level because .0823 > .05 and .0823 > .08. When $.05 < P < .10$, we usually say that there is *some* evidence against H_0.

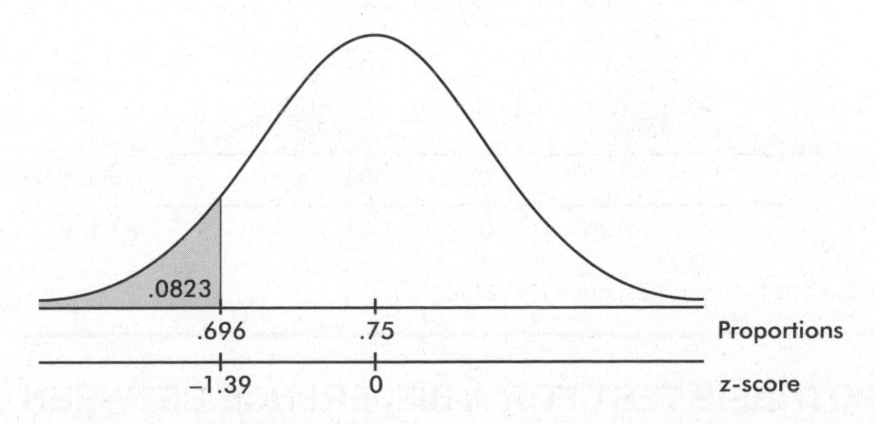

[On the TI-84, under STAT and then TESTS, go to 1-PropZTest. With Po:.75, x:87, n:125, and prop < Po, Calculate gives $P = .0816$.]

(continued)

TIP

Always check assumptions/conditions before proceeding with a test.

TIP

Always give a conclusion in context.

TIP

We never "accept" a null hypothesis; we either do or do not have evidence to reject it.

On the TI-Nspire the result shows as:

zTest_1Prop 0.75,87,125,-1: *stat.results*	"Title"	"1-Prop z Test"
	"Alternate Hyp"	"prop < p0"
	"z"	-1.39427
	"PVal"	0.081617
	"p̂"	0.696
	"n"	125.

EXAMPLE 14.2

A cancer research group surveys 500 women more than 40 years old to test the hypothesis that 28% of women in this age group have regularly scheduled mammograms. Should the hypothesis be rejected at the 5% significance level if 151 of the women respond affirmatively?

Answer: Note that $np_0 = (500)(.28) = 140$ and $n(1 - p_0) = (500)(.72) = 360$ are both >10, we assume an SRS, and clearly 500 < 10% of all women over 40 years old. Since no suspicion is voiced that the 28% claim is low or high, we run a two-sided *z*-test for proportions:

$$H_0:\ p = .28$$
$$H_a:\ p \neq .28$$
$$\alpha = .05$$
$$\sigma_{\hat{p}} = \sqrt{\frac{(.28)(.72)}{500}} = .0201$$

The observed $\hat{p} = \frac{151}{500} = .302$. The *z*-score for .302 is $\frac{.302-.28}{.0201} = 1.09$, which corresponds to a probability of $1 - .8621 = .1379$ in the tail. Doubling this value (because the test is two-sided), we obtain a *P*-value of $2(.1379) = .2758$. Since $.2758 > .05$, there is not sufficient evidence to reject H_0; that is, the cancer research group should not dispute the 28% claim. Note, for example, that the null hypothesis should not be rejected even if α is a relatively large .25. When $P > .10$, we usually say there is little or no evidence to reject H_0.

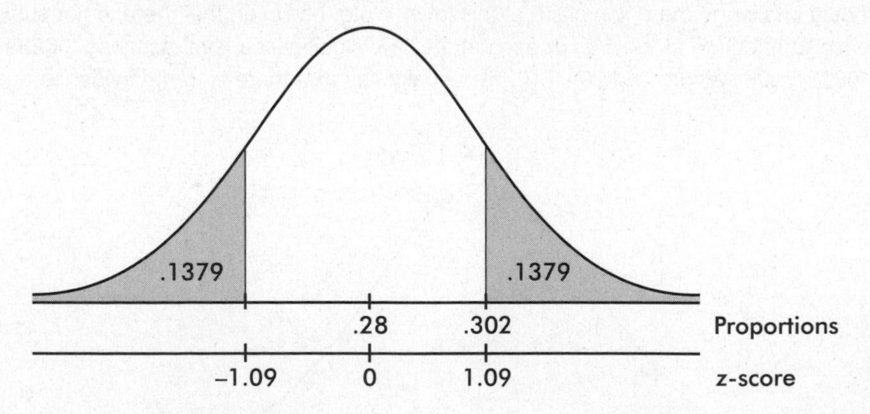

[On the TI-84, 1-PropZTest gives $P = .2732$.]

TIP

The samples must be independent to use these two sample proportion methods.

HYPOTHESIS TEST FOR A DIFFERENCE BETWEEN TWO PROPORTIONS

As with confidence intervals for the difference of two proportions, $n_1\hat{p}_1$, $n_1(1 - \hat{p}_1)$, $n_2\hat{p}_2$, and $n_2(1 - \hat{p}_2)$ should all be at least 10, it is important both that the samples be *simple random samples* and that they be taken *independently* of each other, and the original populations should be large compared to the sample sizes.

The fact that a sample proportion from one population is greater than a sample proportion from a second population does not automatically justify a similar conclusion about the population proportions themselves. Two points need to be stressed. First, sample proportions from the same population can vary from each other. Second, what we are really comparing are confidence interval estimates, not just single points.

For many problems the null hypothesis states that the population proportions are equal or, equivalently, that their difference is 0:

$$H_0\colon p_1 - p_2 = 0$$

The alternative hypothesis is then

$$H_a\colon p_1 - p_2 < 0 \quad H_a\colon p_1 - p_2 > 0 \quad \text{or} \quad H_a\colon p_1 - p_2 \neq 0$$

where the first two possibilities lead to one-sided tests and the third possibility leads to two-sided tests.

Since the null hypothesis is that $p_1 = p_2$, we call this common value p and use it in calculating σ_d:

$$\sigma_d = \sqrt{\frac{p(1-p)}{n_1} + \frac{p(1-p)}{n_2}} = \sqrt{p(1-p)\left(\frac{1}{n_1} + \frac{1}{n_2}\right)}$$

In practice, if $\hat{p}_1 = \frac{x_1}{n_1}$ and $\hat{p}_2 = \frac{x_2}{n_2}$, we use $\hat{p} = \frac{x_1 + x_2}{n_1 + n_2}$ as an estimate of p in calculating σ_d.

EXAMPLE 14.3

Suppose that early in an election campaign a telephone poll of 800 registered voters shows 460 in favor of a particular candidate. Just before election day, a second poll shows only 520 of 1000 registered voters expressing the same preference. At the 10% significance level is there sufficient evidence that the candidate's popularity has decreased?

Answer: Note that $n_1\hat{p}_1 = 460$, $n_1(1 - \hat{p}_1) = 340$, $n_2\hat{p}_2 = 520$, and $n_2(1 - \hat{p}_2) = 480$ are all at least 10, we assume the samples are independent SRSs, and clearly the total population of registered voters is large.

$$H_0: p_1 - p_2 = 0 \quad \hat{p}_1 = \frac{460}{800} = .575$$
$$H_a: p_1 - p_2 > 0 \quad \hat{p}_2 = \frac{520}{1000} = .520$$
$$\alpha = .10 \quad \hat{p} = \frac{460+520}{800+1000} = .544$$

$$\sigma_d \approx \sqrt{(.544)(.456)\left(\frac{1}{800} + \frac{1}{1000}\right)} = .0236$$

The observed difference is .575 − .520 = .055. The *z*-score for .055 is $\frac{.055-0}{.0236} = 2.33$, and so the *P*-value is 1 − .9901 = .0099. [normalcdf(2.33, 100) = .00990 and normalcdf(.055, 1, 0, .0236) = .00989.] Since .0099 < .10, we conclude that at the 10% significance level the candidate's popularity *has* dropped. We note that .0099 < .01, so that the observed difference is statistically significant even at the 1% level.

(continued)

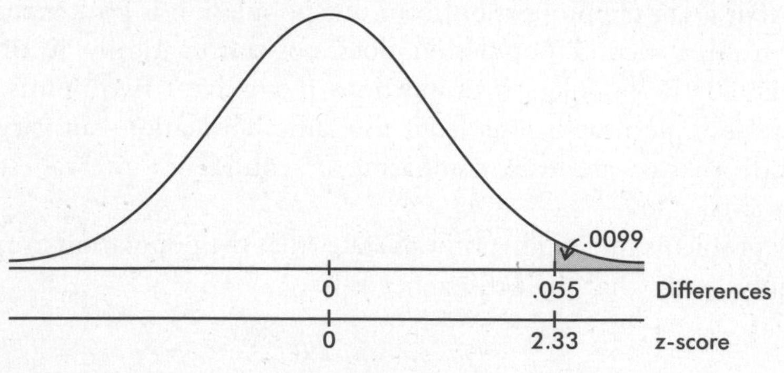

[On the TI-84, 2-PropZTest gives $P = .00995$.]

EXAMPLE 14.4

An automobile manufacturer tries two distinct assembly procedures. In a sample of 350 cars coming off the line using the first procedure there are 28 with major defects, while a sample of 500 autos from the second line shows 32 with defects. Is the difference significant at the 10% significance level?

Answer: Note that $n_1\hat{p}_1 = 28$, $n_1(1 - \hat{p}_1) = 322$, $n_2\hat{p}_2 = 32$, and $n_2(1 - \hat{p}_2) = 468$ are all at least 10, we assume the samples are independent SRSs, and the total population of autos manufactured at the plant is large. Since there is no mention that one procedure is believed to be better or worse than the other, this is a two-sided test.

$$H_0: p_1 - p_2 = 0 \quad \hat{p}_1 = \frac{28}{350} = .080$$
$$H_a: p_1 - p_2 \neq 0 \quad \hat{p}_2 = \frac{32}{500} = .064$$
$$\alpha = .10 \quad \hat{p} = \frac{28+32}{350+500} = .0706$$

$$\sigma_d \approx \sqrt{(.0706)(.9294)\left(\frac{1}{350} + \frac{1}{500}\right)} = .0179$$

The observed difference is .080 – .064 = .016. The *z*-score for .016 is $\frac{.016-0}{.0179} = 0.89$, which corresponds to a tail probability of 1 – .8133 = .1867. [normalcdf(.89, 100) = .1867 and normalcdf(.016, 1, 0, .0179) = .1857.] Doubling this value because the test is two-sided results in a *P*-value of 2(.1867) = .3734. Since .3734 > .10, we conclude that the observed difference is *not* significant at the 10% level. We note that the smallest significance level for which the observed difference is significant is more than 37.34%. When *P* is so large, we can safely say that there is no evidence against H_0. That is, the observed difference in proportions of cars with major defects coming from the two assembly procedures is not significant.

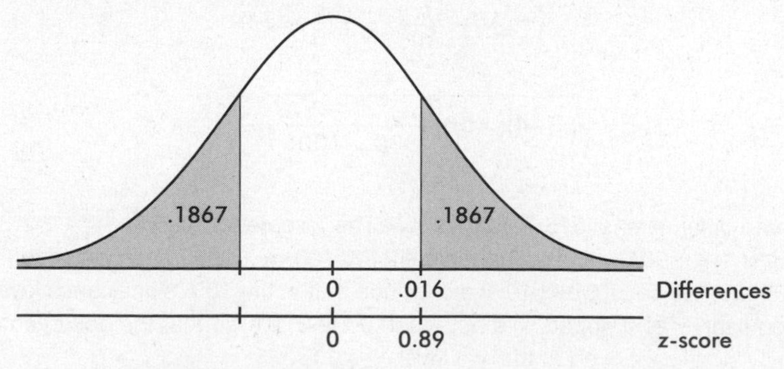

[On the TI-84, 2-PropZTest gives $P = .3701$.]

HYPOTHESIS TEST FOR A MEAN

In the following examples we assume that we have simple random samples from approximately normally distributed populations. (If we are given the sample data we check the normality assumption with a graph such as a histogram or a normal probability plot; the histogram should be unimodal and reasonably symmetric, although this condition can be relaxed the larger the sample size.)

EXAMPLE 14.5

A manufacturer claims that a new brand of air-conditioning unit uses only 6.5 kilowatts of electricity per day. A consumer agency believes the true figure is higher and runs a test on a sample of size 50. If the sample mean is 7.0 kilowatts with a standard deviation of 1.4, should the manufacturer's claim be rejected at a significance level of 5%? Of 1%?

Answer: Assuming an SRS (the normality assumption is less important because $n \geq 40$) we have:

$$H_0: \mu = 6.5$$
$$H_a: \mu > 6.5$$
$$\alpha = .05 \text{ or } .01$$

Here $\sigma_{\bar{x}} = \frac{\sigma}{\sqrt{n}} \approx \frac{s}{\sqrt{n}} = \frac{1.4}{\sqrt{50}} = 0.198$ The t-score of 7.0 is $\frac{7.0-6.5}{0.198} = 2.53$. The critical t-values for the 5% and 1% tests are 1.676 and 2.403, respectively, and we have both 2.53 > 1.676 and 2.53 > 2.403. [On the TI-84, STAT → TESTS → T-Test, then Inpt:Stats, μ_0:6.5, $\bar{x}$:7, Sx:1.4, n:50, μ:>μ_0, Calculate yields $t = 2.525$ and $P = .0074$.]

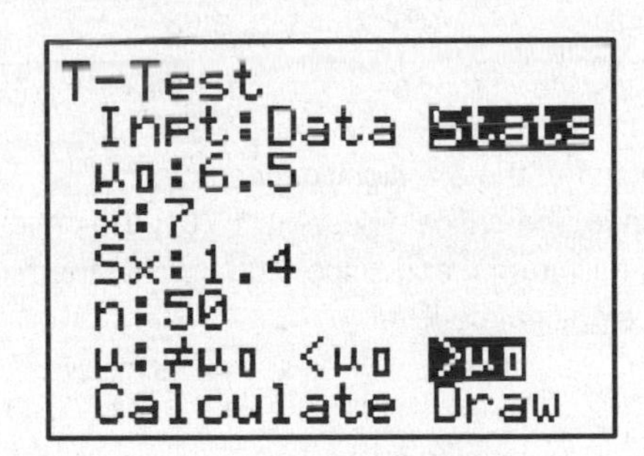

The consumer agency should reject the manufacturer's claim at both the 5% and 1% significance levels. When $P < .01$, we usually say that there is *strong* evidence to reject H_0.

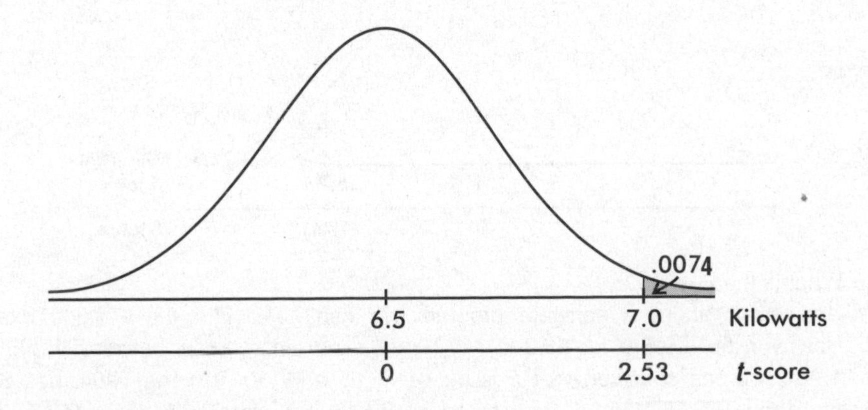

EXAMPLE 14.6

A cigarette industry spokesperson remarks that current levels of tar are no more than 5 milligrams per cigarette. A reporter does a quick check on 15 cigarettes representing a cross section of the market.

a. What conclusion is reached if the sample mean is 5.63 milligrams of tar with a standard deviation of 1.61? Assume a 10% significance level.

Answer: Assuming an SRS from an approximately normally distributed population, we have H_0: $\mu = 5$, H_a: $\mu > 5$, $\alpha = .10$, and

$$\sigma_{\bar{x}} \approx \frac{s}{\sqrt{n}} = \frac{1.61}{\sqrt{15}} = 0.42$$

With $df = 15 - 1 = 14$ and $\alpha = .10$, the critical t-score is 1.345. The critical number of milligrams of tar is $5 + 1.345(0.42) = 5.56$. Since $5.63 > 5.56$, the industry spokesperson's remarks should be rejected at the 10% significance level.

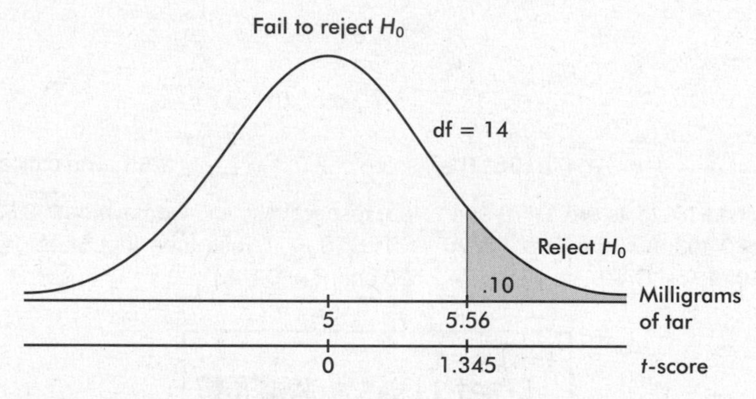

b. What is the conclusion at the 5% significance level?

Answer: In this case, the critical t-score is 1.761, the critical tar level is $5 + 1.761(0.42) = 5.74$ milligrams, and since $5.63 < 5.74$, the remarks cannot be rejected at the 5% significance level.

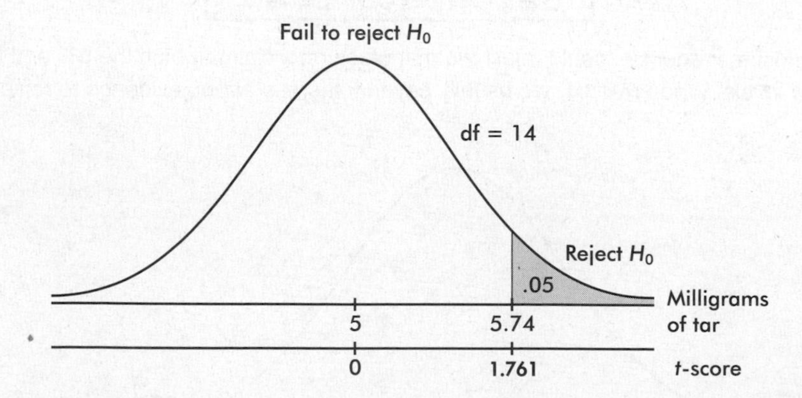

c. What is the P-value?

Answer: Without a software program we can't calculate the P-value exactly. However, the t-score for 5.63 is $\frac{5.63-5}{0.42} = 1.5$, and since 1.5 is between 1.345 and 1.761, we can conclude that P is between .05 and .10. [On the TI-84, use STAT and then TESTS; then T-Test results in $P = .0759$. Note that $.05 < .0759$, while $.0759 < .10$.]

EXAMPLE 14.7

A local chamber of commerce claims that the mean sale price for homes in the city is \$90,000. A real estate salesperson notes eight recent sales of \$75,000, \$102,000, \$82,000, \$87,000, \$77,000, \$93,000, \$98,000, and \$68,000. How strong is the evidence to reject the chamber of commerce claim?

Answer: Assuming an SRS, we check for normality with a histogram, which yields the roughly unimodal, roughly symmetric:

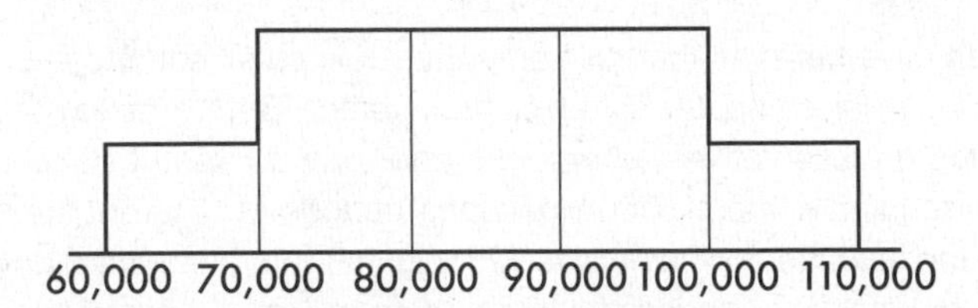

$$H_0: \mu = 90,000 \quad H_a: \mu \neq 90,000$$

A calculator gives $\bar{x} = 85,250$ and $s = 11,877$, and then one calculates $\sigma_{\bar{x}} \approx \frac{s}{\sqrt{n}} = \frac{11877}{\sqrt{8}} = 4199$. The *t*-score of 85,250 is $\frac{85,250-90,000}{4199} = -1.13$. We note that 1.13 is between 1.119 and 1.415, corresponding to .15 and .10, respectively. Doubling these values because this is a two-sided test, we see that the *P*-value is between .20 and .30. With such a high *P*, we conclude that there is no evidence to reject the chamber of commerce claim. [The TI-84 gives $P = .2953$.]

On the TI-Nspire the result shows as:

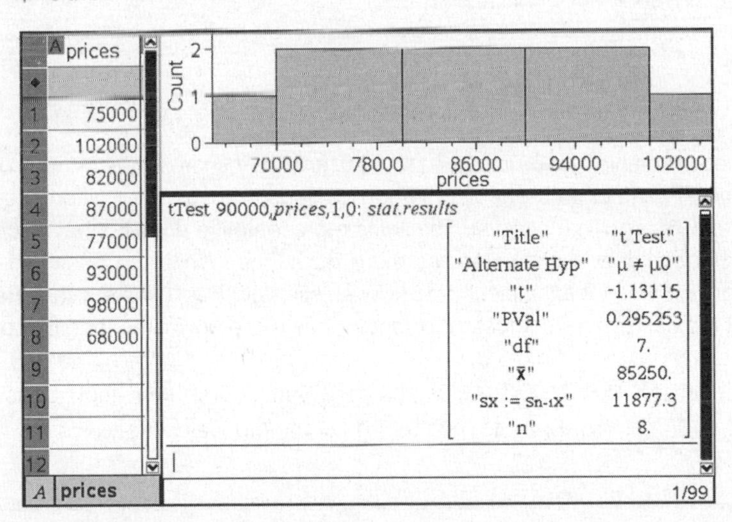

HYPOTHESIS TEST FOR A DIFFERENCE BETWEEN TWO MEANS (UNPAIRED AND PAIRED)

We assume that we have two independent simple random samples, each from an approximately normally distributed population (the normality assumptions can be relaxed if the sample sizes are large).

In this situation the null hypothesis is usually that the means of the populations are the same or, equivalently, that their difference is 0:

$$H_0: \mu_1 - \mu_2 = 0$$

The alternative hypothesis is then

$$H_a\text{: } \mu_1 - \mu_2 < 0 \quad H_a\text{: } \mu_1 - \mu_2 > 0 \quad \text{or} \quad H_a\text{: } \mu_1 - \mu_2 \neq 0$$

The first two possibilities lead to one-sided tests, and the third possibility leads to two-sided tests.

EXAMPLE 14.8

A sales representative believes that his company's computer has more average downtime per week than a similar computer sold by a competitor. Before taking this concern to his director, the sales representative gathers data and runs a hypothesis test. He determines that in a simple random sample of 40 week-long periods at different firms using his company's product, the average downtime was 125 minutes with a standard deviation of 37 minutes. However, 35 week-long periods involving the competitor's computer yield an average downtime of only 115 minutes with a standard deviation of 43 minutes. What conclusion should the sales representative draw assuming a 10% significance level?

Answer: We are given independent SRSs and large enough sample sizes to relax the normality assumptions, so we continue:

H_0: $\mu_1 - \mu_2 = 0$ (where μ_1 = mean down time of company's computers
H_a: $\mu_1 - \mu_2 > 0$ and μ_2 = mean down time of competitor's computers)

$$\alpha = .10$$

$$df = (40 - 1) + (35 - 1) = 73$$

(or the more conservative $df = \min(40–1, 35–1) = 34$)

$$\sigma_{\bar{x}_1-\bar{x}_2} \approx \sqrt{\frac{(37)^2}{40} + \frac{(43)^2}{35}} = 9.33$$

The difference in sample means is 125 – 115 = 10, and the t-score is $\frac{10}{9.33} = 1.07$. Using Table B, we find the P-value to be between .10 and .15. Thus, the observed difference is not significant at the 10% significance level. The sales representative does not have sufficient evidence that his company's computer has more downtime. While the observed difference would be significant at, for example, the 15% level, still, with $P > .1$, we say that the evidence is weak that the salesman's company's computer has more downtime than that of the competitor's.

Remark: On the TI-84, STAT → TESTS → 2-SampTTest, then Inpt : Stats, $\bar{x}1$: 125, Sx1 : 37, n1 : 40, $\bar{x}2$: 115, Sx2 : 43, n2 : 35, $\mu1$: > $\mu2$ and Calculate yields $t = 1.0718$ and $P = .1438$.

TIP

Avoid "calculator speak." For example, do not simply write "2-SampTTest..." or "binomcdf..." There are lots of calculators out there, each with their own abbreviations.

EXAMPLE 14.9

A store manager wishes to determine whether there is a significant difference between two trucking firms with regard to the handling of egg cartons. In a simple random sample of 200 cartons on one firm's truck there was an average of 0.7 broken eggs per carton with a standard deviation of 0.31, while a sample of 300 cartons on the second firm's truck showed an average of 0.775 broken eggs per carton with a standard deviation of 0.42. Is the difference between the averages significant at a significance level of 5%? At a significance level of 1%?

Answer: We are given independent SRSs and large enough sample sizes to relax the normality assumptions, so we continue:

(continued)

H_0: $\mu_1 - \mu_2 = 0$ (where μ_1 = mean number of broken eggs per carton
H_a: $\mu_1 - \mu_2 \neq 0$ for the first firm and μ_2 = mean number of broken eggs per carton for the second firm)

$$\alpha = .05$$

$$df = (200 - 1) + (300 - 1) = 498$$

(or the more conservative $df = \min(200 - 1, 300 - 1) = 199$)

$$\sigma_{\bar{x}_1 - \bar{x}_2} \approx \sqrt{\frac{(0.31)^2}{200} + \frac{(0.42)^2}{300}} = 0.0327$$

Since there is no mention that it is believed that one firm does better than the other, this is a two-sided test. The observed difference of 0.7 – 0.775 = –0.075 has a *t*-score of $\frac{-0.075}{0.0327} = -2.29$. Using Table B, we find that this gives a probability between .01 and .02. The test is two-sided, and so the *P*-value is between .02 and .04. So the observed difference in average number of broken eggs per carton is statistically significant at the 5% level but not at the 1% level. With $.01 < P < .05$, we usually say that there is *moderate* evidence to reject H_0. [On the TI-84, 2-SampTTest gives $P = .0222$.]

EXAMPLE 14.10

A city council member claims that male and female officers wait equal times for promotion in the police department. A women's spokesperson, however, believes women must wait longer than men. If five men waited 8, 7, 10, 5, and 7 years, respectively, for promotion while four women waited 9, 5, 12, and 8 years, respectively, what conclusion should be drawn?

Answer: We must assume that we have two independent SRSs, each from an approximately normally distributed population (the samples are too small for histograms to show anything, but at least a quick calculation indicates no outliers and normal probability plots are fairly linear).

We have H_0: $\mu_1 - \mu_2 = 0$ and H_a: $\mu_1 - \mu_2 < 0$. Using a calculator we find $\bar{x}_1 = 7.4$, $s_1 = 1.82$, $\bar{x}_2 = 8.5$, and $s_2 = 2.89$. Then

$$\sigma_{\bar{x}_1 - \bar{x}_2} = \sqrt{\frac{(1.82)^2}{5} + \frac{(2.89)^2}{4}} = 1.66$$

The *t*-score of the observed difference is $\frac{7.4 - 8.5}{1.66} = -0.66$. With $df = 5 + 4 - 2 = 7$, we note that $0.66 < 0.711$, where 0.711 corresponds to a tail probability of .25. Thus the *P*-value is greater than .25, and we conclude there is no evidence to dispute the council member's claim. (Using the conservative $df = \min(5–1, 4–1) = 3$ leads to the same conclusion.) [On the TI-84, inserting the data into lists and using 2-SampTTest with Data gives $P = .2685$.]

The analysis and procedure described above require that the two samples being compared be independent of each other. However, many experiments and tests involve comparing two populations for which the data naturally occur in pairs. In this case, the proper procedure is to run a one-sample test on a single variable consisting of the differences from the paired data.

EXAMPLE 14.11

Does a particular drug slow reaction times? If so, the government might require a warning label concerning driving a car while taking the medication. An SRS of 30 people are tested before and after taking the drug, and their reaction times (in seconds) to a standard testing procedure are noted. The resulting data are as follows:

Person	Before	After	Person	Before	After
1	1.42	1.48	16	1.83	1.75
2	1.87	1.75	17	1.40	1.51
3	1.34	1.31	18	1.75	1.67
4	0.98	1.22	19	1.56	1.72
5	1.51	1.58	20	1.56	1.63
6	1.43	1.57	21	2.03	1.81
7	1.52	1.48	22	1.38	1.48
8	1.61	1.55	23	1.42	1.35
9	1.37	1.54	24	1.69	1.75
10	1.49	1.37	25	1.50	1.39
11	0.95	1.07	26	1.12	1.24
12	1.32	1.35	27	1.38	1.40
13	1.68	1.77	28	1.71	1.65
14	1.44	1.44	29	0.91	1.11
15	1.17	1.27	30	1.59	1.85

TIP

When the data comes in pairs, you cannot use a two sample *t*-test.

We can calculate the mean reaction times before and after, 1.46 seconds and 1.50 seconds, respectively, and ask if this observed rise is significant. If we performed a two-sample test, we would calculate the *P*-value to be .2708 and would conclude that with such a large *P*, the observed rise is *not* significant. However, this would not be the proper test or conclusion! The two-sample test works for *independent* sets. However, in this case, there is a clear relationship between the data, in pairs, and this relationship is completely lost in the procedure for the two-sample test. The proper procedure is to form the set of 30 differences, being careful with signs, {–0.06, 0.12, 0.03, –0.24, . . .}, and to perform a single sample hypothesis test as follows:

Name the test:

We are using a *paired t-test*, that is, a single sample hypothesis test on the set of differences.

(continued)

State the hypothesis:

H_0: The reaction times of individuals to a standard testing procedure are the same before and after they take a particular drug; the mean difference is zero: $\mu_d = 0$.

H_a: The reaction time is greater after they take the drug; the mean difference is less than zero: $\mu_d < 0$.

Check the conditions:

1. The *data are paired* because they are measurements on the same individuals before and after taking the drug.
2. The reaction times of any individual are independent of the reaction times of the others, so the *differences are independent.*
3. A *random sample* of people are tested.
4. The histogram of the differences looks *nearly normal* (roughly unimodal and symmetric):

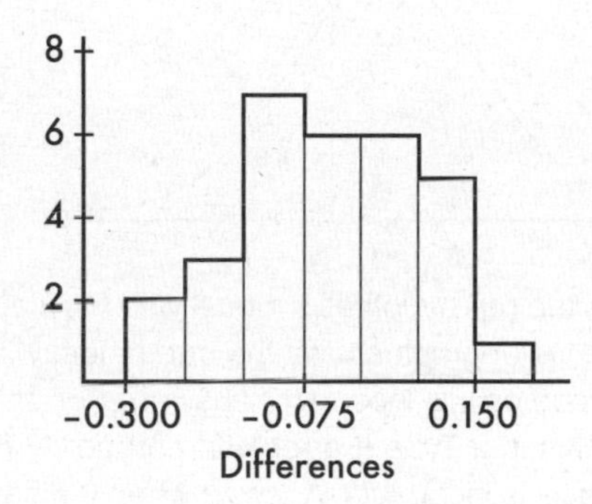

Perform the mechanics:

A calculator readily gives: $n = 30$, $\bar{x} = -0.037667$, and $s = 0.11755$.

Then $\sigma_{\bar{x}} \approx \frac{s}{\sqrt{n}} = \frac{0.11755}{\sqrt{30}} = 0.021462$ and

$$t = \frac{\bar{x} - 0}{\sigma_{\bar{x}}} = \frac{-0.037667}{0.021462} = -1.755$$

With $df = n - 1 = 29$, the *P*-value is $P(t < -1.755) = .0449$.

Give a conclusion in context:

The resulting *P*-value of .0449 is small enough to justify a conclusion (at the 5% significance level) that the observed rise in reaction times after taking the drug is significant.

MORE ON POWER AND TYPE II ERRORS

Given a specific alternative hypothesis, a Type II error is a mistaken failure to reject the false null hypothesis, while the *power* is the probability of rejecting that false null hypothesis.

EXAMPLE 14.12

A candidate claims to have the support of 70% of the people, but you believe that the true figure is lower. You plan to gather an SRS and will reject the 70% claim if in your sample shows 65% or less support. What if in reality only 63% of the people support the candidate?

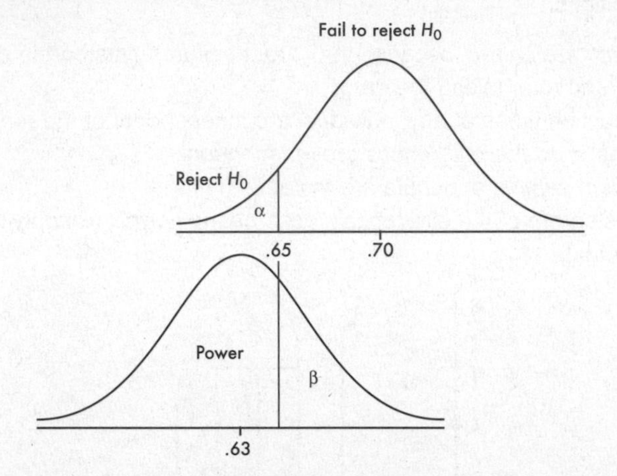

The upper graph shows the null hypothesis model with the claim that $p_0 = 70$, and the plan to reject H_0 if $\hat{p} < .65$. The lower graph shows the true model with $p = .63$. When will we fail to pick up that the null hypothesis is incorrect? Answer: precisely when the sample proportion is greater than .65. This is a Type II error with probability β. When will we rightly conclude that the null hypothesis is incorrect? Answer: when the sample proportion is less than .65. This is the power of the test and has probability 1–β.

The following points should be emphasized:

- Power gives the probability of avoiding a Type II error.
- Power has a different value for different possible correct values of the population parameter; thus it is actually a *function* where the independent variable ranges over specific alternative hypotheses.
- Choosing a smaller α (that is, a tougher standard to reject H_0) results in a higher risk of Type II error and a lower power—observe in the above graphs how making α smaller (in this case moving the critical cutoff value to the left) makes the power less and β more!
- The greater the difference between the null hypothesis p_0 and the true value p, the smaller the risk of a Type II error and the greater the power—observe in the above picture how moving the lower graph to the left makes the power greater and β less. (The difference between p_0 and p is sometimes called the *effect*—thus the greater the effect, the greater is the power to pick it up.)
- A larger sample size n will reduce the standard deviations, making both graphs narrower resulting in smaller α, smaller β, and larger power!

CONFIDENCE INTERVALS VERSUS HYPOTHESIS TESTS

A question sometimes arises as to whether a problem calls for calculating a confidence interval or performing a hypothesis test. Generally, a claim about a population parameter indicates a hypothesis test, while an estimate of a population parameter asks for a confidence interval. However, some confusion may arise because sometimes it is possible to conduct a hypothesis test by constructing a confidence interval estimate of the parameter and then checking whether or not the null parameter value is in the interval. While, when possible, this alternative approach is accepted on the AP exam, it does require very special care in still remembering to state hypotheses, in dealing with one-sided versus two-sided, and in how standard errors are calculated in problems involving proportions. So the recommendation is to conduct a hypothesis test like a hypothesis test. Carefully read the question. If it asks whether or not there is evidence, then this is a hypothesis test; if it involves how much or how effective, then this is a confidence interval calculation. However, it should be noted that there have been AP free-response questions where part (a) calls for a confidence interval calculation and part (b) asks if this calculation provides evidence relating to a hypothesis test.

Summary

- Important assumptions/conditions always must be checked before proceeding with a hypothesis test.
- The null hypothesis is stated in the form of an equality statement about a population parameter, while the alternative hypothesis is in the form of an inequality.
- We attempt to show that a null hypothesis is unacceptable by showing it is improbable.
- The *P*-value is the probability of obtaining a sample statistic as extreme (or more extreme) as the one obtained if the null hypothesis is assumed to be true.
- When the *P*-value is small (typically, less than .05 or .10), we say we have evidence to reject the null hypothesis.
- A Type I error is the probability of mistakenly rejecting a true null hypothesis.
- A Type II error is the probability of mistakenly failing to reject a false null hypothesis.
- The power of a hypothesis test is the probability that a Type II error is not committed.

Questions on Topic Fourteen: Tests of Significance—Proportions and Means

Multiple-Choice Questions

Directions: The questions or incomplete statements that follow are each followed by five suggested answers or completions. Choose the response that best answers the question or completes the statement.

1. Which of the following is a true statement?

   (A) A well-planned hypothesis test should result in a statement either that the null hypothesis is true or that it is false.
   (B) The alternative hypothesis is stated in terms of a sample statistic.
   (C) If a sample is large enough, the necessity for it to be a simple random sample is diminished.
   (D) When the null hypothesis is rejected, it is because it is not true.
   (E) Hypothesis tests are designed to measure the strength of evidence against the null hypothesis.

2. Which of the following is a true statement?

   (A) The *P*-value of a test is the probability of obtaining a result as extreme (or more extreme) as the one obtained assuming the null hypothesis is false.
   (B) If the *P*-value for a test is .015, the probability that the null hypothesis is true is .015.
   (C) The larger the *P*-value, the more evidence there is against the null hypothesis.
   (D) If possible, always examine your data before deciding whether to use a one-sided or two-sided hypothesis test.
   (E) The alternative hypothesis is one-sided if there is interest in deviations from the null hypothesis in only one direction.

Questions 3–4 refer to the following:

Video arcades provide a vehicle for competition and status within teenage peer groups and are recognized as places where teens can hang out and meet their friends. The national PTA organization, concerned about the time and money being spent in video arcades by middle school students, commissions a statistical study to investigate whether or not middle school students are spending an average of over two hours per week in video arcades. Twenty communities are randomly chosen, five middle schools are randomly picked in each of the communities, and ten students are randomly interviewed at each school.

3. What is the parameter of interest?

(A) Video arcades
(B) A particular concern of the national PTA organization
(C) 1000 randomly chosen middle school students
(D) The mean amount of money spent in video arcades by middle school students
(E) The mean number of hours per week spent in video arcades by middle school students

4. What are the null and alternative hypotheses which the PTA is testing?

(A) $H_0: \mu = 2$ $H_a: \mu < 2$
(B) $H_0: \mu = 2$ $H_a: \mu \leq 2$
(C) $H_0: \mu = 2$ $H_a: \mu > 2$
(D) $H_0: \mu = 2$ $H_a: \mu \geq 2$
(E) $H_0: \mu = 2$ $H_a: \mu \neq 2$

5. A hypothesis test comparing two population proportions results in a *P*-value of .032. Which of the following is a proper conclusion?

(A) The probability that the null hypothesis is true is .032.
(B) The probability that the alternative hypothesis is true is .032.
(C) The difference in sample proportions is .032.
(D) The difference in population proportions is .032.
(E) None of the above are proper conclusions.

6. A company manufactures a synthetic rubber (jumping) bungee cord with a braided covering of natural rubber and a minimum breaking strength of 450 kg. If the mean breaking strength of a sample drops below a specified level, the production process is halted and the machinery inspected. Which of the following would result from a Type I error?

   (A) Halting the production process when too many cords break.
   (B) Halting the production process when the breaking strength is below the specified level.
   (C) Halting the production process when the breaking strength is within specifications.
   (D) Allowing the production process to continue when the breaking strength is below specifications.
   (E) Allowing the production process to continue when the breaking strength is within specifications.

7. One ESP test asks the subject to view the backs of cards and identify whether a circle, square, star, or cross is on the front of each card. If p is the proportion of correct answers, this may be viewed as a hypothesis test with H_0: $p = .25$ and H_a: $p > .25$. The subject is recognized to have ESP when the null hypothesis is rejected. What would a Type II error result in?

   (A) Correctly recognizing someone has ESP
   (B) Mistakenly thinking someone has ESP
   (C) Not recognizing that someone really has ESP
   (D) Correctly realizing that someone doesn't have ESP
   (E) Failing to understand the nature of ESP

8. A coffee-dispensing machine is supposed to deliver 12 ounces of liquid into a large paper cup, but a consumer believes that the actual amount is less. As a test he plans to obtain a sample of 5 cups of the dispensed liquid and if the mean content is less than 11.5 ounces, to reject the 12-ounce claim. If the machine operates with a known standard deviation of 0.9 ounces, what is the probability that the consumer will mistakenly reject the 12-ounce claim even though the claim is true?

   (A) .054
   (B) .107
   (C) .214
   (D) .289
   (E) .393

9. A pharmaceutical company claims that a medicine will produce a desired effect for a mean time of 58.4 minutes. A government researcher runs a hypothesis test of 40 patients and calculates a mean of $\overline{x} = 59.5$ with a standard deviation of $s = 8.3$. What is the P-value?

   (A) $P\left(t > \dfrac{59.5 - 58.4}{8.3/\sqrt{40}}\right)$ with $df = 39$

   (B) $P\left(t > \dfrac{59.5 - 58.4}{8.3/\sqrt{40}}\right)$ with $df = 40$

   (C) $2P\left(t > \dfrac{59.5 - 58.4}{8.3/\sqrt{40}}\right)$ with $df = 39$

   (D) $2P\left(t > \dfrac{59.5 - 58.4}{8.3/\sqrt{40}}\right)$ with $df = 40$

   (E) $2P\left(z > \dfrac{59.5 - 58.4}{8.3/\sqrt{40}}\right)$

10. You plan to perform a hypothesis test with a level of significance of $\alpha = .05$. What is the effect on the probability of committing a Type I error if the sample size is increased?

    (A) The probability of committing a Type I error decreases.
    (B) The probability of committing a Type I error is unchanged.
    (C) The probability of committing a Type I error increases.
    (D) The effect cannot be determined without knowing the relevant standard deviation.
    (E) The effect cannot be determined without knowing if a Type II error is committed.

11. A fast food chain advertises that their large bag of french fries has a weight of 150 grams. Some high school students, who enjoy french fries at every lunch, suspect that they are getting less than the advertised amount. With a scale borrowed from their physics teacher, they weigh a random sample of 15 bags. What is the conclusion if the sample mean is 145.8 g and standard deviation is 12.81 g?

    (A) There is sufficient evidence to prove the fast food chain advertisement is true.
    (B) There is sufficient evidence to prove the fast food chain advertisement is false.
    (C) The students have sufficient evidence to reject the fast food chain's claim.
    (D) The students do not have sufficient evidence to reject the fast food chain's claim.
    (E) There is not sufficient data to reach any conclusion.

12. A researcher believes a new diet should improve weight gain in laboratory mice. If ten control mice on the old diet gain an average of 4 ounces with a standard deviation of 0.3 ounces, while the average gain for ten mice on the new diet is 4.8 ounces with a standard deviation of 0.2 ounces, what is the *P*-value?

    (A) $P\left(t < \dfrac{4-4.8}{\sqrt{\dfrac{(0.3)^2}{10}+\dfrac{(0.2)^2}{10}}}\right)$

    (B) $P\left(z < \dfrac{4-4.8}{\sqrt{\dfrac{(0.3)^2}{10}+\dfrac{(0.2)^2}{10}}}\right)$

    (C) $P\left(t < \dfrac{4-4.8}{\dfrac{0.3}{\sqrt{10}}+\dfrac{0.2}{\sqrt{10}}}\right)$

    (D) $P\left(z < \dfrac{4-4.8}{\dfrac{0.3}{\sqrt{10}}+\dfrac{0.2}{\sqrt{10}}}\right)$

    (E) $P\left(t < \dfrac{4-4.8}{\dfrac{0.3}{\sqrt{10-1}}+\dfrac{0.2}{\sqrt{10-1}}}\right)$

13. A school superintendent must make a decision whether or not to cancel school because of a threatening snow storm. What would the results be of Type I and Type II errors for the null hypothesis: The weather will remain dry?

    (A) Type I error: don't cancel school, but the snow storm hits.
    Type II error: weather remains dry, but school is needlessly canceled.
    (B) Type I error: weather remains dry, but school is needlessly canceled.
    Type II error: don't cancel school, but the snow storm hits.
    (C) Type I error: cancel school, and the storm hits.
    Type II error: don't cancel school, and weather remains dry.
    (D) Type I error: don't cancel school, and snow storm hits.
    Type II error: don't cancel school, and weather remains dry.
    (E) Type I error: don't cancel school, but the snow storm hits.
    Type II error: cancel school, and the storm hits.

14. A pharmaceutical company claims that 8% or fewer of the patients taking their new statin drug will have a heart attack in a 5-year period. In a government-sponsored study of 2300 patients taking the new drug, 198 have heart attacks in a 5-year period. Is this strong evidence against the company claim?

    (A) Yes, because the *P*-value is .005657.
    (B) Yes, because the *P*-value is .086087.
    (C) No, because the *P*-value is only .005657.
    (D) No, because the *P*-value is only .086087.
    (E) No, because the *P*-value is over .10.

15. Is Internet usage different in the Middle East and Latin America? In a random sample of 500 adults in the Middle East, 151 claimed to be regular Internet users, while in a random sample of 1000 adults in Latin America, 345 claimed to be regular users. What is the *P*-value for the appropriate hypothesis test?

    (A) $P\left(z < \dfrac{.302 - .345}{\sqrt{\dfrac{(.302)(.698)}{500} + \dfrac{(.345)(.655)}{1000}}}\right)$

    (B) $2P\left(z < \dfrac{.302 - .345}{\sqrt{\dfrac{(.302)(.698)}{500} + \dfrac{(.345)(.655)}{1000}}}\right)$

    (C) $P\left(z < \dfrac{.302 - .345}{\sqrt{(.331)(.669)\left(\dfrac{1}{500} + \dfrac{1}{1000}\right)}}\right)$

    (D) $2P\left(z < \dfrac{.302 - .345}{\sqrt{(.331)(.669)\left(\dfrac{1}{500} + \dfrac{1}{1000}\right)}}\right)$

    (E) 2(.095167)

16. What is the probability of mistakenly failing to reject a false null hypothesis when a hypothesis test is being conducted at the 5% significance level ($\alpha = .05$)?

    (A) .025
    (B) .05
    (C) .10
    (D) .95
    (E) There is insufficient information to answer this question.

17. A research dermatologist believes that cancers of the head and neck will occur most often of the left side, the side next to a window when a person is driving. In a review of 565 cases of head/neck cancers, 305 occurred on the left side. What is the resulting *P*-value?

    (A) $P\left(z > \dfrac{.54-.50}{\sqrt{(.5)(.5)/565}}\right)$

    (B) $2P\left(z > \dfrac{.54-.50}{\sqrt{(.5)(.5)/565}}\right)$

    (C) $P\left(z > \dfrac{.54-.50}{\sqrt{(.54)(.46)/565}}\right)$

    (D) $P\left(z \geq \dfrac{.54-.50}{\sqrt{(.54)(.46)/565}}\right)$

    (E) $2P\left(z > \dfrac{.54-.50}{\sqrt{(.54)(.46)/565}}\right)$

18. Suppose you do five independent tests of the form H_0: $\mu = 38$ versus H_a: $\mu > 38$, each at the $\alpha = .01$ significance level. What is the probability of committing a Type I error and incorrectly rejecting a true null hypothesis with at least one of the five tests?

    (A) .01
    (B) .049
    (C) .05
    (D) .226
    (E) .951

19. Given an experiment with H_0: $\mu = 35$, H_a: $\mu < 35$, and a possible correct value of 32, you obtain a sample statistic of $\bar{x} = 33$. After doing analysis, you realize that the sample size n is actually larger than you first thought. Which of the following results from reworking with the increase in sample size?

   (A) Decrease in probability of a Type I error; decrease in probability of a Type II error; decrease in power.
   (B) Increase in probability of a Type I error; increase in probability of a Type II error; decrease in power.
   (C) Decrease in probability of a Type I error; decrease in probability of a Type II error; increase in power.
   (D) Increase in probability of a Type I error; decrease in probability of a Type II error; decrease in power.
   (E) Decrease in probability of a Type I error; increase in probability of a Type II error; increase in power.

20. Thirty students volunteer to test which of two strategies for taking multiple-choice exams leads to higher average results. Each student flips a coin, and if heads, uses Strategy A on the first exam and then Strategy B on the second, while if tails, uses Strategy B first and then Strategy A. The average of all 30 Strategy A results is then compared to the average of all 30 Strategy B results. What is the conclusion at the 5% significance level if a two-sample hypothesis test, H_0: $\mu_1 = \mu_2$, H_a: $\mu_1 \neq \mu_2$ results in a P-value of .18?

   (A) The observed difference in average scores is significant.
   (B) The observed difference in average scores is not significant.
   (C) A conclusion is not possible without knowing the average scores resulting from using each strategy.
   (D) A conclusion is not possible without knowing the average scores and the standard deviations resulting from using each strategy.
   (E) A two-sample hypothesis test should not be used here.

21. Choosing a smaller level of significance, that is, a smaller α-risk, results in

   (A) a lower risk of Type II error and lower power.
   (B) a lower risk of Type II error and higher power.
   (C) a higher risk of Type II error and lower power.
   (D) a higher risk of Type II error and higher power.
   (E) no change in risk of Type II error or in power.

22. The greater the difference between the null hypothesis claim and the true value of the population parameter,

   (A) the smaller the risk of a Type II error and the smaller the power.
   (B) the smaller the risk of a Type II error and the greater the power.
   (C) the greater the risk of a Type II error and the smaller the power.
   (D) the greater the risk of a Type II error and the greater the power.
   (E) the greater the probability of no change in Type II error or in power.

23. A company selling home appliances claims that the accompanying instruction guides are written at a 6th grade reading level. An English teacher believes that the true figure is higher and with the help of an AP Statistics student runs a hypothesis test. The student randomly picks one page from each of 25 of the company's instruction guides, and the teacher subjects the pages to a standard readability test. The reading levels of the 25 pages are given in the following table:

Reading grade level	5	6	7	8	9	10
Number of pages	6	10	4	2	2	1

Is there statistical evidence to support the English teacher's belief?

(A) No, because the *P*-value is greater than .10.
(B) Yes, the *P*-value is between .05 and .10 indicating some evidence for the teacher's belief.
(C) Yes, the *P*-value is between .01 and .05 indicating evidence for the teacher's belief.
(D) Yes, the *P*-value is between .001 and .01 indicating strong evidence for the teacher's belief.
(E) Yes, the *P*-value is less than .001 indicating very strong evidence for the teacher's belief.

24. Suppose H_0: $p = .4$, and the power of the test for the alternative hypothesis $p = .35$ is .75. Which of the following is a valid conclusion?

(A) The probability of committing a Type I error is .05.
(B) The probability of committing a Type II error is .65.
(C) If the alternative $p = .35$ is true, the probability of failing to reject H_0 is .25.
(D) If the null hypothesis is true, the probability of rejecting it is .25.
(E) If the null hypothesis is false, the probability of failing to reject it is .65.

25. A factory is located close to a city high school. The manager claims that the plant's smokestacks spew forth an average of no more than 350 pounds of pollution per day. As an AP Statistics project, the class plans a one-sided hypothesis test with a critical value of 375 pounds. Suppose the standard deviation in daily pollution poundage is known to be 150 pounds and the true mean is 385 pounds. If the sample size is 100 days, what is the probability that the class will mistakenly fail to reject the factory manager's false claim?

(A) .0475
(B) .2525
(C) .7475
(D) .7514
(E) .9525

Free-Reponse Questions

Directions: You must show all work and indicate the methods you use. You will be graded on the correctness of your methods and on the accuracy of your final answers.

Ten Open-Ended Questions

1. Next to good brakes, proper tire pressure is the most crucial safety issue on your car. Incorrect tire pressure compromises cornering, braking, and stability. Both underinflation and overinflation can lead to problems. The number on the tire is the maximum allowable air pressure—not the recommended pressure for that tire. At a roadside vehicle safety checkpoint, officials randomly select 30 cars for which 35 psi is the recommended tire pressure and calculate the average of the actual tire pressure in the front right tires. What is the parameter of interest, and what are the null and alternative hypotheses that the officials are testing?

2. No vaccinations are 100% risk free, and the theoretical risk of rare complications always have to be balanced against the severity of the disease. Suppose the CDC decides that a risk of one in a million is the maximum acceptable risk of GBS (Guillain-Barre syndrome) complications for a new vaccine for a particularly serious strain of influenza. A large sample study of the new vaccine is conducted with the following hypotheses:

   H_0: The proportion of GBS complications is .000001 (one in a million)

   H_a: The proportion of GBS complications is greater than .000001 (one in a million)

   The *P*-value of the test is .138.

   (a) Interpret the *P*-value in context of this study.

   (b) What conclusion should be drawn at the $\alpha = .10$ significance level?

   (c) Given this conclusion, what possible error, Type I or Type II, might be committed, and give a possible consequence of committing this error.

3. A particular wastewater treatment system aims at reducing the most probable number per ml (MPN/ml) of *E. coli* to 1000 MPN/100 ml. A random study of 40 of these systems in current use is conducted with the data showing a mean of 1002.4 MPN/100 ml and a standard deviation of 7.12 MPN/100 ml. A test of significance is conducted with:

   H_0: The mean concentration of *E. coli* after treatment under this system is 1000 MPN/100 ml.

   H_a: The mean concentration of *E. coli* after treatment under this system is greater than 1000 MPN/100 ml.

   The resulting *P*-value is .0197 with $df = 39$ and $t = 2.132$.

   (a) Interpret the *P*-value in context of this study.

   (b) What conclusion should be drawn at a 5% significance level?

   (c) Given this conclusion, what possible error, Type I or Type II, might be committed, and give a possible consequence of committing this error.

4. A 20-year study of 5000 British adults noted four bad habits: smoking, drinking, inactivity, and poor diet. The study looked to show that there is a higher death rate (proportion who die in a 20-year period) among people with all four bad habits than among people with none of the four bad habits.

   (a) Was this an experiment or observational study? Explain.

   (b) What are the null and alternative hypotheses?

   (c) What would be the result of a Type I error?

   (d) What would be the result of a Type II error?

   Of the 314 people who had all four bad habits, 91 died during the study, while of the 387 people with none of the four bad habits, 32 died during the study.

   (e) Calculate and interpret the *P*-value in context of this study.

5. It is estimated that 17.4% of all U.S. households own a Roth IRA. The American Association of University Professors (AAUP) believes this figure is higher among their members and commissions a study. If 150 out of a random sample of 750 AAUP members own Roth IRAs, is this sufficient evidence to support the AAUP belief?

6. A long accepted measure of the discharge rate (in 1000 ft^3/sec) at the mouth of the Mississippi River is 593. To test if this has changed, ten measurements at random times are taken: 590, 596, 592, 588, 589, 594, 590, 586, 591, 589. Is there statistical evidence of a change?

7. A behavior study of high school students looked at whether a higher proportion of boys than girls met a recommended level of physical activity (increased heart rate for 60 minutes/day for at least 5 days during the 7 days before the survey). What is the proper conclusion if 370 out of a random sample of 850 boys and 218 out of an independent random sample of 580 girls met the recommended level of activity?

8. In a random sample of 35 NFL games the average attendance was 68,729 with a standard deviation of 6,110, while in a random sample of 30 Big 10 Conference football games the average attendance was 70,358 with a standard deviation of 9,139. Is there evidence that the average attendance at Big 10 Conference football games is greater than that at NFL games?

9. A study is proposed to compare two treatments for patients with significantly narrowed neck arteries. Some patients will be treated with surgery to remove built-up plaque, while others will be treated with stents to improve circulation. The response variable will be the proportion of patients who suffer a major complication such as a stroke or heart attack within one month of the treatment. The researchers decide to block on whether or not a patient has had a mini stroke in the previous year.

   (a) There are 1000 patients available for this study, half of whom have had a mini stroke in the previous year. Explain a block design to assign patients to treatments.

   (b) Give two methods other than blocking to increase the power of detecting a difference between using surgery versus stents for patients with this medical condition. Explain your choice of methods.

10. A car simulator was used to compare effect on reaction time between DWI (driving while intoxicated) and DWT (driving while texting). Ten volunteers were instructed to drive at 50 mph and then hit the brakes in response to the sudden image of a child darting into the road. A baseline stopping distance was established for each driver. Then one day each driver was tested for stopping distance while driving while texting, and another day the driver was tested after consuming a quantity of alcohol. For each driver, which test was done on the first day was decided by coin toss. The following table gives the extra number of feet necessary to stop at 50 mph for each driver for DWI and DWT.

DWI	30	26	28	35	42	33	36	28	27	37
DWT	30	31	25	39	45	32	38	30	28	38

The sample means are $\bar{x}_{DWI} = 32.2$, $\bar{x}_{DWT} = 33.6$, and a two-sample *t*-test, H_0: $\mu_{DWI} = \mu_{DWT}$, H_a: $\mu_{DWI} \neq \mu_{DWT}$, gives a *P*-value of 0.590, and a conclusion that there is no evidence of a difference between the effect on reaction time between DWI and DWT. Explain why this is not the proper hypothesis test, and then perform the proper test.

Five Investigative Tasks

1. An exercise electrocardiogram (EKG) checks for changes in your heart during exercise and is useful in diagnosing coronary artery disease. An EKG has fewer potential side effects but is much less precise than thallium tomography. In one EKG study, 500 volunteers with known coronary artery disease and 500 volunteers with healthy arteries underwent EKG checks. The physicians administering and evaluating the tests did not know the physical condition of any volunteer. The following table gives the numbers of volunteers whom the physicians evaluated as "positive" for coronary disease.

Test for coronary disease

	Positive	Negative
Healthy volunteers	100	400
Volunteers with disease	305	195

(a) *Sensitivity* is defined as the probability of a positive test given that the subject has disease. What was the sensitivity of this study?

(b) *Specificity* is defined as the probability of a negative test given that the subject is healthy. What was the specificity of this study?

(c) A valuable tool for assessing the accuracy of such studies is the *positive diagnostic likelihood ratio* (LR^+) which gives the ratio of the probability a positive test result will be observed in a diseased person compared to the probability that the same result will be observed in a healthy person.

$$LR^+ = \frac{\text{sensitivity}}{1\text{-specificity}}$$

What was LR^+ in this study, and explain why the larger the value of LR^+, the more useful the test.

(d) Suppose in one such sample study, $LR^+ = 4.7$. To determine whether or not this is sufficient evidence that the population LR^+ is below the desired value of 5.0, 100 samples from a population with a known LR^+ of 5.0 are generated, and the resulting simulated values of LR^+ are shown in the dotplot:

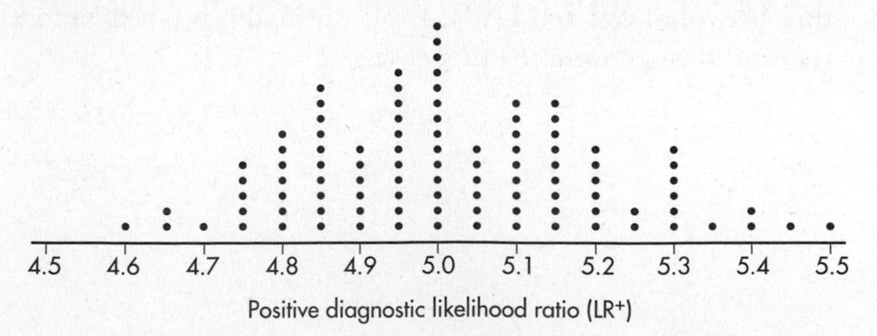

Based on this dotplot and the sample $LR^+ = 4.7$, is there evidence that the population LR^+ is below the desired value of 5.0? Explain.

2. An engineer wishes to test which of two drills can more quickly bore holes in various materials. He assembles a random sampling of ten materials of various hardnesses and thicknesses.

   (a) Given that a drill's efficiency is influenced by how long and how hard it has been operating, the engineer decided to randomly choose the order in which the materials will be tested. Design and implement a scheme to place the materials in random order using the following random number table:

   51844 73424 84380 82259 28273 58102 18727 69708

   (b) Suppose the drilling times (in seconds) are summarized as follows:

Material	A	B	C	D	E	F	G	H	I	J
Drill 1	4.2	5.1	8.8	1.5	0.8	7.4	3.4	4.7	6.2	4.8
Drill 2	4.2	5.4	8.7	1.7	0.9	7.8	3.4	4.4	6.3	4.9

   What is the mean drilling time for each drill? Is the difference significant? Justify your answer.

3. Paul the octopus, who lives in a tank at Sea Life Centre in Oberhausen, Germany, correctly predicted the winner in 12 out of 14 soccer matches (by choosing to eat mussels from the boxes labeled with national flags of the eventual winning teams). Assume these 14 matches represent a random sample of matches which Paul could predict, and assume that all matches end in a win for one of the teams (using sudden death overtime or penalty kicks to settle tie scores from regulation time).

   (a) Why would it be incorrect to perform a one-sample z-test on whether the above is sufficient evidence that Paul can correctly predict the results of more than 50% of soccer matches?

   (b) Perform a proper hypothesis test for the question in (a).

4. A marketing company is interested in whether a particular advertisement will increase interest in a new social networking website. A random sample of 15 teenagers is chosen, and their interest level in the new website is measured on a 1–100 scale before and after seeing the advertisement. The individual results are as follows:

Before ad	33	28	25	37	42	55	41	19	30	51	44	37	25	36	45
After ad	37	30	26	34	40	59	40	17	27	51	47	35	28	38	49

   (a) The ad developer claims that after seeing the ad, the interest level is above 30. Do the data support this conclusion? Explain.

   (b) The ad developer also claims that after seeing the ad, teenagers have increased interest in the new website. Assuming all assumptions for hypothesis testing are met, do the data support this conclusion? Explain.

(c) Can after-ad interest level be predicted from before-ad interest level? Explain.

(d) Can the change in interest level be predicted by before-ad interest level? Explain.

5. According to one national survey, 20% of 18–24-year-olds have passports.

(a) Assuming the 20% figure is correct, use simulation to determine the approximate probability that in a random sample of ten 18–24-year-olds, at least three have passports.

2498346851	4113296825	1485367833	8663018872	7373275392
5062790330	2367029195	4153038298	7360048279	4207598980
9574649262	4488086249	2651769472	9462095309	4072555345
7894788460	2391904958	0201791131	9856022851	1405559336
6003121057	4154811850	7697586849	9644852135	0811348895

(b) Calculate the above probability exactly.

(c) Suppose you believe that the 20% claim is too high and run a hypothesis test. In a simple random sample of 200 18–24-year-olds you find only 33 who have passports. Is this sufficient evidence to dispute the 20% claim?

(d) If the 20% claim is true, what is the probability that the first 18–24-year-old with a passport will be the third one sampled?

(e) A 100-trial simulation is performed to determine the number of 18–24-year-olds sampled before finding one with a passport. The results are as follows:

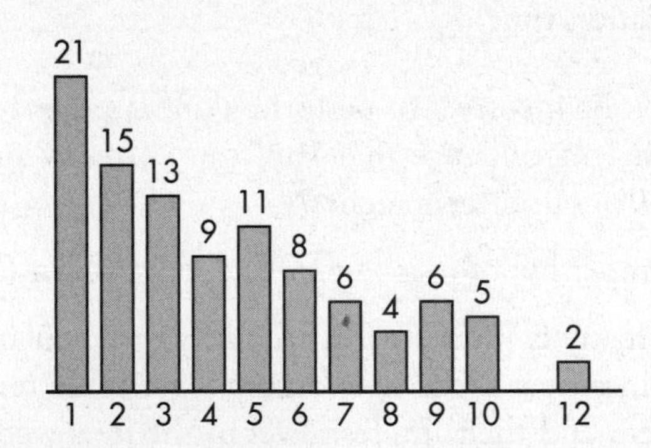

Use the above information and barplot to estimate the mean number of 18–24-year-olds sampled before finding one with a passport.

Answer Key

1. **E**	6. **C**	11. **D**	16. **E**	21. **C**
2. **E**	7. **C**	12. **A**	17. **A**	22. **B**
3. **E**	8. **B**	13. **B**	18. **B**	23. **C**
4. **C**	9. **C**	14. **E**	19. **C**	24. **C**
5. **E**	10. **B**	15. **D**	20. **E**	25. **B**

Answers Explained

Multiple-Choice

1. **(E)** We attempt to show that the null hypothesis is unacceptable by showing that it is improbable; however, we cannot show that it is definitely true or false. Both the null and alternative hypotheses are stated in terms of a population parameter, not a sample statistic. These hypotheses tests assume simple random samples.

2. **(E)** The *P*-value of a test is the probability of obtaining a result as extreme (or more extreme) as the one obtained assuming the null hypothesis is *true*. Small *P*-values are evidence against the null hypothesis. The null and alternative hypotheses are decided upon *before* the data come in.

3. **(E)** The parameter of interest is μ = the mean number of hours per week which middle school students spend in video arcades.

4. **(C)** The alternative hypothesis is always an inequality, either <, or >, or ≠. In this case, the concern is whether middle school students are spending an average of *more than* two hours per week in video arcades.

5. **(E)** The *P*-value is a conditional probability; in this case, there is a .032 probability of an observed difference in sample proportions as extreme (or more extreme) as the one obtained if the null hypothesis is assumed to be true.

6. **(C)** This is a hypothesis test with H_0: breaking strength is within specifications, and H_a: breaking strength is below specifications. A Type I error is committed when a true null hypothesis is mistakenly rejected.

7. **(C)** A Type II error is a mistaken failure to reject a false null hypothesis or, in this case, a failure to realize that a person really does have ESP.

8. **(B)** $P\left(z<\frac{11.5-12}{0.9/\sqrt{5}}\right)=.107$

   [On the TI-84, `normalcdf(–1000,–1.242)` = .107.]

9. **(C)** Medications having an effect shorter or longer than claimed should be of concern, so this is a two-sided test: H_a: $\mu \neq 58.4$, and $df = n - 1 = 40 - 1 = 39$.

10. **(B)** The level of significance is defined to be the probability of committing a Type I error, that is, of mistakenly rejecting a true null hypothesis.

11. **(D)** $t = \frac{145.8 - 150}{12.81/\sqrt{15}} = -1.270$, and with $df = 14$, $P = .112$. With this high P-value ($.112 > .10$), the students do not have sufficient evidence to reject the fast food chain's claim. (On the TI-84, `T-Test` gives $P = .112423$.)

12. **(A)** With unknown population standard deviations, the t-distribution must be used, and $\sigma_{\bar{x}_1 - \bar{x}_2} = \sqrt{\frac{s_1^2}{n_1} + \frac{s_2^2}{n_2}} = \sqrt{\frac{(0.3)^2}{10} + \frac{(0.2)^2}{10}}$

13. **(B)** A Type I error means that the null hypothesis is true (the weather remains dry), but you reject it (thus you needlessly cancel school). A Type II error means that the null hypothesis is wrong (the snow storm hits), but you fail to reject it (so school is not canceled).

14. **(E)** With $\hat{p} = \frac{198}{2300} = .086087$, get $z = \frac{.086087 - .08}{\sqrt{(.08)(.92)/2300}} = 1.076$ and $P = .141$.

    With a P-value this high ($.141 > .10$), the government does not have sufficient evidence to reject the company's claim. (On the TI-84, `1PropZTest` gives $P = .140956$.)

15. **(D)** $\hat{p}_1 = \frac{151}{500} = .302$, $\hat{p}_2 = \frac{345}{1000} = .345$, $\hat{p} = \frac{151 + 345}{500 + 1000} = .331$

    This is a two-sided test (H_0: $p_1 - p_2 = 0$, H_a: $p_1 - p_2 \neq 0$) with

$$\sigma_d = \sqrt{p(1-p)\left(\frac{1}{n_1} + \frac{1}{n_2}\right)}.$$

16. **(E)** There is a different answer for each possible correct value for the population parameter.

17. **(A)** This is a one-sided z-test, H_a: $p > .5$, with $\sigma_{\hat{p}} = \sqrt{(.5)(.5)/565}$.

18. **(B)** P(at least one Type I error) = 1 − P(no Type I errors) = $1 - (.99)^5 = .049$

19. **(C)** A larger sample size n reduces the standard deviation of the sampling distributions resulting in narrower sampling distributions so that for the given sample statistic, the P-value is smaller, and the probabilities of mistakenly rejecting a true null hypothesis or mistakenly failing to reject a false null hypothesis are both decreased. Furthermore, a lower Type II error results in higher power.

20. **(E)** The two-sample hypothesis test is not the proper one and can only be used when the two sets are independent. In this case, there is a clear relationship between the data, in pairs, one pair for each student, and this relationship is completely lost in the procedure for the two-sample test. The proper procedure

is to run a one-sample test on the single variable consisting of the differences from the paired data.

21. **(C)** With a smaller α, that is, with a tougher standard to reject H_0, there is a greater chance of failing to reject a false null hypothesis, that is, there is a greater chance of committing a Type II error. Power is the probability that a Type II error is not committed, so a higher Type II error results in lower power.

22. **(B)** If the null hypothesis is far off from the true parameter value, there is a greater chance of rejecting the false null hypothesis and thus a smaller risk of a Type II error. Power is the probability that a Type II error is not committed, so a lower Type II error results in higher power.

23. **(C)** With $\bar{x} = 6.48$ and $s = 1.388$, $t = \dfrac{6.48-6}{1.388/\sqrt{25}} = 1.729$. Given $df = 29$, $P = .048$.

    [On the TI-84, `T-Test` gives $P = .048322$.]

24. **(C)** If the alternative is true, the probability of failing to reject H_0 and thus committing a Type II error is 1 minus the power, that is, $1 - .75 = .25$.

25. **(B)** $P\left(z < \dfrac{375-385}{150/\sqrt{100}}\right) = .2525$

    [On the TI-84, `normalcdf(–100, –10/15)` = .252492.]

FREE-RESPONSE

1. The parameter of interest is μ = the mean tire pressure in the front right tires of cars with recommended tire pressure of 35 psi. H_0: $\mu = 35$ and H_a: $\mu \neq 35$.

2. (a) The P-value of .138 gives the probability of observing a sample proportion of GBS complications as great or greater as the proportion found in the study if in fact the proportion of GBS complications is .000001.

   (b) Since .138 > .10, there is no evidence to reject H_0, that is, there is no evidence that under the new vaccine the proportion of GBS complications is greater than .000001 (one in a million).

   (c) The null hypothesis is not rejected, so there is the possibility of a Type II error, that is of mistakenly failing to reject a false null hypothesis. A possible consequence is continued use of vaccine with a higher rate of serious complications than is acceptable.

3. (a) The P-value of .0197 gives the probability of observing a sample mean of 1002.4 or greater if in fact this system results in a mean *E. coli* concentration of 1000 MPN/100 ml.

(b) Since .0197 < .05, there is evidence to reject H_0, that is, there is evidence that the mean *E. coli* concentration is greater than 1000 MPN/100 ml, and the system is not working properly.

(c) With rejection of the null hypothesis, there is the possibility of a Type I error, that is of mistakenly rejecting a true null hypothesis. Possible consequences are that sales of the system drop even though the system is doing what it claims, or that the company performs an overhaul to fix the system even though the system is operating properly.

4. (a) This was an observational study as no treatments were imposed. It would have been highly unethical to impose treatments, that is, to instruct randomly chosen volunteers to smoke, drink, skip exercise, and eat poorly.

(b) H_0: $p_4 = p_0$, H_a: $p_4 > p_0$ where p_4 is the proportion of adults with all four bad habits who die during a 20-year period, and p_0 is the proportion of adults with none of the four bad habits who die during a 20-year period. (Note that the hypotheses are about the population of all adults with and with none of the four bad habits, not about the volunteers who took part in the study.)

(c) A Type I error, that is, a mistaken rejection of a true null hypothesis, would result in people being encouraged to not smoke, not drink, exercise, and eat well, when these actions actually will not help decrease 20-year death rates.

(d) A Type II error, that is, a mistaken failure to reject a false null hypothesis, would result in people thinking that smoking, drinking, inactivity, and poor diet don't increase 20-year death rates, when actually they do contribute to higher 20-year death rates.

(e) Calculator software (such as `2-PropZTest` on the TI-84) gives $P = .000$. [For instructional purposes in this review book, we note that:

$\hat{p}_1 = \frac{91}{314} = .290$, $\hat{p}_2 = \frac{32}{387} = .083$, $\hat{p} = \frac{91+32}{314+387}$ and $= .175$, and

$$P\left(z > \frac{.290 - .083}{\sqrt{(.175)(.825)\left(\frac{1}{314} + \frac{1}{387}\right)}}\right) = .000.]$$

If the null hypothesis were true, that is, if there was no difference in the 20-year death rates between people with all four bad habits and people with none of the bad habits, then the probability of sample proportions with a difference as extreme or more extreme than observed is .000 (to three decimals).

5. First, state the hypotheses:

H_0: $p = .174$ and H_a: $p > .174$. (If asked to state the parameter, then state "where p is the proportion of AAUP members who own Roth IRAs.")

Second, identify the test by name or formula and check the assumptions.

This is a one-sample z-test for a population proportion.

Assumptions: Random sample (given), $np = 750(.174) = 130.5 \geq 10$, and $n(1-p) = 750(.826) = 619.5 \geq 10$.

Third, calculate the test statistic z and the P-value. Calculator software (such as `1-PropZTest` on the TI-84) gives $z = 1.878$ and $P = .030$.
[For instructional purposes in this review book, we note that

$$\hat{p} = \frac{250}{750} = .2,\ z = \frac{.2 - .174}{\sqrt{\frac{(.174)(.826)}{750}}} = 1.878, \text{ and } P = P(z > 1.878) = .030.]$$

Fourth, linking to the P-value, give a conclusion in context.

With a P-value this small ($.030 < .05$), there is evidence that more than 17.4% of AAUP members own Roth IRAs.

6. First, state the hypotheses: H_0: $\mu = 593$ and H_a: $\mu \neq 593$ (If asked to state the parameter, then state "where μ is the mean discharge rate (in 1000 ft^3/sec) at the mouth of the Mississippi River.")

Second, identify the test by name or formula and check the assumptions:

This is a one-sample t test for the mean with $t = \frac{\bar{x} - x_0}{s/\sqrt{n}}$. Assumptions: The measurements were taken at random times, and either a dotplot of the sample is roughly unimodal and symmetric or the normal probability plot is roughly linear:

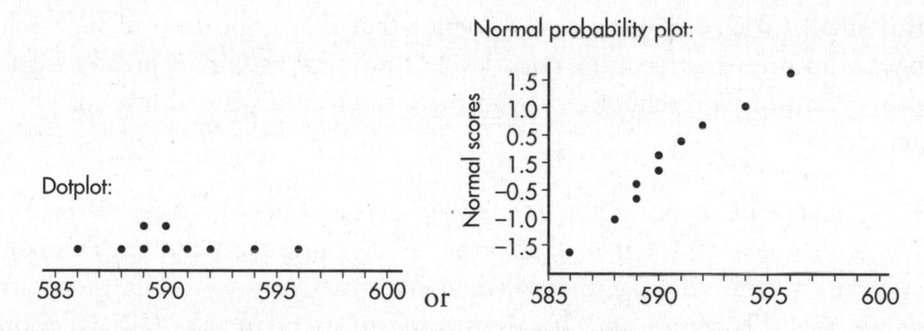

Third, calculate the test statistic t and the P-value:

Calculator software (such as `T-Test` using `Data` on the TI-84) gives $t = -2.7116$ and $P = .0239$. [For instructional purposes in this review book, we note that $\bar{x} = 590.5$ and $s = 2.9155$ which gives

$$t = \frac{590.5 - 593}{2.9155 \neq \sqrt{10}} = -2.7116 \text{ and with } df = 10 - 1 = 9,$$

$$P(t < -2.7116) = .0120.$$

Since this is a two-sided test, we double this value to find the P-value to be $P = .0240$.]

Fourth, linking to the *P*-value, give a conclusion in context: With a *P*-value this small (.0239 < .05), there is evidence that the long accepted measure of the discharge rate at the mouth of the Mississippi River has changed.

7. First, state the hypotheses: H_0: $p_B - p_G = 0$ (or $p_B = p_G$) and H_a: $p_B - p_G > 0$ (or $p_B > p_G$). If asked to state the parameters, then state "where p_B is the proportion of high school boys who meet the recommended level of physical activity, and p_G is the proportion of high school girls who meet the recommended level of physical activity.")

   Second, identify the test by name or formula and check the assumptions:

   This is a two-sample *z*-test for proportions. Assumptions: Independent random samples (given), and $n_B\hat{p}_B = 370 \geq 10$, $n_B(1 - \hat{p}_B) = 480 \geq 10$, $n_G\hat{p}_G = 218 \geq 10$, and $n_G(1 - \hat{p}_G) = 362 \geq 10$.

   Third, calculate the test statistic *z* and the *P*-value. Calculator software (such as `2-PropZTest` on the TI-84) gives $z = 2.2427$ and $P = .012$. (For instructional purposes in this review book, we note that

   $$\hat{p}_B = \frac{370}{850} \approx .4353,\ \hat{p}_G = \frac{218}{580} \approx .3759,\ \hat{p} = \frac{370+218}{850+580} = \frac{588}{1430} \approx .4112,$$

   $$z = \frac{.4353 - .3759}{\sqrt{(.4112)(.5888)\left(\frac{1}{850} + \frac{1}{580}\right)}} \approx 2.2427, \text{ and } P = P(z > 2.2427) = .012.]$$

   Fourth, linking to the *P*-value, give a conclusion in context. With a *P*-value this small (.012 < .05), there is evidence that the proportion of high school boys who meet the recommended level of physical activity is greater than the proportion of high school girls who meet the recommended level of physical activity.

8. First, state the hypotheses: H_0: $\mu_{NFL} - \mu_{10} = 0$ (or $\mu_{NFL} = \mu_{10}$) and H_a: $\mu_{NFL} - \mu_{10} < 0$ (or $\mu_{NFL} < \mu_{10}$) (or ">" depending on choice of variables). (If asked to state the parameters, then state "where μ_{NFL} is the mean attendance at NFL games and μ_{10} is the mean attendance at Big 10 football games."]

   Second, identify the test by name or formula and check the assumptions: This is a two-sample *t*-test for means. Assumptions: Independent random samples (given), and both samples sizes, 35 and 30, are large enough so that by the CLT, the distribution of sample means is approximately normal and a *t*-test may be run.

   Third, calculate the test statistic *t* and the *P*-value. Calculator software (such as `2-SampTTest` on the TI-84) gives $t = -0.8301$, $df = 49.3$, and $P = .2052$. [For instructional purposes in this review book, we note that:

$$t = \frac{68{,}729 - 70{,}358}{\sqrt{\frac{6110^2}{35} + \frac{9139^2}{30}}} = -0.8301,$$ and with the conservative

$df = \min(35 - 1, 30 - 1) = 29$, $P = P(t < -0.8301) = .2066$.]

Fourth, linking to the P-value, give a conclusion in context. With a P-value this large (.2052 > .10), there is no evidence that the average attendance at Big 10 Conference football games is greater than that at NFL games.

9. (a) One block consists of the 500 patients who have had a mini stroke in the previous year, and a second block consists of the 500 patients who have not. In one block, number the patients 1 through 500, and then using a random number generator, pick numbers between 1 and 500, throwing out repeats, until 250 patients are picked. Assign surgery to these patients and stents to the remaining 250. Repeat for the other block.

   (b) One method is to increase the sample size, resulting in a reduction in the standard error of the sampling distribution, which in turn increases the probability of rejecting the null hypothesis if it is false. A second method is to increase α, the significance level, which in turn also increases the probability of rejecting the null hypothesis if it is false. In either case, there is increased probability of detecting any difference in the proportions of patients suffering major complications between the surgery and stent recipients.

10. The data come in pairs, and the two-sample test does not apply the knowledge of what happened to each individual driver (the condition of independence of the two samples is violated). The appropriate test is a one-population, small-sample hypothesis test on the set of differences: {0, –5, 3, –4, –3, 1, –2, –2, –1, –1}. We proceed as follows:

    First, state the hypotheses: H_0: $\mu_D = 0$, H_a: $\mu_D \neq 0$ (If asked to state the parameter, then state "where μ_D is the mean difference in reaction times between DWI and DWT.")

    Second, identify the test and check the assumptions: This is a paired t-test, that is, a single-sample hypothesis test on the set of differences.

    The *data are paired* because they are measurements on the same individuals under DWI and DWT.

    The reaction times of any individual are assumed independent of the reaction times of the others, so the *differences are independent.*

    We must assume that the volunteers are a *representative sample.*

    Either a dotplot of the sample is roughly unimodal and symmetric or the normal probability plot is roughly linear:

Dotplot:

-5 -4 -3 -2 -1 0 1 2 3

Differences

or

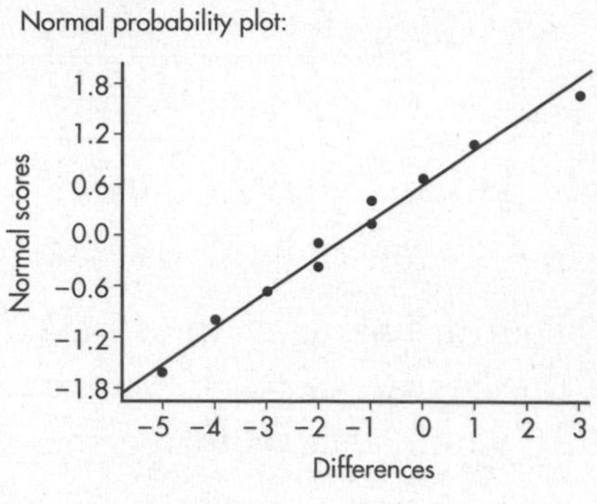

Third, calculate the test statistic t and the P-value: Calculator software (such as `T-Test` and `Data` on the TI-84) gives $t = -1.871$ and $P = .094174$. [For instructional purposes in this review book, we note that

$\bar{x} = -1.4$ and $s = 2.366$ giving $t = \frac{-1.4-4}{2.366/\sqrt{10}} = -1.871$, and with $df = 10 - 1 = 9$, $P(t < -1.871) = .0471$. Since this is a two-sided test, we double this value to find the P-value to be $P = .0942$.]

Fourth, linking to the P-value, give a conclusion in context: With a P-value this small ($.0942 < .10$), there is evidence at the 10% significance level of a difference between the effect on reaction time between DWI and DWT. Or, with a P-value this large ($.0942 > .05$), there is no evidence at the 5% significance level of a difference between the effect on reaction time between DWI and DWT.

Investigative Tasks

1. (a) $P(\textit{positive test} \mid \textit{disease}) = \frac{305}{500} = .61$.

   (b) $P(\textit{negative test} \mid \textit{healthy}) = \frac{400}{500} = .8$.

   (c) $LR^+ = \frac{.61}{1-.8} = 3.05$. A positive test given a diseased subject and a negative test given a healthy subject are both desired outcomes, so higher values for both Sensitivity and Specificity are good. Higher values of Sensitivity, the numerator of LR^+, lead to greater values of LR^+. Higher values of Specificity give lower values of 1 – Specificity, the denominator of LR^+, again leading to greater values of LR^+.

   (d) The estimated P-value is the proportion of the simulated statistics which are less than or equal to the sample statistic of 4.7. Counting values in the dotplot gives a P-value of .04. With a P-value this small, there is evidence that the population LR^+ is below the desired value of 5.0.

2. (a) Different schemes are possible. For example, assign each material a single-digit number, such as A-0, B-1, C-2, D-3, E-4, F-5, G-6, H-7, I-8, J-9. Then read off the digits from the random number list, one at a time, throwing away any repeats, until each of the materials have been picked (or nine have been picked, as the last one left will go last). The order of picking then gives the order of being tested. Using this scheme we would get the following order:

```
FBIE  HD C       A          J                 G
51844 73424 84380 82259 28273 58102 18727 69708
```

(b) The mean drilling times in the ten materials are 4.69 seconds for Drill 1 and 4.77 seconds for Drill 2.

The proper hypothesis test is a matched pairs *t*-test on the set of differences, {0, –0.3, 0.1, –0.2, –0.1, –0.4, 0, 0.3, –0.1, –0.1}

First, state the hypotheses: H_0: $\mu_d = 0$, H_a: $\mu_d \neq 0$. (If asked to state the parameter, then state "where μ_d is the difference in mean drilling times of the two drills through different materials.")

Second, identify the test by name or formula and check the assumptions: This is a one-sample *t*-test for the mean of paired differences.

Assumptions: random samples (given) and either a dotplot of the difference sample is roughly unimodal and symmetric, or the normal probability plot is roughly linear:

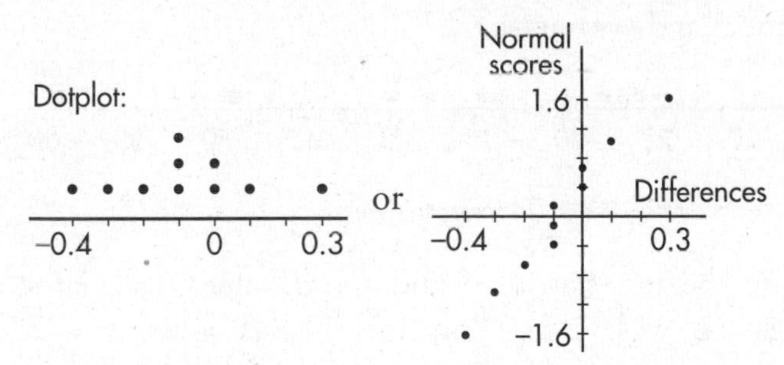

Third, calculate the test statistic *t* and the *P*-value: Calculator software (such as `T-Test` with `Data` on the TI-84) gives $t = -1.272$ and $P = .2353$. [For instructional purposes in this review book, we note that: $\bar{x} = -0.08$ and $s = 0.1989$ which gives $t = \dfrac{-.08 - 0}{0.1989/\sqrt{10}} = -1.272$, and with $df = 10 - 1 = 9$, $P(t < -1.272) = .1176$. Since this is a two-sided test, we double this value to find the *P*-value to be $P = .2352$.]

Fourth, linking to the *P*-value, give a conclusion in context. With a *P*-value this large (.2353 > .10), there is no evidence of a difference in mean drilling times of the two drills through different materials.

3. (a) The conditions for a one-sample *z*-test on a population proportion include that both $n\hat{p} \geq 10$ and $n(1 - \hat{p}) \geq 10$. In this case, $n\hat{p} = 12$ but $n(1 - \hat{p}) = 2$.

(b) First, state the hypotheses: H_0:p = .5 and H_a: p > .5, [If asked to state the parameter, then state "where p is the proportion of soccer matches for which Paul the octopus can correctly predict the winning team."]

Second, identify the test and check the assumptions:

This is a test of a binomial model.

For each trial there are two outcomes (Paul has a choice of two boxes of mussels), the trials are independent, and the probability of Paul picking a winner can be assumed to be the same for each trial. We are given that the 14 matches are a random sample of all matches.

Third, calculate the P-value: With $p = .5$, $P(x \geq 12) = \binom{14}{12}(.5)^{12}(.5)^2 +$

$\binom{14}{13}(.5)^{13}(.5) + (.5)^{14} = .00647$

Fourth, linking to the P-value, give a conclusion in context: With a P-value this small (.00647 < .01), there is strong evidence that Paul can correctly predict the results of more than 50% of soccer matches.

4. (a) First, state the hypotheses: H_0: $\mu = 30$, H_a: $\mu > 30$. (If asked to state the parameter, then state "where μ is the mean interest level of all teenagers after seeing the ad.")

Second, identify the test and check the assumptions: This is a one-sample t-test for the mean. The sample was randomly chosen, and a dotplot is roughly unimodal and symmetric:

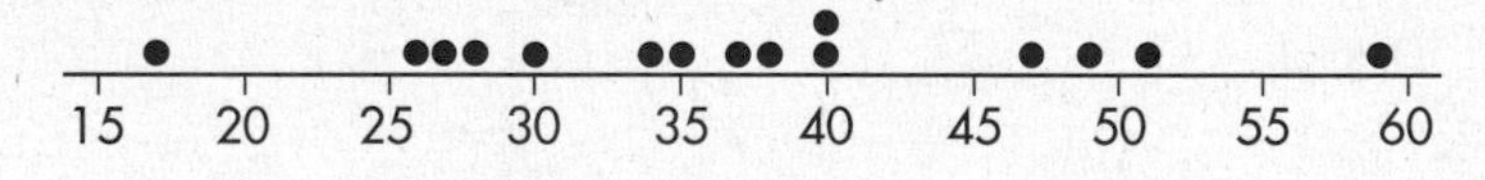

Third, calculate the test statistic t and the P-value: Calculator software (such as `T-Test` with `Data` on the TI-84) gives t = 2.526 and P = .012113. [For instructional purposes in this review book, we note that $\bar{x} = 37.2$ and $s = 11.04$ which gives $t = \dfrac{37.2 - 30}{11.04/\sqrt{15}} = 2.526$ and with $df = 15 - 1 = 14$, $P(t > 2.526) = .0121$.]

Fourth, linking to the P-value, give a conclusion in context: With a P-value this small (.0121 < .05), there is evidence to support the ad developer's claim that after seeing the ad, the interest level is above 30.

(b) First, state the hypotheses: H_0: $\mu_D = 0$, H_a: $\mu_D \neq 0$, where μ_D is the mean difference between after-ad and before-ad interest level.

Second, identify the test and check the assumptions: This is a paired t-test, that is, a single sample hypothesis test on the set of differences, and it is given that all assumptions for hypothesis testing are met.

Third, calculate the test statistic t and the P-value: Calculator software (such as `T-Test` with `Data` on the TI-84) gives $t = 0.9693$ and $P = .174415$. [For instructional purposes in this review book, we note that $\bar{x} = 0.6667$ and $s = 2.664$ giving $t = \frac{0.6667 - 0}{2.664/\sqrt{15}} = 0.9693$ and with $df = 15 - 1 = 14$, $P(t > 0.9693) = .1744$.]

Fourth, linking to the P-value, give a conclusion in context: With a P-value this large ($.1744 > .10$), there is no evidence that after seeing the ad, teenagers have increased interest in the new website.

(c) A least square regression line using the after-ad and before-ad interest levels has a high correlation of $r = .972$, which when combined with the scatterplot, indicates a strong association. Thus, after-ad interest level can be predicted from before-ad interest level.

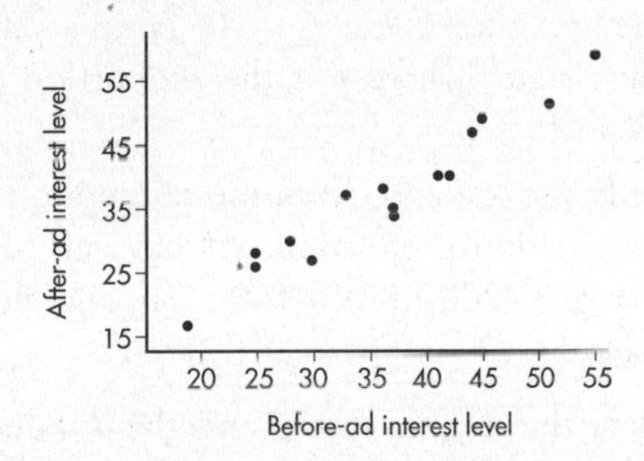

(d) A least square regression line using the before-ad and the change in interest levels yields a very low correlation of $r = .235$, which when combined with the scatterplot, indicates almost no association. Thus, after-ad interest level cannot be predicted from before-ad interest level.

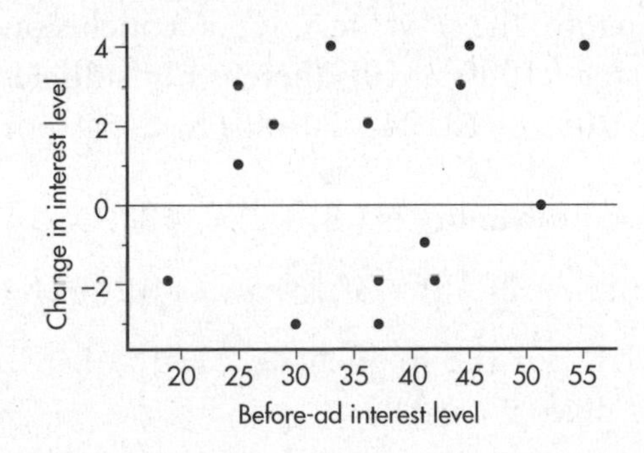

5. (a) One of many ways to proceed: let the digits 1 and 2 represent having a passport, while the remaining single digits represent not having a passport. Read off groups of ten digits, checking for the number of 1s and 2s in each group.

2498346851 4113296825 1485367833 8663018872 7373275392
5062790330 2367029195 4153038298 7360048279 4207598980
9574649262 4488086249 2651769472 9462095309 4072555345
7894788460 2391904958 0201791131 9856022851 1405559336
6003121057 4154811850 7697586849 9644852135 0811348895

Tabulating from the table gives:

No passports (no 1s or 2s): 2
One passport (one 1 or 2): 8
Two passports (two 1s or 2s): 8
Three passports (three 1s or 2s): 5
Four passports (four 1s or 2s): 1
Five passports (five 1s or 2s): 1

The estimated probability of at least three passports among ten 18–24-year-olds is thus $\frac{5+1+1}{25} = .28$.

(b) In a binomial distribution with $n = 10$ and $p = .20$, $P(x \geq 3) = .3222$. (On the TI-84, 1 – binomcdf(10, .20, 2) = 1 – .6778 = .3222.)

(c) First, state the hypotheses: H_0: $p = .20$, H_a: $p < .20$ (If asked to state the parameter, then state "where p is the proportion of all 18–24-year-olds who have passports.")

Second, identify the test and check the assumptions: This is a one-sample z-test for a population proportion, we have an SRS, $np = (200)(.20) = 40 > 10$, $n(1 - p) = (200)(.80) = 160 > 10$, and clearly 200 < 10% of all 18–24-year-olds.

Third, calculate the test statistic z and the P-value: Calculator software (such as `1-PropZTest` on the TI-84) gives $z = -1.2374$ and $P = .107963$. [For instructional purposes in this review book, we note that: $\hat{p} = \frac{33}{200} = .165$, $z = \frac{.165 - .20}{\sqrt{(.2)(.8)/200}} = -1.2374$ and $P(z < -1.2374) = .1080$.]

Fourth, linking to the P-value, give a conclusion in context. With a P-value this large (.1080 > .10), there is not sufficient evidence to dispute the claim that 20% of 18–24-year-olds have passports.

(d) This geometric probability is $(.8)^2(.2) = .128$.

(e) $\sum xP(x) = 1(.21) + 2(.15) + 3(.13) + \ldots + 12(.02) = 4.31$ (On the TI-84, put 1–12 in L1, put the frequencies 21, 15, 13, ... , 2 in L2, and then `1-VarStats` gives $\bar{x} = 4.31$.)

Tests of Significance—Chi-Square and Slope of Least Squares Line

TOPIC 15

- Chi-Square Test for Goodness of Fit
- Chi-Square Test for Independence
- Chi-Square Test for Homogeneity of Proportions
- Hypothesis Test for Slope of Least Squares Line

In this topic we continue our development of tools to analyze data. We learn about inference on distributions of counts using chi-square models. This can be used to solve such problems as "Do test results support Mendel's genetic principles?" (goodness-of-fit test); "Was surviving the Titanic sinking independent of a passenger's status?" (independence test); and "Do students, teachers, and staff show the same distributions in types of cars driven?" (homogeneity test). We then learn about inference with regard to linear association of two variables. This can be used to solve such problems as "Is there a linear relationship between the grade received on a term paper and the number of pages turned in?"

TIP

Unless you have counts, you cannot use χ^2 methods.

CHI-SQUARE TEST FOR GOODNESS OF FIT

A critical question is often whether or not an observed pattern of data fits some given distribution. A perfect fit cannot be expected, and so we must look at discrepancies and make judgments as to the *goodness of fit.*

One approach is similar to that developed earlier. There is the null hypothesis of a good fit, that is, the hypothesis that a given theoretical distribution correctly describes the situation, problem, or activity under consideration. Our observed data consist of one possible sample from a whole universe of possible samples. We ask about the chance of obtaining a sample with the observed discrepancies if the null hypothesis is really true. Finally, if the chance is too small, we reject the null hypothesis and say that the fit is not a good one.

How do we decide about the significance of observed discrepancies? It should come as no surprise that the best information is obtained from squaring the discrepancy values, as this has been our technique for studying variances from the beginning. Furthermore, since, for example, an observed difference of 23 is more significant if the original values are 105 and 128 than if they are 10,602 and 10,625, we must appropriately *weight* each difference. Such weighting is accomplished by dividing each difference by the expected values. The sum of these weighted differences or discrepancies is called *chi-square* and is denoted as χ^2 (χ is the lowercase Greek letter chi):

$$\chi^2 = \sum \frac{(\text{obs} - \text{exp})^2}{\text{exp}}$$

The smaller the resulting χ^2-value, the better the fit. The *P*-value is the probability of obtaining a χ^2 value as extreme as the one obtained if the null hypothesis is assumed true. If the χ^2 value is large enough, that is, if the *P*-value is small enough, we say there is sufficient evidence to reject the null hypothesis and to claim that the fit is poor.

To decide how large a calculated χ^2-value must be to be significant, that is, to choose a critical value, we must understand how χ^2-values are distributed. A χ^2-distribution is not symmetric and is always skewed to the right. There are distinct χ^2-distributions, each with an associated number of degrees of freedom (*df*). The larger the *df* value, the closer the χ^2-distribution to a normal distribution. Note, for example, that squaring the often-used *z*-scores 1.645, 1.96, and 2.576 results in 2.71, 3.84, and 6.63, respectively, which are entries found in the first row of the χ^2-distribution table.

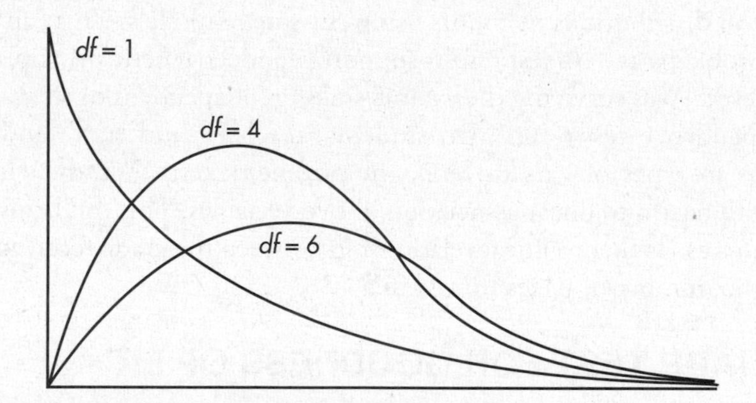

A large χ^2-value may or may not be significant—the answer depends on which χ^2-distribution we are using. A table is given of critical χ^2-values for the more commonly used percentages or probabilities. To use the χ^2-distribution for approximations in goodness-of-fit problems, the individual expected values cannot be too small. An often-used rule of thumb is that no expected value should be less than 5. Finally, as in all hypothesis tests we've looked at, the sample should be randomly chosen from the given population.

EXAMPLE 15.1

In a recent year, at the 6 p.m. time slot, television channels 2, 3, 4, and 5 captured the entire audience with 30%, 25%, 20%, and 25%, respectively. During the first week of the next season, 500 viewers are interviewed.

a. If viewer preferences have not changed, what number of persons is expected to watch each channel?
Answer: .30(500) = 150, .25(500) = 125, .20(500) = 100, and .25(500) = 125, so we have

	Channel 2	3	4	5
Expected number	150	125	100	125

b. Suppose that the actual observed numbers are as follows:

	Channel 2	3	4	5
Observed number	139	138	112	111

Do these numbers indicate a change? Are the differences significant?
Answer: Check the conditions:

1. *Randomization*: We must assume that the 500 viewers are a representative sample.
2. We note that the expected values (150, 125, 100, 125) are all ≥ 5.

H_0: The television audience is distributed over channels 2, 3, 4, and 5 with percentages 30%, 25%, 20%, and 25%, respectively.
H_a: The audience distribution is not 30%, 25%, 20%, and 25%, respectively.

We calculate

$$\chi^2 = \sum \frac{(\text{obs} - \text{exp})^2}{\text{exp}}$$

$$= \frac{(139-150)^2}{150} + \frac{(138-125)^2}{125} + \frac{(112-100)^2}{100} + \frac{(111-125)^2}{125}$$

$$= 5.167$$

Then the *P*-value is $P = P(\chi^2 > 5.167) = .1600$. [With $df = n - 1 = 3$, the TI-84 gives χ^2cdf(5.167,1000,3) = .1600, or a direct test for goodness of fit can be downloaded onto older TI-84+ calculators and comes preloaded on most!]

Conclusion:

With this large a *P*-value (.1600) there is not sufficient evidence to reject H_0. That is, there is not sufficient evidence that viewer preferences have changed.

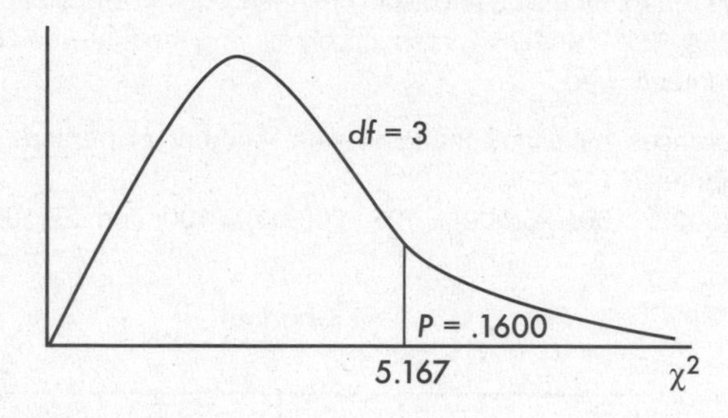

Note: While the TI-83 does not have the goodness of fit download that is available on the TI-84, one can still use a TI-83 to calculate the above χ^2 by putting the observed values in list L1 and the expected values in L2, then calculating $(L1 - L2)^2/L2 \rightarrow L3$ and χ^2 = sum(L3), where "sum" is found under LIST → MATH.

EXAMPLE 15.2

A grocery store manager wishes to determine whether a certain product will sell equally well in any of five locations in the store. Five displays are set up, one in each location, and the resulting numbers of the product sold are noted.

	Location				
	1	2	3	4	5
Actual number sold	43	29	52	34	48

Is there enough evidence that location makes a difference? Test at both the 5% and 10% significance levels.

Answer:

H_0: Sales of the product are uniformly distributed over the five locations.
H_a: Sales are not uniformly distributed over the five locations.

A total of 43 + 29 + 52 + 34 + 48 = 206 units were sold. If location doesn't matter, we would expect $\frac{206}{5}$ = 41.2 units sold per location (uniform distribution).

	Location				
	1	2	3	4	5
Expected number sold	41.2	41.2	41.2	41.2	41.2

Check the conditions:

1. *Randomization:* We must assume that the 206 units sold are a representative sample.
2. We note that the expected values (all 41.2) are all ≥ 5.

Thus

$$\chi^2 = \frac{(43-41.2)^2}{41.2} + \frac{(29-41.2)^2}{41.2} + \frac{(52-41.2)^2}{41.2} + \frac{(34-41.2)^2}{41.2} + \frac{(48-41.2)^2}{41.2} = 8.903$$

The number of degrees of freedom is the number of classes minus 1; that is, $df = 5 - 1 = 4$.

The *P*-value is $P = P(\chi^2 > 8.903) = .0636$. [On the TI-84: χ^2cdf(8.903,1000,4).]

With $P = .0636$ there is sufficient evidence to reject H_0 at the 10% level but not at the 5% level. If the grocery store manager is willing to accept a 10% chance of committing a Type I error, there is enough evidence to claim location makes a difference.

CHI-SQUARE TEST FOR INDEPENDENCE

In the goodness-of-fit problems above, a set of expectations was based on an assumption about how the distribution should turn out. We then tested whether an observed sample distribution could reasonably have come from a larger set based on the assumed distribution.

In many real-world problems we want to compare two or more observed samples without any prior assumptions about an expected distribution. In what is called a *test of independence*, we ask whether the two or more samples might reasonably have come from some larger set. For example, do nonsmokers, light smokers, and heavy smokers all have the same likelihood of being eventually diagnosed with cancer, heart disease, or emphysema? Is there a relationship (association) between smoking status and being diagnosed with one of these diseases?

We classify our observations in two ways and then ask whether the two ways are independent of each other. For example, we might consider several age groups and within each group ask how many employees show various levels of job satisfaction. The null hypothesis is that age and job satisfaction are independent, that is, that the proportion of employees expressing a given level of job satisfaction is the same no matter which age group is considered.

Analysis involves calculating a table of *expected* values, assuming the null hypothesis about independence is true. We then compare these expected values with the observed values and ask whether the differences are reasonable if H_0 is true. The significance of the differences is gauged by the same χ^2-value of weighted squared differences. The smaller the resulting χ^2-value, the more reasonable the null hypothesis of independence. If the χ^2-value is large enough, that is, if the *P*-value is small enough, we can say that the evidence is sufficient to reject the null hypothesis and to claim that there *is* some relationship between the two variables or methods of classification.

In this type of problem,

$$df = (r-1)(c-1)$$

where df is the number of degrees of freedom, r is the number of rows, and c is the number of columns.

A point worth noting is that even if there is sufficient evidence to reject the null hypothesis of independence, we cannot necessarily claim any direct *causal* relationship. In other words, although we can make a statement about some link or relationship between two variables, we are *not* justified in claiming that one causes the other. For example, we may demonstrate a relationship between salary level and job satisfaction, but our methods would not show that higher salaries cause higher job satisfaction. Perhaps an employee's higher job satisfaction impresses his superiors and thus leads to larger increases in pay. Or perhaps there is a third variable, such as training, education, or personality, that has a direct causal relationship to both salary level and job satisfaction.

EXAMPLE 15.3

In a nationwide telephone poll of 1000 adults representing Democrats, Republicans, and Independents, respondents were asked two questions: their party affiliation and if their confidence in the U.S. banking system had been shaken by the savings and loan crisis. The answers, cross-classified by party affiliation, are given in the following *contingency table*.

	Confidence Shaken		
Observed	**Yes**	**No**	**No opinion**
Democrats	175	220	55
Republicans	150	165	35
Independents	75	105	20

Test the null hypothesis that shaken confidence in the banking system is independent of party affiliation. Use a 10% significance level.

Answer:

H_0: Party affiliation and shaken confidence in the banking system are independent.
H_a: Party affiliation and shaken confidence in the banking system are not independent.

The above table gives the observed results. To determine the expected values, we must first determine the row and column totals:

Row totals: 175 + 220 + 55 = 450,
150 + 165 + 35 = 350,
75 + 105 + 20 = 200.

Column totals: 175 + 150 + 75 = 400,
220 + 165 + 105 = 490,
55 + 35 + 20 = 110.

	Yes	No	No opinion	
Democrats				450
Republicans				350
Independents				200
	400	490	110	

To calculate, for example, the expected value in the upper left box, we can proceed in any of several equivalent ways. First, we could note that the proportion of Democrats is $\frac{450}{1000} = .45$; and so, if independent, the expected number of Democrat *yes* responses is .45(400) = 180. Instead, we could note that the proportion of *yes* responses is $\frac{400}{1000} = .4$; and so, if independent, the expected number of Democrat *yes* responses is .4(450) = 180. Finally, we could note that both these calculations simply involve $\frac{(450)(400)}{1000} = 180$.

In other words, *the expected value of any box can be calculated by multiplying the corresponding row total by the appropriate column total and then dividing by the grand total.* Thus, for example, the expected value for the middle box, which corresponds to Republican *no* responses, is $\frac{(350)(490)}{1000} = 171.5$.

Continuing in this manner, we fill in the table as follows:

	Expected			
	Yes	No	No opinion	
Democrats	180	220.5	49.5	450
Republicans	140	171.5	38.5	350
Independents	80	98	22	200
	400	490	110	

[An appropriate check at this point is that each expected cell count is at least 5.]

Next we calculate the value of chi-square:

$$\chi^2 = \frac{(175-180)^2}{180} + \frac{(220-220.5)^2}{220.5} + \frac{(55-49.5)^2}{49.5}$$
$$+ \frac{(150-140)^2}{140} + \frac{(165-171.5)^2}{171.5} + \frac{(35-38.5)^2}{38.5}$$
$$+ \frac{(75-80)^2}{80} + \frac{(100-98)^2}{98} + \frac{(20-22)^2}{22}$$
$$= 3.024$$

[On the TI-84, go to MATRIX and EDIT. Put the data into a matrix. Then STAT, TESTS, χ^2-Test, will give $\chi^2 = 3.0243$. Note also that the expected values are automatically stored in a second matrix.]

Note that, once the 180, 220.5, 140, and 171.5 boxes are calculated, the other expected values can be found by using the row and column totals. Thus the number of degrees of freedom here is 4. Or we calculate

$$df = (r-1)(c-1) = (3-1)(3-1) = 4$$

The *P*-value is calculated to be $P = P(\chi^2 > 3.024) = .5538$.

With such a large *P*-value, there is *no* evidence of any relationship between party affiliation and shaken confidence in the banking system.

On the TI-Nspire the result shows as:

$$\chi^2\text{2way}\begin{bmatrix}175 & 220 & 55\\150 & 165 & 35\\75 & 105 & 20\end{bmatrix} : stat.results \qquad \begin{bmatrix}\text{"Title"} & \text{"}\chi^2\text{ 2-way Test"}\\ \text{"}\chi^2\text{"} & 3.02428\\ \text{"PVal"} & 0.553771\\ \text{"df"} & 4.\\ \text{"ExpMatrix"} & \text{"[...]"}\\ \text{"CompMatrix"} & \text{"[...]"}\end{bmatrix}$$

$$stat.ExpMatrix \qquad \begin{bmatrix}180. & 220.5 & 49.5\\140. & 171.5 & 38.5\\80. & 98. & 22.\end{bmatrix}$$

As for conditions to check for chi-square tests for independence, we should check that the sample is randomly chosen and that the expected values for all cells are at least 5. If a category has one of its expected cell count less than 5, we can combine categories that are logically similar (for example "disagree" and "strongly disagree"), or combine numerically small categories collectively as "other."

EXAMPLE 15.4

To determine whether men with a combination of childhood abuse and a certain abnormal gene are more likely to commit violent crimes, a study is run on a simple random sample of 575 males in the 25 to 35 age group. The data are summarized in the following table:

	Not abused, normal gene	Abused, normal gene	Not abused, abnormal gene	Abused, abnormal gene
Criminal behavior	48	21	32	26
Normal behavior	201	79	118	50

a. Is there evidence of a relationship between the four categories (based on childhood abuse and abnormal genetics) and behavior (criminal versus normal)? Explain.

b. Is there evidence that among men with the normal gene, the proportion of abused men who commit violent crimes is greater than the proportion of nonabused men who commit violent crimes? Explain.

c. Is there evidence that among men who were not abused as children, the proportion of men with the abnormal gene who commit violent crimes is greater than the proportion of men with the normal gene who commit violent crimes? Explain.

d. Is there a contradiction in the above results? Explain.

Answers:

a. A chi-square test for independence is indicated. The expected cell counts are as follows:

55.0	22.1	33.1	16.8
194.0	77.9	116.9	59.2

The condition that all cell counts are greater than 5 is met.

H_0: The four categories (based on childhood abuse and abnormal genetics) and behavior (criminal versus normal) are independent.

H_a: The four categories (based on childhood abuse and abnormal genetics) and behavior (criminal versus normal) are not independent (there is a relationship).

Running a chi-square test gives $\chi^2 = 7.752$. With $df = (r - 1)(c - 1) = 3$, we get $P = .0514$. Since $.0514 < .10$, the data do provide some evidence (at least at the 10% significance level) to reject H_0 and conclude that there is evidence of a relationship between the four categories (based on childhood abuse and abnormal genetics) and behavior (criminal versus normal).

b. A two-proportion *z*-test is indicated. We must check that *n* is large enough: $n_1 \hat{p}_1 = 21 > 10$, $n_1(1 - \hat{p}_1) = 79 > 10$, $n_2\hat{p}_2 = 48 > 10$, $n_2(1 - \hat{p}_2) = 201 > 10$. We must assume simple random samples from the target population, since this is not given.

H_0: $p_1 - p_2 = 0$ (where p_1 is the proportion of abused men with normal genetics who commit violent crimes and p_2 is the proportion of nonabused men with normal genetics who commit violent crimes)

H_a: $p_1 - p_2 > 0$ (the proportion of abused men with normal genetics who commit violent crimes is greater than the proportion of nonabused men with normal genetics who commit violent crimes)

$$\hat{p}_1 = \frac{21}{100} = .21,\ \hat{p}_2 = \frac{48}{249} = .193,\ \text{and}\ \hat{p} = \frac{21+48}{100+249} = .198$$

$$\sigma_d = \sqrt{(.198)(.802)\left(\tfrac{1}{100}+\tfrac{1}{249}\right)} = .0472$$

$$\text{So}\ z = \frac{.21-.193}{.0472} = 0.360\ \text{and}\ P = .359.$$

With such a large P-value there is no evidence to reject H_0, and thus we conclude that there is *no* evidence that the proportion of abused men with normal genetics who commit violent crimes is greater than the proportion of nonabused men with normal genetics who commit violent crimes.

c. A two-proportion z-test is indicated. We must check that n is large enough: $n_1\hat{p}_1 = 32 > 10$, $n_1(1 - \hat{p}_1) = 118 > 10$, $n_2\hat{p}_2 = 48 > 10$, $n_2(1 - \hat{p}_2) = 201 > 10$. We must assume simple random samples from the target population, since this is not given.

H_0: $p_1 - p_2 = 0$ (where p_1 is the proportion of nonabused men with abnormal genetics who commit violent crimes and p_2 is the proportion of nonabused men with normal genetics who commit violent crimes)

H_a: $p_1 - p_2 > 0$ (the proportion of nonabused men with abnormal genetics who commit violent crimes is greater than the proportion of nonabused men with normal genetics who commit violent crimes)

$$\hat{p}_1 = \frac{32}{150} = .213,\ \hat{p}_2 = \frac{48}{249} = .193,\ \text{and}\ \hat{p} = \frac{32+48}{150+249} = .2005$$

$$\sigma_d = \sqrt{(.2005)(.7995)\left(\tfrac{1}{150}+\tfrac{1}{249}\right)} = .0414$$

$$\text{So}\ z = \frac{.213-.193}{.0414} = 0.483\ \text{and}\ P = .315.$$

With such a large P-value there is no evidence to reject H_0, and thus we conclude that there is *no* evidence that the proportion of nonabused men with the abnormal gene who commit violent crimes is greater than the proportion of nonabused men with the normal gene who commit violent crimes.

d. There is no contradiction. It is possible to have evidence of an overall relationship without significant evidence showing in a subset of the categories.

CHI-SQUARE TEST FOR HOMOGENEITY OF PROPORTIONS

In chi-square goodness-of-fit tests we work with a single variable in comparing a single sample to a population model. In chi-square independence tests we work with a single sample classified on two variables. Chi-square procedures can also be used with a single variable to compare samples from two or more populations. It is important that the samples be *simple random samples*, that they be taken *independ-*

ently of each other, that the original populations be large compared to the sample sizes, and that the expected values for all cells be at least 5. The contingency table used has a row for each sample.

EXAMPLE 15.5

In a large city, a group of AP Statistics students work together on a project to determine which group of school employees has the greatest proportion who are satisfied with their jobs. In independent simple random samples of 100 teachers, 60 administrators, 45 custodians, and 55 secretaries, the numbers satisfied with their jobs were found to be 82, 38, 34, and 36, respectively. Is there evidence that the proportion of employees satisfied with their jobs is different in different school system job categories?

Answer:

H_0: The proportion of employees satisfied with their jobs is the same across the various school system job categories.

H_a: At least two of the job categories differ in the proportion of employees satisfied with their jobs.

The observed counts are as follows:

	Satisfied	Not satisfied
Teachers	82	18
Administrators	38	22
Custodians	34	11
Secretaries	36	19

Just as we did in the previous section, we can calculate the expected value of any cell by multiplying the corresponding row total by the appropriate column total and then dividing by the grand total. In this case, this results in the following expected counts:

	Satisfied	Not satisfied	
Teachers	73.1	26.9	100
Administrators	43.8	16.2	60
Custodians	32.9	12.1	45
Secretaries	40.2	14.8	55
	190	70	260

We note that all expected cell counts are >5, and then calculate chi-square:

$$\chi^2 = \sum \frac{(\text{obs} - \text{exp})^2}{\text{exp}} = \frac{(82 - 73.1)^2}{73.1} + \cdots + \frac{(19 - 14.8)^2}{14.8} = 8.640$$

With 4 − 1 = 3 degrees of freedom, we calculate the *P*-value to be $P(\chi^2 > 8.640) = .0345$. [On the TI-84: χ^2cdf(8.640,1000,3).] With this small a *P*-value, there is sufficient evidence to reject H_0, and we can conclude that there is evidence that the proportion of employees satisfied with their jobs is *not* the same across all the school system job categories.

Note: On the TI-84 we could also have put the observed data into a matrix and used χ^2-Test, resulting in $\chi^2 = 8.707$ and $P = .0335$.

The difference between the test for independence and the test for homogeneity can be confusing. When we're simply given a two-way table, it's not obvious which test is being performed. The crucial difference is in the *design* of the study. Did we pick samples from each of two or more populations to compare the distribution of some variable among the different populations? If so, we are doing a test for homogeneity. Did we pick one sample from a single population and cross-categorize on two variables to see if there is an association between the variables? If so, we are doing a test for independence.

For example, if we separately sample Democrats, Republicans, and Independents to determine whether they are for, against, or have no opinion with regard to stem cell research (several samples, one variable), then we do a test for homogeneity with a null hypothesis that the distribution of opinions on stem cell research is the same among Democrats, Republicans, and Independents. However, if we sample the general population, noting the political preference and opinions on stem cell research of the respondents (one sample, two variables), then we do a test for independence with a null hypothesis that political preference is independent of opinion on stem cell research.

Finally, it should be remembered that while we used χ^2 for categorical data, the χ^2 distribution is a *continuous* distribution, and applying it to counting data is just an approximation.

HYPOTHESIS TEST FOR SLOPE OF LEAST SQUARES LINE

In addition to finding a confidence interval for the true slope, one can also perform a hypothesis test for the value of the slope. Often one uses the null hypothesis H_0: $\beta = 0$, that is, that there is no linear relationship between the two variables. Assumptions for inference for the slope of the least squares line include the following: (1) the sample must be randomly selected; (2) The scatterplot should be approximately linear; (3) there should be no apparent pattern in the residuals plot; and (4) the distribution of the residuals should be approximately normal. (This third point can be checked by a histogram, dotplot, stemplot, or normal probability plot of the residuals.)

Note that a low *P*-value tells us that if the two variables did not have some linear relationship, it would be highly unlikely to find such a random sample; however, strong evidence that there is some association does not mean the association is strong.

TIP

Be careful about abbreviations. For example, your teacher might use LOBF (line of best fit), but the grader may have no idea what this refers to.

TIP

If the data don't look straight, do not fit a straight line.

EXAMPLE 15.6

The following table gives serving speeds in mph (using a flat or "cannonball" serve) of ten randomly selected professional tennis players before and after using a newly developed tennis racquet.

With old racquet	125	133	108	128	115	135	125	117	130	121
With new racquet	133	134	112	139	123	142	140	129	139	126

a. Is there evidence of a *straight line* relationship with positive slope between serving speeds of professionals using their old and the new racquets?

(continued)

Answer: Checking the assumptions:

We are told that the data come from a *random* sample of professional players.

Scatterplot is approximately linear.

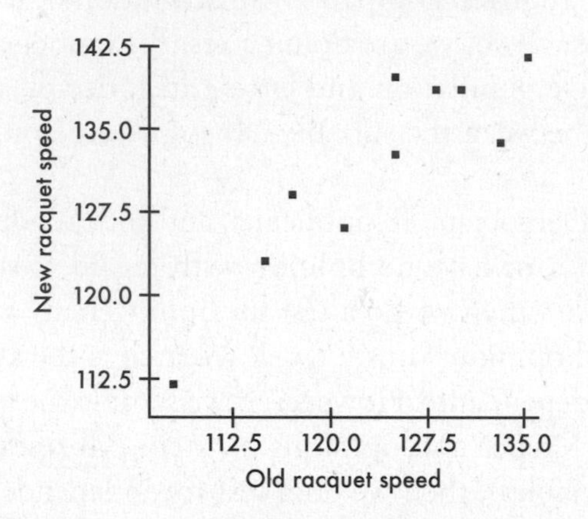

No apparent pattern in residuals plot.

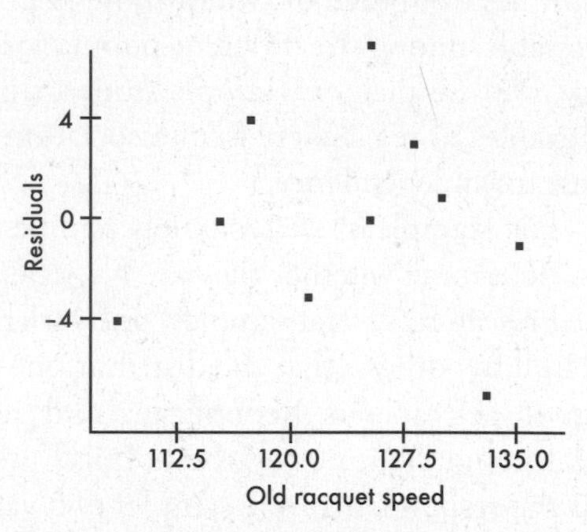

Distribution of residuals is approximately normal.

Checked by histogram. *or* Checked by normal probability plot.

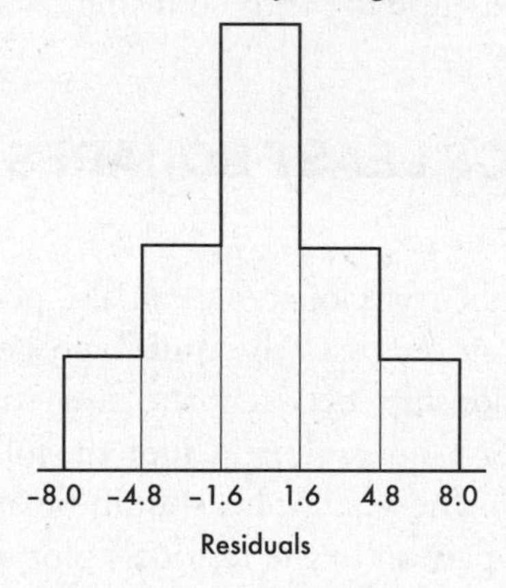

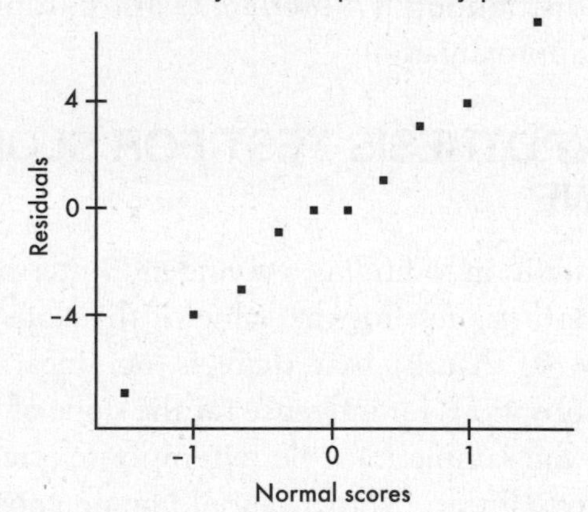

We proceed with the hypothesis test for the slope of the regression line.

H_0: $\beta_1 = 0$ H_a: $\beta_1 > 0$

Using the statistics software on a calculator (for example, `LinRegTTest` on the TI-84), gives:

New speed = 8.76 + 0.99 (Old speed) with $t = 5.853$ and $P = .00019$

With such a small P value, there is very strong evidence of a straight line relationship with positive slope between serving speeds of professionals using their old and the new racquets.

b. Interpret in context the least squares line.

Answer: With a slope of approximately 1 and a y-intercept of 8.76, the regression line indicates that use of the new racquet increases serving speed on the average by 8.76 mph regardless of the old racquet speed. That is, players with lower and higher old racquet speeds experience on the average the same numerical (rather than percentage) increase when using the new racquet.

(continued)

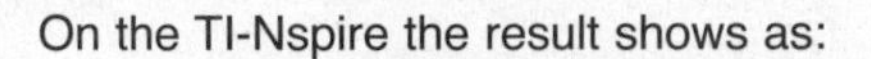

On the TI-Nspire the result shows as:

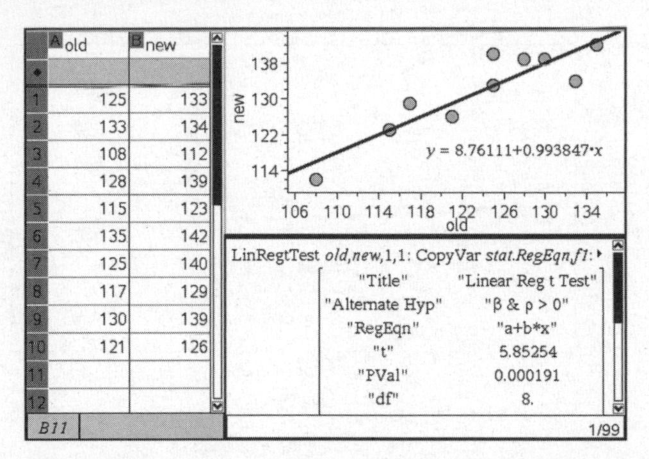

It is important to be able to interpret computer output, and appropriate computer output is helpful in checking the assumptions for inference.

EXAMPLE 15.7

A company offers a 10-lesson program of study to improve students' SAT scores. A survey is made of a random sampling of 25 students. A scatterplot of improvement in total (math + verbal) SAT score versus number of lessons taken is as follows:

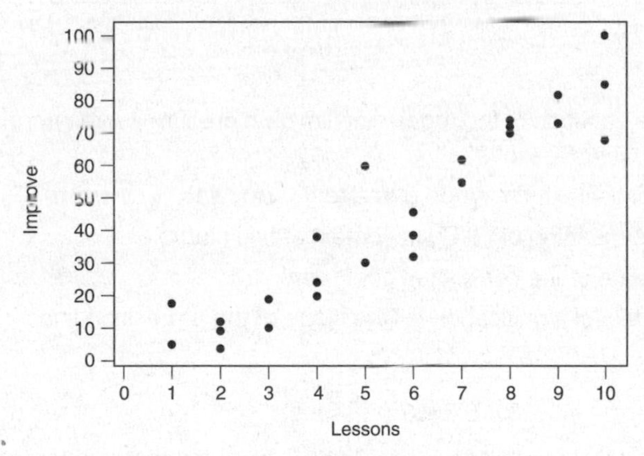

Some computer output of the regression analysis follows, along with a plot and a histogram of the residuals.

Regression Analysis: Improvement Versus Lessons

Dependent variable: Improvement

Predictor	Coef	SE Coef	T	P
Constant	−7.718	4.272	−1.81	0.084
Lessons	9.2854	0.6789	13.68	0.000

S = 9.744 R–Sq = 89.1%

Analysis of Variance

Source	DF	SS	MS	F	P
Regression	1	17761	17761	187.06	0.000
Residual Error	23	2184	95		
Total	24	19945			

(continued)

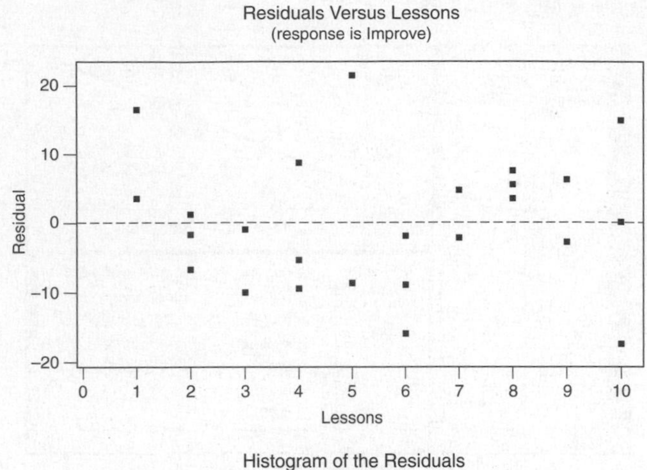

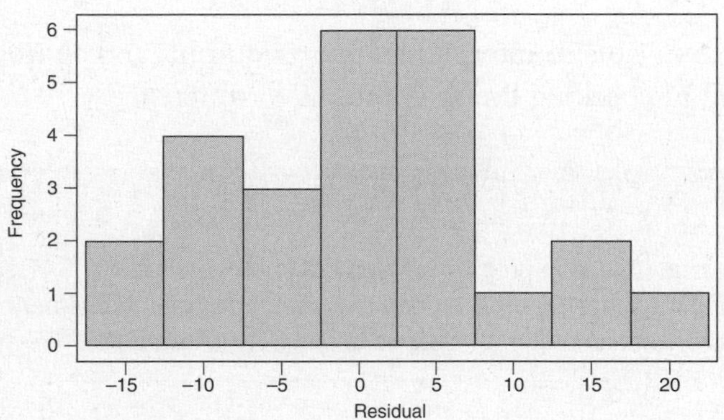

a. What is the equation of the regression line that predicts score improvement as a function of number of lessons?
b. Interpret, in context, the slope of the regression line.
c. What is the meaning of R–Sq in context of this study?
d. Give the value of the correlation coefficient.
e. Perform a test of significance for the slope of the regression line.

Answers:

a. Improvement = −7.718 + 9.2854 (Lessons)
b. The slope of the regression line is 9.2854, meaning that on the average, each additional lesson is predicted to improve one's total SAT score by 9.2854.
c. R-Sq = 89.1% means that 89.1% of the variation in total SAT score improvement can be explained by variation in the number of lessons taken.
d. $r = +\sqrt{.891} = .949$, where the sign is positive because the slope is positive.
e. A test of significance of the slope of the regression line with H_0: $\beta = 0$, H_a: $\beta \neq 0$ (test of significance of correlation) is as follows:

Assumptions: We are told the data came from a random sample, the scatterplot appears to be approximately linear, there is no apparent pattern in the residuals plot, and the histogram of residuals appears to be approximately normal.

$$t = \frac{9.2854 - 0}{0.6789} = 13.68, \text{ and with } df = 25 - 2 = 23, P = .000$$

(Or t and P can simply be read off the computer output.)

With this small a P-value, there is very strong evidence to reject H_0 and conclude that the relationship between improvement in total SAT score and number of lessons taken is significant.

INDEPENDENCE

Throughout this review book the concept of *independence* has arisen many times, with different meanings in different contexts, and a quick review is worthwhile.

1. If the outcome of one event, E, doesn't affect the outcome of another event, F, we say the events are *independent*, and we have $P(E|F) = P(E)$, $P(F|E) = P(F)$, and $P(E \cap F) = P(E)P(F)$.
2. When the distribution of one variable in a contingency table is the same for all categories of another variable, we say the two variables are *independent.*
3. When two random variables are *independent,* we can use the rule for adding variances with regard to the random variables $X + Y$ and $X - Y$.
4. In a binomial distribution, the trials must be *independent;* that is, the probability of success must be the same on every trial, irrespective of what happened on a previous trial. However, when we pick a sample from a population, we do change the size and makeup of the remaining population. We typically allow this as long as the sample is not too large—the usual rule of thumb is that the sample size should be smaller than 10% of the population.
5. In inference we wish to make conclusions about a population parameter by analyzing a sample. A crucial assumption is always that the sampled values are *independent* of each other. This typically involves checking if there was proper randomization in the gathering of data. Also, to minimize the effect on independence from samples being drawn without replacement, we check that less than 10% of the population is sampled.
6. Furthermore, if inference involves the comparison of two groups, the two groups must be *independent* of each other.
7. With paired data, while the observations in each pair are not independent, the differences must be *independent* of each other.
8. With regression, the errors in the regression model must be *independent,* and one may check this by confirming there are no patterns in the residual plot.

Summary

- Chi-square analysis is an important tool for inference on distributions of counts.
- The chi-square statistic is found by summing the weighted squared differences between observed and expected counts.
- A chi-square goodness-of-fit test compares an observed distribution to some expected distribution.
- A chi-square test of independence tests for evidence of an association between two categorical variables from a single sample.
- A chi-square test of homogeneity compares samples from two or more populations with regard to a single categorical variable.
- The sampling distribution for the slope of a regression line can be modeled by a t-distribution with $df = n - 2$.

Questions on Topic Fifteen: Tests of Significance—Chi-Square and Slope of Least Squares Line

Multiple-Choice Questions

Directions: The questions or incomplete statements that follow are each followed by five suggested answers or completions. Choose the response that best answers the question or completes the statement.

1. A random sample of mice is obtained, and each mouse is timed as it moves through a maze to a reward treat at the end. After several days of training, each mouse is timed again. The data should be analyzed using

   (A) a z-test of proportions.
   (B) a two-sample test of means.
   (C) a paired t-test.
   (D) a chi-square test.
   (E) a regression analysis.

2. To test the claim that dogs bite more or less depending upon the phase of the moon, a university hospital counts admissions for dog bites and classifies with moon phase.

	New moon	First quarter	Full moon	Last quarter
Dog bite admissions	32	27	47	38

   Which of the following is the proper conclusion?

   (A) The data prove that dog bites occur equally during all moon phases.
   (B) The data prove that dog bites do not occur equally during all moon phases.
   (C) The data give evidence that dog bites occur equally during all moon phases.
   (D) The data give evidence that dog bites do not occur equally during all moon phases.
   (E) The data do not give sufficient evidence to conclude that dog bites are related to moon phases.

3. For a project, a student randomly picks 100 fellow AP Statistics students to survey on whether each has either a PC or Apple at home (all students in the school have a home computer) and what score (1, 2, 3, 4, 5) each expects to receive on the AP exam. A chi-square test of independence results in a test statistic of 8. How many degrees of freedom are there?

   (A) 1
   (B) 4
   (C) 7
   (D) 9
   (E) 99

4. A random sample of 100 former student-athletes are picked from each of the 16 colleges which are members of the Big East conference. Students are surveyed about whether or not they feel they received a quality education while participating in varsity athletics. A 2×16 table of counts is generated. Which of the following is the most appropriate test to determine whether there is a difference among these schools as to the student-athlete perception of having received a quality education.

   (A) A chi-square goodness-of-fit test for a uniform distribution
   (B) A chi-square test of independence
   (C) A chi-square test of homogeneity of proportions
   (D) A multiple-sample *z*-test of proportions
   (E) A multiple-population *z*-test of proportions

5. According to theory, blood types in the general population occur in the following proportions: 46% O, 40% A, 10% B, and 4% AB. Anthropologists come upon a previously unknown civilization living on a remote island. A random sampling of blood types yields the following counts: 77 O, 85 A, 23 B, and 15 AB. Is there sufficient evidence to conclude that the distribution of blood types found among the island population differs from that which occurs in the general population?

   (A) The data prove that blood type distribution on the island is different from that of the general population.
   (B) The data prove that blood type distribution on the island is not different from that of the general population.
   (C) The data give evidence at the 1% significance level that blood type distribution on the island is different from that of the general population.
   (D) The data give evidence at the 5% significance level, but not at the 1% significance level, that blood type distribution on the island is different from that of the general population.
   (E) The data do not give evidence at the 5% significance level that blood type distribution on the island is different from that of the general population.

6. Is there a relationship between education level and sports interest? A study cross-classified 1500 randomly selected adults in three categories of education level (not a high school graduate, high school graduate, and college graduate) and five categories of major sports interest (baseball, basketball, football, hockey, and tennis). The χ^2-value is 13.95. Is there evidence of a relationship between education level and sports interest?

   (A) The data prove there is a relationship between education level and sports interest.
   (B) The evidence points to a cause-and-effect relationship between education and sports interest.
   (C) There is evidence at the 5% significance level of a relationship between education level and sports interest.
   (D) There is evidence at the 10% significance level, but not at the 5% significance level, of a relationship between education level and sports interest.
   (E) The *P*-value is greater than .10, so there is no evidence of a relationship between education level and sports interest.

7. A disc jockey wants to determine whether middle school students and high school students have similar music tastes. Independent random samples are taken from each group, and each person is asked whether he/she prefers hip-hop, pop, or alternative. A chi-square test of homogeneity of proportions is performed, and the resulting *P*-value is below .05. Which of the following is a proper conclusion?

   (A) There is evidence that for all three music choices the proportion of middle school students who prefer each choice is equal to the corresponding proportion of high school students.
   (B) There is evidence that the proportion of middle school students who prefer hip-hop is different from the proportion of high school students who prefer hip-hop.
   (C) There is evidence that for all three music choices the proportion of middle school students who prefer each choice is different from the corresponding proportion of high school students.
   (D) There is evidence that for at least one of the three music choices the proportion of middle school students who prefer that choice is equal to the corresponding proportion of high school students.
   (E) There is evidence that for at least one of the three music choices the proportion of middle school students who prefer that choice is different from the corresponding proportion of high school students.

8. A geneticist claims that four species of fruit flies should appear in the ratio 1:3:3:9. Suppose that a sample of 2000 flies contained 110, 345, 360, and 1185 flies of each species, respectively. Is there sufficient evidence to reject the geneticist's claim?

   (A) The data prove the geneticist's claim.
   (B) The data prove the geneticist's claim is false.
   (C) The data do not give sufficient evidence to reject the geneticist's claim.
   (D) The data give sufficient evidence to reject the geneticist's claim.
   (E) The evidence from this data is inconclusive.

9. A food biologist surveys people at an ice cream parlor, noting their taste preferences and cross-classifying against the presence or absence of a particular marker in a saliva swab test.

	Presence	Absence
Vanilla	32	12
Chocolate	15	7
Strawberry	24	19

Is there evidence of a relationship between taste preference and the marker presence?

(A) At the 10% significance level, the data prove that there is a relationship between taste preference and the presence of the marker.
(B) At the 10% significance level, the data prove that there is no relationship between taste preference and the presence of the marker.
(C) There is evidence at the 5% significance level of a relationship between taste preference and the presence of the marker.
(D) There is evidence at the 10% significance level, but not at the 5% significance level, of a relationship between taste preference and the presence of the marker.
(E) There is not evidence at the 10% significance level of a relationship between taste preference and the presence of the marker.

10. Random samples of 25 students are chosen from each high school class level, students are asked whether or not they are satisfied with the school cafeteria food, and the results are summarized in the following table:

	Freshmen	Sophomores	Juniors	Seniors
Satisfied	15	12	9	7
Dissatisfied	10	13	16	18

Is there evidence of a difference in cafeteria food satisfaction among the class levels?

(A) The data prove that there is a difference in cafeteria food satisfaction among the class levels.
(B) There is evidence of a linear relationship between food satisfaction and class level.
(C) There is evidence at the 1% significance level of a difference in cafeteria food satisfaction among the class levels.
(D) There is evidence at the 5% significance level, but not at the 1% significance level, of a difference in cafeteria food satisfaction among the class levels.
(E) With $P = .1117$ there is not evidence of a difference in cafeteria food satisfaction among the class levels.

11. Can dress size be predicted from a woman's height? In a random sample of 20 female high school students, dress size versus height (cm) gives the following regression results:

```
The regression equation is
Size = -48.8 + 0.374 Height

Predictor      Coef    SE Coef       T       P
Constant     -48.81      30.57   -1.60   0.128
Height       0.3736     0.1898    1.97   0.065

S = 4.46720        R-Sq = 17.7%       R-Sq(adj) = 13.1%
```

Is there statistical evidence of a linear relationship between dress size and height?

(A) No, because r^2, the coefficient of determination, is too small.
(B) No, because 0.128 is above any reasonable significance level.
(C) Yes, because by any reasonable observation, taller women tend to have larger dress sizes.
(D) Yes, because the computer printout does give a regression equation.
(E) There is evidence at the 10% significance level, but not at the 5% level.

12. To study the relationship between calories (kcal) and fat (g) in pizza, slices of 14 randomly selected major brand pizzas are chemically analyzed. Computer printout for regression:

```
Predictor        Coef     SE Coef       T       P
Constant        1.593       1.422    1.12   0.285
Calories     0.035881    0.004896    7.33   0.000

S = 1.27515        R-Sq = 81.7%       R-Sq(adj) = 80.2%
```

What is measured by `S = 1.27515`?

(A) Variability in calories among slices
(B) Variability in fat among slices
(C) Variability in the slope (g/kcal) of the regression line
(D) Variability in the *y*-intercept of the regression line
(E) Variability in the residuals

Free-Reponse Questions

Directions: You must show all work and indicate the methods you use. You will be graded on the correctness of your methods and on the accuracy of your final answers.

Seven Open-Ended Questions

1. A candy manufacturer advertises that their fruit-flavored sweets have hard sugar shells in five colors with the following distribution: 35% cherry red, 10% vibrant orange, 10% daffodil yellow, 25% emerald green, and 20% royal purple. A random sample of 300 sweets yielded the counts in the following table:

	Cherry red	Vibrant orange	Daffodil yellow	Emerald green	Royal purple
Observed counts	94	34	22	77	73

   Is there evidence that the distribution is different from what is claimed by the manufacturer?

2. You want to study whether smokers and non-smokers have equal fitness levels (low, medium, high) and are considering two survey designs:

   I. Take a random sample of 200 people, asking each whether or not they smoke, and measuring the fitness level of each.

   II. Take a random sample of 100 smokers and measure the fitness level of each, and take a random sample of 100 non-smokers and measure the fitness level of each.

   (a) Which of the designs results in a test of independence and which results in a test of homogeneity? Explain.

   (b) If we are interested in whether the proportion of people who have various fitness levels differs among smokers and non-smokers, which design should be used? Explain.

   (c) If we are interested in the conditional distribution of people with given fitness levels who are smokers or are not smokers, which design should be used? Explain.

3. Is there a difference in happiness levels between busy and idle people? In one study, after filling out a survey, 175 randomly chosen high school students were told they could either sit 15 minutes while the survey was being tabulated, or they could walk 15 minutes to another building where the survey was being tabulated. Then they were given a questionnaire asking how good they felt during the past 15 minutes (on a scale of 1, "not good," to 5, "very good"). The results of the questionnaire were as follows:

	Happiness level				
	1	2	3	4	5
Busy (walking)	8	15	24	26	25
Idle (sitting)	18	20	15	10	14

(a) Does the above data give statistical evidence of a relationship between happiness level and busy/idle choice of the students?

(b) If the answer above is positive for a relationship, is it reasonable to conclude that encouraging high school students to keep more busy will lead to higher happiness levels?

4. A poll, asking a random sample of adults whether or not they eat breakfast and to rate their morning energy level, results in the table:

	Morning energy level		
	Low	Medium	High
Breakfast	22%	24%	24%
No breakfast	12%	10%	8%

(a) If the sample size was $n = 500$, is there evidence of a relationship between eating breakfast and morning energy level?

(b) Does the answer to part (a) change if $n = 1000$ instead? Explain.

5. A survey on acne treatments randomly selected 100 teenagers using topical treatments, 100 using oral medications, and 50 using laser therapy, and asked each subject about satisfaction level. The resulting counts were:

	Topical	Oral	Laser
Very satisfied	61	54	24
Somewhat satisfied	28	21	10
Unsatisfied	11	25	16

(a) Do the different treatments lead to different satisfaction levels? Perform an appropriate hypothesis test.

(b) What is a possible confounding variable which could lead you to jump to a misleading conclusion? Explain.

6. A study of 100 randomly selected teenagers, ages 13–17, looked at number of texts per waking hour versus age, yielding the computer regression output below:

```
Predictor      Coef      SE Coef       T        P
Constant     -1.055        2.815    -0.37    0.709
Age          0.4577       0.1866     2.45    0.016

S = 2.66501        R-Sq = 5.8%      R-Sq(adj) = 4.8%
```

   (a) Interpret the slope in context.

   (b) What three graphs should be checked with regard to conditions for a test of significance for the slope of the regression line?

   (c) Assuming all conditions for inference are met, perform this test of significance.

   (d) Give a conclusion in context, taking into account both your answer to the hypothesis test as well as the value of R-Sq.

7. Can the average women's life expectancy in a country be predicted from the average fertility rate (children/woman) in that country? Following are the most recent data from 11 randomly selected countries:

Country	Fertility rate (children/woman)	Life expectancy (women)
Afghanistan	6.25	45.5
Angola	5.33	51.4
Argentina	2.16	80.0
Australia	1.85	84.4
France	1.85	85.1
Liberia	4.69	61.5
Nepal	2.66	69.0
Netherlands	1.77	82.6
Pakistan	3.58	68.3
Poland	1.29	80.4
Singapore	1.29	83.4

Perform a test of significance for the slope of the regression line that relates fertility rate (children/woman) to life expectancy (women).

Three Investigative Tasks

1. Use of placebos are now the norm for medical studies, but disclosing the compositions of these placebos is unexplicitly not all that common, even though these placebos could have effects. For example, olive oil has been used as a placebo in cholesterol drug tests, and there is some evidence that olive oil can affect cholesterol level. Composition of a placebo injection is disclosed more often than that of a placebo pill, but still it can be difficult to make this finding. In a survey of 800 studies using placebo injections, three researchers independently try to determine the compositions of the placebos, and the counts of successes are as follows:

	Number of researchers able to determine composition of a placebo			
	0	1	2	3
Number of studies	419	314	56	11

   (a) Find the complete binomial distribution (probabilities of 0 through 3 successes) for $n = 3$ and $p = 0.2$.

   (b) Given 800 studies, if the number of researchers who determine composition of a placebo follows a binomial with probability of success $p = 0.2$, what are the expected values for the numbers (0 through 3) of researchers able to determine composition of a placebo?

   (c) Test the null hypothesis that the number of researchers who determine composition of a placebo follows a binomial with probability of success $p = 0.2$.

2. Twelve high schools are randomly selected, and in each, the math teachers, the English teachers, and the students on the honor roll all take a logic puzzle test, and average scores for each group are noted in the following table:

School	1	2	3	4	5	6	7	8	9	10	11	12
English teacher score	9.9	11.7	9.6	9.4	10.2	10.3	10.2	10.9	8.7	9.6	9.9	10.8
Mathematics teacher score	8.5	9.9	9.9	11.1	11.0	10.8	10.9	10.4	9.0	8.5	10.1	10.6
Student score	8.2	11.5	10.6	11.4	11.5	10.7	10.1	11.3	9.5	8.4	10.3	11.7

The regression analysis of the data follows.

Regression Analysis: Student versus English
Dependent variable: Student

```
Predictor   Coef      SE Coef      T       P
Constant    2.315       4.154    0.56    0.590
English     0.8038     0.4102    1.96    0.078

S = 1.06798      R-Sq = 27.8%      R-Sq(adj) = 20.5%
```

Regression Analysis: Student versus Math

Dependent variable: Student

Predictor	Coef	SE Coef	T	P
Constant	-0.330	2.230	-0.15	0.885
Math	1.0701	0.2208	4.85	0.001

S = 0.686661 R-Sq = 70.1% R-Sq(adj) = 67.1%

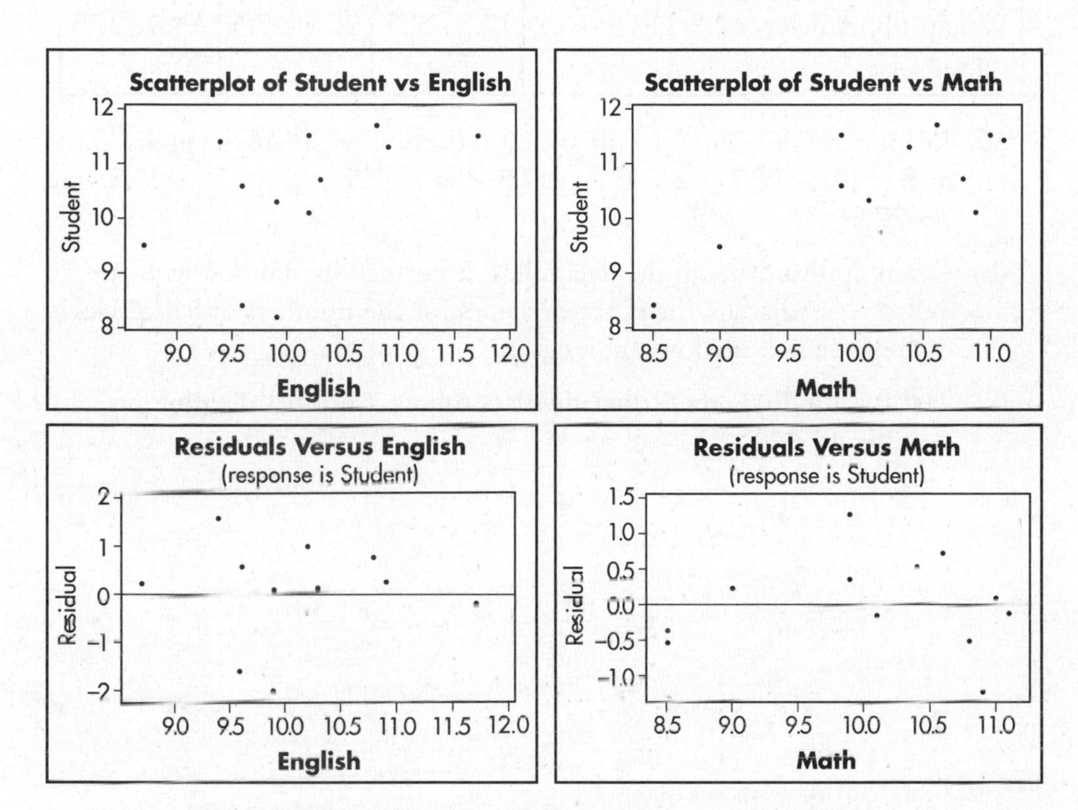

(a) Is there evidence that the average student score is greater than that of the English teachers?

(b) Is there evidence that the average student score is greater than that of the math teachers?

(c) Is the score of the English or math teachers a better predictor of the student score? Explain.

(d) Given the analysis in part (c), predict the average student score in a school where the average English teachers' score is 11.0 and the average math teachers' score is 10.0.

(e) Given the analysis in parts (c) and (d), what is the approximate range of average student scores at schools where the average English teachers' score is 11.0 and the average math teachers' score is 10.0.

3. A heavy backpack can cause chronic shoulder, neck, and back pain. Wide, padded shoulder straps, a waist belt, and avoidance of single-strap bags all help, but weight is the main problem. The recommendation for a safe weight for school backpacks is no more than 10% of body weight. In a study of 500 randomly selected high school students, the weights of their backpacks give rise to the following:

Weight (lb)	Below 12.5	12.5–17.5	17.5–22.5	22.5–27.5	Above 27.5
Observed #	28	134	182	112	44

   (a) In a normal distribution with $\mu = 20$ and $\sigma = 5$, what are the probabilities of $z < 12.5$, $12.5 < z < 17.5$, $17.5 < z < 22.5$, $22.5 < z < 27.5$, and $z > 27.5$?

   (b) Given 500 students, if the data follow a normal distribution with $\mu = 20$ and $\sigma = 5$, what are the expected values for the numbers of backpacks in each of the indicated weight ranges?

   (c) Test the null hypothesis that the data follow a normal distribution with $\mu = 20$ and $\sigma = 5$.

Answer Key

1. **C** 4. **C** 7. **E** 10. **E**
2. **E** 5. **D** 8. **D** 11. **E**
3. **B** 6. **D** 9. **E** 12. **E**

Answers Explained

Multiple-Choice

1. **(C)** There are two observations of each mouse, a *before* time and an *after* time. These two times are dependent so a paired t-test is appropriate, *not* a two-sample test.

2. **(E)** The expected counts if dog bites occur equally during all moon phases are each $\frac{1}{4}(32 + 27 + 47 + 38) = 36$. A chi-square goodness-of-fit test gives

$$\chi^2 = \sum \frac{(obs - exp)^2}{exp} = \frac{(32-36)^2}{36} + \frac{(27-36)^2}{36} + \frac{(47-36)^2}{36} + \frac{(38-36)^2}{36}$$

$= 6.167$, and with $df = 4 - 1 = 3$, $P(\chi^2 > 6.167) = .1038$. With this large a P-value ($.1038 > .10$), there is not sufficient evidence to conclude that dog bites are related to moon phases.

3. **(B)** $df =$ (rows – 1)(columns – 1) = (2 – 1)(5 – 1) = 4

4. **(C)** Picking separate samples from each of 16 populations and classifying according to one variable (perception of quality education) is a survey design which is most appropriately analyzed using a chi-square test of homogeneity of proportions.

5. **(D)** With 77 + 85 + 23 + 15 = 200 samples, the expected counts if the blood type distribution on the island is the same as that of the general population are 46% of 200 = 92, 40% of 200 = 80, 10% of 200 = 20, and 4% of 200 = 8. A chi-square goodness-of-fit test gives

$$\chi^2 = \sum \frac{(obs - exp)^2}{exp} = \frac{(77-92)^2}{92} + \frac{(85-80)^2}{80} + \frac{(23-20)^2}{20} + \frac{(15-8)^2}{8}$$

$= 9.333$, and with $df = 4 - 1 = 3$, $P(\chi^2 > 9.333) = .0252$. With a P-value this small ($.0252 < .05$), there is sufficient evidence at the 5% significance level that blood type distribution on the island is different from that of the general population.

6. **(D)** With $df = (3 - 1)(5 - 1) = 8$, $P(\chi^2 > 13.95) = .083$. Since $.05 < .083 < .10$, there is evidence at the 10% significance level, but not at the 5% significance level, of a relationship between education level and sports interest.

7. **(E)** With a P-value this small (less than .05), there is evidence in support of the alternative hypothesis H_a: the distributions of music preferences are different, that is, they differ for at least one of the proportions.

8. **(D)** With 1 + 3 + 3 + 9 = 16, according to the geneticist the expected number of fruit flies of each species is $\frac{1}{16}(2000) = 125$, $\frac{3}{16}(2000) = 375$, $\frac{3}{16}(2000) = 375$, $\frac{9}{16}(2000) = 1125$. A chi-square goodness-of-fit test gives $\chi^2 =$

$$\sum \frac{(obs - exp)^2}{exp} = \frac{(110-125)^2}{125} + \frac{(345-375)^2}{375} + \frac{(360-375)^2}{375} + \frac{(1185-1125)^2}{1125}$$

= 8, and with $df = 4 - 1 = 3$, $P(\chi^2 > 8) = .0460$. With a *P*-value this small (.0460 < .05), there is sufficient evidence at the 5% significance level to reject the geneticist's claim.

9. **(E)** A chi-square test of independence gives $\chi^2 = 2.852$, and with $df = 2$, we find $P = .2403$, and since .2403 > .10, there is not evidence at the 10% significance level of a relationship between taste preference and the presence of the marker.

10. **(E)** A chi-square test of homogeneity gives $\chi^2 = 5.998$, and with $df = 3$, the *P*-value is .1117. With a *P*-value this large (.1117 > .10) there is not evidence of a difference in cafeteria food satisfaction among the class levels.

11. **(E)** The relevant *P*-value is 0.065 which is less than 0.10 but greater than 0.05.

12. **(E)** In computer printouts of regression analysis, "S" typically gives the standard deviation of the residuals.

Free-Response

1. First, state the hypotheses: H_0: The colors of the sugar shells are distributed according to 35% cherry red, 10% vibrant orange, 10% daffodil yellow, 25% emerald green, and 20% royal purple, and H_a: The colors of the sugar shells are not distributed as claimed by the manufacturer. Or [H_0: $P_{CR} = .35$, $P_{VO} = .10$, $P_{DY} = .10$, $P_{EG} = .25$, $P_{RP} = .20$, and H_a: at least one proportion is different from this distribution.]

   Second, identify the test and check the assumptions: χ^2 goodness-of-fit test. We are given a random sample, and calculate that all expected cells are at least 5: 35% of 300 = 105, 10% of 300 = 30, 25% of 300 = 75, and 20% of 300 = 60.

   Third, calculate the test statistic χ^2 and the *P*-value: A calculator gives $\chi^2 = \sum \frac{(obs - exp)^2}{exp} = 6.689$, and with $df = 5 - 1 = 4$, $P = .153$.

   Fourth, linking to the *P*-value, give a conclusion in context: With a *P*-value this large (.153 > .10), there is not evidence that the distribution is different from what is claimed by the manufacturer.

2. (a) Design I, with a single sample from one population classified on two variables (smoking and fitness), will result in a test of independence. Design II, with independent samples from two populations each with the single variable (fitness), will result in a test of homogeneity.

   (b) Design II, with its test of homogeneity, and using an equal sample size from each of the two populations (smokers and non-smokers), is best for comparing proportions of smokers who have different fitness levels with proportions of non-smokers who have different fitness levels.

   (c) Design I, which classifies one population on the two variables, smoking and fitness, is the only one of these two designs which will give data on the conditional distribution of people with given fitness levels who are smokers or are not smokers.

3. (a) First, state the hypotheses: H_0: Happiness level is independent of busy/idle choice for high school students and H_a: Happiness level is not independent of busy/idle choice for high school students.

   Second, identify the procedure and check the conditions: This is a chi-square test of independence. It is given that there is a random sample, the data are measured as "counts," and the expected counts are all at least 5 (put the observed counts in a matrix; then χ^2`-Test` on the TI-84 gives expected counts of

14.6	19.6	21.8	20.2	21.8
11.4	15.4	17.2	15.8	17.2

   Third, calculate the test statistic and the *P*-value: Calculator software (χ^2`-Test` on the TI-84) gives $\chi^2 = 14.54$ with $P = .0058$ and $df = 4$.

   Fourth, give a conclusion in context with linkage to the *P*-value: With a *P*-value this small ($.0058 < .01$), there is strong evidence of a relationship between happiness level and busy/idle choice for high school students.

   (b) No, it is not reasonable to conclude that encouraging high school students to keep more busy will lead to higher happiness levels. This was not an experiment with students randomly chosen to sit or walk. The students themselves chose whether or not to sit or walk so no cause-and-effect conclusion is possible. For example, it could well be that the happier students choose to walk, whereas the less happy students choose to sit.

4. (a) First, state the hypotheses: H_0: Eating breakfast and morning energy level are independent and H_a: Eating breakfast and morning energy level are not independent.

   Second, identify the test and check the assumptions: This is a χ^2 test of independence on

111	120	120
60	50	40

where we are given a random sample, and a calculator gives that all expected cells are at least 5:

119	119	112
51	51	48

Third, calculate the test statistic χ^2 and the *P*-value: Calculator software (such as `χ2-Test` on the TI-84) gives $\chi^2 = 4.202$ and $P = .1224$.

Fourth, linking to the *P*-value, give a conclusion in context: With this large a *P*-value (.1224 > .10), there is not evidence of a relationship between eating breakfast and morning energy level.

(b) Yes, the conclusion changes. With $n = 1000$, the observed numbers are:

220	240	240
120	100	80

with $\chi^2 = 8.403$ and $P = .0150$. With a *P*-value this small (.0150 < .05), now there is evidence of a relationship between eating breakfast and morning energy level.

5. (a) First, state the hypotheses: H_0: The different treatments lead to the same satisfaction levels and H_a: The different treatments lead to different satisfaction levels.

Second, identify the test and check the assumptions: χ^2 test of homogeneity. We are given a random sample, and a calculator gives that all expected cells are at least 5:

55.6	55.6	27.8
23.6	23.6	11.8
20.8	20.8	10.4

Third, calculate the test statistic χ^2 and the *P*-value: A calculator gives $\chi^2 = 10.9521$, and with $df = 4$, $P = .0271$.

Fourth, linking to the *P*-value, give a conclusion in context: With this small a *P*-value (.0271 < .05), there is evidence (at the 5% significance level) that the different treatments do lead to different satisfaction levels.

(b) An example of a possible confounding variable is severity of the acne outbreak. It could be that those with more severe cases have less satisfaction no matter what the treatment and are also the ones who are encouraged to use oral medications or laser therapy. So it would be wrong to conclude that oral medications or laser therapy are the causes of less satisfaction.

6. (a) Each additional year in age of teenagers is associated with an average of 0.4577 more texts per waking hour.

   (b) The scatterplot of texts per hour versus age should be roughly linear, there should be no apparent pattern in the residual plot, and a histogram of the residuals should be approximately normal.

   (c) First, state the hypotheses: H_0: $\beta = 0$, H_a: $\beta \neq 0$, where β is the slope of the regression line that relates average texts per hour to age.

   Second, identify the test and check the assumptions: This is a test of significance for the slope of the regression line, and we are given that all conditions for inference are met.

   Third, calculate the test statistic t and the P-value: The computer printout gives that $t = 2.45$ and $P = .016$.

   Fourth, linking to the P-value, give a conclusion in context: With this small a P-value ($.016 < .05$), there is evidence of a linear relationship between average texts per hour and age for teenagers ages 13–17.

   (d) R-Sq = 5.8%, so even though there is evidence of a linear relationship between average texts per hour and age for teenagers ages 13–17, only 5.8% of the variability in average texts per hour is explained by this regression model (or "is accounted for by the variation in age.").

7. First, state the hypotheses: H_0: $\beta = 0$, H_a: $\beta \neq 0$. (If asked to state the parameter, then state "where β is the slope of the regression line that relates average fertility rate to women's life expectancy.")

   Second, identify the test and check the assumptions: This is a test of significance for the slope of the regression line. We are given that the data come from a random sample of countries.

Scatterplot is approximately linear

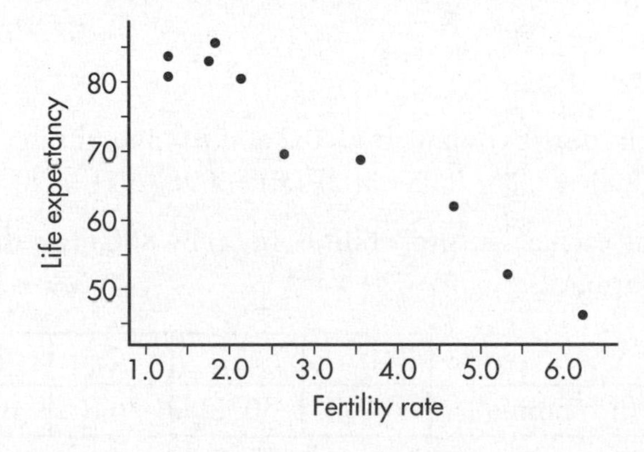

No apparent pattern in residual plot

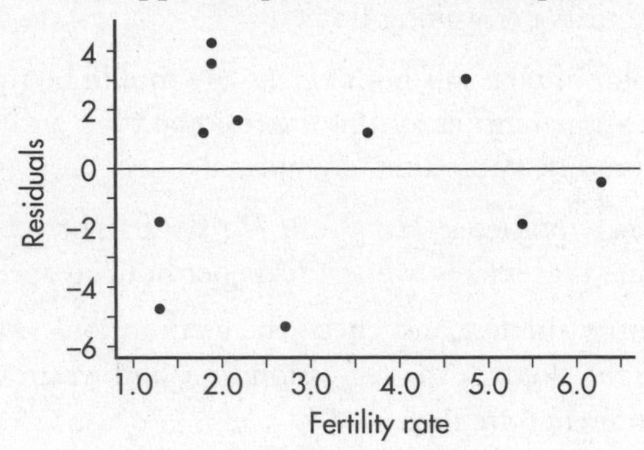

Distribution of residuals is approximately normal
(Checked by normal probability plot)

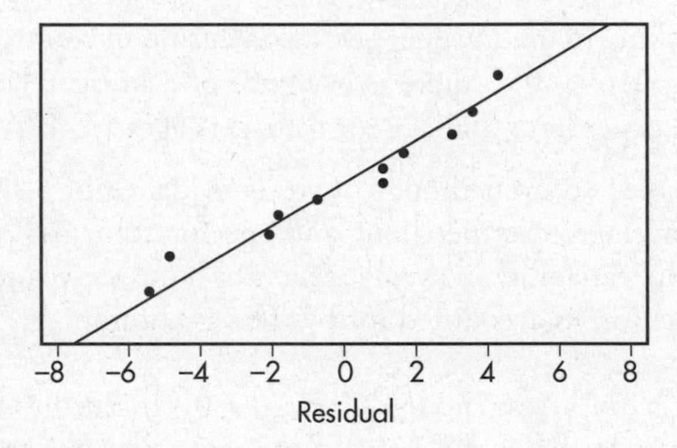

Third, calculate the test statistic t and the P-value: Using a calculator (for example, `LinRegTTest` on the TI-84) gives that $t = -12.53$ and $P = .0000$.

Fourth, linking to the P-value, give a conclusion in context: With a P-value this small ($.0000 < .05$), there is evidence of a linear relationship between average fertility rate (children/woman) and life expectancy (women).

Investigative Task

1. (a) The binomial distribution with $p = 0.2$ and $n = 3$ results in: $P(0) = (.8)^3 = .512$, $P(1) = 3(.2)(.8)^2 = .384$, $P(2) = 3(.2)^2(.8) = .096$, $P(3) = (.8)^3 = .008$.

   (b) Multiplying each of the probabilities in (a) by 800 gives the expected number of occurrences:

	0	1	2	3
Expected (if binomial)	409.6	307.2	76.8	6.4

   (c) First, state the hypotheses: H_0: The number of researchers able to determine the composition of a placebo follows a binomial with $p = 0.2$ and H_a: The number of researchers able to determine the composition of a placebo does not follow a binomial with $p = 0.2$.

Second, identify the test and check the assumptions: This is a chi-square test of goodness-of-fit. We must assume that the 800 studies form a representative sample. We note from (b) that all expected cells are ≥5:

Third, calculate the test statistic χ^2 and the P-value: A calculator gives $\chi^2 = \sum \frac{(obs - exp)^2}{exp} = 9.3058$, and with $df = 4 - 1 = 3$, $P = .02549$.

Fourth, linking to the P-value, give a conclusion in context: With a P-value this small (.02549 < .05), there is evidence that the number of researchers able to determine the composition of a placebo does not follow a binomial with $p = 0.2$.

2. (a) This is a matched-pair t-test on the set of differences {–1.7, –0.2, 1.0, 2.0, 1.3, 0.4, –0.1, 0.4, 0.8, –1.2, 0.4, 0.9}, with H_0: $\mu_d = 0$ and H_a: $\mu_d > 0$. The sample is random (given), and a dotplot of the differences is roughly unimodal and symmetric (so a normal population distribution is a reasonable assumption):

–2.0 –1.0 0 1.0 2.0

Calculator software (such as `T-Test` with `Data` on the TI-84) gives $t = 1.121$ and $P = .143$. [For instructional purposes in this review book, we note that $\bar{x} = 0.3333$, $s = 1.030$, $t = \frac{0.3333 - 0}{1.030/\sqrt{12}} = 1.121$, and with $df = 12 - 1 = 11$, $P = .143$.] With a P-value this large (.143 > .10), there is not evidence that the average student score is greater than that of the English teachers.

(b) This is a matched-pair t-test on the set of differences {–0.3, 1.6, 0.7, 0.3, 0.5, –0.1, –0.8, 0.9, 0.5, –0.1, 0.2, 1.1}, with H_0: $\mu_d = 0$ and H_a: $\mu_d > 0$. The sample is random (given), and a dotplot of the differences is roughly unimodal and symmetric (so a normal population distribution is a reasonable assumption):

–1.0 0 1.0

Calculator software (such as `T-Test` with `Data` on the TI-84) gives $t = 1.974$ and $P = .0370$. [For instructional purposes in this review book, we note that $\bar{x} = 0.375$, $s = 0.6580$, $t = \frac{0.375 - 0}{0.6580/\sqrt{12}} = 1.974$, and with $df = 12 - 1 = 11$, $P = .037$.] With a P-value this small (.037 < .05), there is evidence that the average student score is greater than that of the math teachers.

(c) In both regression analyses, the scatterplots are roughly linear and the residual plots show no apparent pattern; however, while 70.1% of the variance in student scores is explained by the variance in math teacher scores, only

27.8% of the variance in student scores is explained by variance in English teacher scores. Thus, the math teacher scores are a better predictor of student scores.

(d) Using the regression equation giving predicted student score as a function of math teacher score gives –0.330 + 1.0701(10.0) = 10.37.

(e) The regression output gives a standard deviation of s = 0.6867. Using the 10.37 estimate from part (d) gives a range of scores for students whose math teachers' average score is 10.0 to be 10.37 ± 3(0.6867) = 10.37 ± 2.06. Thus, for schools where the math teachers' average score is 10.0, almost all the average student scores will be between 8.31 and 12.4.

3. (a) The z-scores for 22.5 and 27.5 are, respectively, $\frac{22.5-20}{5} = 0.5$ and $\frac{27.5-20}{5} = 1.5$. Similarly, 17.5 and 12.5 have z-scores of –0.5 and –1.5, respectively. Using the normal probability table or a calculator gives the probabilities: $P(z < -1.5) = .0668$, $P(-1.5 < z < -0.5) = .2417$, $P(-0.5 < z < 0.5) = .3830$, $P(0.5 < z < 1.5) = .2417$, $P(z > 1.5) = .0668$.

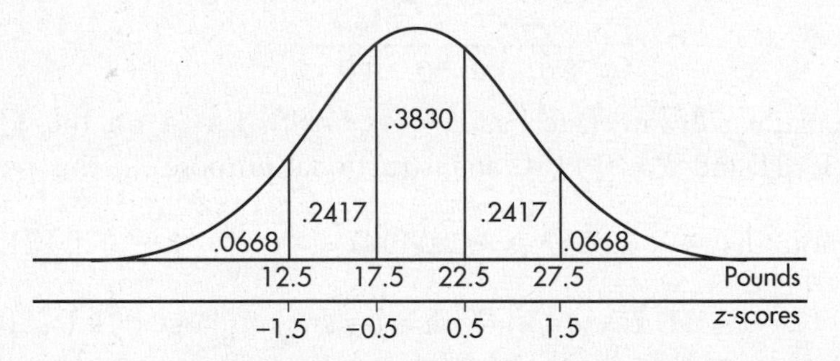

(b) Multiplying each probability by 500: .0668(500) = 33.4, .2417(500) = 120.85, .3830(500) = 191.5, .2417(500) = 120.85, and .0668(500) = 33.4.

Weight (lb)	Below 12.5	12.5–17.5	17.5–22.5	22.5–27.5	Above 27.5
Expected#	33.4	120.85	191.5	120.85	33.4

(c) First, state the hypotheses: H_0: The weights of student backpacks follow a normal distribution with $\mu = 20$ and $\sigma = 5$ and H_a: The weights of student backpacks do not follow a normal distribution with $\mu = 20$ and $\sigma = 5$.

Second, identify the test and check the assumptions: This is a chi-square test of goodness-of-fit. We are given that the weights are from a random sample of students. We note from above that all expected cells are ≥5:

Third, calculate the test statistic χ^2 and the P-value: A calculator gives $\chi^2 = \sum \frac{(obs - exp)^2}{exp} = 6.7874$, and with $df = 5 - 1 = 4$, $P = .1476$.

Fourth, linking to the P-value, give a conclusion in context: With a P-value this large (.1476 > .10), there is no evidence that the data do not follow a normal distribution with $\mu = 20$ and $\sigma = 5$.

Answer Sheet

PRACTICE EXAMINATION 1

1. Ⓐ Ⓑ Ⓒ Ⓓ Ⓔ	11. Ⓐ Ⓑ Ⓒ Ⓓ Ⓔ	21. Ⓐ Ⓑ Ⓒ Ⓓ Ⓔ	31. Ⓐ Ⓑ Ⓒ Ⓓ Ⓔ
2. Ⓐ Ⓑ Ⓒ Ⓓ Ⓔ	12. Ⓐ Ⓑ Ⓒ Ⓓ Ⓔ	22. Ⓐ Ⓑ Ⓒ Ⓓ Ⓔ	32. Ⓐ Ⓑ Ⓒ Ⓓ Ⓔ
3. Ⓐ Ⓑ Ⓒ Ⓓ Ⓔ	13. Ⓐ Ⓑ Ⓒ Ⓓ Ⓔ	23. Ⓐ Ⓑ Ⓒ Ⓓ Ⓔ	33. Ⓐ Ⓑ Ⓒ Ⓓ Ⓔ
4. Ⓐ Ⓑ Ⓒ Ⓓ Ⓔ	14. Ⓐ Ⓑ Ⓒ Ⓓ Ⓔ	24. Ⓐ Ⓑ Ⓒ Ⓓ Ⓔ	34. Ⓐ Ⓑ Ⓒ Ⓓ Ⓔ
5. Ⓐ Ⓑ Ⓒ Ⓓ Ⓔ	15. Ⓐ Ⓑ Ⓒ Ⓓ Ⓔ	25. Ⓐ Ⓑ Ⓒ Ⓓ Ⓔ	35. Ⓐ Ⓑ Ⓒ Ⓓ Ⓔ
6. Ⓐ Ⓑ Ⓒ Ⓓ Ⓔ	16. Ⓐ Ⓑ Ⓒ Ⓓ Ⓔ	26. Ⓐ Ⓑ Ⓒ Ⓓ Ⓔ	36. Ⓐ Ⓑ Ⓒ Ⓓ Ⓔ
7. Ⓐ Ⓑ Ⓒ Ⓓ Ⓔ	17. Ⓐ Ⓑ Ⓒ Ⓓ Ⓔ	27. Ⓐ Ⓑ Ⓒ Ⓓ Ⓔ	37. Ⓐ Ⓑ Ⓒ Ⓓ Ⓔ
8. Ⓐ Ⓑ Ⓒ Ⓓ Ⓔ	18. Ⓐ Ⓑ Ⓒ Ⓓ Ⓔ	28. Ⓐ Ⓑ Ⓒ Ⓓ Ⓔ	38. Ⓐ Ⓑ Ⓒ Ⓓ Ⓔ
9. Ⓐ Ⓑ Ⓒ Ⓓ Ⓔ	19. Ⓐ Ⓑ Ⓒ Ⓓ Ⓔ	29. Ⓐ Ⓑ Ⓒ Ⓓ Ⓔ	39. Ⓐ Ⓑ Ⓒ Ⓓ Ⓔ
10. Ⓐ Ⓑ Ⓒ Ⓓ Ⓔ	20. Ⓐ Ⓑ Ⓒ Ⓓ Ⓔ	30. Ⓐ Ⓑ Ⓒ Ⓓ Ⓔ	40. Ⓐ Ⓑ Ⓒ Ⓓ Ⓔ

Practice Examination 1

SECTION I

Questions 1–40

Spend 90 minutes on this part of the exam.

Directions: The questions or incomplete statements that follow are each followed by five suggested answers or completions. Choose the response that best answers the question or completes the statement.

1. Following is a histogram of home sale prices (in thousands of dollars) in one community:

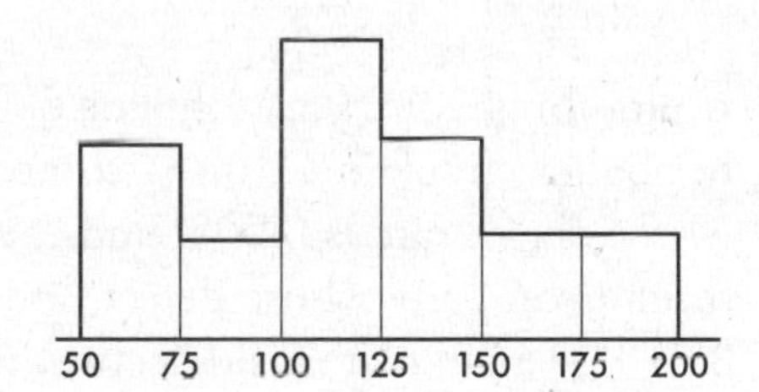

Which of the following conclusions is most correct?

(A) The median price was $125,000.
(B) The mean price was $125,000.
(C) More homes sold for between $100,000 and $125,000 than for over $125,000.
(D) $10,000 is a reasonable estimate of the standard deviation in selling prices.
(E) 1.5×10^9 ($\$^2$) is a reasonable estimate of the variance in selling prices.

2. Which of the following is most useful in establishing cause-and-effect relationships?

(A) A complete census
(B) A least squares regression line showing high correlation
(C) A simple random sample (SRS)
(D) A well-designed, well-conducted survey incorporating chance to ensure a representative sample
(E) An experiment

3. The average yearly snowfall in a city is 55 inches. What is the standard deviation if 15% of the years have snowfalls above 60 inches? Assume yearly snowfalls are normally distributed.

(A) 4.83
(B) 5.18
(C) 6.04
(D) 8.93
(E) The standard deviation cannot be computed from the information given.

GO ON TO THE NEXT PAGE ➤

4. To determine the average cost of running for a congressional seat, a simple random sample of 50 politicians is chosen and the politicians' records examined. The cost figures show a mean of $125,000 with a standard deviation of $32,000. Which of the following is the best interpretation of a 90% confidence interval estimate for the average cost of running for office?

   (A) 90% of politicians running for a congressional seat spend between $117,500 and $132,500.
   (B) 90% of politicians running for a congressional seat spend a mean dollar amount that is between $117,500 and $132,500.
   (C) We are 90% confident that politicians running for a congressional seat spend between $117,500 and $132,500.
   (D) We are 90% confident that politicians running for a congressional seat spend a mean dollar amount between $117,500 and $132,500.
   (E) We are 90% confident that in the chosen sample, the mean dollar amount spent running for a congressional seat is between $117,500 and $132,500.

5. In one study on the effect that eating meat products has on weight level, an SRS of 500 subjects who admitted to eating meat at least once a day had their weights compared with those of an independent SRS of 500 people who claimed to be vegetarians. In a second study, an SRS of 500 subjects were served at least one meat meal per day for 6 months, while an independent SRS of 500 others were chosen to receive a strictly vegetarian diet for 6 months, with weights compared after 6 months.

   (A) The first study is a controlled experiment, while the second is an observational study.
   (B) The first study is an observational study, while the second is a controlled experiment.
   (C) Both studies are controlled experiments.
   (D) Both studies are observational studies.
   (E) Each study is part controlled experiment and part observational study.

6. A plumbing contractor obtains 60% of her boiler circulators from a company whose defect rate is 0.005, and the rest from a company whose defect rate is 0.010. If a circulator is defective, what is the probability that it came from the first company?

   (A) .429
   (B) .500
   (C) .571
   (D) .600
   (E) .750

GO ON TO THE NEXT PAGE ➤

7. A kidney dialysis center periodically checks a sample of its equipment and performs a major recalibration if readings are sufficiently off target. Similarly, a fabric factory periodically checks the sizes of towels coming off an assembly line and halts production if measurements are sufficiently off target. In both situations, we have the null hypothesis that the equipment is performing satisfactorily. For each situation, which is the *more* serious concern, a Type I or Type II error?

   (A) Dialysis center: Type I error, towel manufacturer: Type I error
   (B) Dialysis center: Type I error, towel manufacturer: Type II error
   (C) Dialysis center: Type II error, towel manufacturer: Type I error
   (D) Dialysis center: Type II error, towel manufacturer: Type II error
   (E) This is impossible to answer without making an expected value judgment between human life and accurate towel sizes.

8. A coin is weighted so that the probability of heads is .75. The coin is tossed 10 times and the number of heads is noted. This procedure is repeated a total of 50 times, and the number of heads is recorded each time. What kind of distribution has been simulated?

   (A) The sampling distribution of the sample proportion with $n = 10$ and $p = .75$
   (B) The sampling distribution of the sample proportion with $n = 50$ and $p = .75$
   (C) The sampling distribution of the sample mean with $\bar{x} = 10(.75)$ and $\sigma = \sqrt{10(.75)(.25)}$
   (D) The binomial distribution with $n = 10$ and $p = .75$
   (E) The binomial distribution with $n = 50$ and $p = .75$

9. During the years 1886 through 2000 there were an average of 8.7 tropical cyclones per year, of which an average of 5.1 became hurricanes. Assuming that the probability of any cyclone becoming a hurricane is independent of what happens to any other cyclone, if there are five cyclones in one year, what is the probability that at least three become hurricanes?

   (A) .313
   (B) .345
   (C) .586
   (D) .658
   (E) .686

10. Given the probabilities $P(A) = .3$ and $P(B) = .2$, what is the probability of the union $P(A \cup B)$ if A and B are mutually exclusive? If A and B are independent? If B is a subset of A?

   (A) .44, .5, .2
   (B) .44, .5, .3
   (C) .5, .44, .2
   (D) .5, .44, .3
   (E) 0, .5, .3

GO ON TO THE NEXT PAGE ➤

11. A nursery owner claims that a recent drought stunted the growth of 22% of all her evergreens. A botanist tests this claim by examining an SRS of 1100 evergreens and finds that 268 trees in the sample show signs of stunted growth. With H_0: $p = .22$ and H_a:
$p \neq .22$, what is the value of the test statistic?

(A) $z = \dfrac{.244 - .22}{\sqrt{1100(.22)(1-.22)}}$

(B) $z = 2\dfrac{.244 - .22}{\sqrt{1100(.22)(1-.22)}}$

(C) $z = \dfrac{.244 - .22}{\sqrt{\dfrac{(.22)(1-.22)}{1100}}}$

(D) $z = 2\dfrac{.244 - .22}{\sqrt{\dfrac{(.22)(1-.22)}{1100}}}$

(E) $z = 2\dfrac{.244 - .22}{\sqrt{1100(.244)(1-.244)}}$

12. A medical research team tests for tumor reduction in a sample of patients using three different dosages of an experimental cancer drug. Which of the following is true?

(A) There are three explanatory variables and one response variable.
(B) There is one explanatory variable with three levels of response.
(C) Tumor reduction is the only explanatory variable, but there are three response variables corresponding to the different dosages.
(D) There are three levels of a single explanatory variable.
(E) Each explanatory level has an associated level of response.

13. The graph below shows cumulative proportion plotted against age for a population.

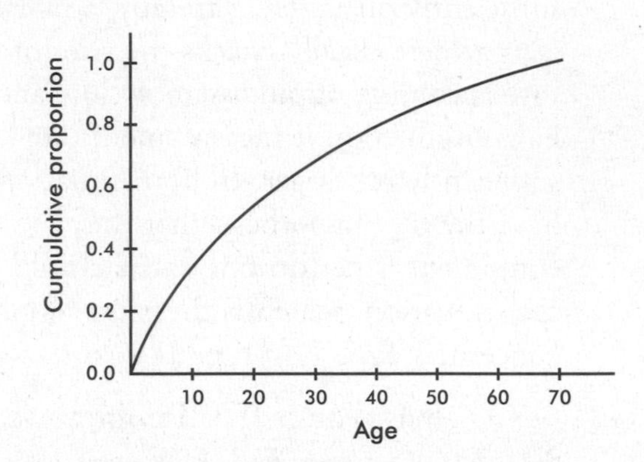

The median is approximately what age?

(A) 17
(B) 30
(C) 35
(D) 40
(E) 45

14. Mr. Bee's statistics class had a standard deviation of 11.2 on a standardized test, while Mr. Em's class had a standard deviation of 5.6 on the same test. Which of the following is the most reasonable conclusion concerning the two classes' performance on the test?

(A) Mr. Bee's class is less heterogeneous than Mr. Em's.
(B) Mr. Em's class is more homogeneous than Mr. Bee's.
(C) Mr. Bee's class performed twice as well as Mr. Em's.
(D) Mr. Em's class did not do as well as Mr. Bee's.
(E) Mr. Bee's class had the higher mean, but this may not be statistically significant.

GO ON TO THE NEXT PAGE ➤

15. A college admissions officer is interested in comparing the SAT math scores of high school applicants who have and have not taken AP Statistics. She randomly pulls the files of five applicants who took AP Statistics and five applicants who did not, and proceeds to run a *t*-test to compare the mean SAT math scores of the two groups. Which of the following is a necessary assumption?

    (A) The population variances from each group are known.
    (B) The population variances from each group are unknown.
    (C) The population variances from the two groups are equal.
    (D) The population of SAT scores from each group is normally distributed.
    (E) The samples must be independent simple random samples, and for each sample *np* and *n*(1 – *p*) must both be at least 10.

16. The appraised values of houses in a city have a mean of $125,000 with a standard deviation of $23,000. Because of a new teachers' contract, the school district needs an extra 10% in funds compared to the previous year. To raise this additional money, the city instructs the assessment office to raise all appraised house values by $5,000. What will be the new standard deviation of the appraised values of houses in the city?

    (A) $23,000
    (B) $25,300
    (C) $28,000
    (D) $30,300
    (E) $30,800

17. A psychologist hypothesizes that scores on an aptitude test are normally distributed with a mean of 70 and a standard deviation of 10. In a random sample of size 100, scores are distributed as in the table below. What is the χ^2 statistic for a goodness-of-fit test?

Score:	below 60	60–70	70–80	above 80
Number of people:	10	40	35	15

    (A) $\frac{(10-16)^2}{10}+\frac{(40-34)^2}{40}+\frac{(35-34)^2}{35}+\frac{(15-16)^2}{15}$
    (B) $\frac{(10-16)^2}{16}+\frac{(40-34)^2}{34}+\frac{(35-34)^2}{34}+\frac{(15-16)^2}{16}$
    (C) $\frac{(10-25)^2}{25}+\frac{(40-25)^2}{25}+\frac{(35-25)^2}{25}+\frac{(15-25)^2}{25}$
    (D) $\frac{(10-25)^2}{10}+\frac{(40-25)^2}{40}+\frac{(35-25)^2}{35}+\frac{(15-25)^2}{15}$
    (E) $\frac{(10-25)^2}{16}+\frac{(40-25)^2}{34}+\frac{(35-25)^2}{34}+\frac{(15-25)^2}{16}$

18. A talk show host recently reported that in response to his on-air question, 82% of the more than 2500 e-mail messages received through his publicized address supported the death penalty for anyone convicted of selling drugs to children. What does this show?

    (A) The survey is meaningless because of voluntary response bias.
    (B) No meaningful conclusion is possible without knowing something more about the characteristics of his listeners.
    (C) The survey would have been more meaningful if he had picked a random sample of the 2500 listeners who responded.
    (D) The survey would have been more meaningful if he had used a control group.
    (E) This was a legitimate sample, randomly drawn from his listeners, and of sufficient size to be able to conclude that most of his listeners support the death penalty for such a crime.

GO ON TO THE NEXT PAGE ➤

19. Define a new measurement as the difference between the 60th and 40th percentile scores in a population. This measurement will give information concerning

(A) central tendency.
(B) variability.
(C) symmetry.
(D) skewness.
(E) clustering.

20. Suppose X and Y are random variables with $E(X) = 37$, $\text{var}(X) = 5$, $E(Y) = 62$, and $\text{var}(Y) = 12$. What are the expected value and variance of the random variable $X + Y$?

(A) $E(X + Y) = 99$, $\text{var}(X + Y) = 8.5$
(B) $E(X + Y) = 99$, $\text{var}(X + Y) = 13$
(C) $E(X + Y) = 99$, $\text{var}(X + Y) = 17$
(D) $E(X + Y) = 49.5$, $\text{var}(X + Y) = 17$
(E) There is insufficient information to answer this question.

21. Which of the following can affect the value of the correlation r?

(A) A change in measurement units
(B) A change in which variable is called x and which is called y
(C) Adding the same constant to all values of the x-variable
(D) All of the above can affect the r value.
(E) None of the above can affect the r value.

22. The American Medical Association (AMA) wishes to determine the percentage of obstetricians who are considering leaving the profession because of the rapidly increasing number of lawsuits against obstetricians. How large a sample should be taken to find the answer to within ±3% at the 95% confidence level?

(A) 6
(B) 33
(C) 534
(D) 752
(E) 1068

23. Suppose you did 10 independent tests of the form H_0: $\mu = 25$ versus H_a: $\mu < 25$, each at the $\alpha = .05$ significance level. What is the probability of committing a Type I error and incorrectly rejecting a true H_0 with at least one of the 10 tests?

(A) .05
(B) .40
(C) .50
(D) .60
(E) .95

24. Which of the following sets has the smallest standard deviation? Which has the largest?

I. 1, 2, 3, 4, 5, 6, 7
II. 1, 1, 1, 4, 7, 7, 7
III. 1, 4, 4, 4, 4, 4, 7

(A) I, II
(B) II, III
(C) III, I
(D) II, I
(E) III, II

25. A researcher plans a study to examine long-term confidence in the U.S. economy among the adult population. She obtains a simple random sample of 30 adults as they leave a Wall Street office building one weekday afternoon. All but two of the adults agree to participate in the survey. Which of the following conclusions is correct?

(A) Proper use of chance as evidenced by the simple random sample makes this a well-designed survey.
(B) The high response rate makes this a well-designed survey.
(C) Selection bias makes this a poorly designed survey.
(D) A voluntary response study like this gives too much emphasis to persons with strong opinions.
(E) Lack of anonymity makes this a poorly designed survey.

GO ON TO THE NEXT PAGE ➤

26. Which among the following would result in the narrowest confidence interval?

    (A) Small sample size and 95% confidence
    (B) Small sample size and 99% confidence
    (C) Large sample size and 95% confidence
    (D) Large sample size and 99% confidence
    (E) This cannot be answered without knowing an appropriate standard deviation.

27. Consider the following three scatterplots:

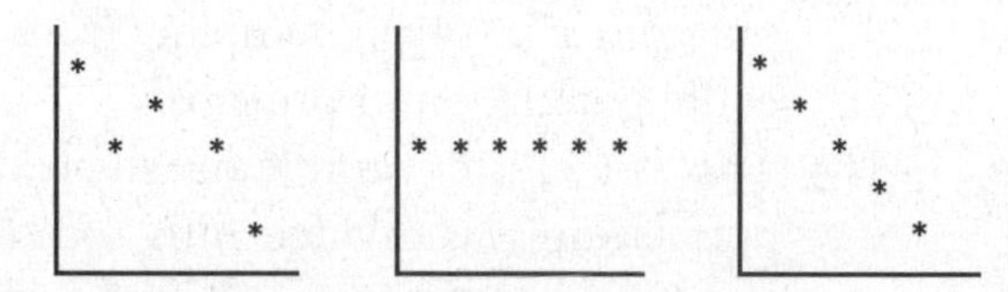

    What is the relationship among r_1, r_2, and r_3, the correlations associated with the first, second, and third scatterplots, respectively?

    (A) $r_1 < r_2 < r_3$
    (B) $r_1 < r_3 < r_2$
    (C) $r_2 < r_3 < r_1$
    (D) $r_3 < r_1 < r_2$
    (E) $r_3 < r_2 < r_1$

28. There are two games involving flipping a coin. In the first game you win a prize if you can throw between 45% and 55% heads. In the second game you win if you can throw more than 80% heads. For each game would you rather flip the coin 30 times or 300 times?

    (A) 30 times for each game
    (B) 300 times for each game
    (C) 30 times for the first game and 300 times for the second
    (D) 300 times for the first game and 30 times for the second
    (E) The outcomes of the games do not depend on the number of flips.

29. City planners are trying to decide among various parking plan options ranging from more on-street spaces to multilevel facilities to spread-out small lots. Before making a decision, they wish to test the downtown merchants' claim that shoppers park for an average of only 47 minutes in the downtown area. The planners have decided to tabulate parking durations for 225 shoppers and to reject the merchants' claim if the sample mean exceeds 50 minutes. If the merchants' claim is wrong and the true mean is 51 minutes, what is the probability that the random sample will lead to a mistaken failure to reject the merchants' claim? Assume that the standard deviation in parking durations is 27 minutes.

    (A) $P\left(z < \dfrac{50-47}{27/\sqrt{225}}\right)$

    (B) $P\left(z > \dfrac{50-47}{27/\sqrt{225}}\right)$

    (C) $P\left(z < \dfrac{50-51}{27/\sqrt{225}}\right)$

    (D) $P\left(z > \dfrac{50-51}{27/\sqrt{225}}\right)$

    (E) $P\left(z > \dfrac{51-47}{27/\sqrt{225}}\right)$

GO ON TO THE NEXT PAGE ➤

30. Consider the following parallel boxplots indicating the starting salaries (in thousands of dollars) for blue collar and white collar workers at a particular production plant:

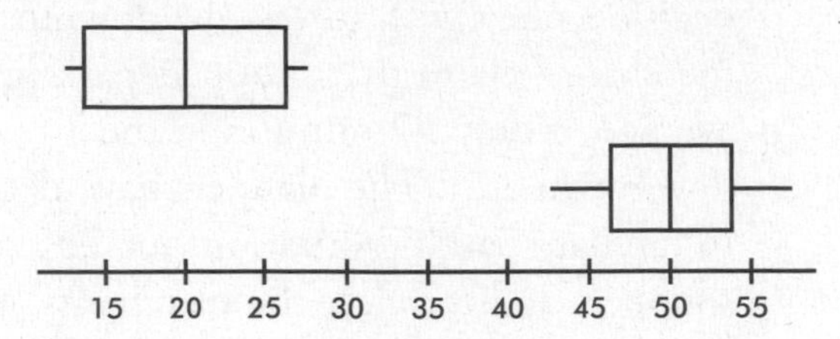

Which of the following is a correct conclusion?

(A) The ranges of the distributions are the same.
(B) In each distribution the mean is equal to the median.
(C) Each distribution is symmetric.
(D) Each distribution is roughly normal.
(E) The distributions are outliers of each other.

31. The mean Law School Aptitude Test (LSAT) score for applicants to a particular law school is 650 with a standard deviation of 45. Suppose that only applicants with scores above 700 are considered. What percentage of the applicants considered have scores below 740? (Assume the scores are normally distributed.)

(A) 13.3%
(B) 17.1%
(C) 82.9%
(D) 86.7%
(E) 97.7%

32. If all the other variables remain constant, which of the following will increase the power of a hypothesis test?

I. Increasing the sample size.
II. Increasing the significance level.
III. Increasing the probability of a Type II error.

(A) I only
(B) II only
(C) III only
(D) I and II
(E) All are true.

33. A researcher planning a survey of school principals in a particular state has lists of the school principals employed in each of the 125 school districts. The procedure is to obtain a random sample of principals from each of the districts rather than grouping all the lists together and obtaining a sample from the entire group. Which of the following is a correct conclusion?

(A) This is a simple random sample obtained in an easier and less costly manner than procedures involving sampling from the entire population of principals.
(B) This is a cluster sample in which the population was divided into heterogeneous groups called clusters.
(C) This is an example of systematic sampling, which gives a reasonable sample as long as the original order of the list is not related to the variables under consideration.
(D) This is an example of proportional sampling based on sizes of the school districts.
(E) This is a stratified sample, which may give comparative information that a simple random sample wouldn't give.

34. A simple random sample is defined by

(A) the method of selection.
(B) examination of the outcome.
(C) both of the above.
(D) how representative the sample is of the population.
(E) the size of the sample versus the size of the population.

35. Changing from a 90% confidence interval estimate for a population proportion to a 99% confidence interval estimate, with all other things being equal,

(A) increases the interval size by 9%.
(B) decreases the interval size by 9%.
(C) increases the interval size by 57%.
(D) decreases the interval size by 57%.
(E) This question cannot be answered without knowing the sample size.

GO ON TO THE NEXT PAGE ➤

36. Which of the following is a true statement about hypothesis testing?

    (A) If there is sufficient evidence to reject a null hypothesis at the 10% level, then there is sufficient evidence to reject it at the 5% level.
    (B) Whether to use a one- or a two-sided test is typically decided after the data are gathered.
    (C) If a hypothesis test is conducted at the 1% level, there is a 1% chance of rejecting the null hypothesis.
    (D) The probability of a Type I error plus the probability of a Type II error always equal 1.
    (E) The power of a test concerns its ability to detect an alternative hypothesis.

37. A population is normally distributed with mean 25. Consider all samples of size 10.

    The variable $\dfrac{\bar{x} - 25}{\frac{s}{\sqrt{10}}}$

    (A) has a normal distribution.
    (B) has a t-distribution with $df = 10$.
    (C) has a t-distribution with $df = 9$.
    (D) has neither a normal distribution or a t-distribution.
    (E) has either a normal or a t-distribution depending on the characteristics of the population standard deviation.

38. A correlation of .6 indicates that the percentage of variation in y that is explained by the variation in x is how many times the percentage indicated by a correlation of .3?

    (A) 2
    (B) 3
    (C) 4
    (D) 6
    (E) There is insufficient information to answer this question.

39. As reported on CNN, in a May 1999 national poll 43% of high school students expressed fear about going to school. Which of the following best describes what is meant by the poll having a margin of error of 5%?

    (A) It is likely that the true proportion of high school students afraid to go to school is between 38% and 48%.
    (B) Five percent of the students refused to participate in the poll.
    (C) Between 38% and 48% of those surveyed expressed fear about going to school.
    (D) There is a .05 probability that the 43% result is in error.
    (E) If similar size polls were repeatedly taken, they would be wrong about 5% of the time.

40. In a high school of 1650 students, 132 have personal investments in the stock market. To estimate the total stock investment by students in this school, two plans are proposed. Plan I would sample 30 students at random, find a confidence interval estimate of their average investment, and then multiply both ends of this interval by 1650 to get an interval estimate of the total investment. Plan II would sample 30 students at random from among the 132 who have investments in the market, find a confidence interval estimate of their average investment, and then multiply both ends of this interval by 132 to get an interval estimate of the total investment. Which is the better plan for estimating the total stock market investment by students in this school?

    (A) Plan I
    (B) Plan II
    (C) Both plans use random samples and so will produce equivalent results.
    (D) Neither plan will give an accurate estimate.
    (E) The resulting data must be seen to evaluate which is the better plan.

If there is still time remaining, you may review your answers.

SECTION II

Part A

Questions 1–5

Spend about 65 minutes on this part of the exam.
Percentage of Section II grade—75

You must show all work and indicate the methods you use. You will be graded on the correctness of your methods and on the accuracy of your results and explanations.

1. Cumulative frequency graphs of the heights (in inches) of athletes playing college baseball (A), basketball (B), and football (C) are given below:

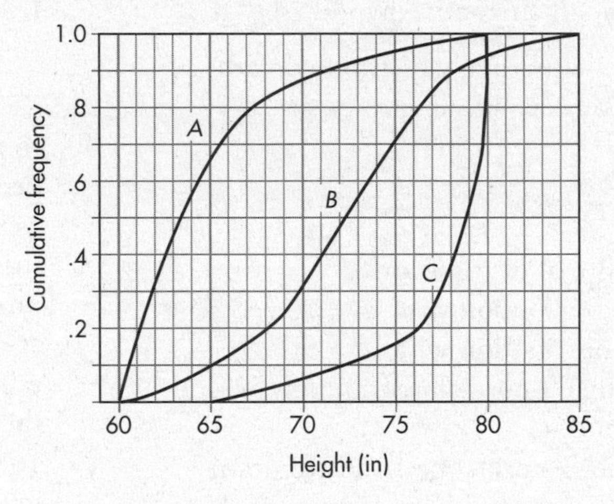

Write a few sentences comparing the distributions of heights of athletes playing the three college sports.

2. When students do not understand an author's vocabulary, there is no way they can fully understand the text; thus, learning vocabulary is recognized as one of the most important skills in all subject areas. A school system is planning a study to see which of two programs is more effective in ninth grade. Program A involves teaching students to apply morphemic analysis, the process of deriving a word's meaning by analyzing its meaningful parts, such as roots, prefixes, and suffixes. Program B involves teaching students to apply contextual analysis, the process of inferring the meaning of an unfamiliar word by examining the surrounding text.

   (a) Explain the purpose of incorporating a control group in this experiment.

GO ON TO THE NEXT PAGE ➤

(b) To comply with informed consent laws, parents will be asked their consent for their children to be randomly assigned to be taught in one of the new programs or by the local standard method. Students of parents who do not return the consent forms will continue to be taught by the local standard method. Explain why these students should not be considered part of the control group.

(c) Given 90 students randomly selected from ninth grade students with parental consent, explain how to assign them to the three groups, Program A, Program B, and Control, for a completely randomized design.

3. An experiment is designed to test a home insulation product. Two side-by-side identical houses are chosen, and in one the insulation product is installed (blown into the attic). Both homes use natural gas for heating, and the thermostat in each house is fixed at 68 degrees Fahrenheit. On 30 randomly chosen winter days, the quantity of gas (in hundreds of cubic feet) used to heat each house is measured. The table below summarizes the data.

	Number	Mean (in 100 cu ft)	Standard Deviation (in 100 cu ft)
Without insulation product	30	7.112	1.579
With insulation product	30	5.555	1.198
Daily differences	30	1.557	1.331

(a) Is there evidence that the insulation product does result in a lower mean quantity of gas being used?

Two least square regression calculations are performed with the following computer output, where Temp is the average outside temperature, and WithOut and With are the quantities of gas (in hundreds of cubic feet) used to heat the two homes on a day with that outside temperature:

```
Regression Analysis: WithOut versus Temp
Predictor        Coef   SE Coef        T        P
Constant      10.0228    0.4009    25.00    0.000
Temp         -0.14926   0.01879    -7.94    0.000
S = 0.890921    R-Sq = 69.3%    R-Sq(adj) = 68.2%

Regression Analysis: With versus Temp
Predictor        Coef   SE Coef        T        P
Constant       7.5850    0.3531    21.48    0.000
Temp         -0.10408   0.01655    -6.29    0.000
S = 0.784653    R-Sq = 58.5%    R-Sq(adj) = 57.1%
```

(b) Compare the slopes of the regression lines in context of this problem.

(c) Compare the y-intercepts of the regression lines in context of this problem.

GO ON TO THE NEXT PAGE ➤

4. A new anti-spam software program is field tested on 1000 e-mails and the results are summarized in the following table (positive test = program labels e-mail as spam):

	Spam	Legitimate	
Positive test	205	90	295
Negative test	45	660	705
	250	750	1000

Using the above empirical results, determine the following probabilities.

(a) (i) What is the *predictive value* of the test? That is, what is the probability that an e-mail is spam and the test is positive?
(ii) What is the *false-positive* rate? That is, what is the probability of testing positive given that the e-mail is legitimate?

(b) (i) What is the *sensitivity* of the test? That is, what is the probability of testing positive given that the e-mail is spam?
(ii) What is the *specificity* of the test? That is, what is the probability of testing negative given that the e-mail is legitimate?

(c) (i) Given a random sample of five legitimate e-mails, what is the probability that the program labels at least one as spam?
(ii) Given a random sample of five spam e-mails, what is the probability that the program correctly labels at least three as spam?

(d) (i) If 35% of the incoming e-mails from one source are spam, what is the probability that an e-mail from that source will be labeled as spam?
(ii) If an e-mail from this source is labeled as spam, what is the probability it really is spam?

5. Cumulative exposure to nitrogen dioxide (NO_2) is a major risk factor for lung disease in tunnel construction workers. In a 2004 study, researchers compared cumulative exposure to NO_2 for a random sample of drill and blast workers and for an independent random sample of concrete workers. Summary statistics are shown below.

Tunnel worker activity	Sample size	Mean cumulative NO_2 exposure (ppm*/yr)	Standard deviation in cumulative NO_2 exposure (ppm*/yr)
Drill and blast	115	4.1	1.8
Concrete	69	4.8	2.4

*Parts per million

(a) Find a 95% confidence interval estimate for the difference in mean cumulative NO_2 exposure for drill and blast workers and concrete workers.

(b) Using this confidence interval, is there evidence of a difference in mean cumulative NO_2 exposure for drill and blast workers and concrete workers? Explain.

(c) It turns out that both data sets were somewhat skewed. Does this invalidate your analysis? Explain.

GO ON TO THE NEXT PAGE ➤

SECTION II

Part B

Question 6

Spend about 25 minutes on this part of the exam.
Percentage of Section II grade—25

6. Do strong magnetic fields have an effect on the early development of mice? A study is performed on 120 one-month-old mice. Each is subjected to a strong magnetic field for one week and no magnetic field for one week, the order being decided randomly for each mouse. The weight gain in grams during each week for each mouse is recorded. The weight gains with no magnetic field are approximately normally distributed with a mean of 25.1 grams and a standard deviation of 3.1 grams, while the weight gains in the magnetic field are approximately normally distributed with a mean of 17.3 grams and a standard deviation of 5.7 grams. Of the 120 mice, 101 gained more weight with no magnetic field.

   (a) Calculate and interpret a 95% confidence interval for the proportion of mice that gain more weight with no magnetic field.
   (b) Suppose two of these mice were chosen at random. What is the probability that the first mouse gained more weight with no magnetic field than the second mouse gained in the magnetic field?
   (c) Based on your answers to (a) and (b), is there evidence that for mice, weight gain with no magnetic field and in a magnetic field are independent?
   (d) In a test for variances on whether or not the variances of the two sets of weight gains are significantly different, $H_0: \sigma_X^2 = \sigma_Y^2$, $H_a: \sigma_X^2 \neq \sigma_Y^2$, with $\alpha = .05$, for samples of this size, the critical cut-off scores for $\frac{S_Y^2}{S_X^2}$ are 0.698 (left tail of .025) and 1.43 (right tail of .025). What is the resulting conclusion about variances?

STOP

If there is still time remaining, you may review your answers.

Answer Key

Section I

1. **E**	9. **D**	17. **B**	25. **C**	33. **E**
2. **E**	10. **D**	18. **A**	26. **C**	34. **A**
3. **A**	11. **C**	19. **B**	27. **D**	35. **C**
4. **D**	12. **D**	20. **E**	28. **D**	36. **E**
5. **B**	13. **A**	21. **E**	29. **C**	37. **C**
6. **A**	14. **B**	22. **E**	30. **A**	38. **C**
7. **C**	15. **D**	23. **B**	31. **C**	39. **A**
8. **D**	16. **A**	24. **E**	32. **D**	40. **B**

Answers Explained

Section I

1. **(E)**

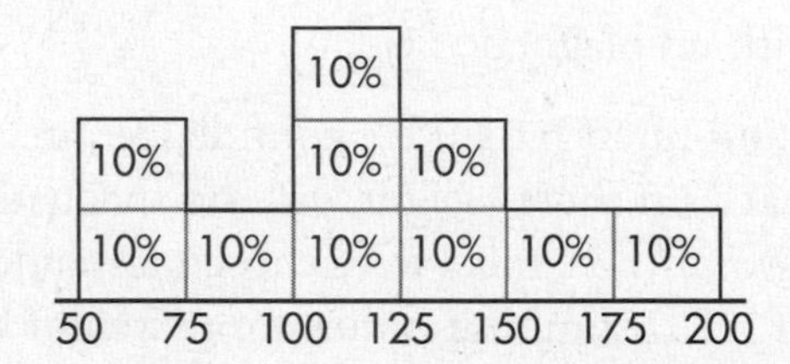

The median price would have 50% of the prices below and 50% of the prices above; however, looking at areas, it is clear that 60% of the prices are below \$125,000. The mean (physically the center of gravity) appears to be less than \$125,000. Area considerations also show that 30% of the prices are between \$100,000 and \$125,000. Many values are too far from the mean for the standard deviation to be only \$10,000. If the distribution were closer to normal, the standard deviation would be around \$25,000; however, the distribution is more spread out than this, and with a SD perhaps between \$35,000 and \$40,000, estimating the variance at 1.5×10^9 is reasonable.

2. **(E)** Regression lines show association, not causation. Surveys suggest relationships, which experiments can help to show to be cause and effect.

3. **(A)** The critical z-score is 1.036. Thus $60 - 55 = 1.036\sigma$ and $\sigma = 4.83$.

4. **(D)** We are 90% confident that the population mean is within the interval calculated using the data from the sample.

5. **(B)** The first study was observational because the subjects were not chosen for treatment.

6. **(A)**

$$\begin{aligned} P(\text{def}) &= P(\text{1st} \cap \text{def}) + P(\text{2nd} \cap \text{def}) \\ &= (.6)(.005) + (.4)(.010) \\ &= .0030 + .0040 = .0070 \end{aligned}$$

$$P\left(\text{1st}\middle|\text{def}\right) = \frac{.0030}{.0070} = .429$$

7. **(C)** At the dialysis center the more serious concern would be a Type II error, which is that the equipment is not performing correctly, yet the check does not pick this up; while at the towel manufacturing plant the more serious concern would be a Type I error, which is that the equipment is performing correctly, yet the check causes a production halt.

8. **(D)** There are two possible outcomes (heads and tails), with the probability of heads always .75 (independent of what happened on the previous toss), and we are interested in the number of heads in 10 tosses. Thus, this is a binomial model with $n = 10$ and $p = .75$. Repeating this over and over (in this case 50 times) simulates the resulting binomial distribution.

9. **(D)** In a binomial distribution with probability of success $p = \frac{5.1}{8.7} = .586$, the probability of at least 3 successes is

$$\binom{5}{3}(.586)^3(.414)^2 + \binom{5}{4}(.586)^4(.414)$$

$+ (.586)^5 = .658$, or using the TI-83, one finds that 1-binomcdf(5, .586, 2) = .658.

10. **(D)** If A and B are mutually exclusive, then $P(A \cap B) = 0$, and so $P(A \cup B) = .3 + .2 - 0 = .5$. If A and B are independent, then $P(A \cap B) = P(A)P(B)$, and so $P(A \cup B) = .3 + .2 - (.3)(.2) = .44$. If B is a subset of A, then $A \cup B = A$, and so $P(A \cup B) = P(A) = .3$.

11. **(C)** The standard deviation of the test statistic is $\sigma_{\hat{p}} = \sqrt{\frac{p(1-p)}{n}}$. Since this is a two-sided test, the P-value will be twice the tail probability of the test statistic; however, the test statistic itself is not doubled.

12. **(D)** Dosage is the only explanatory variable, and it is being tested at three levels. Tumor reduction is the single response variable.

13. **(A)** The median corresponds to a cumulative proportion of 0.5.

14. **(B)** Standard deviation is a measure of variability. The less variability, the more homogeneity.

15. **(D)** Since the sample sizes are small, the samples must come from normally distributed populations. While the samples should be independent simple random samples, np and $n(1 - p)$ refer to conditions for tests involving sample proportions, not means.

16. **(A)** Adding the same constant to all values in a set will increase the mean by that constant, but will leave the standard deviation unchanged.

17. **(B)** With about 68% of the values within 1 standard deviation of the mean, the expected numbers for a normal distribution are as follows:

16	34	34	16

$$\text{Thus } \chi^2 = \sum \frac{(obs - exp)^2}{exp}$$

$$= \frac{(10-16)^2}{16} + \frac{(40-34)^2}{34} + \frac{(35-34)^2}{34} + \frac{(15-16)^2}{16}.$$

18. **(A)** This is a good example of voluntary response bias, which often overrepresents strong or negative opinions. The people who chose to respond were very possibly the parents of children facing drug problems or people who had had bad experiences with drugs being sold in their neighborhoods. There is very little chance that the 2500 respondents were representative of the population. Knowing more about his listeners or taking a sample of the sample would not have helped.

19. **(B)** The range (difference between largest and smallest values), the interquartile range ($Q_3 - Q_1$), and this difference between the 60th and 40th percentile scores all are measures of variability, or how spread out is the population or a subset of the population.

20. **(E)** Without independence we cannot determine var(X + Y) from the information given.

21. **(E)** The correlation coefficient r is not affected by changes in units, by which variable is called x or y, or by adding or multiplying all the values of a variable by the same constant.

22. **(E)** $1.96\left(\frac{.5}{\sqrt{n}}\right) \le .03$ gives $\sqrt{n} \ge 32.67$ and $n \ge 1067.1$.

23. **(B)** P(at least one Type I error) = 1 – P(no Type I errors) = $1 - (.95)^{10} = .40$.

24. **(E)** Note that all three sets have the same mean and the same range. The third set has most of its values concentrated right at the mean, while the second set has most of its values concentrated far from the mean.

25. **(C)** People coming out of a Wall Street office building are a very unrepresentative sample of the adult population, especially given the question under consideration. Using chance and obtaining a high response rate will not change the selection bias and make this into a well-designed survey. This is a convenience sample, not a voluntary response sample.

26. **(C)** Larger samples (so $\frac{\sigma}{\sqrt{n}}$ is smaller) and less confidence (so the critical z or t is smaller) both result in smaller intervals.

27. **(D)** The third scatterplot shows perfect negative association, so $r_3 = -1$. The first scatterplot shows strong, but not perfect, negative correlation, so $-1 < r_1 < 0$. The second scatterplot shows no correlation, so $r_2 = 0$.

28. **(D)** The probability of throwing heads is .5. By the law of large numbers, the more times you flip the coin, the more the relative frequency tends to become closer to this probability. With fewer tosses there is more chance for wide swings in the relative frequency.

29. **(C)** $\sigma_{\bar{x}} = \frac{\sigma}{\sqrt{n}} = \frac{27}{\sqrt{225}} = 1.8$. If the true mean parking duration is 51 minutes, the normal curve should be centered at 51. The critical value of 50 has a z-score of $\frac{50-51}{1.8}$.

30. **(A)** The overall lengths (between tips of whiskers) are the same, and so the ranges are the same. Just because the min and max are equidistant from the median, and Q_1 and Q_3 are equidistant from the median, does not imply that a distribution is symmetric or that the mean and median are equal. And even if a distribution is symmetric, this does not imply that it is roughly normal. Particular values, not distributions, may be outliers.

31. **(C)** Critical z-scores are $\frac{700-650}{45} = 1.11$ and $\frac{740-650}{45} = 2$ with right tail probabilities of .1335 and .0228, respectively. The percentage below 740 given that the scores are above 700 is $\frac{.1335-.0228}{.1335} = 82.9\%$.

32. **(D)** There is a different probability of Type II error for each possible correct value of the population parameter, and 1 minus this probability is the power of the test against the associated correct value.

33. **(E)** This is not a simple random sample because all possible sets of the required size do not have the same chance of being picked. For example, a set of principals all from just half the school districts has no chance of being picked to be the sample. This is not a cluster sample in that there is no reason to believe that each school district resembles the population as a whole, and furthermore, there was no random sample taken of the school districts. This is not systematic sampling as the districts were not put in some order with every *n*th district chosen. Stratified samples are often easier and less costly to obtain and also make comparative data available. In this case responses can be compared among various districts.

34. **(A)** A simple random sample can be any size and may or may not be representative of the population. It is a method of selection in which every possible sample of the desired size has an equal chance of being selected.

35. **(C)** The critical z-scores go from ±1.645 to ±2.576, resulting in an increase in the interval size: $\frac{2.576}{1.645} = 1.57$ or an increase of 57%.

36. **(E)** If the *P*-value is less than .10, it does not follow that it is less than .05. Decisions such as whether a test should be one- or two-sided are made before the data are gathered. If $\alpha = .01$, there is a 1% chance of rejecting the null hypothesis *if* the null hypothesis is true. There is one probability of a Type I error, the significance level, while there is a different probability of a Type II error associated with each possible correct alternative, so the sum does not equal 1.

37. **(C)** $\frac{\bar{x} - \mu}{\frac{s}{\sqrt{n}}}$ has a t-distribution with $df = n - 1$.

38. **(C)** A correlation of .6 explains $(.6)^2$ or 36% of the variation in y, while a correlation of .3 explains only $(.3)^2$ or 9% of the variation in y.

39. **(A)** Using a measurement from a sample, we are never able to say *exactly* what a population proportion is; rather we always say we have a certain *confidence* that the population proportion lies in a particular *interval.* In this case that interval is 43% ± 5% or between 38% and 48%.

40. **(B)** With Plan I the expected number of students with stock investments is only 2.4 out of 30. Plan II allows an estimate to be made using a full 30 investors.

SECTION II

1. A complete answer compares shape, center, and spread.

   Shape: The baseball players, (A), for which the cumulative frequency plot rises steeply at first, include more shorter players, and thus the distribution is skewed to the right (toward the greater heights). The football players, (C), for which the cumulative frequency plot rises slowly at first, and then steeply toward the end, include more taller players, and thus the distribution is skewed to the left (toward the lower heights). The basketball players, (B), for which the cumulative frequency plot rises slowly at each end, and steeply in the middle, have a more bell-shaped distribution of heights.

   Center: The medians correspond to relative frequencies of 0.5. Reading across from 0.5 and then down to the x-axis shows the median heights to be about 63.5 inches for baseball players, about 72.5 inches for basketball players, and about 79 inches for football players. Thus, the center of the baseball height distribution is the least, and the center of the football height distribution is the greatest.

   Spread: The range of the football players is the smallest, 80 – 65 = 15 inches, then comes the range of the baseball players, 80 – 60 = 20 inches, and finally the range of the basketball players is the greatest, 85 – 60 = 25 inches.

Scoring

The discussion of shape is essentially correct for correctly identifying which distribution is skewed left, skewed right, and more bell-shaped, and for giving a correct justification based on the cumulative frequency plots. The discussion of shape is partially correct for correctly identifying which distribution is skewed left, skewed right, and more bell-shaped without giving a good explanation.

The discussion of center is essentially correct for correctly noting that the baseball players have the lowest median height and the football players have the greatest median height, and giving some numerical justification. The discussion of center is partially correct for correctly noting that the baseball players have the lowest median height and the football players have the greatest median height but without giving a good explanation.

The discussion of spread is essentially correct for correctly noting that the football players have the smallest range for their heights and the basketball players have the greatest range for their heights, and giving some numerical justification. The discussion of spread is partially correct for correctly noting that the football players have the smallest range for their heights and the basketball players have the greatest range for their heights but without giving a good explanation.

4 Complete Answer	All three parts essentially correct.
3 Substantial Answer	Two parts essentially correct and one part partially correct.
2 Developing Answer	Two parts essentially correct OR one part essentially correct and one or two parts partially correct OR all three parts partially correct.
1 Minimal Answer	One part essentially correct OR two parts partially correct.

2. (a) A control group would allow the school system to compare the effectiveness of each of the new programs to the local standard method currently being used.

   (b) Parents who fail to return the consent form are a special category who may well place less priority on education. The effect of using their children may distort results, since their children could only be placed in the control group.

   (c) Assign each student a unique number 01–90. Using a random number table or a random number generator on a calculator, pick numbers between 01 and 90, throwing out repeats. The students corresponding to the first 30 such numbers picked will be assigned to Program A, the next 30 picked to Program B, and the remaining to the control group.

Scoring

Part (a) is essentially correct if the purpose is given for using a control group in this study. Part (a) is partially correct if a correct explanation for the use of a control group is given, but not in context of this study.

Part (b) is essentially correct for a clear explanation in context. Part (b) is partially correct if the explanation is weak.

Part (c) is essentially correct if randomization is used correctly and the method is clear. Part (c) is partially correct if randomization is used but the method is not clearly explained.

4 Complete Answer	All three parts essentially correct.
3 Substantial Answer	Two parts essentially correct and one part partially correct.
2 Developing Answer	Two parts essentially correct OR one part essentially correct and one or two parts partially correct OR all three parts partially correct.
1 Minimal Answer	One part essentially correct OR two parts partially correct.

3. (a) This is a paired data test, not a two-sample test. There are four parts to a complete solution.

   Part 1: Must state a correct pair of hypotheses.

   H_0: $\mu_d = 0$ and H_a: $\mu_d > 0$, where μ_d is the mean daily difference in quantity of gas used (in 100 cu ft) between the house without insulation and the house with insulation.

   Part 2: Must name the test and check the conditions.

   This is a paired *t*-test, that is, a single sample hypothesis test on the set of differences.
   Conditions: We are given that the sample of days was randomly chosen. The population may not be normal, but with $n = 30$ it is OK to proceed with the *t*-test.

   Part 3: Must find the test statistic t and the *P*-value.

   A calculator quickly gives $t = 6.41$ and $P = .000$.

   Or with $\bar{x}_d = 1.557$and $s_d = 1.331$ we have $t = \dfrac{1.557 - 0}{\frac{1.331}{\sqrt{30}}} = 6.407$ and with $df = 29$, $P = .000$.

   Part 4: Must state the conclusion in context with linkage to the *P*-value.

With this small a *P*-value (.000), there is very strong evidence to reject H_0. That is, there is very strong evidence that the mean daily gas usage in a home without the insulation product is greater than the mean daily gas usage in a home with the insulation product.

(b) In the house without the insulation product an estimate for the average *increase* in quantity of gas used when the outside temperature goes *down* 1 degree Fahrenheit is 0.15 hundred cubic feet, while in the house with the insulation product an estimate for the average *increase* in quantity of gas used when the outside temperature goes *down* 1 degree Fahrenheit is 0.10 hundred cubic feet. So when the temperature goes down, the increase in gas used is less in the house with the insulation product than in the house without the insulation product.

(c) In the house without the insulation product an estimate for the average quantity of gas used when the outside temperature is 0 degrees Fahrenheit is 10.02 hundred cubic feet, while in the house with the insulation product an estimate for the average quantity of gas used when the outside temperature is 0 degrees Fahrenheit is 7.59 hundred cubic feet. So at an outside temperature of 0 degrees Fahrenheit, the average quantity of gas used is less in the house with the insulation product than in the house without the insulation product.

Scoring

In Part (a), Parts 1–2 are essentially correct for a correct statement of the hypotheses (in terms of *population* means), together with naming the test and checking the conditions, and partially correct for one of Part 1 or Part 2 completely correct.

Parts 3–4 are essentially correct for a correct calculation of both the test statistic *t* and the *P*-value, together with a correct conclusion in context linked to the *P*-value, and partially correct for one of Part 3 or Part 4 completely correct.

Part (b) is essentially correct for giving the correct values for the slopes, in context of the problem, and making a direct comparison. Part (b) is partially correct for giving the correct slopes but missing the context or the direct comparison.

Part (c) is essentially correct for giving the correct values for the *y*-intercepts, in context of the problem, and making a direct comparison. Part (c) is partially correct for giving the correct *y*-intercepts but missing the context or the direct comparison.

Count essentially correct answers as one point and partially correct answers as one-half point.

4 Complete Answer Four points

3 Substantial Answer Three points

2 Developing Answer Two points

1 Minimal Answer One point

Use a holistic approach to decide a score totaling between two numbers.

4. (a) (i) $P(spam \cap positive) = \frac{205}{1000} = .205$

(ii) $P(positive \mid legitimate) = \frac{90}{750} = .12$

(b) (i) $P(positive \mid spam) = \frac{205}{250} = .82$

(ii) $P(negative \mid legitimate) = \frac{660}{750} = .88$

(c) (i) $1 - (.88)^5 = .4723$

(ii) $10(.82)^3(.18)^2 + 5(.82)^4(.18) + (.82)^5 = .9563$

(d) (i) $P(positive) = P(positive \cap spam) + P(positive \cap legitimate) =$
$P(spam)\ P(positive \mid spam) + P(legitimate) P(positive \mid legitimate) =$
$(.35)(.82) + (.65)(.12) = .365$

(ii) $P(spam \mid positive) = \frac{P(spam \cap positive)}{P(positive)} = \frac{(.35)(.82)}{.365} = .7863$

Scoring

There are two probabilities to calculate in each Part (a)–(d). Each Part is essentially correct for both probabilities correctly calculated and partially correct for one probability correctly calculated. In Parts that use results from previous parts, full credit is given for correctly using the results of the earlier Part, whether that earlier calculation was correct or not. For credit for Part (d), a correct methodology must also be shown.

Count partially correct answers as one-half an essentially correct answer.

4 Complete Answer	Four essentially correct answers.
3 Substantial Answer	Three essentially correct answers.
2 Developing Answer	Two essentially correct answers.
1 Minimal Answer	One essentially correct answer.

Use a holistic approach to decide a score totaling between two numbers.

5. (a) Identify the confidence interval and check conditions: two-sample *t*-interval for $\mu_{db} - \mu_c$, the difference in mean cumulative NO_2 exposure for drill and blast workers and concrete workers. We are given that we have independent random samples, and we note that the sample sizes are large ($n_{db} = 115 > 30$ and $n_c = 69 > 30$).

Calculate the confidence interval: with $df = \min(115–1,69–1) = 68$,

$$(4.1 - 4.8) \pm 1.995\sqrt{\frac{(1.8)^2}{115} + \frac{(2.4)^2}{69}} = -0.7 \pm 0.667.$$

Interpretation in context: We are 95% confident that the true difference in the mean cumulative NO_2 exposure for drill and blast workers and concrete workers is between –1.367 and –0.033 ppm/yr. [On the TI-84, 2-SampTInt gives (–1.362,–.0381).]

(b) Zero is not in the above 95% confidence interval, so at the α = .05 significance level, there is evidence to reject H_0: $\mu_{db} - \mu_c = 0$ in favor of H_a: $\mu_{db} - \mu_c \neq 0$. That is, there is evidence of a difference in mean cumulative NO_2 exposure for drill and blast workers and for concrete workers.

(c) With sample sizes this large, the central limit theorem applies and our analysis is valid.

Scoring

Part (a) has two components. The first component, identifying the confidence interval and checking conditions, is essentially correct for naming the confidence interval procedure, noting independent random samples, and noting the large sample sizes. This component is partially correct for correctly noting two of the three points.

The second component of Part (a) is essentially correct for correct mechanics in calculating the confidence interval and for a correct (based on the shown mechanics) interpretation in context, and is partially correct for one of these two features.

Part (b) is essentially correct for noting that zero is not in the interval so the observed difference is significant and stating this in context of the problem. Part (b) is partially correct for noting that zero is not in the interval so the observed difference is significant, but failing to put this conclusion in context of the problem.

Part (c) is essentially correct for relating the central limit theorem (CLT) to the samples being large. Part (c) is partially correct for referring to the CLT or to the large samples, but not linking the two.

Count partially correct answers as one-half an essentially correct answer.

4 Complete Answer	Four essentially correct answers.
3 Substantial Answer	Three essentially correct answers.
2 Developing Answer	Two essentially correct answers.
1 Minimal Answer	One essentially correct answer.

Use a holistic approach to decide a score totaling between two numbers.

6. (a) Identify the confidence interval by name or formula.

95% confidence interval for the population proportion

$$\hat{p} \pm 1.96\sqrt{\frac{\hat{p}(1-\hat{p})}{n}}$$

Check the assumptions.

$$n\hat{p} = 120\left(\frac{101}{120}\right) = 101 > 10 \text{ and } n(1-\hat{p}) = 120\left(\frac{19}{120}\right) = 19 > 10.$$

Calculate the confidence interval.

$$\hat{p} = \frac{101}{120} = .842 \text{ and } .842 \pm 1.96\sqrt{\frac{(.842)(.158)}{120}} = .842 \pm .065$$

Interpret the confidence interval in context.

We are 95% confident that between 77.7% and 90.7% of mice will gain more weight with no magnetic field.

(b) The distribution of the set of all possible differences of weight gains without and with magnetic fields is approximately normal, with mean 25.1 – 17.3 = 7.8 and standard deviation $\sqrt{(3.1)^2 + (5.7)^2} = 6.49$. Then the probability that the first mouse gained more weight than the second equals

$$P\left(z > \frac{0-7.8}{6.49}\right) = P\left(z > -1.202\right) = .885$$

(c) From the answer to (b), if weight gains without or with the magnetic field are independent, we would expect approximately 88.5% of mice will gain more weight with no magnetic field. From the answer to (a) we estimate the percent of mice who gain more weight with no magnetic field to be between 77.7% and 90.7%. Since 88.5% is in this interval, there is no evidence to suggest that weight gain with no magnetic field and in a magnetic field are not independent.

(d) Calculating $\frac{S_Y^2}{S_X^2} = \frac{(3.1)^2}{(5.7)^2} = 0.296$ or $\frac{(5.7)^2}{(3.1)^2} = 3.38$, the test statistic is outside the critical cutoff scores of 0.698 and 1.43. Thus there *is* evidence that the variances of the distributions of weight gains without and with magnetic fields are different.

Scoring

Part (a) has four parts: 1) identifying the confidence interval; 2) checking assumptions; 3) calculating the confidence interval; and 4) interpreting the confidence interval in context. Part (a) is essentially correct if three or four of these parts are correct and partially correct if one or two of these parts are correct.

Part (b) is essentially correct for a complete answer, and partially correct for finding the mean and SD of the set of differences but incorrectly finding the probability, OR making a mistake in finding the mean or standard deviation but using these correctly in finding the probability.

Part (c) is essentially correct for a complete answer and is partially correct if the conclusion is right but the links to (a) and (b) are unclear.

Part (d) is essentially correct for a correct calculation of the quotient of sample variances together with a correct conclusion in context, and is partially correct for one of these two parts correct.

Count partially correct answers as one-half an essentially correct answer.

4 Complete Answer	Four essentially correct answers.
3 Substantial Answer	Three essentially correct answers.
2 Developing Answer	Two essentially correct answers.
1 Minimal Answer	One essentially correct answer.

Use a holistic approach to decide a score totaling between two numbers.

Answer Sheet

PRACTICE EXAMINATION 2

1. (A) (B) (C) (D) (E)	11. (A) (B) (C) (D) (E)	21. (A) (B) (C) (D) (E)	31. (A) (B) (C) (D) (E)
2. (A) (B) (C) (D) (E)	12. (A) (B) (C) (D) (E)	22. (A) (B) (C) (D) (E)	32. (A) (B) (C) (D) (E)
3. (A) (B) (C) (D) (E)	13. (A) (B) (C) (D) (E)	23. (A) (B) (C) (D) (E)	33. (A) (B) (C) (D) (E)
4. (A) (B) (C) (D) (E)	14. (A) (B) (C) (D) (E)	24. (A) (B) (C) (D) (E)	34. (A) (B) (C) (D) (E)
5. (A) (B) (C) (D) (E)	15. (A) (B) (C) (D) (E)	25. (A) (B) (C) (D) (E)	35. (A) (B) (C) (D) (E)
6. (A) (B) (C) (D) (E)	16. (A) (B) (C) (D) (E)	26. (A) (B) (C) (D) (E)	36. (A) (B) (C) (D) (E)
7. (A) (B) (C) (D) (E)	17. (A) (B) (C) (D) (E)	27. (A) (B) (C) (D) (E)	37. (A) (B) (C) (D) (E)
8. (A) (B) (C) (D) (E)	18. (A) (B) (C) (D) (E)	28. (A) (B) (C) (D) (E)	38. (A) (B) (C) (D) (E)
9. (A) (B) (C) (D) (E)	19. (A) (B) (C) (D) (E)	29. (A) (B) (C) (D) (E)	39. (A) (B) (C) (D) (E)
10. (A) (B) (C) (D) (E)	20. (A) (B) (C) (D) (E)	30. (A) (B) (C) (D) (E)	40. (A) (B) (C) (D) (E)

Practice Examination 2

SECTION I

Questions 1–40

Spend 90 minutes on this part of the exam.

> **Directions:** The questions or incomplete statements that follow are each followed by five suggested answers or completions. Choose the response that best answers the question or completes the statement.

1. Suppose that the regression line for a set of data, $y = mx + 3$, passes through the point (2, 7). If $\bar{x}$ and $\bar{y}$ are the sample means of the x- and y-values, respectively, then $\bar{y}$ =

   (A) $\bar{x}$.
   (B) $\bar{x} - 2$.
   (C) $\bar{x} + 3$.
   (D) $2\bar{x} + 3$.
   (E) $3.5\bar{x} + 3$.

2. A study is made to determine whether more hours of academic studying leads to higher point scoring by basketball players. In surveying 50 basketball players, it is noted that the 25 who claim to study the most hours have a higher point average than the 25 who study less. Based on this study, the coach begins requiring the players to spend more time studying. Which of the following is a correct statement?

   (A) While this study may have its faults, it still does prove causation.
   (B) There could well be a confounding variable responsible for the seeming relationship.
   (C) While this is a controlled experiment, the conclusion of the coach is not justified.
   (D) To get the athletes to study more, it would be more meaningful to have them put in more practice time on the court to boost their point averages, as higher point averages seem to be associated with more study time.
   (E) No proper conclusion is possible without somehow introducing *blinding*.

3. The longevity of people living in a certain locality has a standard deviation of 14 years. What is the mean longevity if 30% of the people live longer than 75 years? Assume a normal distribution for life spans.

   (A) 61.00
   (B) 67.65
   (C) 74.48
   (D) 82.35
   (E) The mean cannot be computed from the information given.

GO ON TO THE NEXT PAGE ➤

4. Which of the following is a correct statement about correlation?

   (A) If the slope of the regression line is exactly 1, then the correlation is exactly 1.
   (B) If the correlation is 0, then the slope of the regression line is undefined.
   (C) Switching which variable is called x and which is called y changes the sign of the correlation.
   (D) The correlation r is equal to the slope of the regression line when z-scores for the y-variable are plotted against z-scores for the x-variable.
   (E) Changes in the measurement units of the variables may change the correlation.

5. Which of the following are affected by outliers?

   I. Mean
   II. Median
   III. Standard deviation
   IV. Range
   V. Interquartile range

   (A) I, III, and V
   (B) II and IV
   (C) I and V
   (D) III and IV
   (E) I, III, and IV

6. An engineer wishes to determine the quantity of heat being generated by a particular electronic component. If she knows that the standard deviation is 2.4, how many of these components should she consider to be 99% sure of knowing the mean quantity to within ±0.6?

   (A) 27
   (B) 87
   (C) 107
   (D) 212
   (E) 425

7. A company that produces facial tissues continually monitors tissue strength. If the mean strength from sample data drops below a specified level, the production process is halted and the machinery inspected. Which of the following would result from a Type I error?

   (A) Halting the production process when sufficient customer complaints are received.
   (B) Halting the production process when the tissue strength is below specifications.
   (C) Halting the production process when the tissue strength is within specifications.
   (D) Allowing the production process to continue when the tissue strength is below specifications.
   (E) Allowing the production process to continue when the tissue strength is within specifications.

8. Suppose that for a certain Caribbean island in any 3-year period the probability of a major hurricane is .25, the probability of water damage is .44, and the probability of both a hurricane and water damage is .22. What is the probability of water damage given that there is a hurricane?

   (A) $.25 + .44 - .22$
   (B) $\frac{.22}{.44}$
   (C) $.25 + .44$
   (D) $\frac{.22}{.25}$
   (E) $.25 + .44 + .22$

GO ON TO THE NEXT PAGE ➤

9. Two possible wordings for a questionnaire on a proposed school budget increase are as follows:

 I. This school district has one of the highest per student expenditure rates in the state. This has resulted in low failure rates, high standardized test scores, and most students going on to good colleges and universities. Do you support the proposed school budget increase?
 II. This school district has one of the highest per student expenditure rates in the state. This has resulted in high property taxes, with many people on fixed incomes having to give up their homes because they cannot pay the school tax. Do you support the proposed school budget increase?

 One of these questions showed that 58% of the population favor the proposed school budget increase, while the other question showed that only 13% of the population support the proposed increase. Which produced which result and why?

 (A) The first showed 58% and the second 13% because of the lack of randomization as evidenced by the wording of the questions.
 (B) The first showed 13% and the second 58% because of a placebo effect due to the wording of the questions.
 (C) The first showed 58% and the second 13% because of the lack of a control group.
 (D) The first showed 13% and the second 58% because of response bias due to the wording of the questions.
 (E) The first showed 58% and the second 13% because of response bias due to the wording of the questions.

10. A union spokesperson is trying to encourage a college faculty to join the union. She would like to argue that faculty salaries are not truly based on years of service as most faculty believe. She gathers data and notes the following scatterplot of salary versus years of service.

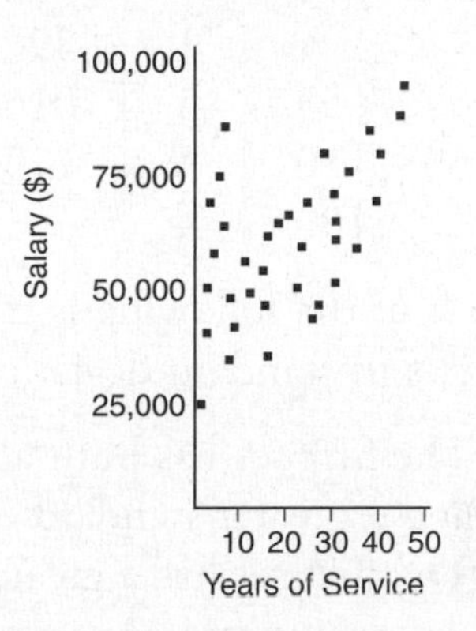

 Which of the following most correctly interprets the overall scatterplot?

 (A) The faculty member with the fewest years of service makes the lowest salary, and the faculty member with the most service makes the highest salary.
 (B) A faculty member with more service than another has the greater salary than the other.
 (C) There is a strong positive correlation with little deviation.
 (D) There is no clear relationship between salary and years of service.
 (E) While there is a strong positive correlation, there is a distinct deviation from the overall pattern for faculty with fewer than ten years of service.

11. Two random samples of students are chosen, one from those taking an AP Statistics class and one from those not. The following back-to-back stemplots compare the GPAs.

AP Statistics		No AP Statistics
	1	89
97653	2	015688
98775332110	3	133344777888
1100	4	

Which of the following is true about the ranges and standard deviations?

(A) The first set has both a greater range and a greater standard deviation.
(B) The first set has a greater range, while the second has a greater standard deviation.
(C) The first set has a greater standard deviation, while the second has a greater range.
(D) The second set has both a greater range and a greater standard deviation.
(E) The two sets have equal ranges and equal standard deviations.

12. In a group of 10 scores, the largest score is increased by 40 points. What will happen to the mean?

(A) It will remain the same.
(B) It will increase by 4 points.
(C) It will increase by 10 points.
(D) It will increase by 40 points.
(E) There is not sufficient information to answer this question.

13. Suppose X and Y are random variables with $\mu_x = 32$, $\sigma_x = 5$, $\mu_y = 44$, and $\sigma_y = 12$. Given that X and Y are independent, what are the mean and standard deviation of the random variable $X + Y$?

(A) $\mu_{x+y} = 76$, $\sigma_{x+y} = 8.5$
(B) $\mu_{x+y} = 76$, $\sigma_{x+y} = 13$
(C) $\mu_{x+y} = 76$, $\sigma_{x+y} = 17$
(D) $\mu_{x+y} = 38$, $\sigma_{x+y} = 17$
(E) There is insufficient information to answer this question.

14. Suppose you toss a fair die three times and it comes up an even number each time. Which of the following is a true statement?

(A) By the law of large numbers, the next toss is more likely to be an odd number than another even number.
(B) Based on the properties of conditional probability the next toss is more likely to be an even number given that three in a row have been even.
(C) Dice actually do have memories, and thus the number that comes up on the next toss will be influenced by the previous tosses.
(D) The law of large numbers tells how many tosses will be necessary before the percentages of evens and odds are again in balance.
(E) The probability that the next toss will again be even is .5.

GO ON TO THE NEXT PAGE ➤

15. A pharmaceutical company is interested in the association between advertising expenditures and sales for various over-the-counter products. A sales associate collects data on nine products, looking at sales (in $1000) versus advertising expenditures (in $1000). The results of the regression analysis are shown below.

```
Dependent variable: Sales
Source          df              Sum of Squares      Mean Square       F-ratio
Regression      1                   9576.1             9576.1         1118.45
Residual        7                     59.9                8.6

Variable        Coefficient         SE Coef            t-ratio           P
Constant          123.800            1.798              68.84          0.000
Advertising        12.633            0.378              33.44          0.000
R-Sq = 99.4%     R-Sq (adj)  = 99.3%
s = 2.926 with 9-2 = 7 degrees of freedom
```

Which of the following gives a 90% confidence interval for the slope of the regression line?

(A) 12.633 ± 1.415(0.378)
(B) 12.633 ± 1.895(0.378)
(C) 123.800 ± 1.414(1.798)
(D) 123.800 ± 1.895(1.798)
(E) 123.800 ± 1.645(1.798/ $\sqrt{9}$)

16. Suppose you wish to compare the AP Statistics exam results for the male and female students taking AP Statistics at your high school. Which is the most appropriate technique for gathering the needed data?

(A) Census
(B) Sample survey
(C) Experiment
(D) Observational study
(E) None of these is appropriate.

17. Jonathan obtained a score of 80 on a statistics exam, placing him at the 90th percentile. Suppose five points are added to everyone's score. Jonathan's new score will be at the

(A) 80th percentile.
(B) 85th percentile.
(C) 90th percentile.
(D) 95th percentile.
(E) There is not sufficient information to answer this quesiton.

18. To study the effect of music on piecework output at a clothing manufacturer, two experimental treatments are planned: day-long classical music for one group versus day-long light rock music for another. Which one of the following groups would serve best as a control for this study?

(A) A third group for which no music is played
(B) A third group that randomly hears either classical or light rock music each day
(C) A third group that hears day-long R & B music
(D) A third group that hears classical music every morning and light rock every afternoon
(E) A third group in which each worker has earphones and chooses his or her own favorite music

19. Suppose H_0: $p = .6$, and the power of the test for H_a: $p = .7$ is .8. Which of the following is a valid conclusion?

(A) The probability of committing a Type I error is .1.
(B) If H_a is true, the probability of failing to reject H_0 is .2.
(C) The probability of committing a Type II error is .3.
(D) All of the above are valid conclusions.
(E) None of the above are valid conclusions.

GO ON TO THE NEXT PAGE ➤

20. Following is a histogram of the numbers of ties owned by bank executives.

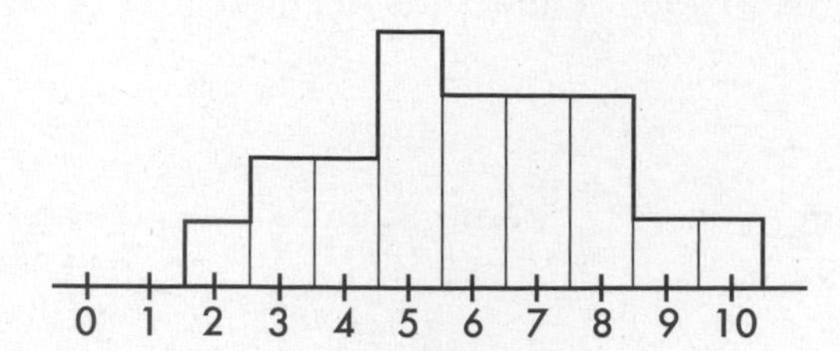

Which of the following is a correct statement ?

(A) The median number of ties is five.
(B) More than four executives own over eight ties each.
(C) An executive is equally likely to own fewer than five ties or more than seven ties.
(D) One tie is a reasonable estimate for the standard deviation.
(E) Removing all the executives with three, nine, and ten ties may change the median.

21. Which of the following is a binomial random variable?

(A) The number of tosses before a "5" appears when tossing a fair die.
(B) The number of points a hockey team receives in 10 games, where two points are awarded for wins, one point for ties, and no points for losses.
(C) The number of hearts out of five cards randomly drawn from a deck of 52 cards, without replacement.
(D) The number of motorists not wearing seat belts in a random sample of five drivers.
(E) None of the above.

22. Company I manufactures bomb fuses that burn an average of 50 minutes with a standard deviation of 10 minutes, while company II advertises fuses that burn an average of 55 minutes with a standard deviation of 5 minutes. Which company's fuse is more likely to last at least 1 hour? Assume normal distributions of fuse times.

(A) Company I's, because of its greater standard deviation
(B) Company II's, because of its greater mean
(C) For both companies, the probability that a fuse will last at least 1 hour is 15.9%
(D) For both companies, the probability that a fuse will last at least 1 hour is 84.1%
(E) The problem cannot be solved from the information given.

23. Which of the following is *not* important in the design of experiments?

(A) Control of confounding variables
(B) Randomization in assigning subjects to different treatments
(C) Use of a lurking variable to control the placebo effect
(D) Replication of the experiment using sufficient numbers of subjects
(E) All of the above are important in the design of experiments.

Practice Examination 2

GO ON TO THE NEXT PAGE ➤

24. The travel miles claimed in weekly expense reports of the sales personnel at a corporation are summarized in the following boxplot.

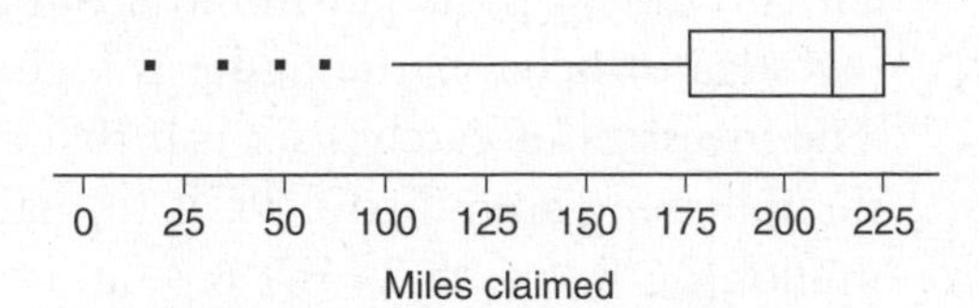

Which of the following is the most reasonable conclusion?

(A) The mean and median numbers of travel miles are roughly equal.
(B) The mean number of travel miles is greater than the median number.
(C) Most of the claimed numbers of travel miles are in the [0, 200] interval.
(D) Most of the claimed numbers of travel miles are in the [200, 240] interval.
(E) The left and right whiskers contain the same number of values from the set of personnel travel mile claims.

25. Which of the following statements about residuals is true?

(A) Influential scores have large residuals.
(B) If the linear model is good, the number of positive residuals will be the same as the number of negative residuals.
(C) The mean of the residuals is always zero.
(D) If the correlation is 0, there will be a distinct pattern in the residual plot.
(E) If the correlation is 1, there will not be a distinct pattern in the residual plot.

26. Four pairs of data are used in determining a regression line $y = 3x + 4$. If the four values of the independent variable are 32, 24, 29, and 27, respectively, what is the mean of the four values of the dependent variable?

(A) 68
(B) 84
(C) 88
(D) 100
(E) The mean cannot be determined from the given information.

27. According to one poll, 12% of the public favor legalizing all drugs. In a simple random sample of six people, what is the probability that at least one person favors legalization?

(A) $6(.12)(.88)^5$
(B) $(.88)^6$
(C) $1 - (.88)^6$
(D) $1 - 6(.12)(.88)^5$
(E) $6(.12)(.88)^5 + (.88)^6$

28. Sampling error occurs

(A) when interviewers make mistakes resulting in bias.
(B) because a sample statistic is used to estimate a population parameter.
(C) when interviewers use judgment instead of random choice in picking the sample.
(D) when samples are too small.
(E) in all of the above cases.

29. A telephone executive instructs an associate to contact 104 customers using their service to obtain their opinions in regard to an idea for a new pricing package. The associate notes the number of customers whose names begin with *A* and uses a random number table to pick four of these names. She then proceeds to use the same procedure for each letter of the alphabet and combines the $4 \times 26 = 104$ results into a group to be contacted. Which of the following is a correct conclusion?

    (A) Her procedure makes use of chance.
    (B) Her procedure results in a simple random sample.
    (C) Each customer has an equal probability of being included in the survey.
    (D) Her procedure introduces bias through *sampling error.*
    (E) With this small a sample size, it is better to let the surveyor pick representative customers to be surveyed based on as many features such as gender, political preference, income level, race, age, and so on, as are in the company's data banks.

30. The graph below shows cumulative proportions plotted against GPAs for high school seniors.

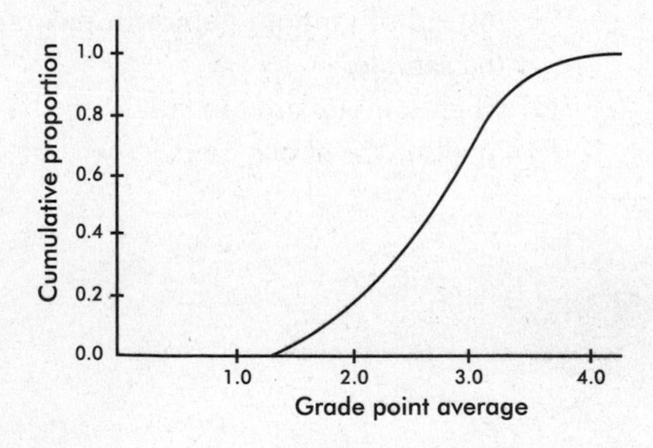

What is the approximate interquartile range?

    (A) 0.85
    (B) 2.25
    (C) 2.7
    (D) 2.75
    (E) 3.1

31. PCB contamination of a river by a manufacturer is being measured by amounts of the pollutant found in fish. A company scientist claims that the fish contain only 5 parts per million, but an investigator believes the figure is higher. The investigator catches six fish that show the following amounts of PCB (in parts per million): 6.8, 5.6, 5.2, 4.7, 6.3, and 5.4. In performing a hypothesis test with H_0: $\mu = 5$ and H_a: $\mu > 5$, what is the test statistic?

    (A) $t = \dfrac{5.67 - 5}{0.763}$
    (B) $t = \dfrac{5.67 - 5}{\sqrt{0.763/5}}$
    (C) $t = \dfrac{5.67 - 5}{\sqrt{0.763/6}}$
    (D) $t = \dfrac{5.67 - 5}{0.763/\sqrt{5}}$
    (E) $t = \dfrac{5.67 - 5}{0.763/\sqrt{6}}$

32. The distribution of weights of 16-ounce bags of a particular brand of potato chips is approximately normal with a standard deviation of 0.28 ounce. How does the weight of a bag at the 40th percentile compare with the mean weight?

    (A) 0.40 ounce above the mean
    (B) 0.25 ounce above the mean
    (C) 0.07 ounce above the mean
    (D) 0.07 ounce below the mean
    (E) 0.25 ounce below the mean

GO ON TO THE NEXT PAGE ➤

33. In general, how does tripling the sample size change the confidence interval size?

(A) It triples the interval size.
(B) It divides the interval size by 3.
(C) It multiples the interval size by 1.732.
(D) It divides the interval size by 1.732.
(E) This question cannot be answered without knowing the sample size.

34. Which of the following statements is *false?*

(A) Like the normal distribution, the *t*-distributions are symmetric.
(B) The *t*-distributions are lower at the mean and higher at the tails, and so are more spread out than the normal distribution.
(C) The greater the *df*, the closer the *t*-distributions are to the normal distribution.
(D) The smaller the *df*, the better the 68–95–99.7 Rule works for *t*-models.
(E) The area under all *t*-distribution curves is 1.

35. A study on school budget approval among people with different party affiliations resulted in the following segmented bar chart:

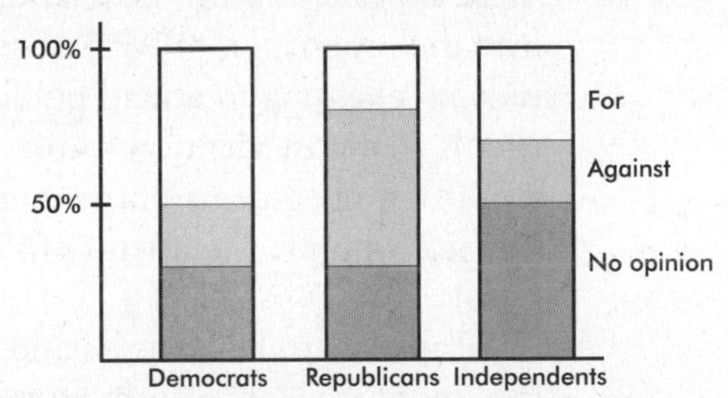

Which of the following is greatest?

(A) Number of Democrats who are for the proposed budget
(B) Number of Republicans who are against the budget
(C) Number of Independents who have no opinion on the budget
(D) The above are all equal.
(E) The answer is impossible to determine without additional information.

36. The sampling distribution of the sample mean is close to the normal distribution

(A) only if both the original population has a normal distribution and n is large.
(B) if the standard deviation of the original population is known.
(C) if n is large, no matter what the distribution of the original population.
(D) no matter what the value of n or what the distribution of the original population.
(E) only if the original population is not badly skewed and does not have outliers.

37. What is the probability of a Type II error when a hypothesis test is being conducted at the 10% significance level ($\alpha = .10$)?

(A) .05
(B) .10
(C) .90
(D) .95
(E) There is insufficient information to answer this question.

38.

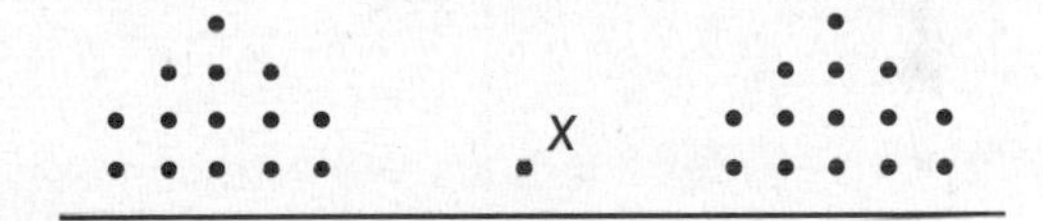

Above is the dotplot for a set of numbers. One element is labeled X. Which of the following is a correct statement?

(A) X has the largest z-score, in absolute value, of any element in the set.
(B) A modified boxplot will plot an outlier like X as an isolated point.
(C) A stemplot will show X isolated from two clusters.
(D) Because of X, the mean and median are different.
(E) The IQR is exactly half the range.

Practice Examination 2

GO ON TO THE NEXT PAGE ➤

39. A 2008 survey of 500 households concluded that 82% of the population uses grocery coupons. Which of the following best describes what is meant by the poll having a margin of error of 3%?

(A) Three percent of those surveyed refused to participate in the poll.
(B) It would not be unexpected for 3% of the population to begin using coupons or stop using coupons.
(C) Between 395 and 425 of the 500 households surveyed responded that they used grocery coupons.
(D) If a similar survey of 500 households were taken weekly, a 3% change in each week's results would not be unexpected.
(E) It is likely that between 79% and 85% of the population use grocery coupons.

40.

	College Plans	
	Public	Private
Taking AP Statistics:	18	27
Not taking AP Statistics:	26	40

The above two-way table summarizes the results of a survey of high school seniors conducted to determine if there is a relationship between whether or not a student is taking AP Statistics and whether he or she plans to attend a public or a private college after graduation. Which of the following is the most reasonable conclusion about the relationship between taking AP Statistics and the type of college a student plans to attend?

(A) There appears to be no association since the proportion of AP Statistics students planning to attend public schools is almost identical to the proportion of students not taking AP Statistics who plan to attend public schools.
(B) There appears to be an association since the proportion of AP Statistics students planning to attend public schools is almost identical to the proportion of students not taking AP Statistics who plan to attend public schools.
(C) There appears to be an association since more students plan to attend private than public schools.
(D) There appears to be an association since fewer students are taking AP Statistics than are not taking AP Statistics.
(E) These data do not address the question of association.

STOP

If there is still time remaining, you may review your answers.

SECTION II

PART A

Questions 1–5

Spend about 65 minutes on this part of the exam.
Percentage of Section II grade—75

You must show all work and indicate the methods you use. You will be graded on the correctness of your methods and on the accuracy of your results and explanations.

1. Ten volunteer male subjects are to be used for an experiment to study four drugs (aloe, camphor, eucalyptus oil, and benzocaine) and a placebo with regard to itching relief. Itching is induced on a forearm with an itch stimulus (cowage), a drug is topically administered, and the duration of itching is recorded.

   (a) If the experiment is to be done in one sitting, how would you assign treatments for a completely randomized design?
   (b) If 5 days are set aside for the experiment, with one sitting a day, how would you assign treatments for a randomized block design where the subjects are the blocks?
   (c) What limits are there on generalization of any results?

2. A comprehensive study of more than 3,000 baseball games results in the graph below showing relative frequencies of runs scored by home teams.

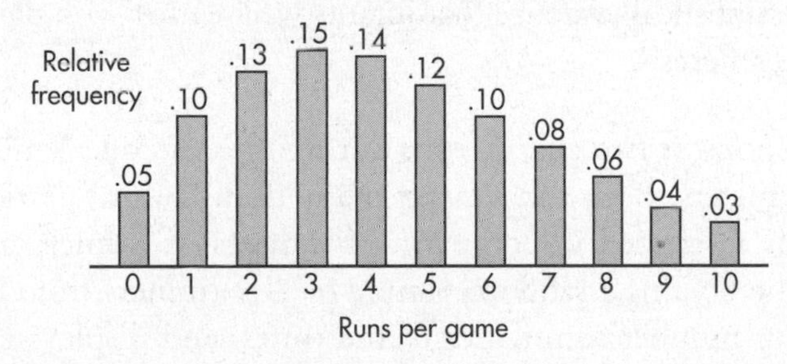

   (a) Calculate the mean and the median.
   (b) Between the mean and the median, is the one that is greater what was expected? Explain.
   (c) What is the probability that in 4 randomly selected games, the home team is shut out (scores no runs) in at least one of the games?
   (d) If X is the random variable for the runs scored per game by home teams, its standard deviation is 2.578. Suppose 200 games are selected at random, and $\overline{X}$, the mean number of runs scored by the home teams, is calculated. Describe the sampling distribution of $\overline{X}$.

GO ON TO THE NEXT PAGE ➤

3. A study is performed to explore the relationship, if any, between 24-hour urinary metabolite 3-methoxy-4-hydroxyphenylglycol (MHPG) levels and depression in bipolar patients. The MHPG level is measured in micrograms per 24 hours while the manic-depression (MD) scale used goes from 0 (manic delirium), through 5 (euthymic), up to 10 (depressive stupor). A partial computer printout of regression analysis with MHPG as the independent variable follows:

```
Average MHPG = 1243.1 with SD = 384.9
Average MD = 5.4 with SD = 2.875
95% confidence interval for slope b1 = (−.0096, −.0016)
```

(a) Calculate the slope of the regression line and interpret it in the context of this problem.
(b) Find the equation of the regression line.
(c) Calculate and interpret the value of r^2 in the context of this problem.
(d) What does the correlation say about causation in the context of this problem.

4. An experiment is run to test whether daily stimulation of specific reflexes in young infants will lead to earlier walking. Twenty infants were recruited through a pediatrician's service, and were randomly split into two groups of ten. One group received the daily stimulation while the other was considered a control group. The ages (in months) at which the infants first walked alone were recorded.

	Ages in months at which first steps alone were taken
With stimulation:	10, 12, 11, 10.5, 11, 11.5, 11.5, 11, 12, 11.5
Control:	10, 13, 12, 11, 11.5, 11.5, 12.5, 12, 11.5, 11

Is there statistical evidence that infants walk earlier with daily stimulation of specific reflexes?

5. A city sponsors two charity runs during the year, one 5 miles and the other 10 kilometers. Both runs attract many thousands of participants. A statistician is interested in comparing the types of runners attracted to each race. She obtains a random sample of 100 runners from each race, calculates five-number summaries of the times, and displays the results in the parallel boxplots below.

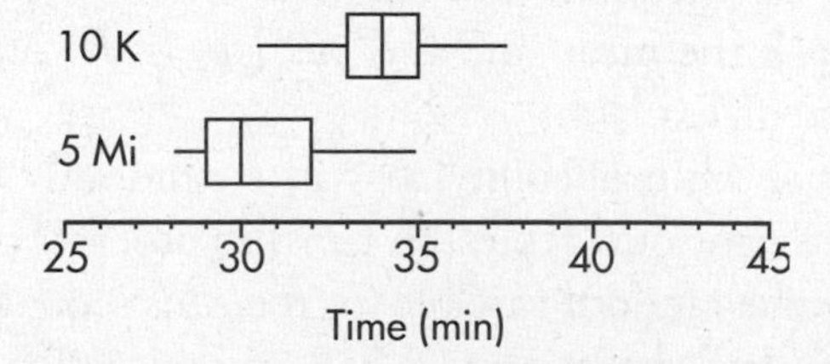

GO ON TO THE NEXT PAGE ➤

Practice Examination 2

(a) Compare the distributions of times above.

The statistician notes that 5 miles equals 8 kilometers and so decides to multiply the times from the 5-mile event by $\frac{10}{8}$ and then compare box plots. This display is below.

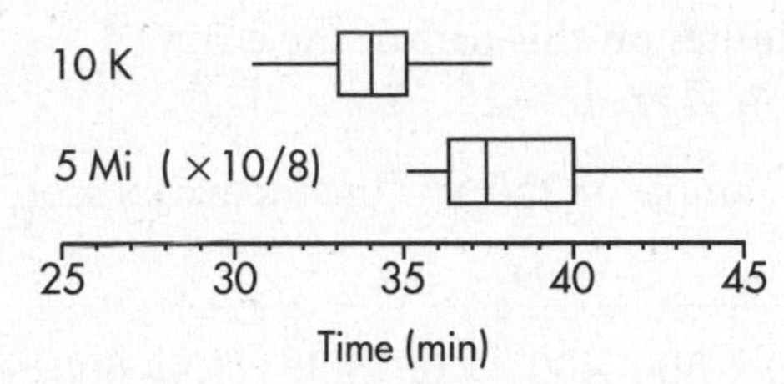

(b) How do the distributions now compare?
(c) Given that the 10-kilometer race was a longer distance than the 5-mile race, was the change from the first set of boxplots to the second set as expected? Explain.
(d) Based on the boxplots, would you expect the difference in mean times in the first set of parallel boxplots to be less than, greater than, or about the same as the difference in mean times in the second set of parallel boxplots? Explain.

SECTION II

PART B

Question 6

Spend about 25 minutes on this part of the exam.
Percentage of Section II grade—25

6. A random sample of 250 SAT mathematics scores tabulates as follows:

200–300	300–400	400–500	500–600	600–700	700–800
7	39	74	82	36	12

(a) Find a 95% confidence interval for the proportion of scores over 500.

(b) Test the null hypothesis that the data follow a *normal distribution* with $\mu = 500$ and $\sigma = 100$.

(c) Assume the above data comes from a normally distributed population with unknown μ and σ. In this case, the sampling distribution of $\frac{(n-1)s^2}{\sigma^2}$ follows a chi-square distribution. If computer output of the above data yields $n = 250$, $\bar{x} = 504.8$, and $s = 114.4$, and if in this problem the χ^2-values for tails of .025 and .975 are 294.6 and 207.2, respectively, find a 95% confidence interval for the standard deviation σ.

STOP

If there is still time remaining, you may review your answers.

Answer Key

Section I

1. **D**	9. **E**	17. **C**	25. **C**	33. **D**
2. **B**	10. **E**	18. **A**	26. **C**	34. **D**
3. **B**	11. **D**	19. **B**	27. **C**	35. **E**
4. **D**	12. **B**	20. **C**	28. **B**	36. **C**
5. **E**	13. **B**	21. **D**	29. **A**	37. **E**
6. **C**	14. **E**	22. **C**	30. **A**	38. **C**
7. **C**	15. **B**	23. **C**	31. **E**	39. **E**
8. **D**	16. **A**	24. **D**	32. **D**	40. **A**

Answers Explained

Section I

1. **(D)** Since (2, 7) is on the line $y = mx + 3$, we have $7 = 2m + 3$ and $m = 2$. Thus the regression line is $y = 2x + 3$. The point $(\bar{x}, \bar{y})$ is always on the regression line, and so we have $\bar{y} = 2\bar{x} + 3$.

2. **(B)** It could well be that conscientious students are the same ones who both study and do well on the basketball court. If students could be randomly assigned to study or not study, the results would be more meaningful. Of course, ethical considerations might make it impossible to isolate the confounding variable in this way.

3. **(B)** The critical *z*-score is 0.525. Thus $75 - \mu = 0.525(14)$ and $\mu = 67.65$.

4. **(D)** The slope of the regression line and the correlation are related by $b_1 = r\frac{s_y}{s_x}$. When using *z*-scores, the standard deviations s_x and s_y are 1. If $r = 0$, then $b_1 = 0$. Switching which variable is *x* and which is *y*, or changing units, will not change the correlation.

5. **(E)** The median and interquartile range are specifically used when outliers are suspected of unduly influencing the mean, range, or standard deviation.

6. **(C)** $2.576\left(\frac{2.4}{\sqrt{n}}\right) \le 0.6$, which gives $\sqrt{n} \ge 10.304$ and $n \ge 106.2$.

7. **(C)** This is a hypothesis test with H_0: tissue strength is within specifications, and H_a: tissue strength is below specifications. A Type I error is committed when a true null hypothesis is mistakenly rejected.

8. **(D)**

$$P(\text{water}|\text{hurricane}) = \frac{P(\text{water} \cap \text{hurricane})}{P(\text{hurricane})} = \frac{.22}{.25} = .88$$

9. **(E)** The wording of questions can lead to response bias. The neutral way of asking this question would simply have been, "Do you support the proposed school budget increase?"

10. **(E)** While it is important to look for basic patterns, it is also important to look for deviations from these patterns. In this case, there is an overall positive correlation; however, those faculty with under ten years of service show little relationship between years of service and salary. While (A) is a true statement, it does not give an overall interpretation of the scatterplot.

11. **(D)** The second set has a greater range, 3.8 – 1.8 = 2.0 as compared to 4.1 – 2.3 = 1.8, and with its skewness it also has a greater standard deviation.

12. **(B)** With $n = 10$, increasing Σx by 40 increases $\frac{\Sigma x}{n}$ by 4.

13. **(B)** The means and the variances can be added. Thus the new variance is $5^2 + 12^2 = 169$, and the new standard deviation is 13.

14. **(E)** Dice have no memory, so the probability that the next toss will be an even number is .5 and the probability that it will be an odd number is .5. The law of large numbers says that as the number of tosses becomes larger, the proportion of even numbers tends to become closer to .5.

15. **(B)** The critical *t*-scores for 90% confidence with $df = 7$ are ±1.895.

16. **(A)** Either directly or anonymously, you should be able to obtain the test results for *every* student.

17. **(C)** Percentile ranking is a measure of relative position. Adding five points to everyone's score will not change the relative positions.

18. **(A)** The control group should have experiences identical to those of the experimental groups except for the treatment under examination. They should not be given a new treatment.

19. **(B)** If H_a is true, the probability of failing to reject H_0 and thus committing a Type II error is 1 minus the power, that is, 1 – .8 = .2.

20. **(C)** Five does not split the area in half, so 5 is not the median. Histograms such as these show relative frequencies, not actual frequencies. The area from 1.5 to 4.5 is the same as that between 7.5 and 10.5, each being about 25% of the total. Given the spread, 1 is too small an estimate of the standard deviation. The area above 3 looks to be the same as the area above 9 and 10, so the median won't change.

21. **(D)** There must be a fixed number of trials, which rules out (A); only two possible outcomes, which rules out (B); and a constant probability of success on any trial, which rules out (C).

22. **(C)** In both cases 1 hour is one standard deviation from the mean with a right tail probability of .1587.

23. **(C)** Control, randomization, and replication are all important aspects of well-designed experiments. We try to control lurking variables, not to use them to control something else.

24. **(D)** The data are strongly skewed to the left, indicating that the mean is less than the median. The median appears to be roughly 215, indicating that the interval [200, 240] probably has more than 50% of the values. While in a standard boxplot each whisker contains 25% of the values, this is a modified boxplot showing four outliers, and so the left whisker has four fewer values than the right whisker.

25. **(C)** For the regression line, the sum and thus the mean of the residuals are always zero. An influential score may have a small residual but still have a great effect on the regression line. If the correlation is 1, all the residuals would be 0, resulting in a very distinct pattern.

26. **(C)** $\bar{x} = \frac{32 + 24 + 29 + 27}{4} = 28$. Since $(\bar{x}, \bar{y})$ is a point on the regression line,

    $\bar{y} = 3(28) + 4 = 88$.

27. **(C)** $P(\text{at least } 1) = 1 - P(\text{none}) = 1 - (.88)^6 = .536$

28. **(B)** Different samples give different sample statistics, all of which are estimates for the same population parameter, and so error, called *sampling error*, is naturally present.

29. **(A)** While the associate does use chance, each customer would have the same chance of being selected only if the same number of customers had names starting with each letter of the alphabet. This selection does not result in a simple random sample because each possible set of 104 customers does not have the same chance of being picked as part of the sample. For example, a group of customers whose names all start with *A* will not be chosen. Sampling error, the natural variation inherent in a survey, is always present and is not a source of bias. Letting the surveyor have free choice in selecting the sample, rather than incorporating chance in the selection process, is a recipe for disaster!

30. **(A)** Corresponding to cumulative proportions of 0.25 and 0.75 are $Q_1 = 2.25$ and $Q_3 = 3.1$, respectively, and so the interquartile range is $3.1 - 2.25 = 0.85$.

31. **(E)** The standard deviation of the test statistic is $\sigma_{\bar{x}} = \frac{\sigma}{\sqrt{n}} \approx \frac{s}{\sqrt{n}}$.

32. **(D)** From a table or a calculator (for example `invNorm` on the TI-84), the 40th percentile corresponds to a *z*-score of -0.2533, and $-0.2533(.28) = -0.0709$.

33. **(D)** Increasing the sample size by a multiple of d divides the interval estimate by $\sqrt{d}$.

34. **(D)** The t-distributions are symmetric; however, they are lower at the mean and higher at the tails and so are more spread out than the normal distribution. The greater the *df*, the closer the t-distributions are to the normal distribution. The 68–95–99.7 Rule applies to the z-distribution and will work for t-models with very large *df*. All probability density curves have an area of 1 below them.

35. **(E)** The given bar chart shows percentages, not actual numbers.

36. **(C)** This follows from the central limit theorem.

37. **(E)** There is a different Type II error for each possible correct value for the population parameter.

38. **(C)** X is close to the mean and so will have a z-score close to 0. Modified boxplots show only outliers that are far from the mean. X and the two clusters are clearly visible in a stemplot of these data. In symmetric distributions the mean and median are equal. The IQR here is close to the range.

39. **(E)** Using a measurement from a sample, we are never able to say *exactly* what a population proportion is; rather we always say we have a certain *confidence* that the population proportion lies in a particular *interval*. In this case that interval is 82% ± 3% or between 79% and 85%.

40. **(A)** Whether or not students are taking AP Statistics seems to have no relationship to which type of school they are planning to go to. Chi-square is close to 0.

SECTION II

1. (a) Number the volunteers 1 through 10. Use a random number generator to pick numbers between 1 and 10, throwing out repeats. The volunteers corresponding to the first two numbers chosen will receive aloe, the next two will receive camphor, the next two eucalyptus oil, the next two benzocaine, and the remaining two a placebo.
(b) Each volunteer (the volunteers are "blocks") should receive all five treatments, one a day, with the time-order randomized. For example, label aloe 1, camphor 2, eucalyptus oil 3, benzocaine 4, and the placebo 5. Then for each volunteer use a random number generator to pick numbers between 1 and 5, throwing away repeats. The order picked gives the day on which each volunteer receives each treatment.
(c) Results cannot be generalized to women.

Scoring

Part (a) is essentially correct for giving a procedure which randomly assigns 2 volunteers to each of the five treatments. Part (a) is partially correct for giving a procedure which randomly assigns one of the treatments for each volunteer, but may not result in *two* volunteers receiving each treatment.

Part (b) is essentially correct for giving a procedure which assigns a random order for each of the volunteers to have all 5 treatments. Part (b) is partially correct for having each volunteer take all five treatments, one a day, but not clearly randomizing the time-order.

Part (c) is essentially correct for stating that the results cannot be generalized to women and is incorrect otherwise.

4 Complete Answer	All three parts essentially correct.
3 Substantial Answer	Two parts essentially correct and one part partially correct.
2 Developing Answer	Two parts essentially correct OR one part essentially correct and one or two parts partially correct OR all three parts partially correct.
1 Minimal Answer	One part essentially correct OR two parts partially correct.

2. (a) $\bar{X} = \sum xp(x) = 0(.05) + 1(.10) + 2(.13) + 3(.15) + 4(.14) + 5(.12) + 6(.10) + 7(.08) + 8(.06) + 9(.04) + 10(.03) = 4.27$.
With $(.05 + .10 + .13 + .15) = .43$ below 4 runs, and $(.12 + .10 + .08 + .06 + .04 + .03) = .43$ above 4 runs, the median must be 4.

(b) The mean is greater than the median, as was to be expected because the distribution is skewed to the right.

(c) P(at least one shutout in 4 games) $= 1 - P$(no shutouts in the 4 games) $= 1 - (1 - .05)^4 = 1 - (.95)^4 = .1855$

(d) The distribution of $\bar{X}$ is approximately normal with mean $\mu_{\bar{x}} = 4.27$ (from above) and standard deviation $\sigma_{\bar{x}} = \dfrac{2.578}{\sqrt{200}} = 0.1823$.

Scoring

Part (a) is essentially correct for correctly calculating both the mean and median. Part (a) is partially correct for correctly calculating one of these two measures.

Part (b) is essentially correct for noting that the mean is greater than the median and relating this to the skew.

Part (c) is essentially correct for recognizing this as a binomial probability calculation and making the correct calculation. Part (c) is partially correct for recognizing this as a binomial probability calculation, but with an error such as $1 - (.05)^4$ or $4(.05)(.95)^3$.

Part (d) is essentially correct for "approximately normal," $\mu_{\bar{x}} = 4.27$, and $\sigma_{\bar{x}} = 0.1823$, and partially correct for two of these three answers.

Count partially correct answers as one-half an essentially correct answer.

4 Complete Answer	Four essentially correct answers.
3 Substantial Answer	Three essentially correct answers.
2 Developing Answer	Two essentially correct answers.
1 Minimal Answer	One essentially correct answer.

Use a holistic approach to decide a score totaling between two numbers.

3. (a) The slope b_1 is in the center of the confidence interval, so $b_1 = \frac{-.0096-.0016}{2} = -0.0056$. In context, 0.0056 *estimates* the *average* decrease in the manic-depressive scale score for each 1-microgram increase in the level of urinary MHPG. (Thus high levels of MHPG are associated with increased mania, and conversely, low levels of MHPG are associated with increased depression.)

(b) Recalling that the regression line goes through the point $(\bar{x}, \bar{y})$ or using the AP Exam formula page, $b_o = \bar{y} - b_1\bar{x} = 5.4 - (-0.0056)(1243.1) = 12.36$, and thus the equation of the regression line is $\widehat{MD} = 12.36 - 0.0056(MHPG)$ (or $\hat{y} = 12.36 - 0.0056x$, where x is the level of urinary MHPG in micrograms per 24 hours, and y is the score on a 0-10 manic-depressive scale).

(c) Recalling that on the regression line, each one SD increase in the independent variable corresponds to an increase of r SD in the dependent variable, or using the AP Exam formula page, $b_1 = r\frac{S_y}{S_x}$, so

$$r = b_1\frac{S_x}{S_y} = -0.0056\left(\frac{384.9}{2.875}\right) = -.750$$

and $r^2 = 56.2\%$. Thus, 56.2% of the variation in the manic-depression scale level is explained by urinary MHPG levels.

(d) Correlation never proves causation. It could be that depression causes biochemical changes leading to low levels of urinary MHPG, or it could be that low levels of urinary MHPG cause depression, or it could be that some other variable (a lurking variable) simultaneously affects both urinary MHPG levels and depression.

Scoring

Part (a) is essentially correct if the slope is correctly calculated and correctly interpreted in context. Part (a) is partially correct if the slope is not correctly calculated but a correct interpretation is given using the incorrect value for the slope.

Part (b) is essentially correct if the regression equation is correctly calculated (using the slope found in part (a)) and it is clear what the variables stand for. Part (b) is partially correct if the correct equation is found (using the slope found in part (a)) but it is unclear what the variables stand for.

Part (c) is essentially correct if the coefficient of determination, r^2, is correctly calculated and correctly interpreted in context. Part (c) is partially correct if r^2 is not correctly calculated but a correct interpretation is given using the incorrect value for r^2.

Part (d) is essentially correct for noting that correlation never proves causation, and referring to context. Part (d) is partially correct for a correct statement about correlation and causation, but with no reference to context.

A partially correct answer counts half of an essentially correct answer.

4 Complete Answer	4 essentially correct answers.
3 Substantial Answer	3 essentially correct answers.
2 Developing Answer	2 essentially correct answers.
1 Minimal Answer	1 essentially correct answer.

Use a holistic approach to decide a score totaling between two numbers.

4. There are four parts to this solution.

(a) State the hypotheses.

$$H_0: \mu_1 - \mu_2 = 0 \text{ and } H_a: \mu_1 - \mu_2 < 0$$

where μ_1 = mean number of months at which first steps alone are taken for infants receiving daily stimulation, and μ_2 = mean number of months at which first steps are taken for infants in the control group. [Other notations are also possible, for example, $H_0: \mu_1 = \mu_2$ and $H_a: \mu_1 < \mu_2$.]

(b) Identify the test by name or formula and check the assumptions.

Two-sample t-test OR $t = \dfrac{\bar{x}_1 - \bar{x}_2}{\sqrt{\dfrac{s_1^2}{n_1} + \dfrac{s_2^2}{n_2}}}$

Assumptions:

1. *Randomization:* We are given that the infants were randomly split into the two groups.
2. *Nearly normal populations:* Dotplots of the two samples show no outliers and are roughly bell-shaped.

10 11 12 13
Age (mo)
(with stimulation)

10 11 12 13
Age (mo)
(control)

(c) Demonstrate correct mechanics.

$$t = \frac{11.2 - 11.6}{\sqrt{\frac{(0.6235)^2}{10} + \frac{(0.8433)^2}{10}}} = -1.206$$

and with $df = 9$, we have a P-value = .1293
[with a calculator's $df = 16.69$, we get $P = .1234$]

(d) State the conclusion in context with linkage to the P-value.

With this large a P-value (.1293 > .10), there is no evidence that infants walk earlier with daily stimulation of specific reflexes.

Scoring

Part (a) is essentially correct for stating the hypotheses and identifying the variables. Part (a) is partially correct for correct hypotheses, but missing identification of the variables.

Part (b) is essentially correct for identifying the test and checking the assumptions. Part (b) is partially correct for only one of these two elements.

Part (c) is essentially correct for correctly calculating both the t-score and the P-value. Part (c) is partially correct for only one of these two elements.

Part (d) is essentially correct for a conclusion in context with linkage to the P-value. Part (d) is partially correct for only one of these two elements.

Count partially correct answers as one-half an essentially correct answer.

4 Complete Answer	Four essentially correct answers.
3 Substantial Answer	Three essentially correct answers.
2 Developing Answer	Two essentially correct answers.
1 Minimal Answer	One essentially correct answer.

Use a holistic approach to decide a score totaling between two numbers.

5. (a) A complete answer compares shape, center, and spread.
Shape: The 10-kilometer distribution appears symmetric, while the 5-mile distribution is skewed right.
Center: The median of the 10-kilometer distribution is 4 minutes greater than the median of the 5-mile distribution.
Spread: The ranges of the two distributions are equal, both about 7 minutes.

(b) *Shape:* Again, the 10-kilometer distribution appears symmetric, while the 5-mile distribution is still skewed right.
Center: Now the median of the 10-kilometer distribution is about 3.5 minutes less than the median of the 5-mile distribution.
Spread: Now the range of the 10-kilometer distribution is less than the range of the 5-mile distribution.

(c) One possible answer is that this was not as expected with the following explanation: One would expect that for 5 miles, the shorter run, the speeds would be faster, so adjusting the 5-mile run times would result in faster times (less minutes) than the 10-kilometer times, but this was not the case. [Perhaps slower, less serious runners participate in the shorter race, so when their times are adjusted, the minutes are greater than the times for the 10-kilometer run.]

(d) In each set of parallel boxplots, the 10-kilometer distribution appears symmetric, so that the mean will be about the same as the median, while the 5-mile distribution is skewed right so that the mean will in all likelihood be greater than the median. In the first set of boxplots (where the 5-mile median < 10-kilometer median) this will result in the means being closer, while in the second set of boxplots (where the 5-mile median > 10-kilometer median) this will result in the means being further apart. Thus we would expect the difference in mean times in the first set of parallel boxplots to be less than the difference in mean times in the second set of parallel boxplots.

Scoring

Part (a) is essentially correct for correctly comparing shape, center, and spread. Part (a) is partially correct for correctly comparing two of the three features.

Part (b) is essentially correct for correctly comparing shape, center, and spread. Part (b) is partially correct for correctly comparing two of the three features.

Part (c) is essentially correct for a reasonable statement about the change between the two sets of parallel boxplots together with a correct explanation to go along with the statement. Part (c) is partially correct if the explanation is weak.

Part (d) is essentially correct for a correct prediction about the means together with a reasonable justification based on symmetry of the 10-kilometer distribution and skewness of the 5-mile distribution. Part (d) is partially correct if the justification is weak.

Count partially correct answers as one-half an essentially correct answer.

4 Complete Answer	Four essentially correct answers.
3 Substantial Answer	Three essentially correct answers.
2 Developing Answer	Two essentially correct answers.
1 Minimal Answer	One essentially correct answer.

Use a holistic approach to decide a score totaling between two numbers.

6. (a) Identify the confidence interval by name or formula.

95% confidence interval for the population proportion

$$\hat{p} \pm 1.96\sqrt{\frac{\hat{p}(1-\hat{p})}{n}}$$

Check the assumptions.

We are given that this is a *random* sample.

$n\hat{p} = 250\left(\frac{82+36+12}{250}\right) = 130 > 10$ and $n(1-\hat{p}) = 250\left(\frac{120}{250}\right) = 120 > 10.$

Calculate the confidence interval.

$\hat{p} = \left(\frac{130}{250}\right) = .52$ and $.52 \pm 1.96\sqrt{\frac{(.52)(.48)}{250}} = .52 \pm .062$

Interpret the confidence interval in context.

We are 95% confident that between 45.8% and 58.2% of SAT mathematics scores are over 500.

(b) State the hypotheses.

H_0: The distribution of SAT mathematics scores is normal with $\mu = 500$ and $\sigma = 100$.
H_a: The distribution of SAT mathematics scores is not normal with $\mu = 500$ and $\sigma = 100$.

Identify the test by name or formula and check the assumptions.

Chi-square goodness-of-fit test

$$\chi^2 = \sum \frac{(\text{observed} - \text{expected})^2}{\text{expected}}$$

Check assumptions

We have a random sample.
A normal distribution has 34.13% on each side of the mean and within one SD, 13.59% on each side between one and two SDs from the mean, and 2.28% on each side more than two SDs from the mean. This gives expected cell counts of (.0228)250 = 5.7, (.1359)500 = 34.0, and (.3413)500 = 85.3, each of which is at least 5.

Demonstrate correct mechanics.

$$\chi^2 = \frac{(7-5.7)^2}{5.7} + \frac{(39-34.0)^2}{34.0} + \frac{(74-85.3)^2}{85.3} +$$

$$\frac{(82-85.3)^2}{85.3} + \frac{(36-34.0)^2}{34.0} + \frac{(12-5.7)^2}{5.7} = 9.737$$

With $df = 6 - 1 = 5$, the P-value is .083

State the conclusion in context with linkage to the *P*-value.

Either with this *small* a *P*-value (.083 < α = .10) there *is* some evidence that the distribution of SAT mathematics scores is not normal, OR with this *large* a *P*-value (.083 > α = .05) there is *no* evidence that the distribution of SAT mathematics scores is not normal.

(c) $207.2 < \frac{(250-1)(114.4)^2}{\sigma^2} < 294.6$

$207.2 < \frac{3{,}259{,}000}{\sigma^2} < 294.6$

$\frac{1}{207.2} > \frac{\sigma^2}{3{,}259{,}000} > \frac{1}{294.6}$

$15{,}729 > \sigma^2 > 11{,}062$

$125.4 > \sigma > 105.2$

Thus we are 95% confident that the standard deviation for the distribution of SAT mathematics scores is between 105.2 and 125.4.

Scoring

Part (a) has four parts: 1) identifying the confidence interval; 2) checking assumptions; 3) calculating the confidence interval; and 4) interpreting the confidence interval in context. Part (a) is essentially correct if three or four of these parts are correct and partially correct if one or two of these parts are correct.

Part (b) has four parts: 1) stating the hypotheses; 2) identifying the test and checking assumptions; 3) calculating the test statistic and the *P*-value; and 4) giving a conclusion in context with linkage to the *P*-value. Part (b) is essentially correct if three or four of these parts are correct and partially correct if one or two of these parts are correct.

Part (c) is essentially correct for calculating the confidence interval and interpreting it in context. Part (c) is partially correct for one of these two parts correct.

4 Complete Answer	All three parts essentially correct.
3 Substantial Answer	Two parts essentially correct and one part partially correct.
2 Developing Answer	Two parts essentially correct OR one part essentially correct and one or two parts partially correct OR all three parts partially correct.
1 Minimal Answer	One part essentially correct OR two parts partially correct.

Answer Sheet

PRACTICE EXAMINATION 3

1. Ⓐ Ⓑ Ⓒ Ⓓ Ⓔ	11. Ⓐ Ⓑ Ⓒ Ⓓ Ⓔ	21. Ⓐ Ⓑ Ⓒ Ⓓ Ⓔ	31. Ⓐ Ⓑ Ⓒ Ⓓ Ⓔ
2. Ⓐ Ⓑ Ⓒ Ⓓ Ⓔ	12. Ⓐ Ⓑ Ⓒ Ⓓ Ⓔ	22. Ⓐ Ⓑ Ⓒ Ⓓ Ⓔ	32. Ⓐ Ⓑ Ⓒ Ⓓ Ⓔ
3. Ⓐ Ⓑ Ⓒ Ⓓ Ⓔ	13. Ⓐ Ⓑ Ⓒ Ⓓ Ⓔ	23. Ⓐ Ⓑ Ⓒ Ⓓ Ⓔ	33. Ⓐ Ⓑ Ⓒ Ⓓ Ⓔ
4. Ⓐ Ⓑ Ⓒ Ⓓ Ⓔ	14. Ⓐ Ⓑ Ⓒ Ⓓ Ⓔ	24. Ⓐ Ⓑ Ⓒ Ⓓ Ⓔ	34. Ⓐ Ⓑ Ⓒ Ⓓ Ⓔ
5. Ⓐ Ⓑ Ⓒ Ⓓ Ⓔ	15. Ⓐ Ⓑ Ⓒ Ⓓ Ⓔ	25. Ⓐ Ⓑ Ⓒ Ⓓ Ⓔ	35. Ⓐ Ⓑ Ⓒ Ⓓ Ⓔ
6. Ⓐ Ⓑ Ⓒ Ⓓ Ⓔ	16. Ⓐ Ⓑ Ⓒ Ⓓ Ⓔ	26. Ⓐ Ⓑ Ⓒ Ⓓ Ⓔ	36. Ⓐ Ⓑ Ⓒ Ⓓ Ⓔ
7. Ⓐ Ⓑ Ⓒ Ⓓ Ⓔ	17. Ⓐ Ⓑ Ⓒ Ⓓ Ⓔ	27. Ⓐ Ⓑ Ⓒ Ⓓ Ⓔ	37. Ⓐ Ⓑ Ⓒ Ⓓ Ⓔ
8. Ⓐ Ⓑ Ⓒ Ⓓ Ⓔ	18. Ⓐ Ⓑ Ⓒ Ⓓ Ⓔ	28. Ⓐ Ⓑ Ⓒ Ⓓ Ⓔ	38. Ⓐ Ⓑ Ⓒ Ⓓ Ⓔ
9. Ⓐ Ⓑ Ⓒ Ⓓ Ⓔ	19. Ⓐ Ⓑ Ⓒ Ⓓ Ⓔ	29. Ⓐ Ⓑ Ⓒ Ⓓ Ⓔ	39. Ⓐ Ⓑ Ⓒ Ⓓ Ⓔ
10. Ⓐ Ⓑ Ⓒ Ⓓ Ⓔ	20. Ⓐ Ⓑ Ⓒ Ⓓ Ⓔ	30. Ⓐ Ⓑ Ⓒ Ⓓ Ⓔ	40. Ⓐ Ⓑ Ⓒ Ⓓ Ⓔ

Practice Examination 3

SECTION I

Questions 1–40

Spend 90 minutes on this part of the exam.

Directions: The questions or incomplete statements that follow are each followed by five suggested answers or completions. Choose the response that best answers the question or completes the statement.

1. Which of the following is a true statement?
   (A) While properly designed experiments can strongly suggest cause-and-effect relationships, a complete census is the only way of establishing such a relationship.
   (B) If properly designed, observational studies can establish cause-and-effect relationships just as strongly as properly designed experiments.
   (C) Controlled experiments are often undertaken later to establish cause-and-effect relationships first suggested by observational studies.
   (D) A useful approach to overcome bias in observational studies is to increase the sample size.
   (E) In an experiment, the control group is a self-selected group who choose not to receive a designated treatment.

2. Two classes take the same exam. Suppose a certain score is at the 40th percentile for the first class and at the 80th percentile for the second class. Which of the following is the most reasonable conclusion?
   (A) Students in the first class generally scored higher than students in the second class.
   (B) Students in the second class generally scored higher than students in the first class.
   (C) A score at the 20th percentile for the first class is at the 40th percentile for the second class.
   (D) A score at the 50th percentile for the first class is at the 90th percentile for the second class.
   (E) One of the classes has twice the number of students as the other.

GO ON TO THE NEXT PAGE ➤

3. In an experiment, the control group should receive

   (A) treatment opposite that given the experimental group.
   (B) the same treatment given the experimental group without knowing they are receiving the treatment.
   (C) a procedure identical to that given the experimental group except for receiving the treatment under examination.
   (D) a procedure identical to that given the experimental group except for a random decision on receiving the treatment under examination.
   (E) none of the procedures given the experimental group.

4. In a random sample of Toyota car owners, 83 out of 112 said they were satisfied with the Toyota front-wheel drive, while in a similar survey of Subaru owners, 76 out of 81 said they were satisfied with the Subaru four-wheel drive. A 90% confidence interval estimate for the difference in proportions between Toyota and Subaru car owners who are satisfied with their drive systems is reported to be $-.197 \pm .081$. Which is a proper conclusion?

   (A) The interval is invalid because probabilities cannot be negative.
   (B) The interval is invalid because it does not contain zero.
   (C) Subaru owners are approximately 19.7% more satisfied with their drive systems than are Toyota owners.
   (D) 90% of Subaru owners are approximately 19.7% more satisfied with their drive systems than are Toyota owners.
   (E) We are 90% confident that the difference in proportions between Toyota and Subaru car owners who are satisfied with their drive systems is between $-.278$ and $-.116$.

5. In a study on the effect of music on worker productivity, employees were told that a different genre of background music would be played each day and the corresponding production outputs noted. Every change in music resulted in an increase in production. This is an example of

   (A) the effect of a treatment unit.
   (B) the placebo effect.
   (C) the control group effect.
   (D) sampling error.
   (E) voluntary response bias.

6. A computer manufacturer sets up three locations to provide technical support for its customers. Logs are kept noting whether or not calls about problems are solved successfully. Data from a sample of 1000 calls are summarized in the following table:

Location

	1	2	3	Total
Problem solved	325	225	150	700
Problem not solved	125	100	75	300
Total	450	325	225	1000

Assuming there is no association between location and whether or not a problem is resolved successfully, what is the expected number of successful calls (problem solved) from location 1?

   (A) $\frac{(325)(450)}{700}$
   (B) $\frac{(325)(700)}{450}$
   (C) $\frac{(325)(450)}{1000}$
   (D) $\frac{(325)(700)}{1000}$
   (E) $\frac{(450)(700)}{1000}$

Practice Examination 3

GO ON TO THE NEXT PAGE ➤

7. Which of the following statements about the correlation coefficient is true?

   (A) The correlation coefficient and the slope of the regression line may have opposite signs.
   (B) A correlation of 1 indicates a perfect cause-and-effect relationship between the variables.
   (C) Correlations of +.87 and –.87 indicate the same degree of clustering around the regression line.
   (D) Correlation applies equally well to quantitative and categorical data.
   (E) A correlation of 0 shows little or no association between two variables.

8. Suppose X and Y are random variables with $E(X) = 780$, $\text{var}(X) = 75$, $E(Y) = 430$, and $\text{var}(Y) = 25$. Given that X and Y are independent, what is the variance of the random variable $X - Y$?

   (A) $75 - 25$
   (B) $75 + 25$
   (C) $\sqrt{75-25}$
   (D) $\sqrt{75+25}$
   (E) $\sqrt{75} - \sqrt{25}$

9. What is a sample?

   (A) A measurable characteristic of a population
   (B) A set of individuals having a characteristic in common
   (C) A value calculated from raw data
   (D) A subset of a population
   (E) None of the above

10. A histogram of the cholesterol levels of all employees at a large law firm is as follows:

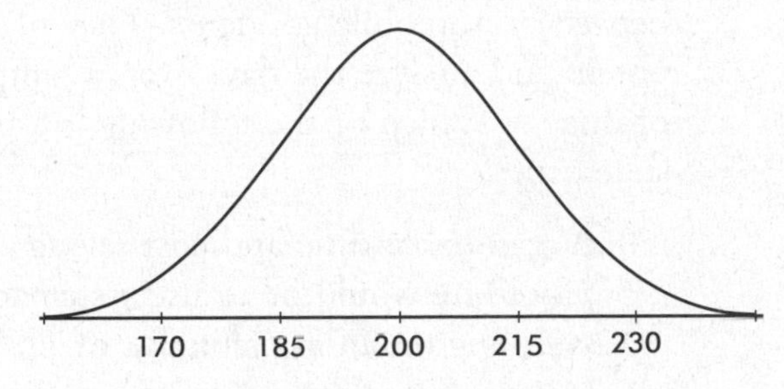

   Which of the following is the best estimate of the standard deviation of this distribution?

   (A) $\frac{(230-170)}{6} = 10$
   (B) 15
   (C) $\frac{200}{6} = 33.3$
   (D) $230 - 170 = 60$
   (E) $245 - 155 = 90$

11. A soft drink dispenser can be adjusted to deliver any fixed number of ounces. If the machine is operating with a standard deviation in delivery equal to 0.3 ounce, what should be the mean setting so that a 12-ounce cup will overflow less than 1% of the time? Assume a normal distribution for ounces delivered.

   (A) 11.23 ounces
   (B) 11.30 ounces
   (C) 11.70 ounces
   (D) 12.70 ounces
   (E) 12.77 ounces

GO ON TO THE NEXT PAGE ➤

12. An insurance company wishes to study the number of years drivers in a large city go between automobile accidents. They plan to obtain and analyze the data from a sample of drivers. Which of the following is a true statement?

    (A) A reasonable time-and-cost-saving procedure would be to use systematic sampling on an available list of all AAA (Automobile Association of America) members in the city.
    (B) A reasonable time-and-cost-saving procedure would be to randomly choose families and include all drivers in each of these families in the sample.
    (C) To determine the mean number of years between accidents, randomness in choosing a sample of drivers is not important as long as the sample size is very large.
    (D) The larger a simple random sample, the more likely its standard deviation will be close to the population standard deviation divided by the square root of the sample size.
    (E) None of the above are true statements.

13. The probability that a person will show a certain gene-transmitted trait is .8 if the father shows the trait and .06 if the father doesn't show the trait. Suppose that the children in a certain community come from families in 25% of which the father shows the trait. Given that a child shows the trait, what is the probability that her father shows the trait?

    (A) .245
    (B) .250
    (C) .750
    (D) .816
    (E) .860

14. Given an experiment with $H_0: \mu = 10$, $H_a: \mu > 10$, and a possible correct value of 11, which of the following increases as n increases?

    I. The probability of a Type I error.
    II. The probability of a Type II error.
    III. The power of the test.

    (A) I only
    (B) II only
    (C) III only
    (D) II and III
    (E) None will increase.

15. If all the values of a data set are the same, all of the following must equal zero except for which one?

    (A) Mean
    (B) Standard deviation
    (C) Variance
    (D) Range
    (E) Interquartile range

16. To determine the average number of minutes it takes to manufacture one unit of a new product, an assembly line manager tracks a random sample of 15 units and records the number of minutes it takes to make each unit. The assembly times are assumed to have a normal distribution. If the mean and standard deviation of the sample are 3.92 and 0.45 minutes respectively, which of the following gives a 90% confidence interval for the mean assembly time, in minutes, for units of the new product?

    (A) $3.92 \pm 1.645\frac{0.45}{\sqrt{14}}$
    (B) $3.92 \pm 1.753\frac{0.45}{\sqrt{14}}$
    (C) $3.92 \pm 1.761\frac{0.45}{\sqrt{14}}$
    (D) $3.92 \pm 1.753\frac{0.45}{\sqrt{15}}$
    (E) $3.92 \pm 1.761\frac{0.45}{\sqrt{15}}$

GO ON TO THE NEXT PAGE ➤

17. A company has 1000 employees evenly distributed throughout five assembly plants. A sample of 30 employees is to be chosen as follows. Each of the five managers will be asked to place the 200 time cards of their respective employees in a bag, shake them up, and randomly draw out six names. The six names from each plant will be put together to make up the sample. Will this method result in a simple random sample of the 1000 employees?

    (A) Yes, because every employee has the same chance of being selected.
    (B) Yes, because every plant is equally represented.
    (C) Yes, because this is an example of stratified sampling, which is a special case of simple random sampling.
    (D) No, because the plants are not chosen randomly.
    (E) No, because not every group of 30 employees has the same chance of being selected.

18. Given that $P(E) = .32$, $P(F) = .15$, and $P(E \cap F) = .048$, which of the following is a correct conclusion?

    (A) The events E and F are both independent and mutually exclusive.
    (B) The events E and F are neither independent nor mutually exclusive.
    (C) The events E and F are mutually exclusive but not independent.
    (D) The events E and F are independent but not mutually exclusive.
    (E) The events E and F are independent, but there is insufficient information to determine whether or not they are mutually exclusive.

19. The number of leasable square feet of office space available in a city on any given day has a normal distribution with mean 640,000 square feet and standard deviation 18,000 square feet. What is the interquartile range for this distribution?

    (A) 652,000 – 628,000
    (B) 658,000 – 622,000
    (C) 667,000 – 613,000
    (D) 676,000 – 604,000
    (E) 694,000 – 586,000

20. Consider the following back-to-back stemplot:

Left	Stem	Right
	0	348
	1	01256
843	2	29
65210	3	2557
92	4	
7552	5	6
	6	1458
6	7	09
8541	8	
90	9	

Which of the following is a correct statement?

(A) The distributions have the same mean.
(B) The distributions have the same median.
(C) The interquartile range of the distribution to the left is 20 greater than the interquartile range of the distribution to the right.
(D) The distributions have the same variance.
(E) None of the above is correct.

GO ON TO THE NEXT PAGE ➤

21. Which of the following is a correct statement?

(A) A study results in a 99% confidence interval estimate of (34.2, 67.3). This means that in about 99% of all samples selected by this method, the sample means will fall between 34.2 and 67.3.
(B) A high confidence level may be obtained no matter what the sample size.
(C) The central limit theorem is most useful when drawing samples from normally distributed populations.
(D) The sampling distribution for a mean has standard deviation $\frac{\sigma}{\sqrt{n}}$ only when n is sufficiently large (typically one uses $n \geq 30$).
(E) The center of any confidence interval is the population parameter.

22. The binomial distribution is an appropriate model for which of the following?

(A) The number of minutes in an hour for which the Dow-Jones average is above its beginning average for the day.
(B) The number of cities among the 10 largest in New York State for which the weather is cloudy for most of a given day.
(C) The number of drivers wearing seat belts if 10 consecutive drivers are stopped at a police roadblock.
(D) The number of A's a student receives in his/her five college classes.
(E) None of the above.

23. Suppose two events, E and F, have nonzero probabilities p and q, respectively. Which of the following is impossible?

(A) $p + q > 1$
(B) $p - q < 0$
(C) $p / q > 1$
(D) E and F are neither independent nor mutually exclusive.
(E) E and F are both independent and mutually exclusive.

24. An inspection procedure at a manufacturing plant involves picking four items at random and accepting the whole lot if at least three of the four items are in perfect condition. If in reality 90% of the whole lot are perfect, what is the probability that the lot will be accepted?

(A) $(.9)^4$
(B) $1 - (.9)^4$
(C) $4(.9)^3(.1)$
(D) $.1 - 4(.9)^3(.1)$
(E) $4(.9)^3(.1) + (.9)^4$

25. A town has one high school, which buses students from urban, suburban, and rural communities. Which of the following samples is recommended in studying attitudes toward tracking of students in honors, regular, and below-grade classes?

(A) Convenience sample
(B) Simple random sample (SRS)
(C) Stratified sample
(D) Systematic sample
(E) Voluntary response sample

26. Suppose there is a correlation of $r = 0.9$ between number of hours per day students study and GPAs. Which of the following is a reasonable conclusion?

(A) 90% of students who study receive high grades.
(B) 90% of students who receive high grades study a lot.
(C) 90% of the variation in GPAs can be explained by variation in number of study hours per day.
(D) 10% of the variation in GPAs cannot be explained by variation in number of study hours per day.
(E) 81% of the variation in GPAs can be explained by variation in number of study hours per day.

GO ON TO THE NEXT PAGE ➤

27. To determine the average number of children living in single-family homes, a researcher picks a simple random sample of 50 such homes. However, even after one follow-up visit the interviewer is unable to make contact with anyone in 8 of these homes. Concerned about nonresponse bias, the researcher picks another simple random sample and instructs the interviewer to keep trying until contact is made with someone in a total of 50 homes. The average number of children is determined to be 1.73. Is this estimate probably too low or too high?

(A) Too low, because of undercoverage bias.
(B) Too low, because convenience samples overestimate average results.
(C) Too high, because of undercoverage bias.
(D) Too high, because convenience samples overestimate average results.
(E) Too high, because voluntary response samples overestimate average results.

28. The graph below shows cumulative proportions plotted against land values (in dollars per acre) for farms on sale in a rural community.

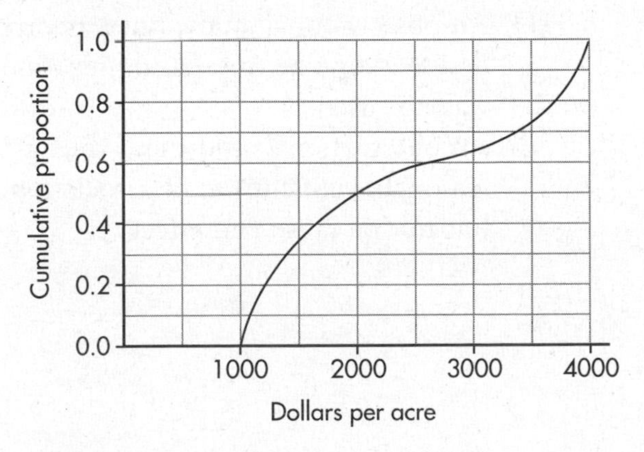

What is the median land value?

(A) $2000
(B) $2250
(C) $2500
(D) $2750
(E) $3000

29. An experiment is to be conducted to determine whether taking fish oil capsules or garlic capsules has more of an effect on cholesterol levels. In past studies it was noted that daily exercise intensity (low, moderate, high) is associated with cholesterol level, but average sleep length (< 5, $5 - 8$, > 8 hours) is not associated with cholesterol level. This experiment should be done

(A) by blocking on exercise intensity
(B) by blocking on sleep length
(C) by blocking on cholesterol level
(D) by blocking on capsule type
(E) without blocking

30. A confidence interval estimate is determined from the monthly grocery expenditures in a random sample of n families. Which of the following will result in a smaller margin of error?

I. A smaller confidence level
II. A smaller sample standard deviation
III. A smaller sample size

(A) II only
(B) I and II
(C) I and III
(D) II and III
(E) I, II, and III

GO ON TO THE NEXT PAGE ➤

31. A medical research team claims that high vitamin C intake increases endurance. In particular, 1000 milligrams of vitamin C per day for a month should add an average of 4.3 minutes to the length of maximum physical effort that can be tolerated. Army training officers believe the claim is exaggerated and plan a test on an SRS of 400 soldiers in which they will reject the medical team's claim if the sample mean is less than 4.0 minutes. Suppose the standard deviation of added minutes is 3.2. If the true mean increase is only 4.2 minutes, what is the probability that the officers will fail to reject the false claim of 4.3 minutes?

    (A) $P\left(z < \dfrac{4.0 - 4.3}{3.2/\sqrt{400}}\right)$

    (B) $P\left(z > \dfrac{4.0 - 4.3}{3.2/\sqrt{400}}\right)$

    (C) $P\left(z < \dfrac{4.3 - 4.2}{3.2/\sqrt{400}}\right)$

    (D) $P\left(z > \dfrac{4.3 - 4.2}{3.2/\sqrt{400}}\right)$

    (E) $P\left(z > \dfrac{4.0 - 4.2}{3.2/\sqrt{400}}\right)$

32. Consider the two sets $X = \{10, 30, 45, 50, 55, 70, 90\}$ and $Y = \{10, 30, 35, 50, 65, 70, 90\}$. Which of the following is false?

    (A) The sets have identical medians.
    (B) The sets have identical means.
    (C) The sets have identical ranges.
    (D) The sets have identical boxplots.
    (E) None of the above are false.

33. The weight of an aspirin tablet is 300 milligrams according to the bottle label. An FDA investigator weighs a simple random sample of seven tablets, obtains weights of 299, 300, 305, 302, 299, 301, and 303, and runs a hypothesis test of the manufacturer's claim. Which of the following gives the *P*-value of this test?

    (A) $P(t > 1.54)$ with $df = 6$
    (B) $2P(t > 1.54)$ with $df = 6$
    (C) $P(t > 1.54)$ with $df = 7$
    (D) $2P(t > 1.54)$ with $df = 7$
    (E) $0.5P(t > 1.54)$ with $df = 7$

34. A teacher believes that giving her students a practice quiz every week will motivate them to study harder, leading to a greater overall understanding of the course material. She tries this technique for a year, and everyone in the class achieves a grade of at least C. Is this an experiment or an observational study?

    (A) An experiment, but with no reasonable conclusion possible about cause and effect
    (B) An experiment, thus making cause and effect a reasonable conclusion
    (C) An observational study, because there was no use of a control group
    (D) An observational study, but a poorly designed one because randomization was not used
    (E) An observational study, and thus a reasonable conclusion of association but not of cause and effect

35. Which of the following is *not* true with regard to contingency tables for chi-square tests for independence?

    (A) The categories are not numerical for either variable.
    (B) Observed frequencies should be whole numbers.
    (C) Expected frequencies should be whole numbers.
    (D) Expected frequencies in each cell should be at least 5, and to achieve this, one sometimes combines categories for one or the other or both of the variables.
    (E) The expected frequency for any cell can be found by multiplying the row total by the column total and dividing by the sample size.

36. Which of the following is a correct statement?

    (A) The probability of a Type II error does not depend on the probability of a Type I error.
    (B) In conducting a hypothesis test, it is possible to simultaneously make both a Type I and a Type II error.
    (C) A Type II error will result if one incorrectly assumes the data are normally distributed.
    (D) In medical disease testing with the null hypothesis that the patient is healthy, a Type I error is associated with a *false negative;* that is, the test incorrectly indicates that the patient is disease free.
    (E) When you choose a significance level α, you're setting the probability of a Type I error to exactly α.

37.

X

Above is a scatterplot with one point labeled *X*. Suppose you find the least squares regression line. Which of the following is a correct statement?

    (A) *X* has the largest residual, in absolute value, of any point on the scatterplot.
    (B) *X* is an influential point.
    (C) The residual plot will show a curved pattern.
    (D) The association between the *x* and *y* variables is very weak.
    (E) If the point *X* were removed, the correlation would be 1.

38. A banking corporation advertises that 90% of the loan applications it receives are approved within 24 hours. In a random sample of 50 applications, what is the expected number of loan applications that will be turned down?

    (A) 50(.90)
    (B) 50(.10)
    (C) 50(.90)(.10)
    (D) $\sqrt{50(.90)(.10)}$
    (E) $\sqrt{\frac{(.90)(.10)}{50}}$

GO ON TO THE NEXT PAGE ➤

39. The parallel boxplots below show monthly rainfall summaries for Liberia, West Africa.

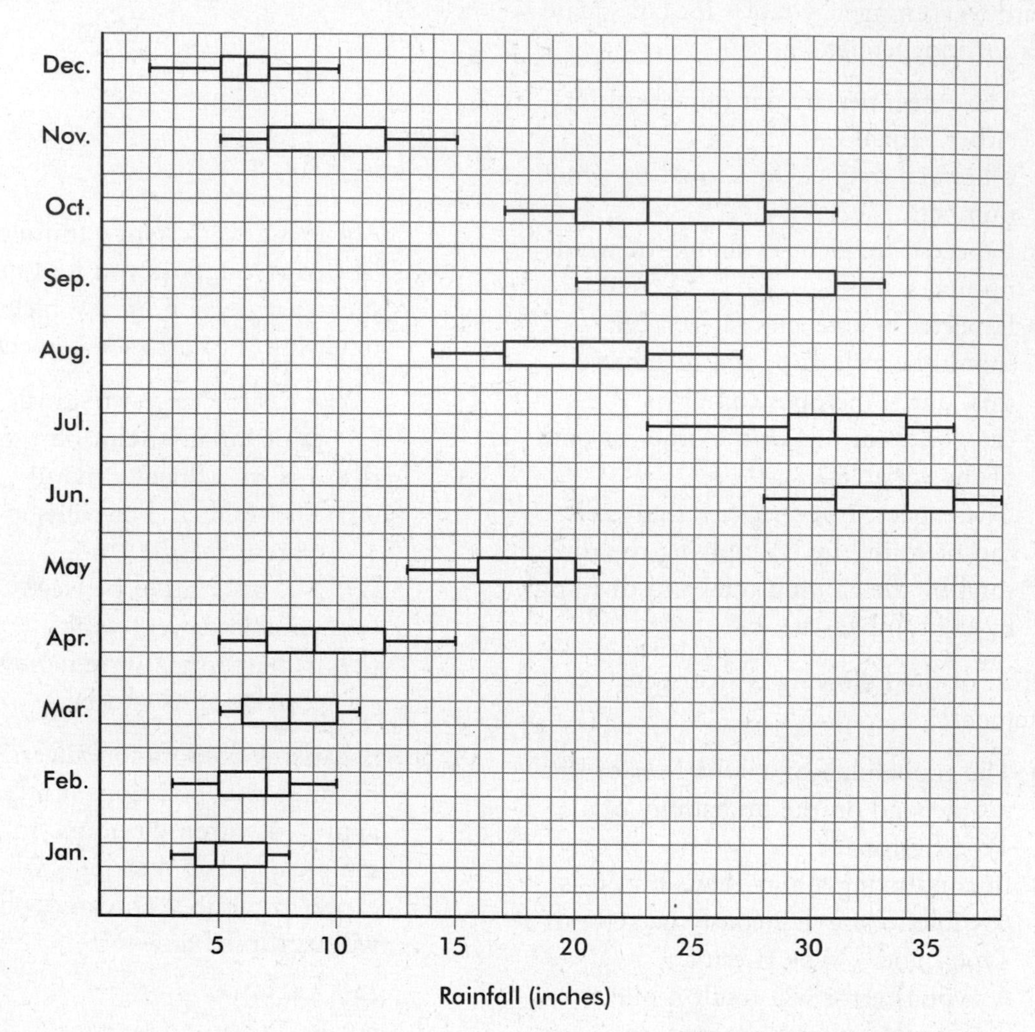

Which of the following months has the least variability as measured by *interquartile range*?

(A) January
(B) February
(C) March
(D) May
(E) December

GO ON TO THE NEXT PAGE ➤

40. In comparing the life expectancies of two models of refrigerators, the average years before complete breakdown of 10 model A refrigerators is compared with that of 15 model B refrigerators. The 90% confidence interval estimate of the difference is (6, 12). Which of the following is the most reasonable conclusion?

(A) The mean life expectancy of one model is twice that of the other.
(B) The mean life expectancy of one model is 6 years, while the mean life expectancy of the other is 12 years.
(C) The probability that the life expectancies are different is .90.
(D) The probability that the difference in life expectancies is greater than 6 years is .90.
(E) We should be 90% confident that the difference in life expectancies is between 6 and 12 years.

STOP

If there is still time remaining, you may review your answers.

SECTION II

PART A

Questions 1–5

Spend about 65 minutes on this part of the exam.
Percentage of Section II grade—75

> You must show all work and indicate the methods you use. You will be graded on the correctness of your methods and on the accuracy of your results and explanations.

1. The Information Technology Services division at a university is considering installing a new spam filter software product on all campus computers to combat unwanted advertising and spyware. A sample of 60 campus computers was randomly divided into two groups of 30 computers each. One group of 30 was considered to be a control group, while each computer in the other group had the spam filter software installed. During a two-week period each computer user was instructed to keep track of the number of unwanted spam e-mails received. The back-to-back stemplot below shows the distribution of such e-mails received for the control and treatment groups.

Control		Treatment
9 8 7 5	0	2 2 3 5 5 6 6 6 7 8 8 9 9 9
9 8 8 7 5 5 5 4 2 2 0	1	0 0 1 1 2 2 5 7 9
8 7 5 4 4 3 2 1 1 0	2	0 2 4 6
4 3 2 0 0	3	2 5
	4	1

(a) Compare the distribution of spam e-mails from the control and treatment groups.

(b) The standard deviation of the numbers of e-mails in the control group is 8.1. How does this value summarize variability for the control group data?

(c) A researcher in Information Technology Services calculates a 95% confidence interval for the difference in mean number of spam e-mails received between the control group and the treatment group with the new software and obtains (1.5, 10.9). Assuming all conditions for a two-sample *t*-interval are met, comment on whether or not there is evidence of a difference in the means for the number of spam e-mails received during a two-week period by computers with and without the software.

(d) The computer users on campus fall into four groups: administrators, staff, faculty, and students. Explain why a researcher might decide to use blocking in setting up this experiment.

2. A game contestant flips three fair coins and receives a score equal to the absolute value of the difference between the number of heads and number of tails showing.

   (a) Construct the probability distribution table for the possible scores in this game.

   (b) Calculate the expected value of the score for a player.

   (c) What is the probability that if a player plays this game three times the total score will be 3?

   (d) Suppose a player wins a major prize if he or she can average a score of at least 2. Given the choice, should he or she try for this average by playing 10 times or by playing 15 times? Explain.

3. A high school has a room set aside after school for students to play games. The attendance data are summarized in the following table.

Week	1	2	3	4	5	6	7	8
Attendance	73	78	84	88	29	35	39	44

A student plugs these data into his calculator and comes up with:

$$\widehat{Attendance} = 91.1 - 7.19\,(Week) \text{ with } r = -.726$$

   (a) Interpret the slope of the regression line in context. Does this seem to adequately explain the data? Why or why not?

   Another student points out that during the first 4 weeks the school allowed computer games, but then the policy changed.

   (b) Give a more appropriate regression analysis modeling attendance in terms of week. Include correlation and interpretation of slope in your analysis.

   (c) Explain the above results in terms of a scatterplot.

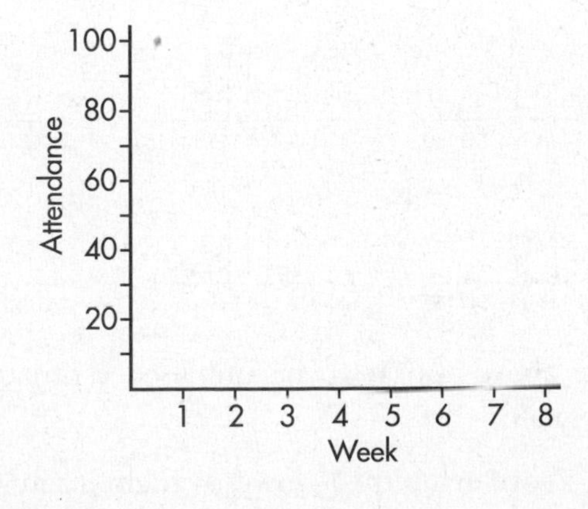

GO ON TO THE NEXT PAGE ➤

4. State investigators believe that a particular auto repair facility is fraudulently charging customers for repairs they don't need. As part of their investigation they pick a random sample of ten damaged cars, do their own cost estimate for repair work, and then send the cars to the facility under suspicion for an estimate. The data obtained are shown in the table below.

Car

	1	2	3	4	5	6	7	8	9	10
Investigator Estimate ($)	2585	3040	560	8250	3800	1575	3590	2830	6550	1820
Facility Estimate ($)	2250	3600	800	9100	4675	1920	3710	4050	6300	2200

Is the mean estimate of the facility under suspicion significantly greater than the mean estimate by the investigators? Justify your answer.

5. Five new estimators are being evaluated with regard to quality control in manufacturing professional baseballs of a given weight. Each estimator is tested every day for a month on samples of sizes $n = 10$, $n = 20$, and $n = 40$. The baseballs actually produced that month had a consistent mean weight of 146 grams. The distributions given by each estimator are as follows:

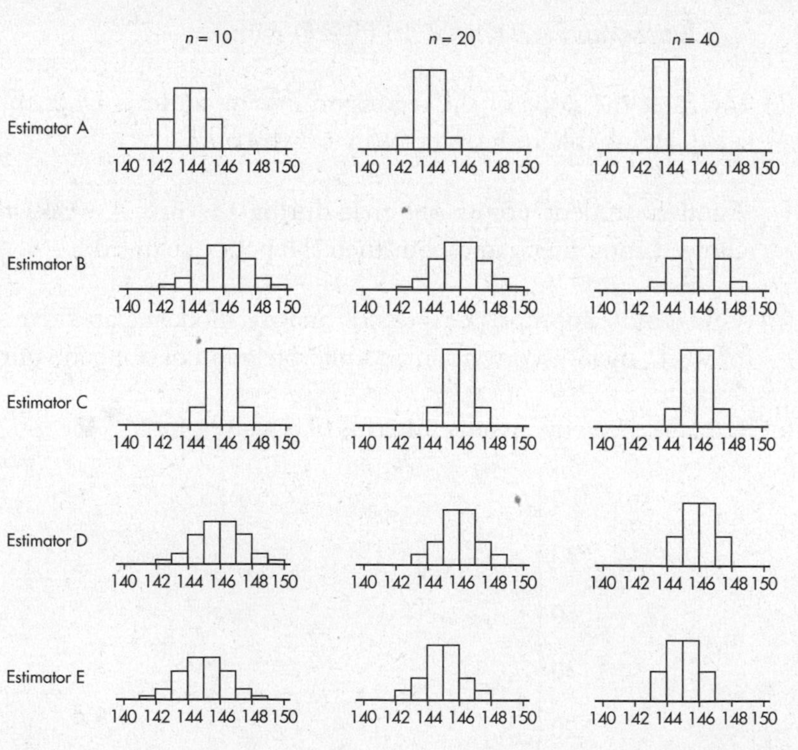

(a) Which of the above appear to be unbiased estimators of the population parameter? Explain.

(b) Which of the above exhibits the lowest variability for $n = 40$? Explain.

(c) Which of the above is the best estimator if the selected estimator will eventually be used with a sample of size $n = 100$? Explain.

GO ON TO THE NEXT PAGE ➤

SECTION II

PART B

Question 6

Spend about 25 minutes on this part of the exam.
Percentage of Section II grade—25

6. Dose intensity in chemotherapy is a balance between minimizing adverse side effects and maximizing therapeutic effect of the treatment. A study is performed on 800 patients with stage 4 colon cancer to determine the relationship between nausea (a side effect of most types of chemotherapy) and dosage of an experimental chemotherapy agent. Each patient is randomly placed in one of eight groups, each group receiving a different dose intensity, and the numbers of patients experiencing severe nausea are noted. The table below summarizes the data collected.

Dose intensity (mg/week)	**Number of patients**	**Number experiencing severe nausea**	**Proportion experiencing severe nausea**
10	110	34	.31
15	105	15	.14
20	85	22	.26
25	95	17	.18
50	100	47	.47
55	95	54	.57
60	110	42	.38
65	100	62	.62

(a) A medical researcher notes that a total of 205 out of 405 patients given higher doses ($\geq$ 50 mg/week) experienced severe nausea, while only 88 of the 395 patients given lower doses ($\leq$ 25 mg/week) experienced severe nausea. Do these data support that a greater proportion of patients receiving higher doses experience severe nausea than patients receiving lower doses?

(b) Another researcher thinks there is more of a relation between dose intensity and proportion experiencing severe nausea than one can see by simply looking at low and high doses. She runs a regression analysis with the following computer output.

```
Predictor      Coef       SE Coef      T       P
Constant     0.11652     0.07548     1.54    0.174
Dose        0.006659    0.001761     3.78    0.009
S = 0.103431     R-Sq = 70.4%
```

GO ON TO THE NEXT PAGE ➤

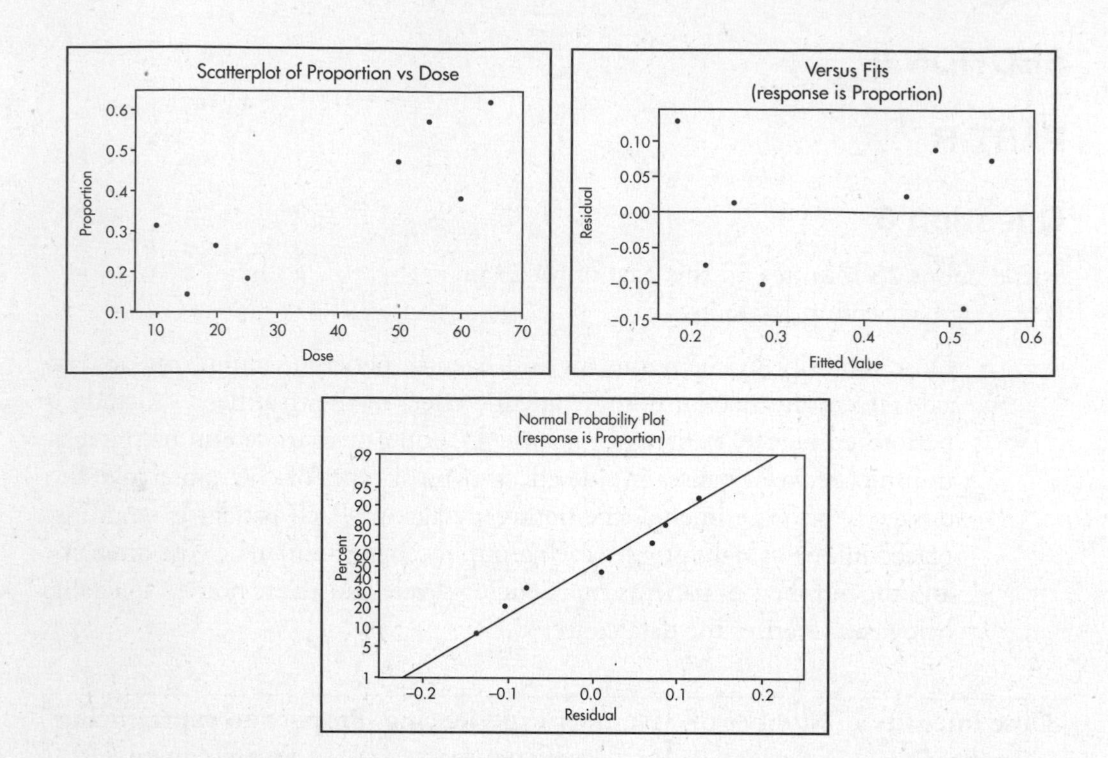

Find a 95% confidence interval for the slope of the regression line, and interpret your answer in context.

(c) A researcher plans to interview five patients randomly chosen from among the 800 participants in this experiment. What is the probability that a majority of those chosen will have experienced severe nausea?

STOP

If there is still time remaining, you may review your answers.

Answer Key

Section I

1. **C**	9. **D**	17. **E**	25. **C**	33. **B**
2. **A**	10. **B**	18. **D**	26. **E**	34. **A**
3. **C**	11. **B**	19. **A**	27. **C**	35. **C**
4. **E**	12. **E**	20. **D**	28. **A**	36. **E**
5. **B**	13. **D**	21. **B**	29. **A**	37. **C**
6. **E**	14. **C**	22. **E**	30. **B**	38. **B**
7. **C**	15. **A**	23. **E**	31. **E**	39. **E**
8. **B**	16. **E**	24. **E**	32. **E**	40. **E**

Answers Explained

Section I

1. **(C)** A complete census can give much information about a population, but it doesn't necessarily establish a cause-and-effect relationship among seemingly related population parameters. While the results of well-designed observational studies might suggest relationships, it is difficult to conclude that there is cause and effect without running a well-designed experiment. If bias is present, increasing the sample size simply magnifies the bias. The control group is selected by the researchers making use of chance procedures.

2. **(A)** In the first class only 40% of the students scored below the given score, while in the second class 80% scored below the same score.

3. **(C)** The control group should have experiences identical to those of the experimental groups except for the treatment under examination. They should not be given a new treatment.

4. **(E)** The negative sign comes about because we are dealing with the difference of proportions. The confidence interval estimate means that we have a certain *confidence* that the difference in population proportions lies in a particular *interval.*

5. **(B)** The desire of the workers for the study to be successful led to a placebo effect.

6. **(E)** The proportion of successful calls (problem solved) is $\frac{700}{1000}$, so $\frac{700}{1000}(450)$ is the expected number of calls from location 1 that are successful. Alternatively, the proportion of calls from location 1 is $\frac{450}{1000}$, so $\frac{450}{1000}(700)$ gives the expected number of successful calls from location 1.

7. **(C)** The slope and the correlation always have the same sign. Correlation shows association, not causation. Correlation does not apply to categorical data. Correlation measures linear association, so even with a correlation of 0, there may be very strong nonlinear association.

8. **(B)** If two random variables are independent, the mean of the difference of the two random variables is equal to the difference of the two individual means; however, the variance of the difference of the two random variables is equal to the *sum* of the two individual variances.

9. **(D)** A sample is simply a subset of a population.

10. **(B)** The markings, spaced 15 apart, clearly look like the standard deviation spacings associated with a normal curve.

11. **(B)** With a right tail having probability .01, the critical *z*-score is 2.326. Thus $\mu + 2.326(.3) = 12$, giving $\mu = 11.3$.

12. **(E)** There is no reason to think that AAA members are representative of the city's drivers. Family members may have similar driving habits and the independence condition would be violated. Random selection is important regardless of the sample size. The larger a random sample, the closer its standard deviation will be to the population standard deviation.

13. **(D)**

$$P\begin{pmatrix}\text{child}\\\text{shows}\end{pmatrix} = P\begin{pmatrix}\text{father}\\\text{shows}\end{pmatrix. \cap \begin{pmatrix}.\text{child}\\\text{shows}\end{pmatrix} + P\begin{pmatrix}\text{father}\\\text{doesn't}\end{pmatrix. \cap \begin{pmatrix}.\text{child}\\\text{shows}\end{pmatrix}$$

$$= (.25)(.8) + (.75)(.06)$$
$$= .200 + .045$$
$$= .245$$

$$P\left(\begin{matrix}\text{father}\\\text{shows}\end{matrix}\middle|\begin{matrix}\text{child}\\\text{shows}\end{matrix}\right) = \frac{.200}{.245} = .816$$

14. **(C)** As *n* increases the probabilities of Type I and Type II errors both decrease.

15. **(A)** The mean equals the common value of all the data elements. The other terms all measure variability, which is zero when all the data elements are equal.

16. **(E)** $\mu_{\bar{x}} \approx \frac{s}{\sqrt{n}} = \frac{0.45}{\sqrt{15}}$ and with $df = 15 - 1 = 14$, the critical *t*-scores are ± 1.761.

17. **(E)** In a simple random sample, every possible group of the given size has to be equally likely to be selected, and this is not true here. For example, with this procedure it is impossible for the employees in the final sample to all be from a single plant. This method is an example of stratified sampling, but stratified sampling does not result in simple random samples.

18. **(D)** $(.32)(.15) = .048$ so $P(E \cap F) = P(E)P(F)$ and thus E and F are independent. $P(E \cap F) \neq 0$, so E and F are not mutually exclusive.

19. **(A)** The quartiles Q_1 and Q_3 have *z*-scores of ± 0.67, so $Q_1 = 640{,}000 - (0.67)18{,}000 \approx 628{,}000$, while $Q_3 = 640{,}000 + (0.67)18{,}000 \approx 652{,}000$. The interquartile range is the difference $Q_3 - Q_1$.

20. **(D)** One set is a shift of 20 units from the other, so they have different means and medians, but they have identical shapes and thus the same variability including IQR, standard deviation, and variance.

21. **(B)** A 99% confidence interval estimate means that in about 99% of all samples selected by this method, the population mean will be included in the confidence interval. The wider the confidence interval, the higher the confidence level. The central limit theorem applies to any population, no matter if it is normally distributed or not. The sampling distribution for a mean always has standard deviation $\frac{\sigma}{\sqrt{n}}$; large enough sample size n refers to the closer the distribution will be to a normal distribution. The center of a confidence interval is the sample statistic, not the population parameter.

22. **(E)** In none of these are the trials independent. For example, as each consecutive person is stopped at a roadblock, the probability the next person has a seat belt on will quickly increase; if a student has one A, the probability is increased that he or she has another A.

23. **(E)** Independence implies $P(E \cap F) = P(E)P(F)$, while mutually exclusive implies $P(E \cap F) = 0$.

24. **(E)** In a binomial with $n = 4$ and $p = .9$, $P(\text{at least 3 successes}) = P(\text{exactly 3 successes}) + P(\text{exactly 4 successes}) = 4(.9)^3(.1) + (.9)^4$.

25. **(C)** In stratified sampling the population is divided into representative groups, and random samples of persons from each group are chosen. In this case it might well be important to be able to consider separately the responses from each of the three groups—urban, suburban, and rural.

26. **(E)** r^2, the coefficient of determination, indicates the percentage of variation in y that is explained by variation in x.

27. **(C)** It is most likely that the homes at which the interviewer had difficulty finding someone home were homes with fewer children living in them. Replacing these homes with other randomly picked homes will most likely replace homes with fewer children with homes with more children.

28. **(A)** The median corresponds to the 0.5 cumulative proportion.

29. **(A)** Blocking divides the subjects into groups of similar individuals, in this case individuals with similar exercise habits, and runs the experiment on each separate group. This controls the known effect of variation in exercise level on cholesterol level.

30. **(B)** The margin of error varies directly with the critical *z*-value and directly with the standard deviation of the sample, but inversely with the square root of the sample size.

31. **(E)** $\sigma_{\bar{x}} = \frac{3.2}{\sqrt{400}} = 0.16$. With a true mean increase of 4.2, the *z*-score for 4.0 is $\frac{4.0-4.2}{0.16} = -1.25$ and the officers fail to reject the claim if the sample mean has *z*-score greater than this.

32. **(E)** Both have 50 for their means and medians, both have a range of $90 - 10 = 80$, and both have identical boxplots, with first quartile 30 and third quartile 70.

33. **(B)** Since we are not told that the investigator suspects that the average weight is over 300 mg or is under 300 mg, and since a tablet containing too little or too much of a drug clearly should be brought to the manufacturer's attention, this is a two-sided test. Thus the *P*-value is twice the tail probability obtained (using the *t*-distribution with $df = n - 1 = 6$.)

34. **(A)** This study was an experiment because a treatment (weekly quizzes) was imposed on the subjects. However, it was a poorly designed experiment with no use of randomization and no control over lurking variables.

35. **(C)** The expected frequencies, as calculated by the rule in (E), may not be whole numbers.

36. **(E)** The probabilities of Type I and Type II error are related; for example, lowering the Type I error increases the probability of a Type II error. A Type I error can be made only if the null hypothesis is true, while a Type II error can be made only if the null hypothesis is false. In medical testing, with the usual null hypothesis that the patient is healthy, a Type I error is that a healthy patient is diagnosed with a disease, that is, a *false positive.* We reject H_0 when the *P*-value falls below α, and when H_0 is true this rejection will happen precisely with probability α.

37. **(C)** *X* is probably very close to the least squares regression line and so has a small residual. Removing *X* will change the regression line very little if at all, and so it is not an influential point. The association between the *x* and *y* variables is very strong, just not linear. Correlation measures the strength of a *linear* relationship, which is very weak regardless of whether the point *X* is present or not.

38. **(B)** The probability of an application being turned down is $1 - .90 = .10$, and the expected value of a binomial with $n = 50$ and $p = .10$ is $np = 50(.10)$.

39. **(E)** A boxplot gives a five-number summary: smallest value, 25th percentile (Q_1), median, 75th percentile (Q_3), and largest value. The interquartile range is given by $Q_3 - Q_1$, or the total length of the two "boxes" minus the "whiskers."

40. **(E)** Using a measurement from a sample, we are never able to say *exactly* what a population mean is; rather we always say we have a certain *confidence* that the population mean lies in a particular *interval.*

SECTION II

1. (a) A complete answer compares shape, center, and spread.
Shape. The control group distribution is somewhat bell-shaped and symmetric, while the treatment group distribution is somewhat skewed right.
Center. The center of the control group distribution is around 20, which is greater than the center of the treatment group distribution, which is somewhere around 10 to 12.
Spread. The spread of the control group distribution, 5 to 34, is less than the spread of the treatment group distribution, which is 2 to 41.

(b) For computers in the control group (no spam software), the number of spam e-mails received varies an "average" amount of 8.1 from the mean number of spam e-mails received in the control group.

(c) Since the 95% confidence interval for the difference does not contain zero, the researcher can conclude the observed difference in mean numbers of spam e-mails received between the control group and the treatment group that received spam software is significant.

(d) It may well be that the four groups—administrators, staff, faculty, and students—are each exposed to different kinds of spam e-mail risks, and possibly the software will be more or less of a help to each group. In that case, the researcher should in effect run four separate experiments on the homogeneous groups, called blocks. Conclusions will be more specific.

Scoring

Part (a) is essentially correct for correctly comparing shape, center, and spread. Part (a) is partially correct for correctly comparing two of the three features.

Part (b) is essentially correct if standard deviation is explained correctly in context of this problem. Part (b) is partially correct if there is a correct explanation of SD but no reference to context.

Part (c) is essentially correct for noting that zero is not in the interval so the observed difference is significant, and stating this in context of the problem. Part (c) is partially correct for noting that zero is not in the interval so the observed difference is significant, but failing to put this conclusion in context of the problem.

Part (d) is essentially correct if the purpose of blocking is correctly explained in context of this problem. Part (d) is partially correct if the general purpose of blocking is correctly explained but not in context of this problem.

Count partially correct answers as one-half an essentially correct answer.

4 Complete Answer	Four essentially correct answers.
3 Substantial Answer	Three essentially correct answers.
2 Developing Answer	Two essentially correct answers.
1 Minimal Answer	One essentially correct answer.

Use a holistic approach to decide a score totaling between two numbers.

2. (a) Listing the eight possibilities: {HHH, HHT, HTH, HTT, THH, THT, TTH, TTT} clearly shows that the only possibilities for the absolute value of the difference are 1 with probability 6/8 = .75 and 3 with probability 2/8 = .25. The table is:

Absolute difference	Probability
1	.75
3	.25

(b) $E = \sum xP(x) = 1(.75) + 3(.25) = 1.5$

(c) Since the only possible scores for each game are 1 and 3, the only way to have a total score of 3 in three games is to score 1 in each game. The probability of this is $(.75)^3 = .421875$ [or $\left(\frac{3}{4}\right)^3 = \frac{27}{64}$].

(d) The more times the game is played, the closer the average score will be to the expected value of 1.5. The player does not want to average close to 1.5, so should prefer playing 10 times rather than 15 times.

Scoring

Part (a) is essentially correct for the correct probability distribution table. Part (a) is partially correct for one minor error.

Part (b) is essentially correct for the correct calculation of expected value based on the answer given in Part (a). Part (b) is partially correct for the correct formula for expected value but an incorrect calculation based on the answer given in Part (a).

Part (c) is essentially correct for the correct probability calculation with some indication of where the answer is coming from. Part (c) is partially correct for the correct probability with no work shown.

Part (d) is essentially correct for choosing 10 and giving a clear explanation. Part (d) is partially correct for choosing 10 and giving a weak explanation. Part (d) is incorrect for choosing 10 with no explanation or with an incorrect explanation.

Count partially correct answers as one-half an essentially correct answer.

4 Complete Answer	Four essentially correct answers.
3 Substantial Answer	Three essentially correct answers.
2 Developing Answer	Two essentially correct answers.
1 Minimal Answer	One essentially correct answer.

Use a holistic approach to decide a score totaling between two numbers.

3. (a) The slope of –7.19 says that the attendance *dropped* an average of 7.19 students per week. No, this does not seem to adequately explain the data, because the attendance *increased* every week during the first 4 weeks, and again *increased* every week during the final 4 weeks.

(b) Modeling the first 4 weeks (computer games allowed) gives *Attendance* = 68 + 5.1*(Week)* with r = .997, and modeling the final 4 weeks (no computer games) gives *Attendance* = 4.9 + 4.9*(Week)* with r = .997. These models give an average increase in attendance each week of 5.1 and 4.9, respectively, as well as much higher correlations.

(c) 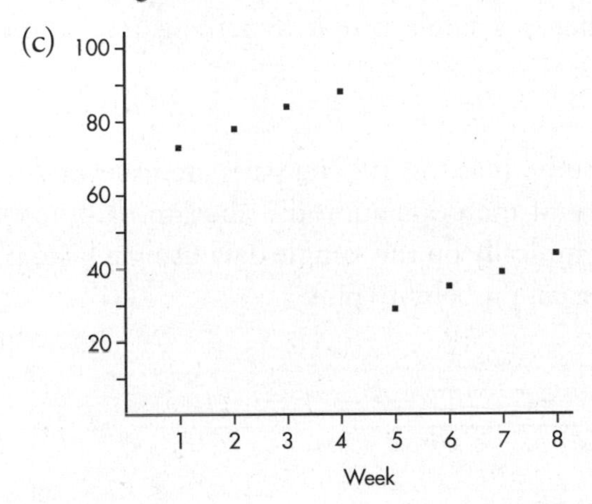

The scatterplot clearly shows the linearity in the first 4 weeks and in the final 4 weeks, and also the nonlinearity of the full set of data.

Scoring

Part (a) is essentially correct if the slope is correctly interpreted in context and a reasonable explanation is given as to why this slope does not explain the data. Part (a) is partially correct for one of these two components correct.

Part (b) is essentially correct for two separate regression models, correct interpretations of the slopes, and noting the increased correlation. Part (b) is partially correct if one of these components is missing.

Part (c) is essentially correct if the scatterplot is correctly drawn and the strong linearity in the first four weeks and separately in the final four weeks is noted. Part (c) is partially correct for a correctly drawn scatterplot but no observations on linearity.

4 Complete Answer All three parts essentially correct.

3 Substantial Answer Two parts essentially correct and one part partially correct.

2 Developing Answer Two parts essentially correct OR one part essentially correct and one or two parts partially correct OR all three parts partially correct.

1 Minimal Answer One part essentially correct OR two parts partially correct.

4. This is a paired data test, not a two-sample test, with four parts to a complete solution.

Part 1: Must state a correct pair of hypotheses.

Either $H_0 : \mu_d = 0$ and $H_a : \mu_d < 0$ where μ_d is the mean difference between the investigator and facility estimates; or

$H_0 : \mu_1 - \mu_2 = 0$ and $H_a : \mu_1 - \mu_2 < 0$ where μ_1 is the mean estimate of the investigator and μ_2 is the mean estimate of the facility.

Part 2: Must name the test and check the conditions.

This is a paired *t*-test, that is, a single sample hypothesis test on the set of differences.

Conditions:

It is reasonable to assume that the 10 data pairs are independent of each other. Normality of the population distribution of differences should be checked graphically on the sample data using a histogram, or a boxplot, or a normal probability plot:

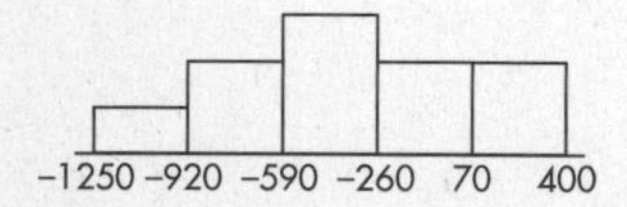

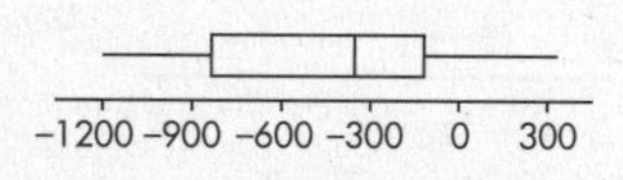

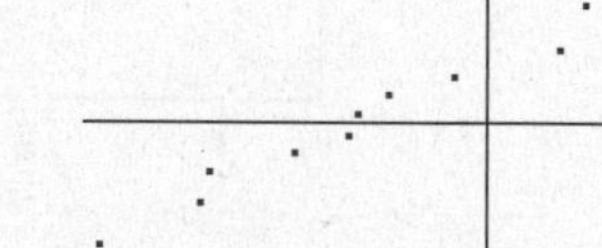

Part 3: Must find the test statistic *t* and the *P*-value.

A calculator quickly gives $t = -2.56$ and $P = .015$.

Or, with $\bar{x}_d = -400.5$ and $s_d = 494.0$ we have $t = \frac{-400.5 - 0}{\frac{494.0}{\sqrt{10}}} = -2.56$ and with $df = 9$, $P = .015$.

Part 4: Linking to the *P*-value, give a correct conclusion in context.

With this small a *P*-value (.015), there is evidence to reject H_0. That is, there is evidence that the mean estimate of the facility under suspicion is significantly greater than the mean estimate by the investigators.

Scoring

Part 1 is essentially correct for a correct statement of the hypotheses (in terms of *population* means).

Part 2 is essentially correct if the test is correctly identified by name or formula and a graphical check of the normality condition is given.

Part 3 is essentially correct for a correct calculation of both the test statistic t and the P-value.

Part 4 is essentially correct for a correct conclusion in context, linked to the P-value.

4 Complete Answer	All four parts essentially correct.
3 Substantial Answer	Three parts essentially correct.
2 Developing Answer	Two parts essentially correct.
1 Minimal Answer	One part essentially correct.

5. (a) A statistic used to estimate a population parameter is unbiased if the mean of the sampling distribution of the statistic is equal to the true value of the parameter being estimated. Estimators B, C, and D appear to have means equal to the population mean of 146.

   (b) For $n = 40$, estimator A exhibits the lowest variability, with a range of only 2 grams compared to the other ranges of 6 grams, 4 grams, 4 grams, and 4 grams.

   (c) The estimator should have a distribution centered at 146, thus eliminating A and E. As n increases, D shows tighter clustering around 146 than does B. Finally, while C looks better than D for $n = 40$, the estimator will be used with $n = 100$, and the D distribution is clearly converging as the sample size increases while the C distribution remains the same. Choose D.

Scoring

Part (a) is essentially correct for a correct answer with a good explanation of what unbiased means, and is partially correct for a correct answer with a weak explanation. Part (a) is incorrect for a correct answer with no explanation or with an incorrect explanation.

Part (b) is essentially correct for a correct answer together with some numerical justification, and is partially correct for a correct answer with a weak explanation. Part (b) is incorrect for a correct answer with no explanation or with an incorrect explanation.

Part (c) is essentially correct for a correct answer with a good explanation, and is partially correct for a correct answer with a weak explanation. Part (c) is incorrect for a correct answer with no explanation or with an incorrect explanation.

4 Complete Answer	All three parts essentially correct.
3 Substantial Answer	Two parts essentially correct and one part partially correct.
2 Developing Answer	Two parts essentially correct OR one part essentially correct and one or two parts partially correct OR all three parts partially correct.
1 Minimal Answer	One part essentially correct OR two parts partially correct.

6. (a) State the hypotheses:

$H_0 : p_H - p_L = 0$ and $H_a : p_H - p_L > 0$ where p_H is the proportion of patients receiving higher doses who experience severe nausea, and p_L is the proportion of patients receiving lower doses who experience severe nausea [other possible expressions include $H_0 : p_H = p_L$ and $H_a : p_H > p_L$].

Identify the test by name or formula and check the assumptions:

Two-sample test for proportions

$$z = \frac{\hat{p}_H - \hat{p}_L}{\sqrt{\hat{p}(1-\hat{p})\left(\frac{1}{n_H} + \frac{1}{n_L}\right)}}$$

Assumptions: Patients were randomly placed in different groups, so it is reasonable to assume independence of samples. Then we note that with $\hat{p}_H = \frac{205}{405} = .506$ and $\hat{p}_L = \frac{88}{395} = .223$, we have $n_H\hat{p}_H = 205$, $n_H(1 - \hat{p}_H) = 200$, $n_L\hat{p}_L = 88$, and $n_L(1 - \hat{p}_L) = 307$. These are all greater than 10.

Calculate the test statistic and the P-value:

$$\hat{p} = \frac{205+88}{405+395} = \frac{293}{800} = .366$$

$$z = \frac{.506-.223}{\sqrt{(.366)(.634)\left(\frac{1}{405}+\frac{1}{395}\right)}} = \frac{.283}{.034} = 8.32$$

P-value = .000

[Note: a graphing calculator will give $z = 8.31792$ with $P = 4.5279\text{E}{-17}$]

State the conclusion in context with linkage to the P-value:

With this small a P-value ($P < .001$), there is very strong evidence to reject H_0. That is, there is very strong evidence that a greater proportion of patients receiving higher doses experience severe nausea than patients receiving lower doses.

(b) Identify the confidence interval by name or formula:

95% confidence interval for the slope of the regression line $b \pm ts_b$

Check the assumptions:

The scatterplot is roughly linear, there is no apparent pattern in the residuals plot, and the distribution of the residuals is approximately normal (because the normal probability plot is roughly linear).

Calculate the confidence interval:

$df = n - 2 = 8 - 2 = 6$

$0.006659 \pm 2.447(0.001761) = 0.006659 \pm 0.004309$

$(0.00235, 0.010968)$

Interpret the confidence interval in context:

We are 95% confident that the mean proportion of patients experiencing severe nausea goes up between 0.00235 and 0.010968 with each increase of 1 milligram per week in dose intensity.

(c) The probability that a randomly chosen patient experienced severe nausea is $\frac{293}{800} = .366$, so the probability that at least 3 out of 5 experienced severe nausea is $\binom{5}{3}(.366)^3(.634)^2 + \binom{5}{4}(.366)^4(.634) + \binom{5}{5}(.366)^5 = 10(.366)^3(.634)^2 + 5(.366)^4(.634) + (.366)^5 = .261$ [Or 1 – binomcdf(5,.366,2)]

Scoring

Part (a) has four parts: 1) stating the hypotheses; 2) identifying the test and checking assumptions; 3) calculating the test statistic and the P-value; and 4) giving a conclusion in context with linkage to the P-value. Part (a) is essentially correct if three or four of these parts are correct and partially correct if one or two of these parts are correct.

Part (b) has four parts: 1) identifying the confidence interval; 2) checking assumptions; 3) calculating the confidence interval; and 4) interpreting the confidence interval in context. Part (b) is essentially correct if three or four of these parts are correct and partially correct if one or two of these parts are correct.

Part (c) is essentially correct if the correct probability is calculated and the derivation is clear. Part (c) is partially correct for indicating a binomial with $n = 5$ and $p = .366$, but then calculating incorrectly.

4 Complete Answer	All three parts essentially correct.
3 Substantial Answer	Two parts essentially correct and one part partially correct.
2 Developing Answer	Two parts essentially correct OR one part essentially correct and one or two parts partially correct OR all three parts partially correct.
1 Minimal Answer	One part essentially correct OR two parts partially correct.

Answer Sheet

PRACTICE EXAMINATION 4

1. Ⓐ Ⓑ Ⓒ Ⓓ Ⓔ	11. Ⓐ Ⓑ Ⓒ Ⓓ Ⓔ	21. Ⓐ Ⓑ Ⓒ Ⓓ Ⓔ	31. Ⓐ Ⓑ Ⓒ Ⓓ Ⓔ
2. Ⓐ Ⓑ Ⓒ Ⓓ Ⓔ	12. Ⓐ Ⓑ Ⓒ Ⓓ Ⓔ	22. Ⓐ Ⓑ Ⓒ Ⓓ Ⓔ	32. Ⓐ Ⓑ Ⓒ Ⓓ Ⓔ
3. Ⓐ Ⓑ Ⓒ Ⓓ Ⓔ	13. Ⓐ Ⓑ Ⓒ Ⓓ Ⓔ	23. Ⓐ Ⓑ Ⓒ Ⓓ Ⓔ	33. Ⓐ Ⓑ Ⓒ Ⓓ Ⓔ
4. Ⓐ Ⓑ Ⓒ Ⓓ Ⓔ	14. Ⓐ Ⓑ Ⓒ Ⓓ Ⓔ	24. Ⓐ Ⓑ Ⓒ Ⓓ Ⓔ	34. Ⓐ Ⓑ Ⓒ Ⓓ Ⓔ
5. Ⓐ Ⓑ Ⓒ Ⓓ Ⓔ	15. Ⓐ Ⓑ Ⓒ Ⓓ Ⓔ	25. Ⓐ Ⓑ Ⓒ Ⓓ Ⓔ	35. Ⓐ Ⓑ Ⓒ Ⓓ Ⓔ
6. Ⓐ Ⓑ Ⓒ Ⓓ Ⓔ	16. Ⓐ Ⓑ Ⓒ Ⓓ Ⓔ	26. Ⓐ Ⓑ Ⓒ Ⓓ Ⓔ	36. Ⓐ Ⓑ Ⓒ Ⓓ Ⓔ
7. Ⓐ Ⓑ Ⓒ Ⓓ Ⓔ	17. Ⓐ Ⓑ Ⓒ Ⓓ Ⓔ	27. Ⓐ Ⓑ Ⓒ Ⓓ Ⓔ	37. Ⓐ Ⓑ Ⓒ Ⓓ Ⓔ
8. Ⓐ Ⓑ Ⓒ Ⓓ Ⓔ	18. Ⓐ Ⓑ Ⓒ Ⓓ Ⓔ	28. Ⓐ Ⓑ Ⓒ Ⓓ Ⓔ	38. Ⓐ Ⓑ Ⓒ Ⓓ Ⓔ
9. Ⓐ Ⓑ Ⓒ Ⓓ Ⓔ	19. Ⓐ Ⓑ Ⓒ Ⓓ Ⓔ	29. Ⓐ Ⓑ Ⓒ Ⓓ Ⓔ	39. Ⓐ Ⓑ Ⓒ Ⓓ Ⓔ
10. Ⓐ Ⓑ Ⓒ Ⓓ Ⓔ	20. Ⓐ Ⓑ Ⓒ Ⓓ Ⓔ	30. Ⓐ Ⓑ Ⓒ Ⓓ Ⓔ	40. Ⓐ Ⓑ Ⓒ Ⓓ Ⓔ

Practice Examination 4

SECTION I

Questions 1–40

Spend 90 minutes on this part of the exam.

Directions: The questions or incomplete statements that follow are each followed by five suggested answers or completions. Choose the response that best answers the question or completes the statement.

1. A company wishes to determine the relationship between the number of days spent training employees and their performances on a job aptitude test. Collected data result in a least squares regression line, $\hat{y} = 12.1 + 6.2x$, where x is the number of training days and $\hat{y}$ is the predicted score on the aptitude test. Which of the following statements best interprets the slope and y-intercept of the regression line?

   (A) The base score on the test is 12.1, and for every day of training one would expect, on average, an increase of 6.2 on the aptitude test.
   (B) The base score on the test is 6.2, and for every day of training one would expect, on average, an increase of 12.1 on the aptitude test.
   (C) The mean number of training days is 12.1, and for every additional 6.2 days of training one would expect, on average, an increase of one unit on the aptitude test.
   (D) The mean number of training days is 6.2, and for every additional 12.1 days of training one would expect, on average, an increase of one unit on the aptitude test.
   (E) The mean number of training days is 12.1, and for every day of training one would expect, on average, an increase of 6.2 on the aptitude test.

GO ON TO THE NEXT PAGE ➤

2. To survey the opinions of the students at your high school, a researcher plans to select every twenty-fifth student entering the school in the morning. Assuming there are no absences, will this result in a simple random sample of students attending your school?

(A) Yes, because every student has the same chance of being selected.
(B) Yes, but only if there is a single entrance to the school.
(C) Yes, because the 24 out of every 25 students who are not selected will form a control group.
(D) Yes, because this is an example of systematic sampling, which is a special case of simple random sampling.
(E) No, because not every sample of the intended size has an equal chance of being selected.

3. Consider a hypothesis test with $H_0: \mu = 70$ and $H_a: \mu < 70$. Which of the following choices of significance level and sample size results in the greatest power of the test when $\mu = 65$?

(A) $\alpha = 0.05,\ n = 15$
(B) $\alpha = 0.01,\ n = 15$
(C) $\alpha = 0.05,\ n = 30$
(D) $\alpha = 0.01,\ n = 30$
(E) There is no way of answering without knowing the strength of the given power.

4. Suppose that 60% of a particular electronic part last over 3 years, while 70% last less than 6 years. Assuming a normal distribution, what are the mean and standard deviation with regard to length of life of these parts?

(A) $\mu = 3.677,\ \sigma = 3.561$
(B) $\mu = 3.977,\ \sigma = 3.861$
(C) $\mu = 4.177,\ \sigma = 3.561$
(D) $\mu = 4.377,\ \sigma = 3.261$
(E) The mean and standard deviation cannot be computed from the information given.

5. The graph below shows cumulative proportions plotted against numbers of employees working in mid-sized retail establishments.

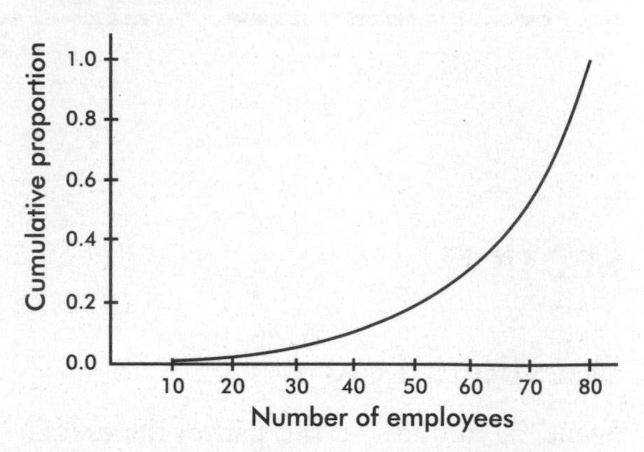

What is the approximate interquartile range?

(A) 18
(B) 35
(C) 57
(D) 68
(E) 75

6. The label on a package of cords claims that the breaking strength of a cord is 3.5 pounds, but a hardware store owner believes the real value is less. She plans to test 36 such cords; if their mean breaking strength is less than 3.25 pounds, she will reject the claim on the label. If the standard deviation for the breaking strengths of all such cords is 0.9 pounds, what is the probability of mistakenly rejecting a true claim?

(A) .05
(B) .10
(C) .15
(D) .45
(E) .94

7. What is a placebo?

(A) A method of selection
(B) An experimental treatment
(C) A control treatment
(D) A parameter
(E) A statistic

GO ON TO THE NEXT PAGE ➤

8. To study the effect of alcohol on reaction time, subjects were randomly selected and given three beers to consume. Their reaction time to a simple stimulus was measured before and after drinking the alcohol. Which of the following is a correct statement?

   (A) This study was an observational study.
   (B) Lack of blocking makes this a poorly designed study.
   (C) The placebo effect is irrelevant in this type of study.
   (D) This study was an experiment with no controls.
   (E) This study was an experiment in which the subjects were used as their own controls.

9. Suppose that the regression line for a set of data, $y = 7x + b$, passes through the point $(-2, 4)$. If $\bar{x}$ and $\bar{y}$ are the sample means of the x- and y-values, respectively, then $\bar{y} =$

   (A) $\bar{x}$
   (B) $\bar{x} + 2$
   (C) $\bar{x} - 4$
   (D) $7\bar{x}$
   (E) $7\bar{x} + 18$

10. Which of the following is a false statement about simple random samples?

    (A) A sample must be reasonably large to be properly considered a simple random sample.
    (B) Inspection of a sample will give no indication of whether or not it is a simple random sample.
    (C) Attributes of a simple random sample may be very different from attributes of the population.
    (D) Every element of the population has an equal chance of being picked.
    (E) Every sample of the desired size has an equal chance of being picked.

11. A local school has seven math teachers and seven English teachers. When comparing their mean salaries, which of the following is most appropriate?

    (A) A two-sample z-test of population means.
    (B) A two-sample t-test of population means.
    (C) A one-sample z-test on a set of differences.
    (D) A one-sample t-test on a set of differences.
    (E) None of the above are appropriate.

12. Following is a histogram of ages of people applying for a particular high school teaching position.

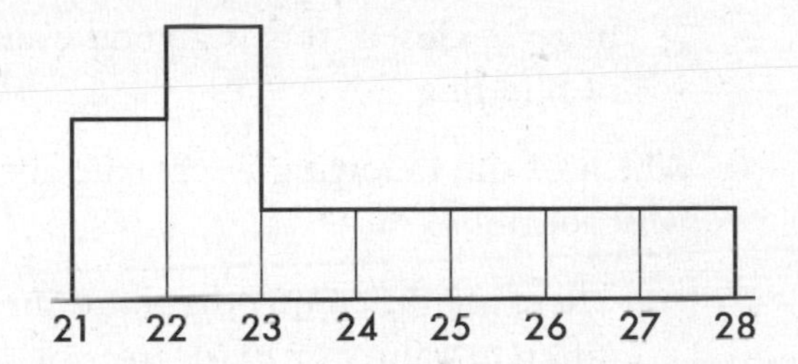

    Which of the following is a correct statement?

    (A) The median age is between 24 and 25.
    (B) The mean age is between 22 and 23.
    (C) The mean age is greater than the median age.
    (D) More applicants are under 23 years of age than are over 23.
    (E) There are a total of 10 applicants.

GO ON TO THE NEXT PAGE ➤

13. To conduct a survey of which long distance carriers are used in a particular locality, a researcher opens a telephone book to a random page, closes his eyes, puts his finger down on the page, and then calls the next 75 names. Which of the following is a correct statement?

(A) The procedure results in a simple random sample.
(B) While the survey design does incorporate chance, the procedure could easily result in selection bias.
(C) This is an example of cluster sampling with 26 clusters.
(D) This is an example of stratified sampling with 26 strata.
(E) Given that the researcher truly keeps his eyes closed, this is a good example of blinding.

14. Which of the following is *not* true about *t*-distributions?

(A) There are different *t*-distributions for different values of *df* (degrees of freedom).
(B) *t*-distributions are bell-shaped and symmetric.
(C) *t*-distributions always have mean 0 and standard deviation 1.
(D) *t*-distributions are more spread out than the normal distribution.
(E) The larger the *df* value, the closer the distribution is to the normal distribution.

15. Suppose the probability that a person picked at random has lung cancer is .035 and the probability that the person both has lung cancer and is a heavy smoker is .014. Given that someone picked at random has lung cancer, what is the probability that the person is a heavy smoker?

(A) .035 – .014
(B) .035 + .014
(C) .035 + .014 – (.035)(.014)
(D) $\frac{.035}{1-.014}$
(E) $\frac{.014}{.035}$

16. Taxicabs in a metropolitan area are driven an average of 75,000 miles per year with a standard deviation of 12,000 miles. What is the probability that a randomly selected cab has been driven less than 100,000 miles if it is known that it has been driven over 80,000 miles? Assume a normal distribution of miles per year among cabs.

(A) .06
(B) .34
(C) .66
(D) .68
(E) .94

17. A plant manager wishes to determine the difference in number of accidents per day between two departments. How many days' records should be examined to be 90% certain of the difference in daily averages to within 0.25 accidents per day? Assume standard deviations of 0.8 and 0.5 accidents per day in the two departments, respectively.

(A) 39
(B) 55
(C) 78
(D) 109
(E) 155

18. Suppose the correlation between two variables is $r = .19$. What is the new correlation if .23 is added to all values of the *x*-variable, every value of the *y*-variable is doubled, and the two variables are interchanged?

(A) .19
(B) .42
(C) .84
(D) –.19
(E) –.84

GO ON TO THE NEXT PAGE ➤

19. Two commercial flights per day are made from a small county airport. The airport manager tabulates the number of on-time departures for a sample of 200 days.

Number of on-time departures	0	1	2
Observed number of days	10	80	110

What is the χ^2 statistic for a goodness-of-fit test that the distribution is binomial with probability equal to .8 that a flight leaves on time?

(A) $\frac{(10-8)^2}{8}+\frac{(80-64)^2}{64}+\frac{(110-128)^2}{128}$

(B) $\frac{(10-8)^2}{10}+\frac{(80-64)^2}{80}+\frac{(110-128)^2}{110}$

(C) $\frac{(10-10)^2}{10}+\frac{(80-30)^2}{30}+\frac{(110-160)^2}{160}$

(D) $\frac{(10-10)^2}{10}+\frac{(80-30)^2}{80}+\frac{(110-160)^2}{110}$

(E) $\frac{(10-66)^2}{10}+\frac{(80-67)^2}{80}+\frac{(110-67)^2}{110}$

20. A company has a choice of three investment schemes. Option I gives a sure \$25,000 return on investment. Option II gives a 50% chance of returning \$50,000 and a 50% chance of returning \$10,000. Option III gives a 5% chance of returning \$100,000 and a 95% chance of returning nothing. Which option should the company choose?

(A) Option II if it wants to maximize expected return
(B) Option I if it needs at least \$20,000 to pay off an overdue loan
(C) Option III if it needs at least \$80,000 to pay off an overdue loan
(D) All of the above answers are correct.
(E) Because of chance, it really doesn't matter which option it chooses.

21. Suppose $P(X) = .35$ and $P(Y) = .40$. If $P(X|Y) = .28$, what is $P(Y|X)$?

(A) $\frac{(.28)(.35)}{.40}$

(B) $\frac{(.28)(.40)}{.35}$

(C) $\frac{(.35)(.40)}{.28}$

(D) $\frac{(.28)}{.40}$

(E) $\frac{(.28)}{.35}$

22. To test whether extensive exercise lowers the resting heart rate, a study is performed by randomly selecting half of a group of volunteers to exercise 1 hour each morning, while the rest are instructed to perform no exercise. Is this study an experiment or an observational study?

(A) An experiment with a control group and blinding
(B) An experiment with blocking
(C) An observational study with comparison and randomization
(D) An observational study with little if any bias
(E) None of the above

23. The waiting times for a new roller coaster ride are normally distributed with a mean of 35 minutes and a standard deviation of 10 minutes. If there are 150,000 riders the first summer, which of the following is the shortest time interval associated with 100,000 riders?

(A) 0 to 31.7 minutes
(B) 31.7 to 39.3 minutes
(C) 25.3 to 44.7 minutes
(D) 25.3 to 35 minutes
(E) 39.3 to 95 minutes

GO ON TO THE NEXT PAGE ➤

24. A medical researcher, studying the effective durations of three over-the-counter pain relievers, obtains the following boxplots from three equal-sized groups of patients, one group using each of the pain relievers.

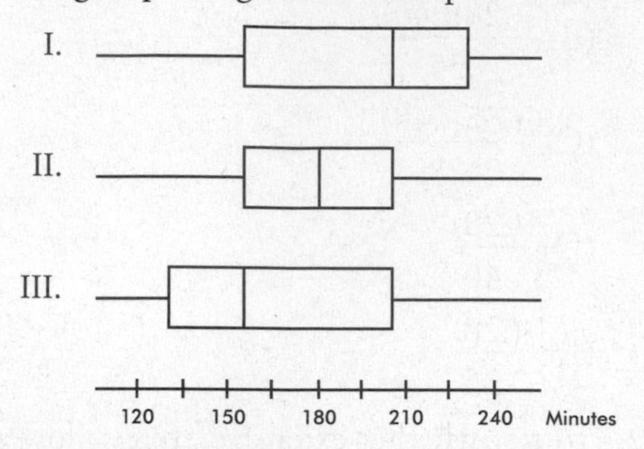

Which of the following is a correct statement with regard to comparing the effective durations in minutes of the three pain relievers?

(A) All three have the same interquartile range.
(B) More patients had over 210 minutes of pain relief in the (I) group than in either of the other two groups.
(C) More patients had over 240 minutes of pain relief in the (I) group than in either of the other two groups.
(D) More patients had less than 120 minutes of pain relief in the (III) group than in either of the other two groups.
(E) The durations of pain relief in the (II) group form a normal distribution.

25. For their first exam, students in an AP Statistics class studied an average of 4 hours with a standard deviation of 1 hour. Almost everyone did poorly on the exam, and so for the second exam every student studied 10 hours. What is the correlation between the numbers of hours students studied for each exam?

(A) −1
(B) 0
(C) .4
(D) 1
(E) There is not sufficient information to answer this question.

26. Suppose that 54% of the graduates from your high school go on to 4-year colleges, 20% go on to 2-year colleges, 19% find employment, and the remaining 7% search for a job. If a randomly selected student is not going on to a 2-year college, what is the probability she will be going on to a 4-year college?

(A) .460
(B) .540
(C) .630
(D) .675
(E) .730

27. We are interested in the proportion p of people who are unemployed in a large city. Eight percent of a simple random sample of 500 people are unemployed. What is the midpoint for a 95% confidence interval estimate of p?

(A) .012
(B) .025
(C) .475
(D) p
(E) None of the above.

28. Random samples of size n are drawn from a population. The mean of each sample is calculated, and the standard deviation of this set of sample means is found. Then the procedure is repeated, this time with samples of size $4n$. How does the standard deviation of the second group compare with the standard deviation of the first group?

(A) It will be the same.
(B) It will be twice as large.
(C) It will be four times as large.
(D) It will be half as large.
(E) It will be one-quarter as large.

GO ON TO THE NEXT PAGE ➤

29. Leech therapy is used in traditional medicine for treating localized pain. In a double blind experiment on 50 patients with osteoarthritis of the knee, half are randomly selected to receive injections of leech saliva while the rest receive a placebo. Pain levels 7 days later among those receiving the saliva show a mean of 19.5, while pain levels among those receiving the placebo show a mean of 25.6 (higher numbers indicate more pain). Partial calculator output is shown below.

```
2-SampTTest
 μ1<μ2
 t=-3.939503313
 df=43.43159286
 x̄1=19.5
 x̄2=25.6
 Sx1=4.5
 Sx2=6.3
 n1=25
 n2=25
```

Which of the following is a correct conclusion?

(A) After 7 days, the mean pain level with the leech treatment is significantly lower than the mean pain level with the placebo at the 0.01 significance level.
(B) After 7 days, the mean pain level with the leech treatment is significantly lower than the mean pain level with the placebo at the 0.05 significance level, but not at the 0.01 level.
(C) After 7 days, the mean pain level with the leech treatment is significantly lower than the mean pain level with the placebo at the 0.10 significance level, but not at the 0.05 level.
(D) After 7 days, the mean pain level with the leech treatment is not significantly lower than the mean pain level with the placebo at the 0.10 significance level.
(E) The proper test should be a one-sample t-test on a set of differences.

30. A survey was conducted to determine the percentage of parents who would support raising the legal driving age to 18. The results were stated as 67% with a margin of error of ±3%. What is meant by ±3%?

(A) Three percent of the population were not surveyed.
(B) In the sample, the percentage of parents who would support raising the driving age is between 64% and 70%.
(C) The percentage of the entire population of parents who would support raising the driving age is between 64% and 70%.
(D) It is unlikely that the given sample proportion result could be obtained unless the true percentage was between 64% and 70%.
(E) Between 64% and 70% of the population were surveyed.

31. It is estimated that 30% of all cars parked in a metered lot outside City Hall receive tickets for meter violations. In a random sample of 5 cars parked in this lot, what is the probability that at least one receives a parking ticket?

(A) $1 - (.3)^5$
(B) $1 - (.7)^5$
(C) $5(.3)(.7)^4$
(D) $5(.3)^4 (.7)$
(E) $5(.3)^4(.7) + 10(.3)^3(.7)^2 + 10(.3)^2(.7)^3 + 5(.3)(.7)^4 + (.7)^5$

32. Data are collected on income levels x versus number of bank accounts y. Summary calculations give $\bar{x} = 32{,}000$; $s_x = 11{,}500$; $\bar{y} = 1.7$; $s_y = 0.4$, and $r = .42$. What is the slope of the least squares regression line of number of bank accounts on income level?

(A) 0.000015
(B) 0.000022
(C) 0.000035
(D) 0.000053
(E) 0.000083

GO ON TO THE NEXT PAGE ➤

33. Given that the sample has a standard deviation of zero, which of the following is a true statement?

    (A) The standard deviation of the population is also zero.
    (B) The sample mean and sample median are equal.
    (C) The sample may have outliers.
    (D) The population has a symmetric distribution.
    (E) All samples from the same population will also have a standard deviation of zero.

34. In one study half of a class were instructed to watch exactly 1 hour of television per day, the other half were told to watch 5 hours per day, and then their class grades were compared. In a second study students in a class responded to a questionnaire asking about their television usage and their class grades.

    (A) The first study was an experiment without a control group, while the second was an observational study.
    (B) The first study was an observational study, while the second was a controlled experiment.
    (C) Both studies were controlled experiments.
    (D) Both studies were observational studies.
    (E) Each study was part controlled experiment and part observational study.

35. All of the following statements are true for all discrete random variables except for which one?

    (A) The possible outcomes must all be numerical.
    (B) The possible outcomes must be mutually exclusive.
    (C) The mean (expected value) always equals the sum of the products obtained by multiplying each value by its corresponding probability.
    (D) The standard deviation of a random variable can never be negative.
    (E) Approximately 95% of the outcomes will be within two standard deviations of the mean.

36. In leaving for school on an overcast April morning you make a judgment on the null hypothesis: The weather will remain dry. What would the results be of Type I and Type II errors?

    (A) Type I error: get drenched
    Type II error: needlessly carry around an umbrella
    (B) Type I error: needlessly carry around an umbrella
    Type II error: get drenched
    (C) Type I error: carry an umbrella, and it rains
    Type II error: carry no umbrella, but weather remains dry
    (D) Type I error: get drenched
    Type II error: carry no umbrella, but weather remains dry
    (E) Type I error: get drenched
    Type II error: carry an umbrella, and it rains

37. The mean thrust of a certain model jet engine is 9500 pounds. Concerned that a production process change might have lowered the thrust, an inspector tests a sample of units, calculating a mean of 9350 pounds with a *z*-score of −2.46 and a *P*-value of .0069. Which of the following is the most reasonable conclusion?

    (A) 99.31% of the engines produced under the new process will have a thrust under 9350 pounds.
    (B) 99.31% of the engines produced under the new process will have a thrust under 9500 pounds.
    (C) 0.69% of the time an engine produced under the new process will have a thrust over 9500 pounds.
    (D) There is evidence to conclude that the new process is producing engines with a mean thrust under 9350 pounds.
    (E) There is evidence to conclude that the new process is producing engines with a mean thrust under 9500 pounds.

GO ON TO THE NEXT PAGE ➤

38. In a group of 10 third graders, the mean height is 50 inches with a median of 47 inches, while in a group of 12 fourth graders, the mean height is 54 inches with a median of 49 inches. What is the median height of the combined group?

    (A) 48 inches
    (B) 52 inches
    (C) $\frac{10(47)+12(49)}{22}$ inches
    (D) $\frac{10(50)+12(54)}{22}$ inches
    (E) The median of the combined group cannot be determined from the given information.

39. A reading specialist in a large public school system believes that the more time students spend reading, the better they will do in school. She plans a middle school experiment in which an SRS of 30 eighth graders will be assigned four extra hours of reading per week, an SRS of 30 seventh graders will be assigned two extra hours of reading per week, and an SRS of 30 sixth graders with no extra assigned reading will be a control group. After one school year, the mean GPAs from each group will be compared. Is this a good experimental design?

    (A) Yes
    (B) No, because while this design may point out an association between reading and GPA, it cannot establish a cause-and-effect relationship.
    (C) No, because without blinding, there is a strong chance of a placebo effect.
    (D) No, because any conclusion would be flawed because of blocking bias.
    (E) No, because grade level is a lurking variable which may well be confounded with the variables under consideration.

40. A study at 35 large city high schools gives the following back-to-back stemplot of the percentages of students who say they have tried alcohol.

School Year 1995–96		**School Year 1998–99**
	0	
2	1	
9	2	
6 6 3	3	1 8
4 3 1 1	4	0 2 9
9 9 8 6 5 3 2 2 0	5	3 3 4 6
9 8 7 6 6 5 1 1	6	1 2 2 2 7
5 4 3 2 2	7	0 1 3 3 5 5 6 7 8 9
5 4 0	8	2 3 4 5 8 8 9
0	9	0 1 1 2

Which of the following does *not* follow from the above data?

(A) In general, the percentage of students trying alcohol seems to have increased from 1995–96 to 1998–99.
(B) The median alcohol percentage among the 35 schools increased from 1995–96 to 1998–99.
(C) The spread between the lowest and highest alcohol percentages decreased from 1995–96 to 1998–99.
(D) For both school years in most of the 35 schools, most of the students said they had tried alcohol.
(E) The percentage of students trying alcohol increased in each of the schools between 1995–96 and 1998–99.

STOP

If there is still time remaining, you may review your answers.

SECTION II

PART A

Questions 1–5

Spend about 65 minutes on this part of the exam.
Percentage of Section II grade—75

You must show all work and indicate the methods you use. You will be graded on the correctness of your methods and on the accuracy of your results and explanations.

1. A student is interested in estimating the average length of words in a 600-page textbook, and plans the following three-stage sampling procedure:
   (1) Noting that each of the 12 chapters has a different author, the student decides to obtain a sample of words from each chapter.
   (2) Each chapter is approximately 50 pages long. The student uses a random number generator to pick three pages from each chapter.
   (3) On each chosen page the student notes the length of every tenth word.
   (a) The first stage above represents what kind of sampling procedure? Give an advantage in using it in this context.
   (b) The second stage above represents what kind of sampling procedure? Give an advantage in using it in this context.
   (c) The third stage above represents what kind of sampling procedure? Give a *disadvantage* to using it dependent upon an author's writing style.

2. (a) Suppose that nationwide, 69% of all registered voters would answer "Yes" to the question "Do you consider yourself highly focused on this year's presidential election?" Which of the following is more likely: an SRS of 50 registered voters having over 75% answer "Yes," or an SRS of 100 registered voters having over 75% answer "Yes" to the given question? Explain.

   (b) A particular company with 95 employees wishes to survey their employees with regard to interest in the presidential election. They pick an SRS of 30 employees and ask "Do you consider yourself highly focused on this year's presidential election?" Suppose that in fact 60% of all 95 employees would have answered "Yes." Explain why it is not reasonable to say that the distribution for the count in the sample who say "Yes" is a binomial with $n = 30$ and $p = .6$.

GO ON TO THE NEXT PAGE ➤

(c) Suppose that nationwide, 78% of all registered voters would answer "Yes" to the question "Do you consider yourself highly focused on this year's presidential election?" You plan to interview an SRS of 20 registered voters. Explain why it is not reasonable to say that the distribution for the proportion in the sample who say "Yes" is approximately a normal distribution.

3. An instructor takes an anonymous survey and notes exam score, hours studied, and gender for the first exam in a large college statistics class ($n = 250$). A resulting regression model is:

$$Score = 50.90 + 9.45(Hours) + 4.40(Gender)$$

where *Gender* takes the value 0 for males and 1 for females.

(a) Provide an interpretation in context for each of the three numbers appearing in the above model formula.

(b) Sketch the separate prediction lines for males and females resulting from using 0 or 1 in the above model.

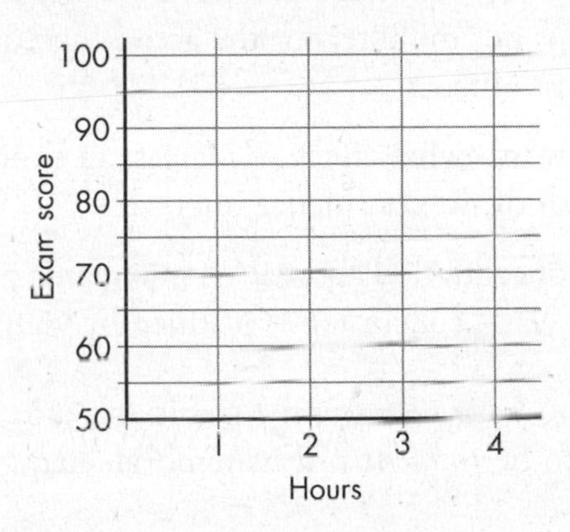

Looking at the data separately by gender results in the following two regression models:

For males: $Score = 51.8 + 8.5(Hours)$
For females: $Score = 54.4 + 10.4(Hours)$

(c) What comparative information do the slope coefficients from these two models give that does not show in the original model?

GO ON TO THE NEXT PAGE ➤

(d) Sketch the prediction lines given by these last two models.

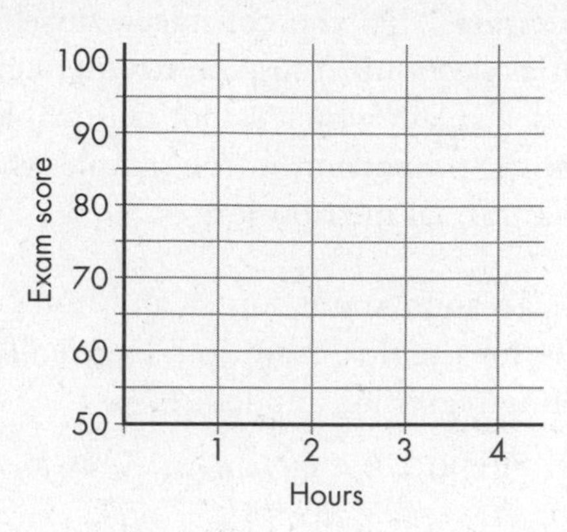

4. A laboratory is testing the concentration level in milligrams per milliliter for the active ingredient found in a pharmaceutical product. In a random sample of five vials of the product, the concentrations were measured at 2.46, 2.57, 2.70, 2.64, and 2.54 mg/ml.

   (a) Determine a 95% confidence interval for the mean concentration level in milligrams per milliliter for the active ingredient found in this pharmaceutical product.

   (b) Explain in words what effect an increase in confidence level would have on the width of the confidence interval.

   (c) Suppose a concentration above 2.70 milligrams per milliliter is considered dangerous. What conclusion is justified by your answers to (a) and (b)?

5. A point is said to have *high leverage* if it is an outlier in the x-direction, and a point is said to be *influential* if its removal sharply changes the regression line.

   (a) In the scatterplots below, compare points A and B with regard to having high leverage and with regard to being influential.

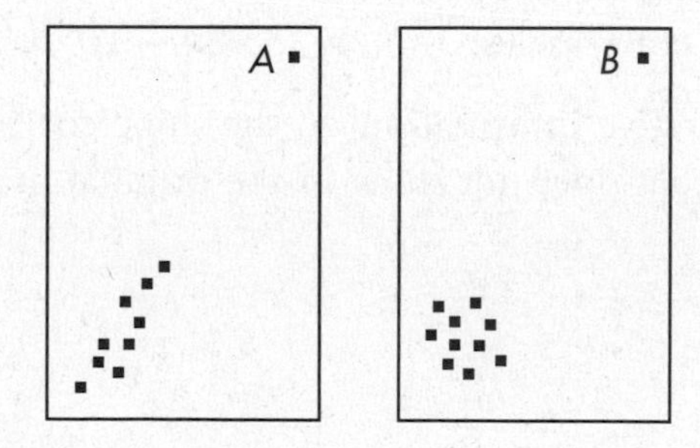

GO ON TO THE NEXT PAGE ➤

(b) In the scatterplots below, compare points *C* and *D* with regard to residuals and influence on β_1, the slope of the regression line.

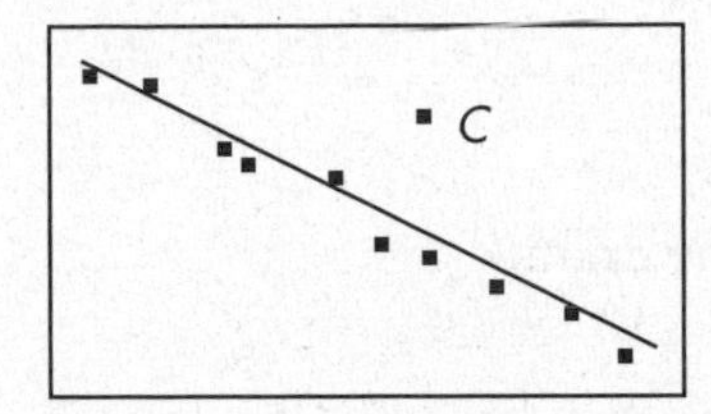

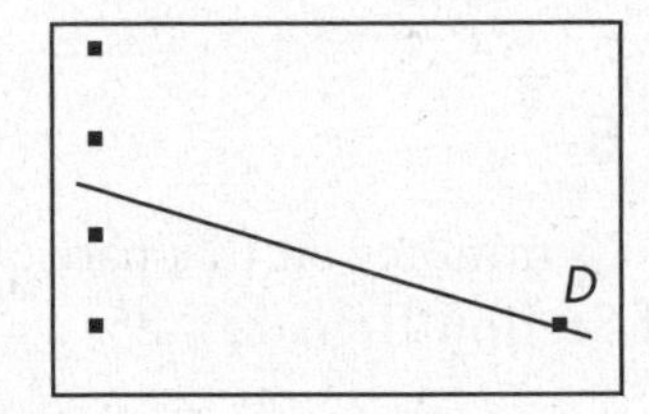

(c) In the scatterplot below, compare the effect of removing point *E* or point *F* to that of removing both *E* and *F*.

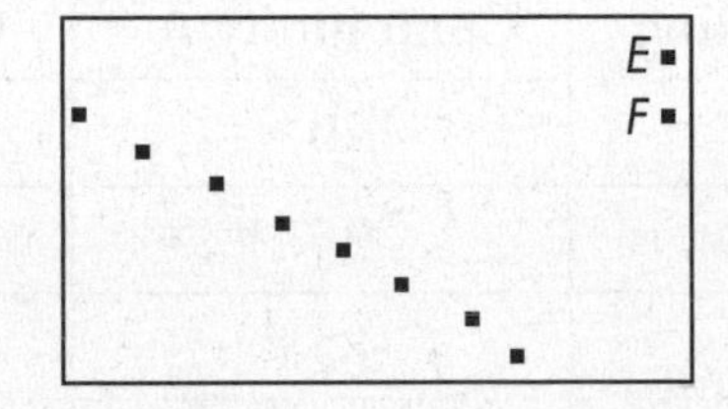

(d) In the scatterplot below, compare the effect of removing point *G* or point *H* to that of removing both *G* and *H*.

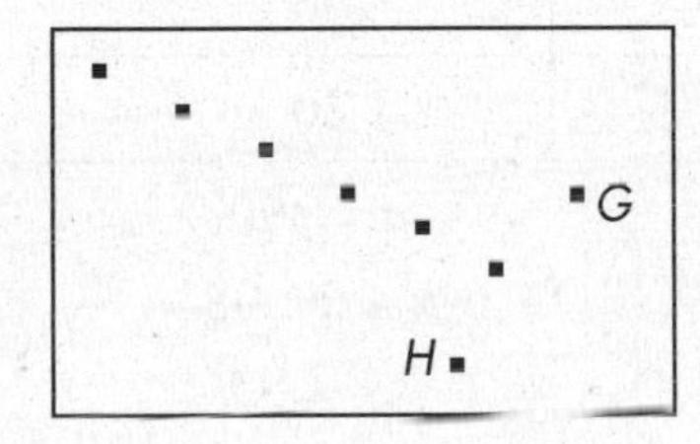

GO ON TO THE NEXT PAGE ➤

SECTION II

PART B

Question 6

Spend about 25 minutes on this part of the exam.
Percentage of Section II grade—25

6. In two random samples of adults, one from each of two communities, counts were made of the number of people with different years of schooling and tabulated as follows:

Years of schooling	Community A	Community B
8	20	15
10	35	40
12	85	40
14	100	20
16	90	40
18	40	55
20	20	25
	$n = 390$	$n = 235$
	$\bar{x} = 14.08$	$\bar{x} = 14.51$
	$s = 2.93$	$s = 3.71$

(a) Is there evidence of a difference in the mean years of schooling between the two communities?

(b) Is there evidence of a difference in the distribution of years of schooling in the two communities?

(c) Use histograms to help explain the above results.

STOP

If there is still time remaining, you may review your answers.

Answer Key

Section I

1. **A**	9. **E**	17. **A**	25. **B**	33. **B**
2. **E**	10. **A**	18. **A**	26. **D**	34. **A**
3. **C**	11. **E**	19. **A**	27. **E**	35. **E**
4. **B**	12. **C**	20. **D**	28. **D**	36. **B**
5. **A**	13. **B**	21. **B**	29. **A**	37. **E**
6. **A**	14. **C**	22. **E**	30. **D**	38. **E**
7. **C**	15. **E**	23. **C**	31. **B**	39. **E**
8. **E**	16. **E**	24. **B**	32. **A**	40. **E**

Answers Explained

Section I

1. **(A)** The slope, 6.2, gives the predicted increase in the *y*-variable for each unit increase in the *x*-variable.

2. **(E)** For a simple random sample, every possible group of the given size has to be equally likely to be selected, and this is not true here. For example, with this procedure it will be impossible for all the early arrivals to be together in the final sample. This procedure is an example of systematic sampling, but systematic sampling does not result in simple random samples.

3. **(C)** Power = $1 - \beta$, and β is smallest when α is more and n is more.

4. **(B)** The critical *z*-scores for 60% to the right and 70% to the left are −0.253 and 0.524, respectively. Then $\{\mu - 0.253\sigma = 3, \mu + 0.524\sigma = 6\}$ gives $\mu = 3.977$ and $\sigma = 3.861$.

5. **(A)** The cumulative proportions of 0.25 and 0.75 correspond to $Q_1 = 57$ and $Q_3 = 75$, respectively, and so the interquartile range is $75 - 57 = 18$.

6. **(A)** We have H_0: $\mu = 3.5$ and H_a: $\mu < 3.5$. Then $\sigma_{\bar{x}} = \frac{0.9}{\sqrt{36}} = 0.15$, the *z*-score of 3.25 is $\frac{3.25-3.5}{0.15} = -1.67$, and Table A gives .0475. (A *t*-test on the TI-83 gives .0522.)

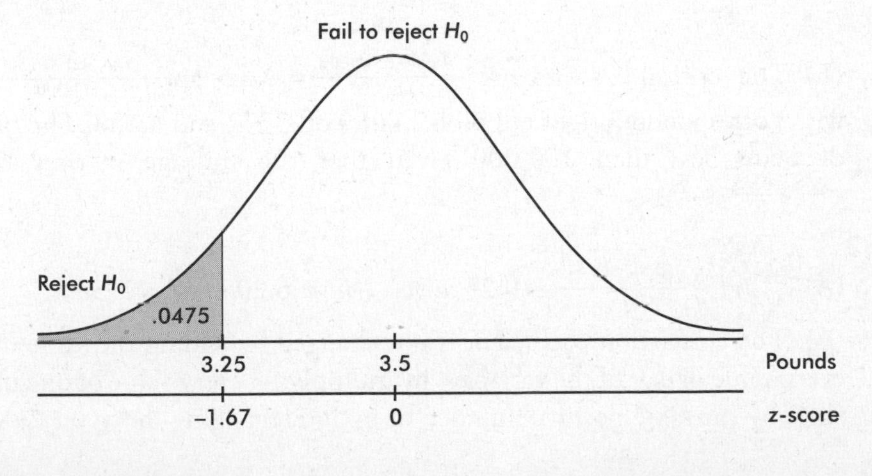

7. **(C)** A placebo is a control treatment in which members of the control group do not realize whether or not they are receiving the experimental treatment.

8. **(E)** In experiments on people, subjects can be used as their own controls, with responses noted before and after the treatment. However, with such designs there is always the danger of a placebo effect. In this case, subjects might well have slower reaction times after drinking the alcohol because they think they should. Thus the design of choice would involve a separate control group to use for comparison. Blocking is not necessary for a well-designed experiment, and there is no indication that it would be useful here.

9. **(E)** Since $(-2, 4)$ is on the line $y = 7x + b$, we have $4 = -14 + b$ and $b = 18$. Thus the regression line is $y = 7x + 18$. The point $(\bar{x}, \bar{y})$ is always on the regression line, and so we have $\bar{y} = 7\bar{x} + 18$.

10. **(A)** A simple random sample can be any size.

11. **(E)** With such small populations, censuses instead of samples are used, and there is no resulting probability statement about the difference.

12. **(C)** Half the area is on either side of 23, so 23 is the median. The distribution is skewed to the right, and so the mean is greater than the median. With half the area to each side of 23, half the applicants' ages are to each side of 23. Histograms such as this show relative frequencies, not actual frequencies.

13. **(B)** While the procedure does use some element of chance, all possible groups of size 75 do not have the same chance of being picked, so the result is not a simple random sample. There is a real chance of selection bias. For example, a number of relatives with the same name and all using the same long-distance carrier might be selected.

14. **(C)** While *t*-distributions do have mean 0, their standard deviations are greater than 1.

15. **(E)**

$$P(\text{smoker}|\text{cancer}) = \frac{P(\text{smoker} \cap \text{cancer})}{P(\text{cancer})} = \frac{.014}{.035} = .4$$

16. **(E)** The critical *z*-scores are $\frac{80{,}000 - 75{,}000}{12{,}000} = 0.42$ and $\frac{100{,}000 - 75{,}000}{12{,}000} = 2.08$, with corresponding right tail probabilities of .3372 and .0188. The probability of being less than 100,000 given that the mileage is over 80,000 is $\frac{.3372 - .0188}{.3372} = .94$.

17. **(A)** $(1.645)\frac{\sqrt{(0.8)^2 + (0.5)^2}}{\sqrt{n}} \leq 0.25$ gives $\sqrt{n} \geq 6.208$ and $n \geq 38.5$.

18. **(A)** The correlation coefficient is not changed by adding the same number to every value of one of the variables, by multiplying every value of one of the variables by the same positive number, or by interchanging the *x*- and *y*-variables.

19. **(A)** The binomial distribution with $n = 2$ and $p = .8$ is $P(0) = (.2)^2 = .04$, $P(1) = 2(.2)(.8) = .32$, and $P(2) = (.8)^2 = .64$, resulting in expected numbers of $.04(200) = 8$, $.32(200) = 64$, and $.64(200) = 128$. Thus,

$$\chi^2 = \sum \frac{(obs - exp)^2}{exp} = \frac{(10-8)^2}{8} + \frac{(80-64)^2}{64} + \frac{(110-128)^2}{128}$$

20. **(D)** Option II gives the highest expected return: $(50{,}000)(.5) + (10{,}000)(.5) = 30{,}000$, which is greater than 25,000 and is also greater than $(100{,}000)(.05) = 5000$. Option I guarantees that the \$20,000 loan will be paid off. Option III provides the only chance of paying off the \$80,000 loan. The moral is that the highest expected value is not automatically the "best" answer.

21. **(B)** $P(X \cap Y) = P(X|Y)P(Y) = (.28)(.40) = .112$. Then $P(Y|X) = \frac{P(X \cap Y)}{P(X)} = \frac{.112}{.35} = .32$.

22. **(E)** This study is an experiment because a treatment (extensive exercise) is imposed. There is no blinding because subjects clearly know whether or not they are exercising. There is no blocking because subjects are not divided into blocks before random assignment to treatments. For example, blocking would have been used if subjects had been separated by gender or age before random assignment to exercise or not.

23. **(C)** From the shape of the normal curve, the answer is in the middle. The middle two-thirds is between z-scores of ± 0.97, and $35 \pm 0.97(10)$ gives (25.3, 44.7).

24. **(B)** The interquartile range is the length of the box, so they are not all equal. More than 25% of the patients in the A group had over 210 minutes of pain relief, which is not the case for the other two groups. There is no way to positively conclude a normal distribution from a boxplot.

25. **(B)** A scatterplot would be horizontal; the correlation is zero.

26. **(D)** $\frac{.54}{.54+.19+.07} = .675$

27. **(E)** The midpoint of the confidence interval is .08.

28. **(D)** While the sample proportion is between 64% and 70% (more specifically, it is 67%), this is not the meaning of ±3%. While the percentage of the entire population is likely to be between 64% and 70%, this is not known for certain.

29. **(A)** With $df = 43.43$ and $t = -3.94$, the P-value is $.000146 < .01$. [On the TI-84, use `tcdf`.]

30. **(D)** Increasing the sample size by a multiple of d^2 divides the standard deviation of the set of sample means by d.

31. **(B)** In this binomial situation, the probability that a car does not receive a ticket is 1 – 3 = .7, the probability that none of the five cars receives a ticket is $(.7)^5$, and thus the probability that at least one receives a ticket is $1 - (.7)^5$.

32. **(A)** $b_1 = r\frac{s_y}{s_x} = .42\frac{1.7}{11,500} = 0.000015$

33. **(B)** If the standard deviation of a set is zero, all the values in the set are equal. The mean and median would both equal this common value and so would equal each other. If all the values are equal, there are no outliers. Just because the sample happens to have one common value, there is no reason for this to be true for the whole population. Statistics from one sample can be different from statistics from any other sample.

34. **(A)** The first study is an experiment with two treatment groups and no control group. The second study is observational; the researcher did not randomly divide the subjects into groups and have each group watch a designated number of hours of television per night.

35. **(E)** This refers only to very particular random variables, for example, random variables whose values are the numbers of successes in a binomial probability distribution with large *n*.

36. **(B)** A Type I error means that the null hypothesis is correct (the weather will remain dry), but you reject it (thus you needlessly carry around an umbrella). A Type II error means that the null hypothesis is wrong (it will rain), but you fail to reject it (thus you get drenched).

37. **(E)** If the sample statistic is far enough away from the claimed population parameter, we say that there is sufficient evidence to reject the null hypothesis. In this case the null hypothesis is that $\mu = 9500$. The *P*-value is the probability of obtaining a sample statistic as extreme as the one obtained if the null hypothesis is assumed to be true. The smaller the *P*-value, the more significant the difference between the null hypothesis and the sample results. With $P = .0069$, there is strong evidence to reject H_0.

38. **(E)** There are 10 + 12 = 22 students in the combined group. In ascending order, where are the two middle scores? At least 5 third graders and 6 fourth graders have heights less than or equal to 49 inches, so at most 11 students have heights greater than or equal to 49 and thus the median is less than or equal to 49. At least 5 third graders and 6 fourth graders have heights greater than or equal to 47 inches, so at most 11 students have heights less than or equal to 47 and thus the median is greater than or equal to 47. All that can be said about the median of the combined group is that it is between 47 and 49 inches.

39. **(E)** Good experimental design aims to give each group the same experiences except for the treatment under consideration, Thus, all three SRSs should be picked from the same grade level.

40. **(E)** The stemplot does not indicate what happened for any individual school.

SECTION II

1. (a) This is an example of *stratified sampling,* where the chapters are strata. The advantage is that the student is ensuring that the final sample will represent the 12 different authors, who may well use different average word lengths.
   (b) This is an example of *cluster sampling,* where for each chapter the three chosen pages are clusters. It is reasonable to assume that each page (cluster) resembles the author's overall pattern. The advantage is that using these clusters is much more practical than trying to sample from among all an author's words.
   (c) This is an example of *systematic sampling,* which is quicker and easier than many other procedures. A possible disadvantage is that if ordering is related to the variable under consideration, this procedure will likely result in an unrepresentative sample. For example, in this study if an author's word length is related to word order in sentences, the student could end up with words of particular lengths.

Scoring

Part (a) is essentially correct for identifying the procedure and giving a correct advantage. Part (a) is partially correct for one of these two elements.

Part (b) is essentially correct for identifying the procedure and giving a correct advantage. Part (b) is partially correct for one of these two elements.

Part (c) is essentially correct for identifying the procedure and giving a correct disadvantage. Part (c) is partially correct for one of these two elements.

4 Complete Answer	All three parts essentially correct.
3 Substantial Answer	Two parts essentially correct and one part partially correct.
2 Developing Answer	Two parts essentially correct OR one part essentially correct and one or two parts partially correct OR all three parts partially correct.
1 Minimal Answer	One part essentially correct OR two parts partially correct.

2. (a) The SRS with $n = 50$ is more likely to have a sample proportion greater than 75%. In each case, the sampling distribution of $\hat{p}$ is approximately normal with a mean of .69 and a standard deviation of $\sigma_{\hat{p}} = \sqrt{\frac{pq}{n}} = \sqrt{\frac{(.69)(.31)}{n}}$ $= \frac{.462}{\sqrt{n}}$. Thus the sampling distribution with $n = 50$ will have more variability than the sampling distribution with $n = 100$. Thus the tail area ($\hat{p} > .75$) will be larger for $n = 50$.

(b) The size of the sample, 30, is much too large compared to the size of the population, 95. With a sample size this close to the population size, the necessary assumption of independence does not follow. A commonly accepted condition is that the sample should be no more than 10% of the population.

(c) A normal distribution may be used to approximate a binomial distribution only if the sample size n is not too small. The commonly used condition check is that both np and nq are at least 10. In this case $nq = (20)(1 - .78) = 4.4$.

Scoring

Part (a) is essentially correct for correctly giving $n = 50$ and linking in context to variability in the sampling distributions. Part (a) is partially correct for a correct answer missing comparison of variability in the sampling distributions.

Part (b) is essentially correct for a clear explanation in context. Part (b) is partially correct for saying the sample size is too close to the population size but not linking to this context (sample size 30 and population size 95).

Part (c) is essentially correct for a clear explanation in context. Part (c) is partially correct for saying the sample size is too small but not linking to this context ($nq = 4.4 < 10$).

4 Complete Answer	All three parts essentially correct.
3 Substantial Answer	Two parts essentially correct and one part partially correct.
2 Developing Answer	Two parts essentially correct OR one part essentially correct and one or two parts partially correct OR all three parts partially correct.
1 Minimal Answer	One part essentially correct OR two parts partially correct.

3. (a) The value 50.90 estimates the average score of males who spend 0 hours studying. The value 9.45 estimates the average increase in score for each additional hour study time. For any fixed number of hours study time, the value 4.40 estimates the average number of points that females score higher than males.

(b)

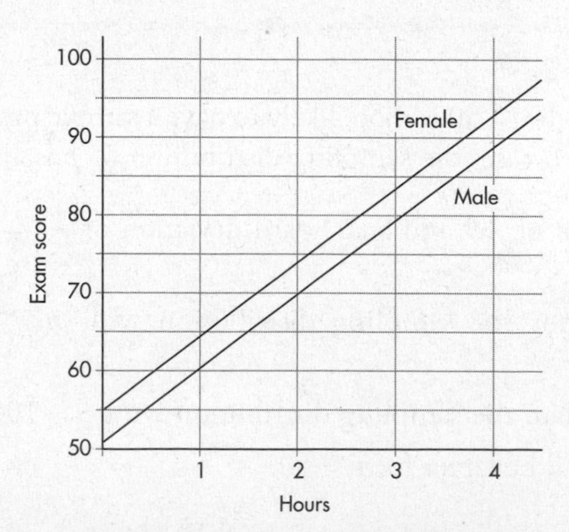

(c) For each additional hour of study time, the scores of males increase an average of 8.5 while those of females increase an average of 10.4. Additional hours of study time appear to benefit females more than males, something that does not show in the original model.

(d)

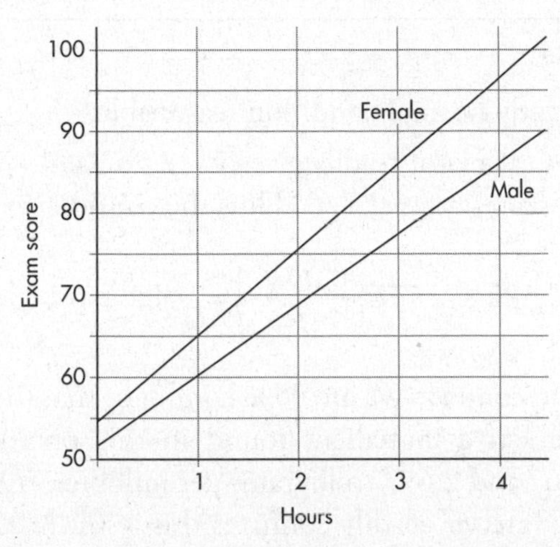

Scoring

Part (a) is essentially correct for correctly interpreting all three numbers in context and partially correct for correctly interpreting two of the three numbers.

Part (b) is essentially correct for correct graphs, clearly parallel and labeled as to which is which. Part (b) is partially correct for correct parallel graphs but missing labels, or for labeled, parallel graphs with incorrect y-intercepts.

Part (c) is essentially correct for correctly interpreting both slopes and for some comparative statement about additional hours study time appearing to benefit females more than males. Part (c) is partially correct for correctly interpreting both slopes but failing to make a comparative statement, or for making a correct comparative statement without interpreting the slopes.

Part (d) is essentially correct for correct graphs, clearly showing different slopes and labeled as to which is which. Part (d) is partially correct for graphs, clearly showing different slopes but missing labels, or for labeled graphs with different slopes but with incorrect y-intercepts.

Count partially correct answers as one-half an essentially correct answer.

4 Complete Answer Four essentially correct answers.

3 Substantial Answer Three essentially correct answers.

2 Developing Answer Two essentially correct answers.

1 Minimal Answer One essentially correct answer.

Use a holistic approach to decide a score totaling between two numbers.

4. (a) Name the procedure: This is a one-sample *t*-interval for the mean.

Check the conditions: We are given that this is a random sample, and a dot-plot.

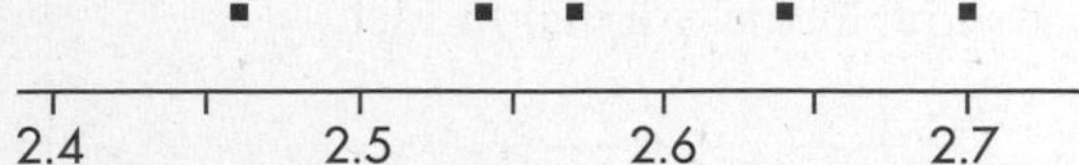

makes the nearly normal condition reasonable.

Mechanics: A calculator readily gives $\bar{x} = 2.582$ and $s = 0.0923$. With $df = 4$, the critical *t*-scores are $\pm$ 2.776. Thus the confidence interval is

$$2.582 \pm 2.776\left(\frac{0.0923}{\sqrt{5}}\right) = 2.582 \pm 0.115.$$

Conclusion in context: We are 95% confident that the mean concentration level for the active ingredient found in this pharmaceutical product is between 2.467 and 2.697 milligrams per milliliter. [On the TI-84, with the data in L1, TInterval readily confirms this result.]

(b) Raising the confidence level would increase the width of the confidence interval.

(c) Since the whole confidence interval (2.467, 2.697) is below the critical 2.7, at the 95% confidence level the mean concentration is at a safe level. However, with 2.697 so close to 2.7, based on the statement in (c), if the confidence level is raised we are no longer confident that the mean concentration is at a safe level.

Scoring

Part (a) has two components. The first component is essentially correct for correctly naming the procedure and checking the conditions, and is partially correct for one of these two. The second component is essentially correct for correct mechanics and a correct conclusion in context, and is partially correct for one of these two.

Part (b) is either essentially correct or incorrect.

Part (c) is essentially correct for a correct conclusion in context for both the 95% interval and for if the confidence is raised, both conclusions referencing the answers from (a) and (b). Part (c) is partially correct for a correct conclusion in context for either of the confidence intervals or for both but with a weak explanation.

Count partially correct answers as one-half an essentially correct answer.

4 Complete Answer	Four essentially correct answers.
3 Substantial Answer	Three essentially correct answers.
2 Developing Answer	Two essentially correct answers.
1 Minimal Answer	One essentially correct answer.

Use a holistic approach to decide a score totaling between two numbers.

5. (a) Points A and B both have high leverage; that is, both their x-coordinates are outliers in the x-direction. However, point B is influential (its removal sharply changes the regression line), while point A is not influential (it lies directly on the regression line, so its removal will not change the line).
   (b) Point C lies off the regression line so its residual is much greater than that of point D, whose residual is 0 (point D lies on the regression line). However, the removal of point C will very minimally affect the slope of the regression line, if at all, while the removal of point D dramatically affects the slope of the regression line.
   (c) Removal of either point E or point F minimally affects the regression line, while removal of both has a dramatic effect.
   (d) Removing either point G or point H will definitely affect the regression line (pulling the line toward the remaining of the two points), while removing both will have little, if any, effect on the line.

Scoring

Each of parts (a), (b), (c), and (d) has two components and is scored essentially correct for both components correct and partially correct for one component correct.

Give 1 point for each essentially correct part and $\frac{1}{2}$ point for each partially correct part.

4 Complete Answer	4 points
3 Substantial Answer	3 points
2 Developing Answer	2 points
1 Minimal Answer	1 point

Use a holistic approach to decide a score totaling between two numbers.

6. (a) State the hypotheses: H_0: $\mu_A - \mu_B = 0$ and H_a: $\mu_A - \mu_B \neq 0$
   Identify the test and check conditions: two-sample t-test. It is given that the samples are random. The populations may not be normal, but since n_A = 390 and n_B = 235 are both large, it is OK to proceed with the t-test.

   Calculate the test statistic t and the P-value: $t = \dfrac{14.08 - 14.51}{\sqrt{\dfrac{2.93^2}{390} + \dfrac{3.71^2}{235}}} = -1.51$

   and with df = 199, we have P = 2(.066) = .133. [Or use df = 623. Or a calculator using the given sample statistics gives t = −1.51, df = 408.25, and P = .131. Or a calculator using the sample data gives t = −1.53, df = 408.27, and P = .127.]

   Conclusion in context: With this large a P-value, there is no evidence of a difference in the mean years of schooling between the two communities.

(b) State the hypotheses: H_0: The distribution of years of schooling is the same in the two communities. H_a: The distribution of years of schooling is not the same in the two communities.

Identify the test and check conditions: Chi-square test (of homogeneity). All expected counts are > 5.

Calculate the test statistic and the *P*-value: $\chi^2 = 57.85$ and with $df = (7 - 1)(2 - 1) = 6$, we have $P = .000$.

State the conclusion in context: With this small a *P*-value, there is strong evidence that the distribution of years of schooling is not the same in the two communities.

(c)

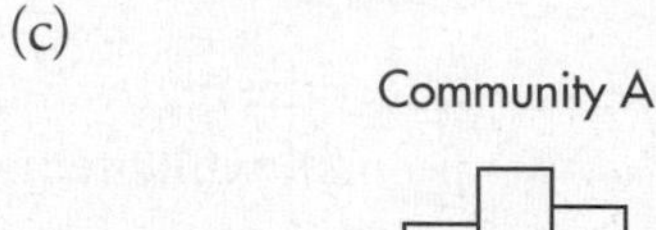

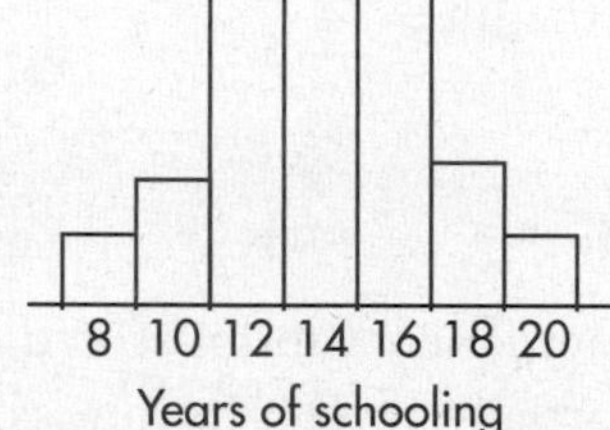

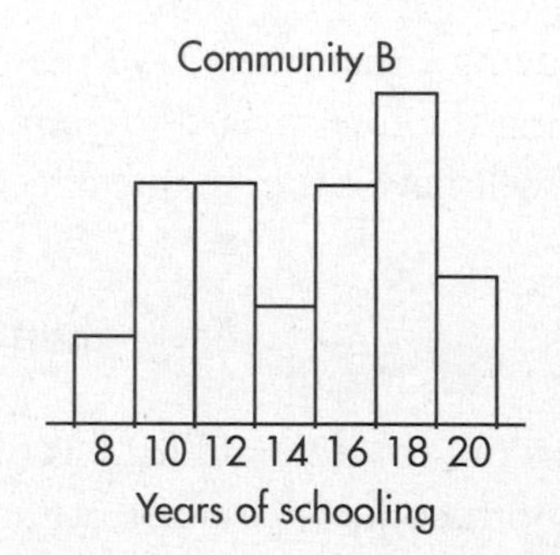

Although the two distributions have roughly the same center, their shapes are different. The distribution of years of schooling for Community A is unimodal, symmetric, and bell-shaped, while that of Community B is roughly bimodal with few scores in the center.

Scoring

Part (a) is essentially correct if all four steps for a two-sample *t*-test are correct and partially correct if three steps are correct.

Part (b) is essentially correct if all four steps for a chi-square test (of homogeneity) are correct and partially correct if three steps are correct.

Part (c) is essentially correct for an accurate comparison of the distributions linked to correctly drawn histograms. Part (c) is partially correct if only one of the two parts (comparison and histograms) is correct or if both are correct but there is no linkage.

4 Complete Answer	All three parts essentially correct.
3 Substantial Answer	Two parts essentially correct and one part partially correct.
2 Developing Answer	Two parts essentially correct OR one part essentially correct and one or two parts partially correct OR all three parts partially correct.
1 Minimal Answer	One part essentially correct OR two parts partially correct.

Answer Sheet

PRACTICE EXAMINATION 5

1. Ⓐ Ⓑ Ⓒ Ⓓ Ⓔ	11. Ⓐ Ⓑ Ⓒ Ⓓ Ⓔ	21. Ⓐ Ⓑ Ⓒ Ⓓ Ⓔ	31. Ⓐ Ⓑ Ⓒ Ⓓ Ⓔ
2. Ⓐ Ⓑ Ⓒ Ⓓ Ⓔ	12. Ⓐ Ⓑ Ⓒ Ⓓ Ⓔ	22. Ⓐ Ⓑ Ⓒ Ⓓ Ⓔ	32. Ⓐ Ⓑ Ⓒ Ⓓ Ⓔ
3. Ⓐ Ⓑ Ⓒ Ⓓ Ⓔ	13. Ⓐ Ⓑ Ⓒ Ⓓ Ⓔ	23. Ⓐ Ⓑ Ⓒ Ⓓ Ⓔ	33. Ⓐ Ⓑ Ⓒ Ⓓ Ⓔ
4. Ⓐ Ⓑ Ⓒ Ⓓ Ⓔ	14. Ⓐ Ⓑ Ⓒ Ⓓ Ⓔ	24. Ⓐ Ⓑ Ⓒ Ⓓ Ⓔ	34. Ⓐ Ⓑ Ⓒ Ⓓ Ⓔ
5. Ⓐ Ⓑ Ⓒ Ⓓ Ⓔ	15. Ⓐ Ⓑ Ⓒ Ⓓ Ⓔ	25. Ⓐ Ⓑ Ⓒ Ⓓ Ⓔ	35. Ⓐ Ⓑ Ⓒ Ⓓ Ⓔ
6. Ⓐ Ⓑ Ⓒ Ⓓ Ⓔ	16. Ⓐ Ⓑ Ⓒ Ⓓ Ⓔ	26. Ⓐ Ⓑ Ⓒ Ⓓ Ⓔ	36. Ⓐ Ⓑ Ⓒ Ⓓ Ⓔ
7. Ⓐ Ⓑ Ⓒ Ⓓ Ⓔ	17. Ⓐ Ⓑ Ⓒ Ⓓ Ⓔ	27. Ⓐ Ⓑ Ⓒ Ⓓ Ⓔ	37. Ⓐ Ⓑ Ⓒ Ⓓ Ⓔ
8. Ⓐ Ⓑ Ⓒ Ⓓ Ⓔ	18. Ⓐ Ⓑ Ⓒ Ⓓ Ⓔ	28. Ⓐ Ⓑ Ⓒ Ⓓ Ⓔ	38. Ⓐ Ⓑ Ⓒ Ⓓ Ⓔ
9. Ⓐ Ⓑ Ⓒ Ⓓ Ⓔ	19. Ⓐ Ⓑ Ⓒ Ⓓ Ⓔ	29. Ⓐ Ⓑ Ⓒ Ⓓ Ⓔ	39. Ⓐ Ⓑ Ⓒ Ⓓ Ⓔ
10. Ⓐ Ⓑ Ⓒ Ⓓ Ⓔ	20. Ⓐ Ⓑ Ⓒ Ⓓ Ⓔ	30. Ⓐ Ⓑ Ⓒ Ⓓ Ⓔ	40. Ⓐ Ⓑ Ⓒ Ⓓ Ⓔ

Practice Examination 5

SECTION I

Questions 1–40

Spend 90 minutes on this part of the exam.

Directions: The questions or incomplete statements that follow are each followed by five suggested answers or completions. Choose the response that best answers the question or completes the statement.

1. The mean and standard deviation of the population {1, 5, 8, 11, 15} are $\mu = 8$ and $\sigma = 4.8$, respectively. Let S be the set of the 125 *ordered* triples (repeats allowed) of elements of the original population. Which of the following is a correct statement about the mean $\mu_{\bar{x}}$ and standard deviation $\sigma_{\bar{x}}$ of the means of the triples in S?

   (A) $\mu_{\bar{x}} = 8,\ \sigma_{\bar{x}} = 4.8$
   (B) $\mu_{\bar{x}} = 8,\ \sigma_{\bar{x}} < 4.8$
   (C) $\mu_{\bar{x}} = 8,\ \sigma_{\bar{x}} > 4.8$
   (D) $\mu_{\bar{x}} < 8,\ \sigma_{\bar{x}} = 4.8$
   (E) $\mu_{\bar{x}} > 8,\ \sigma_{\bar{x}} > 4.8$

2. A survey to measure job satisfaction of high school mathematics teachers was taken in 1993 and repeated 5 years later in 1998. Each year a random sample of 50 teachers rated their job satisfaction on a 1-to-100 scale with higher numbers indicating greater satisfaction. The results are given in the following back-to-back stemplot.

1993		1998
	0	
98775	1	
98530	2	1
65210	3	3589
96430	4	01122233455667889999
87421	5	035667899
99877555322100	6	1344789
976442	7	22689
7511	8	138
0	9	7

What is the trend from 1993 to 1998 with regard to the standard deviation and range of the two samples?

(A) Both the standard deviation and range increased.
(B) The standard deviation increased, while the range decreased.
(C) The range increased, while the standard deviation decreased.
(D) Both the standard deviation and range decreased.
(E) Both the standard deviation and range remained unchanged.

GO ON TO THE NEXT PAGE ➤

3. Consider the following studies being run by three different AP Statistics instructors.

   I. One rewards students every day with lollipops for relaxation, encouragement, and motivation to learn the material.
   II. One promises that all students will receive A's as long as they give their best efforts to learn the material.
   III. One is available every day after school and on weekends so that students with questions can come in and learn the material.

   (A) None of these studies use randomization.
   (B) None of these studies use control groups.
   (C) None of these studies use blinding.
   (D) Important information can be found from all these studies, but none can establish causal relationships.
   (E) All of the above.

4. The number of days it takes to build a new house has a variance of 386. A sample of 40 new homes shows an average building time of 83 days. With what confidence can we assert that the average building time for a new house is between 80 and 90 days?

   (A) 15.4%
   (B) 17.8%
   (C) 20.0%
   (D) 38.8%
   (E) 82.2%

5. A shipment of resistors have an average resistance of 200 ohms with a standard deviation of 5 ohms, and the resistances are normally distributed. Suppose a randomly chosen resistor has a resistance under 194 ohms. What is the probability that its resistance is greater than 188 ohms?

   (A) .07
   (B) .12
   (C) .50
   (D) .93
   (E) .97

6. Suppose 4% of the population have a certain disease. A laboratory blood test gives a positive reading for 95% of people who have the disease and for 5% of people who do not have the disease. If a person tests positive, what is the probability the person has the disease?

   (A) .038
   (B) .086
   (C) .442
   (D) .558
   (E) .950

7. For which of the following is it appropriate to use a census?

   (A) A 95% confidence interval of mean height of teachers in a small town.
   (B) A 95% confidence interval of the proportion of students in a small town who are taking some AP class.
   (C) A two-tailed hypothesis test where the null hypothesis was that the mean expenditure on entertainment by male students at a high school is the same as that of female students.
   (D) Calculation of the standard deviation in the number of broken eggs per carton in a truckload of eggs.
   (E) All of the above.

8. On the same test, Mary and Pam scored at the 64th and 56th percentiles, respectively. Which of the following is a true statement?

   (A) Mary scored eight more points than Pam.
   (B) Mary's score is 8% higher than Pam's.
   (C) Eight percent of those who took the test scored between Pam and Mary.
   (D) Thirty-six people scored higher than both Mary and Pam.
   (E) None of the above.

GO ON TO THE NEXT PAGE ➤

9. Which of the following is a true statement?
   (A) While observational studies gather information on an already existing condition, they still often involve intentionally forcing some treatment to note the response.
   (B) In an experiment, researchers decide on the treatment but typically allow the subjects to self-select into the control group.
   (C) If properly designed, either observational studies or controlled experiments can easily be used to establish cause and effect.
   (D) Wording to disguise hidden interests in observational studies is the same idea as blinding in experimental design.
   (E) Stratifying in sampling is the same idea as blocking for experiments.

10. The random variable describing the number of minutes high school students spend in front of a computer daily has a mean of 200 minutes. Samples of two different sizes result in sampling distributions with the two graphs below.

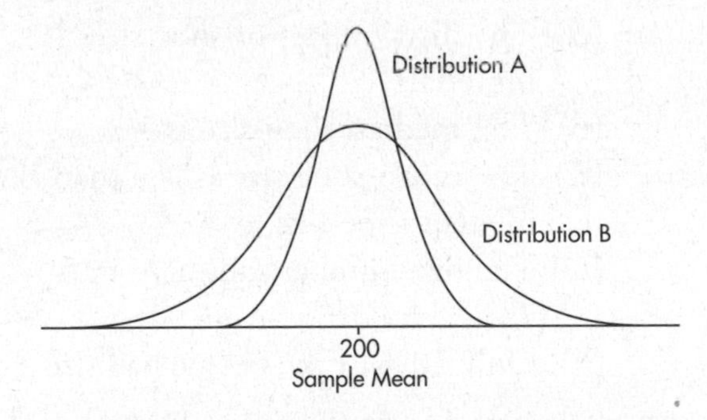

Which of the following is a true statement?
   (A) Based on these graphs, no comparison between the two sample sizes is possible.
   (B) More generally, sample sizes have no effect on sampling distributions.
   (C) The sample size in A is the same as the sample size in B.
   (D) The sample size in A is less than the sample size in B.
   (E) The sample size in A is greater than the sample size in B.

11. To determine the mean cost of groceries in a certain city, an identical grocery basket of food is purchased at each store in a random sample of ten stores. If the average cost is \$47.52 with a standard deviation of \$1.59, find a 98% confidence interval estimate for the cost of these groceries in the city.
   (A) $47.52 \pm 2.33\sqrt{1.59}$
   (B) $47.52 \pm 2.33\left(\frac{1.59}{\sqrt{10}}\right)$
   (C) $47.52 \pm 2.33\left(\sqrt{\frac{1.59}{10}}\right)$
   (D) $47.52 \pm 2.821\left(\frac{1.59}{\sqrt{10}}\right)$
   (E) $47.52 \pm 2.821\sqrt{\frac{1.59}{10}}$

12. A set consists of four numbers. The largest value is 200, and the range is 50. Which of the following statements is true?
   (A) The mean is less than 185.
   (B) The mean is greater than 165.
   (C) The median is less than 195.
   (D) The median is greater than 155.
   (E) The median is the mean of the second and third numbers if the set is arranged in ascending order.

13. A telephone survey of 400 registered voters showed that 256 had not yet made up their minds 1 month before the election. How sure can we be that between 60% and 68% of the electorate were still undecided at that time?
   (A) 2.4%
   (B) 8.0%
   (C) 64.0%
   (D) 90.5%
   (E) 95.3%

GO ON TO THE NEXT PAGE ➤

14. Suppose we have a random variable X where the probability associated with the value k is

$$\binom{15}{k}(.29)^k(.71)^{15-k} \text{ for } k = 0, \ldots, 15.$$

What is the mean of X?

(A) 0.29
(B) 0.71
(C) 4.35
(D) 10.65
(E) None of the above

15. The financial aid office at a state university conducts a study to determine the total student costs per semester. All students are charged $4500 for tuition. The mean cost for books is $350 with a standard deviation of $65. The mean outlay for room and board is $2800 with a standard deviation of $380. The mean personal expenditure is $675 with a standard deviation of $125. Assuming independence among categories, what is the standard deviation of the total student costs?

(A) $24
(B) $91
(C) $190
(D) $405
(E) $570

16. Suppose X and Y are random variables with $E(X) = 312$, $\text{var}(X) = 6$, $E(X) = 307$, and $\text{var}(Y) = 8$. What are the expected value and variance of the random variable $X + Y$?

(A) $E(X + Y) = 619$, $\text{var}(X + Y) = 7$
(B) $E(X + Y) = 619$, $\text{var}(X + Y) = 10$
(C) $E(X + Y) = 619$, $\text{var}(X + Y) = 14$
(D) $E(X + Y) = 309.5$, $\text{var}(X + Y) = 14$
(E) There is insufficient information to answer this question.

17. In sample surveys, what is meant by *bias*?

(A) A systematic error in a sampling method that leads to an unrepresentative sample.
(B) Prejudice, for example in ethnic and gender related studies.
(C) Natural variability seen between samples.
(D) Tendency for some distributions to be skewed.
(E) Tendency for some distributions to vary from normality.

18. The following histogram gives the shoe sizes of people in an elementary school building one morning.

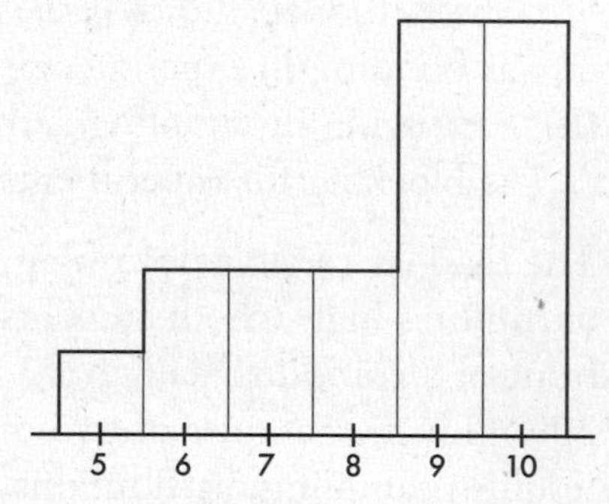

Which of the following is a true statement?

(A) The distribution of shoe sizes is bimodal.
(B) The median shoe size is $7\frac{1}{2}$.
(C) The mean shoe size is less than the median shoe size.
(D) The five-number summary is: $5\frac{1}{2} - 6\frac{1}{2} - 7\frac{1}{2} - 8\frac{1}{2} - 9\frac{1}{2}$.
(E) Only 10% of the people had size 5 shoes.

19. When comparing the standard normal (z) distribution to the t-distribution with $df = 30$, which of (A)–(D), if any, are false?

(A) Both are symmetric.
(B) Both are bell-shaped.
(C) Both have center 0.
(D) Both have standard deviation 1.
(E) All the above are true statements.

GO ON TO THE NEXT PAGE ➤

20. Given a probability of .65 that interest rates will jump this year, and a probability of .72 that if interest rates jump the stock market will decline, what is the probability that interest rates will jump and the stock market will decline?

(A) $.72 + .65 - (.72)(.65)$
(B) $(.72)(.65)$
(C) $1 - (.72)(.65)$
(D) $\frac{.65}{.72}$
(E) $1-\frac{.65}{.72}$

21. Sampling error is

(A) the mean of a sample statistic.
(B) the standard deviation of a sample statistic.
(C) the standard error of a sample statistic.
(D) the result of bias.
(E) the difference between a population parameter and an estimate of that parameter.

22. Suppose that the weights of trucks traveling on the interstate highway system are normally distributed. If 70% of the trucks weigh more than 12,000 pounds and 80% weigh more than 10,000 pounds, what are the mean and standard deviation for the weights of trucks traveling on the interstate system?

(A) $\mu = 14{,}900$; $\sigma = 6100$
(B) $\mu = 15{,}100$; $\sigma = 6200$
(C) $\mu = 15{,}300$; $\sigma = 6300$
(D) $\mu = 15{,}500$; $\sigma = 6400$
(E) The mean and standard deviation cannot be computed from the information given.

23. If the correlation coefficient $r = .78$, what percentage of variation in y is explained by variation in x?

(A) 22%
(B) 39%
(C) 44%
(D) 61%
(E) 78%

24. Consider the following scatterplot:

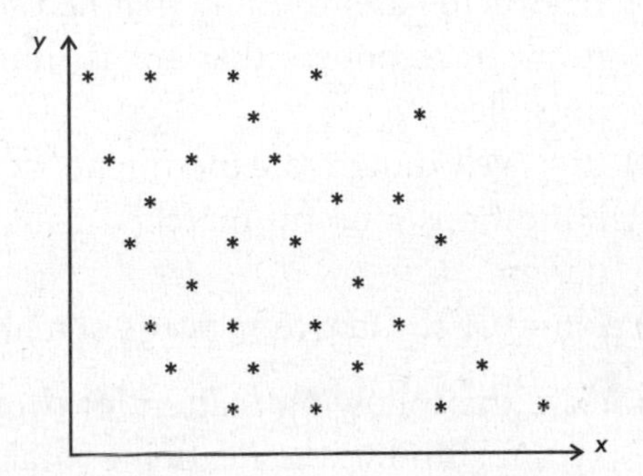

Which of the following is the best estimate of the correlation between x and y?

(A) −.95
(B) −.15
(C) 0
(D) .15
(E) .95

25. For one NBA playoff game the actual percentage of the television viewing public who watched the game was 24%. If you had taken a survey of 50 television viewers that night and constructed a confidence interval estimate of the percentage watching the game, which of the following would have been true?

I. The center of the interval would have been 24%.
II. The interval would have contained 24%.
III. A 99% confidence interval estimate would have contained 24%.

(A) I and II
(B) I and III
(C) II and III
(D) All are true.
(E) None is true.

GO ON TO THE NEXT PAGE ➤

26. Which of the following is a true statement?

    (A) In a well-designed, well-conducted sample survey, sampling error is effectively eliminated.
    (B) In a well-designed observational study, responses are influenced through an orderly, carefully planned procedure during the collection of data.
    (C) In a well-designed experiment, the treatments are carefully planned to result in responses that are as similar as possible.
    (D) In a well-designed experiment, double-blinding is a useful matched pairs design.
    (E) None of the above is a true statement.

27. Consider the following scatterplot showing the relationship between caffeine intake and job performance.

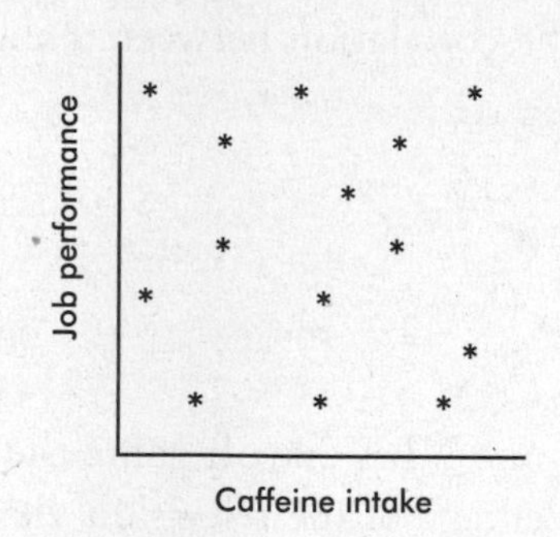

Which of the following is a reasonable conclusion?

    (A) Low caffeine intake is associated with low job performance.
    (B) Low caffeine intake is associated with high job performance.
    (C) High caffeine intake is associated with low job performance.
    (D) High caffeine intake is associated with high job performance.
    (E) Job performance cannot be predicted from caffeine intake.

28. An author of a new book claims that anyone following his suggested diet program will lose an average of 2.8 pounds per week. A researcher believes that the true figure will be lower and plans a test involving a random sample of 36 overweight people. She will reject the author's claim if the mean weight loss in the volunteer group is less than 2.5 pounds per week. Assume that the standard deviation among individuals is 1.2 pounds per week. If the true mean value is 2.4 pounds per week, what is the probability that the researcher will mistakenly fail to reject the author's false claim of 2.8 pounds?

    (A) $P\left(z > \frac{2.5-2.4}{1.2/\sqrt{36}}\right)$
    (B) $P\left(z < \frac{2.5-2.4}{1.2/\sqrt{36}}\right)$
    (C) $P\left(z < \frac{2.8-2.5}{1.2/\sqrt{36}}\right)$
    (D) $P\left(z > \frac{2.8-2.4}{1.2/\sqrt{36}}\right)$
    (E) $P\left(z < \frac{2.8-2.4}{1.2/\sqrt{36}}\right)$

GO ON TO THE NEXT PAGE ➤

29. Which of the following is the central limit theorem?

(A) No matter how the population is distributed, as the sample size increases, the mean of the sample means becomes closer to the mean of the population.
(B) No matter how the population is distributed, as the sample size increases, the standard deviation of the sample means becomes closer to the standard deviation of the population divided by the square root of the sample size.
(C) If the population is normally distributed, then as the sample size increases, the sampling distribution of the sample mean becomes closer to a normal distribution.
(D) All of the above together make up the central limit theorem.
(E) The central limit theorem refers to something else.

30. What is a sampling distribution?

(A) A distribution of all the statistics that can be found in a given sample
(B) A histogram, or other such visual representation, showing the distribution of a sample
(C) A normal distribution of some statistic
(D) A distribution of all the values taken by a statistic from all possible samples of a given size
(E) All of the above

31. A judge chosen at random reaches a just decision roughly 80% of the time. What is the probability that in randomly chosen cases at least two out of three judges reach a just decision?

(A) $3(.8)^2(.2)$
(B) $1 - 3(.8)^2(.2)$
(C) $(.8)^3$
(D) $1 - (.8)^3$
(E) $3(.8)^2(.2) + (.8)^3$

32. Miles per gallon versus speed (miles per hour) for a new model automobile is fitted with a least squares regression line. Following is computer output of the statistical analysis of the data.

```
Dependent variable: Miles per gallon

Source       df     Sum of Squares   Mean Square   F-ratio
Regression    1         199.34          199.34       3.79
Residual      6         315.54            5.59

Variable    Coefficient       SE Coef         t-ratio        P
Constant      38.929          5.651            6.89        0.000
Speed         -0.2179         0.112           -1.95        0.099

R-Sq = 38.7%        R-Sq(adj) = 28.5%
s = 7.252 with 8 - 2 = 6 degrees of freedom
```

Which of the following gives a 99% confidence interval for the slope of the regression line?

(A) $-0.2179 \pm 3.707(0.112)$
(B) $-0.2179 \pm 3.143(0.112/\sqrt{8})$
(C) $-0.2179 \pm 3.707(0.112/\sqrt{8})$
(D) $38.929 \pm 3.143(3.651/\sqrt{8})$
(E) $38.929 \pm 3.707(5.651\sqrt{8})$

GO ON TO THE NEXT PAGE ➤

33. What fault do all these sampling designs have in common?

    I. The Parent-Teacher Association (PTA), concerned about rising teenage pregnancy rates at a high school, randomly picks a sample of high school students and interviews them concerning unprotected sex they have engaged in during the past year.

    II. A radio talk show host asks people to phone in their views on whether the United States should keep troops in Bosnia indefinitely to enforce the cease-fire.

    III. The *Ladies Home Journal* plans to predict the winner of a national election based on a survey of its readers.

    (A) All the designs make improper use of stratification.
    (B) All the designs have errors that can lead to strong bias.
    (C) All the designs confuse association with cause and effect.
    (D) All the designs suffer from sampling error.
    (E) None of the designs makes use of chance in selecting a sample.

34. Hospital administrators wish to determine the average length of stay for all surgical patients. A statistician determines that for a 95% confidence level estimate of the average length of stay to within ±0.50 days, 100 surgical patients' records would have to be examined. How many records should be looked at for a 95% confidence level estimate to within ±0.25 days?

    (A) 25
    (B) 50
    (C) 200
    (D) 400
    (E) There is not enough information given to determine the necessary sample size.

35. A chess master wins 80% of her games, loses 5%, and draws the rest. If she receives 1 point for a win, $\frac{1}{2}$ point for a draw, and no points for a loss, what is true about the sampling distribution X of the points scored in two independent games?

    (A) X takes on the values 0, 1, and 2 with respective probabilities .10, .26, and .64.
    (B) X takes on the values 0, $\frac{1}{2}$, 1, $1\frac{1}{2}$, and 2 with respective probabilities .0025, .015, .1025, .24, and .64.
    (C) X takes on values according to a binomial distribution with $n = 2$ and $p = .8$.
    (D) X takes on values according to a binomial distribution with mean $1(.8) + \frac{1}{2}(.15) + 0(.05)$.
    (E) X takes on values according to a distribution with mean $(2)(.8)$ and standard deviation $\sqrt{2(.8)(.2)}$.

36. Which of the following is a true statement?

    (A) The P-value is a conditional probability.
    (B) The P-value is usually chosen before an experiment is conducted.
    (C) The P-value is based on a specific test statistic and thus should not be used in a two-sided test.
    (D) P-values are more appropriately used with t-distributions than with z-distributions.
    (E) If the P-value is less than the level of significance, then the null hypothesis is proved false.

GO ON TO THE NEXT PAGE ➤

37. An assembly line machine is supposed to turn out ball bearings with a diameter of 1.25 centimeters. Each morning the first 30 bearings produced are pulled and measured. If their mean diameter is under 1.23 centimeters or over 1.27 centimeters, the machinery is stopped and an engineer is called to make adjustments before production is resumed. The quality control procedure may be viewed as a hypothesis test with the null hypothesis H_0: $\mu = 1.25$ and the alternative hypothesis H_a: $\mu \neq 1.25$. The engineer is asked to make adjustments when the null hypothesis is rejected. In test terminology, what would a Type II error result in?

(A) A warranted halt in production to adjust the machinery
(B) An unnecessary stoppage of the production process
(C) Continued production of wrong size ball bearings
(D) Continued production of proper size ball bearings
(E) Continued production of ball bearings that randomly are the right or wrong size

38. Both over-the-counter niacin and the prescription drug Lipitor are known to lower blood cholesterol levels. In one double-blind study Lipitor outperformed niacin. The 95% confidence interval estimate of the difference in mean cholesterol level lowering was (18, 41). Which of the following is a reasonable conclusion?

(A) Niacin lowers cholesterol an average of 18 points, while Lipitor lowers cholesterol an average of 41 points.
(B) There is a .95 probability that Lipitor will outperform niacin in lowering the cholesterol level of any given individual.
(C) There is a .95 probability that Lipitor will outperform niacin by at least 23 points in lowering the cholesterol level of any given individual.
(D) We should be 95% confident that Lipitor will outperform niacin as a cholesterol-lowering drug.
(E) None of the above.

39. The following parallel boxplots show the average daily hours of bright sunshine in Liberia, West Africa:

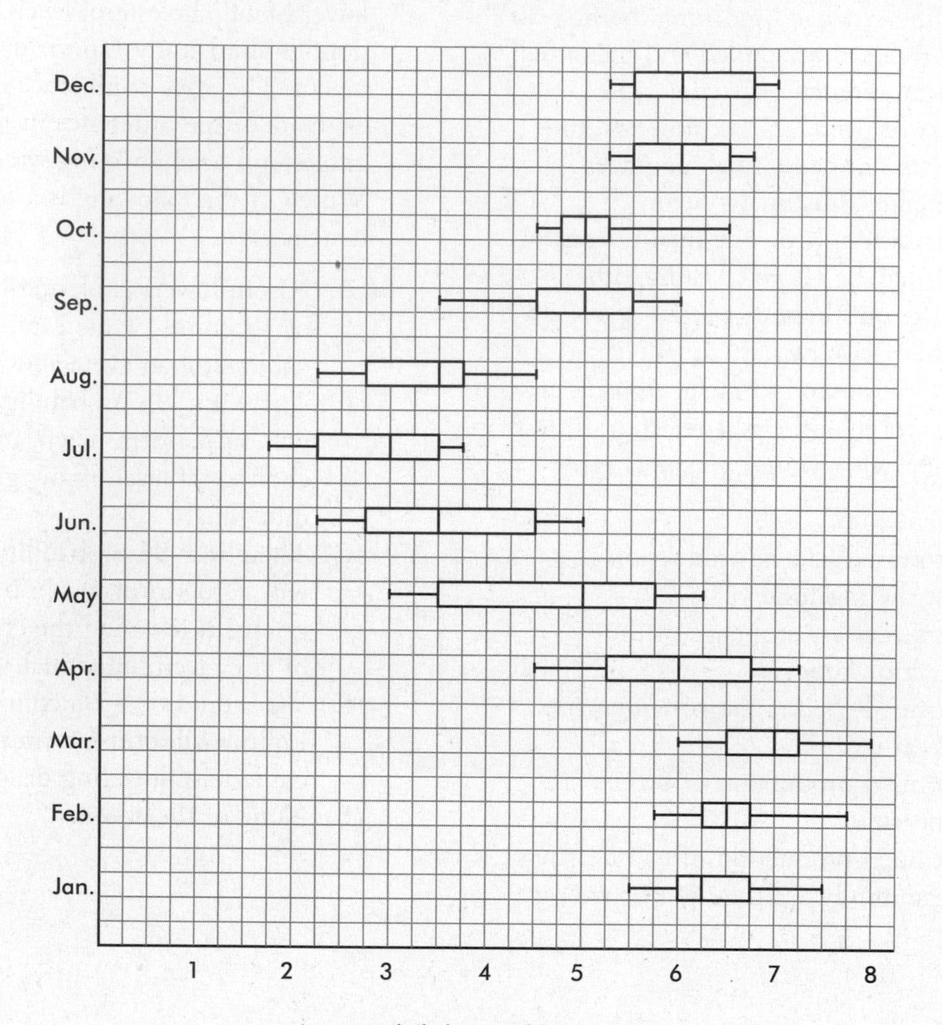

For how many months is the median below 4 hours?

(A) One
(B) Two
(C) Three
(D) Four
(E) Five

40. Following is a cumulative probability graph for the number of births per day in a city hospital.

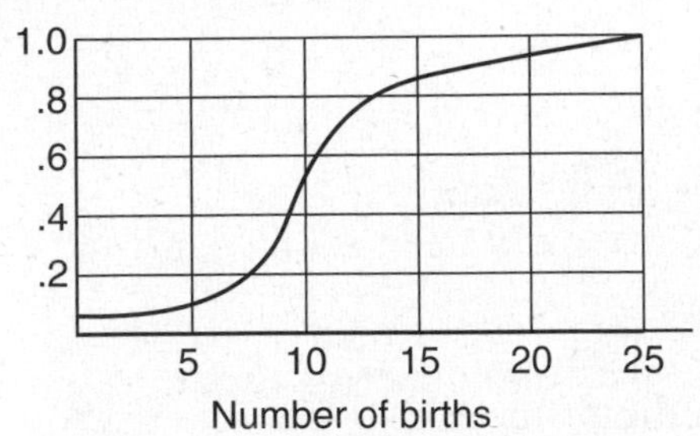

Assuming that a birthing room can be used by only one woman per day, how many rooms must the hospital have available to be able to meet the demand at least 90 percent of the days?

(A) 5
(B) 10
(C) 15
(D) 20
(E) 25

STOP

If there is still time remaining, you may review your answers.

SECTION II

PART A

Questions 1–5

Spend about 65 minutes on this part of the exam.
Percentage of Section II grade—75

> You must show all work and indicate the methods you use. You will be graded on the correctness of your methods and on the accuracy of your results and explanations.

1. An experiment is being planned to study urban land use practices aimed at reviving and sustaining native bird populations. Vegetation types A and B are to be compared, and eight test sites are available. After planting, volunteer skilled birdwatchers will collect data on the abundance of bird species making each of the two habitat types their home. The east side of the city borders a river, while the south side borders an industrial park.

(a) Suppose the decision is made to block using the scheme below (one block is white, one gray). How would you use randomization, and what is the purpose of the randomization?

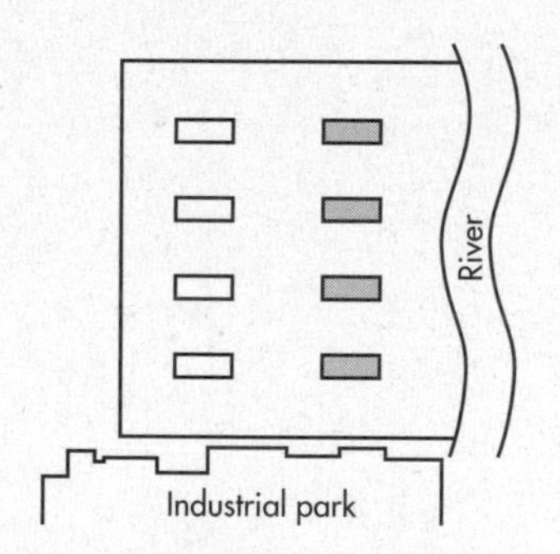

(b) Comment on the strength and weakness of the above scheme as compared to the following blocking scheme (one block is white, one gray).

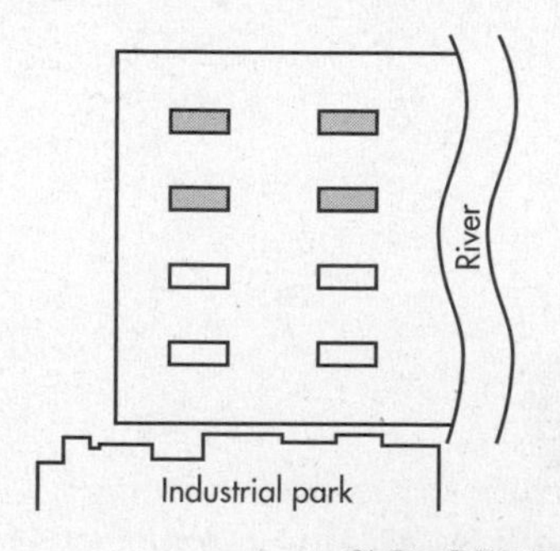

2. A world organization report gives the following percentages for primary-school-age children enrolled in school in the 17 countries of each of two geographic regions.

Region A: 36, 45, 52, 56, 56, 58, 60, 63, 65, 66, 69, 71, 72, 74, 77, 82, 92
Region B: 35, 37, 41, 43, 43, 48, 50, 54, 65, 71, 78, 82, 83, 87, 89, 91, 92

(a) Draw a back-to-back stemplot of this data.

(b) The report describes both regions as having the same median percentage (for primary-school-age children enrolled in school) among their 17 countries and approximately the same range. What about the distributions is missed by the report?

(c) If the organization has education funds to help only one region, give an argument for which should be helped.

(d) A researcher plans to run a two-sample *t*-test to study the difference in means between the percentages from each region. Comment on his plan.

3. Data were collected from a random sample of 100 student athletes at a large state university. The plot below shows grade point average (GPA) versus standard normal value (*z*-score) corresponding to the percentile of each GPA (when arranged in order). Also shown is the *normal line,* that is, a line passing through expected values for a normal distribution with the mean and SD of the given data.

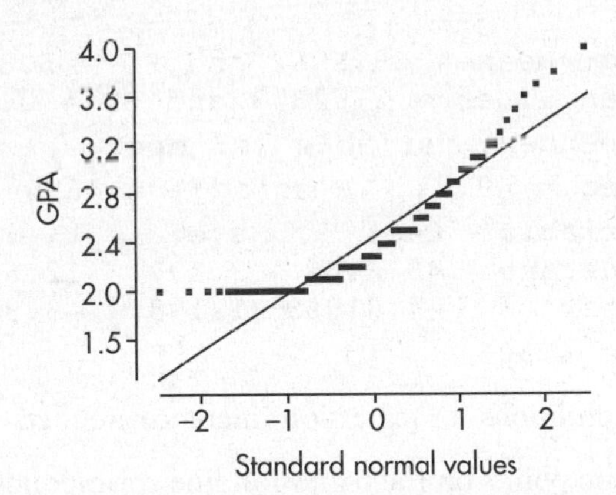

(a) What is the shape of this distribution? Explain.

(b) What is the 95th percentile of the data? Explain.

(c) In a normal distribution with the mean and SD of this data, what is the 95th percentile? Explain.

Practice Examination 5

GO ON TO THE NEXT PAGE ➤

4. Although blood type frequencies in the United States are in the ratios of 9:8:2:1 for types O, A, B, and AB, respectively, local differences are often found depending upon a variety of demographic characteristics. Two researchers are independently assigned to determine if patients at a particular large city general hospital exhibit blood types supporting the above model. The table below gives the data results from what each researcher claims to be random samples of 500 patient lab results.

	O	A	B	AB
Researcher 1	253	194	38	15
Researcher 2	230	196	48	26

(a) Do the data reported by Researcher 1 support the 9:8:2:1 model for blood types of patients at the particular hospital? Justify your answer.

(b) The editorial board of a medical publication rejects the findings of Researcher 2, claiming that his data are suspicious in that they are too good to be true. Give a statistical justification for the board's decision.

5. In a random sample of automobiles the highway mileage (in mpg) and the engine size (in liters) are measured, and the following computer output for regression is obtained:

```
mpg: Mean = 28.5882 and SD = 5.88493
Mean liter = 2.52353 and SD = 0.898978
Dependent variable is: mpg
R-sq = 49.8%   R-sq(adj) = 46.5%
Variable    Coef       s.e.    t       p
Constant    40.2462    3.197   12.6    0.0001
liter       −4.61969   1.198   −3.86   0.0015
s = 4.306
```

Assume all conditions for regression inference are met.

(a) One of the points on the regression line corresponds to an auto with an engine size of 1.5 liters and highway mileage of 35 miles per gallon. What is the residual of this point?

Suppose another auto with an engine size of 5.0 liters and highway mileage of 17 mpg is added to the data set.

(b) Explain whether the new slope will be greater than, less than, or about the same as the slope given by the output above.

(c) Explain whether the new correlation will be greater than, less than, or about the same in absolute value as the correlation given by the output above.

GO ON TO THE NEXT PAGE ➤

SECTION II

PART B

Question 6

Spend about 25 minutes on this part of the exam.
Percentage of Section II grade—25

6. A demographer randomly selects five northern and five southern U.S. states and notes the populations in 1990 and 2000:

Population (1000s)

Northern states			Southern states		
	1990	2000		1990	2000
Connecticut	3300	3400	Alabama	4000	4400
Idaho	1000	1300	Florida	12900	16000
Illinois	11400	12400	Louisiana	4200	4500
Michigan	9300	9940	New Mexico	1500	1800
New York	18000	19000	Texas	17000	20900

She calculates the following summary statistics:

	1990	2000	Difference (2000–1990)
Northern states	$\bar{x} = 8600$	$\bar{x} = 9208$	$\bar{x} = 608$
	$s = 6755$	$s = 7122$	$s = 407$
Southern states	$\bar{x} = 7920$	$\bar{x} = 9520$	$\bar{x} = 1600$
	$s = 6665$	$s = 8404$	$s = 1758$

Assuming that all conditions for inference are met,

(a) at the $\alpha = .05$ significance level, test the null hypothesis that southern states did not grow between 1990 and 2000

(b) find a 95% confidence interval for the difference in populations between northern and southern states in 1990

(c) test the hypothesis that southern states grew faster than northern states between 1990 and 2000

If there is still time remaining, you may review your answers.

Answer Key

Section I

1. **B**	9. **E**	17. **A**	25. **E**	33. **B**
2. **C**	10. **E**	18. **C**	26. **E**	34. **D**
3. **E**	11. **D**	19. **D**	27. **E**	35. **B**
4. **E**	12. **E**	20. **B**	28. **A**	36. **A**
5. **D**	13. **D**	21. **E**	29. **E**	37. **C**
6. **C**	14. **C**	22. **C**	30. **D**	38. **E**
7. **D**	15. **D**	23. **D**	31. **E**	39. **C**
8. **C**	16. **E**	24. **B**	32. **A**	40. **D**

Answers Explained

Section I

1. **(B)** $\mu_{\bar{x}} = \mu = 8$, and $\sigma_{\bar{x}} = \frac{\sigma}{\sqrt{n}} = \frac{4.8}{\sqrt{3}} < 4.8$

2. **(C)** The range increased from 90 − 15 = 75 to 97 − 21 = 76, while the standard deviation decreased (note how the values are bunched together more closely in 1998).

3. **(E)** None of the studies have any controls such as randomization, a control group, or blinding, and so while they may give valuable information, they cannot establish cause and effect.

4. **(E)** $\sigma_{\bar{x}} = \frac{\sqrt{386}}{\sqrt{40}} = 3.106$. The critical *z*-scores are $\frac{80-83}{3.106} = -0.97$ and $\frac{90-83}{3.106} = 2.25$, resulting in a probability of .3340 + .4878 = .8218.

5. **(D)** The critical *z*-scores are $\frac{188-200}{5} = -2.4$ and $\frac{194-200}{5} = -1.2$, with corresponding left tail probabilities of .0082 and .1151, respectively. The probability of being greater than 188 given that it is less than 194 is $\frac{.1151-.0082}{.1151} = .93$.

6. **(C)**

$$\begin{aligned} P(\text{pos test}) &= P(\text{disease} \cap \text{pos}) + P(\text{healthy} \cap \text{pos}) \\ &= (.04)(.95) + (.96)(.05) \\ &= .038 + .048 = .086 \end{aligned}$$

$$P(\text{disease} \mid \text{pos test}) = \frac{.038}{.086} = .442$$

7. **(D)** Given a census, the population parameter is known, and there is no need to use the techniques of inference.

Practice Examination 5

8. **(C)** Sixty-four percent of the students scored below Mary and 56% scored below Pam, and so 8% must have scored between them.

9. **(E)** Intentionally forcing some treatment to note the response is associated with controlled experiments, not with observational studies. In experiments, the researchers decide how people are placed in different groups; self-selection is associated with observational studies. Results of observational studies may suggest cause-and-effect relationships; however, controlled studies are used to establish such relationships.

10. **(E)** With $\sigma_{\bar{x}} = \frac{\sigma}{\sqrt{n}}$, the greater the sample size n, the smaller the standard deviation $\sigma_{\bar{x}}$.

11. **(D)** $df = 9$, and $47.52 \pm 2.821\left(\frac{1.59}{\sqrt{10}}\right)$.

12. **(E)** The set could be {150, 150, 150, 200} with mean 162.5 and median 150. It might also be {150, 200, 200, 200} with mean 187.5 and median 200. The median of a set of four elements is the mean of the two middle elements.

13. **(D)** $\hat{p} = \frac{256}{400} = .64$, and $\sigma_{\hat{p}} = \sqrt{\frac{(.64)(.36)}{400}} = .024$. The z-scores of .60 and .68 are $\pm\frac{.04}{.024} = \pm 1.67$ for a probability of $.9525 - .0475 = .9050$.

14. **(C)** This is a binomial with $n = 15$ and $p = .29$, and so the mean is $np = 15(.29) = 4.35$.

15. **(D)** With independence, variances add, so $\sqrt{65^2 + 380^2 + 125^2} = 405$.

16. **(E)** Without independence we cannot determine var($X + Y$) from the information given.

17. **(A)** Bias is the tendency to favor the selection of certain members of a population. It has nothing to do with the shape of distributions. The natural variability between samples is called *sampling error.*

18. **(C)** There are not two distinct peaks, so the distribution is not bimodal. Relative frequency is given by relative area. The distribution is skewed to the left, and so the mean is less than the median.

19. **(D)** While the standard deviation of the z-distribution is 1, the standard deviation of the t-distribution is greater than 1.

20. **(B)**

$$P\left(\begin{array}{c}\text{market}\\\text{decline}\end{array} \cap \begin{array}{c}\text{interest}\\\text{jumps}\end{array}\right) = P\left(\begin{array}{c}\text{market}\\\text{decline}\end{array} \middle| \begin{array}{c}\text{interest}\\\text{jumps}\end{array}\right) P\left(\begin{array}{c}\text{interest}\\\text{jumps}\end{array}\right)$$
$$= (.72)(.65)$$
$$= .468$$

21. **(E)** Different samples give different sample statistics, all of which are estimates for the same population parameter, and so error, called *sampling error*, is naturally present.

22. **(C)** With four-digit accuracy as found on the TI-83, the critical z-scores for 70% to the right and 80% to the right are -0.5244 and -0.8416, respectively. Then $\{\mu - 0.5244\sigma = 12{,}000,\ \mu - 0.8416\sigma = 10{,}000\}$ gives $\mu = 15{,}306$ and $\sigma = 6305$. Using two-digit accuracy as found in Table A, that is, -0.52 and -0.84, results in $\mu = 15{,}250$ and $\sigma = 6250$.

23. **(D)** The percentage of the variation in y explained by the variation in x is given by the coefficient of determination r^2. In this example, $(.78)^2 = .61$.

24. **(B)** There is a weak negative correlation, and so $-.15$ is the only reasonable possibility among the choices given.

25. **(E)** There is no guarantee that 24 is anywhere near the interval, and so none of the statements is true.

26. **(E)** Sampling error relates to natural variation between samples, and it can never be eliminated. In good observational studies, responses are not influenced during the collection of data. In good experiments, treatments are compared as to the differences in responses.

27. **(E)** The scatterplot suggests a zero correlation.

28. **(A)** $\sigma_{\bar{x}} = \frac{1.2}{\sqrt{36}} = 0.2$. The z-score of 2.5 is $\frac{2.5-2.4}{0.2} = 0.5$, and to the right of 2.5 is the probability of failing to reject the false claim.

29. **(E)** The central limit theorem says that no matter how the original population is distributed, as the sample size increases, the sampling distribution of the sample mean becomes closer to a normal distribution.

30. **(D)** A sampling distribution is the distribution of all the values taken by a statistic, such as sample mean or sample proportion, from all possible samples of a given size.

31. **(E)** $P(\text{at least 2}) = P(\text{exactly 2}) + P(\text{exactly 3}) = 3(.8)^2(.2) + (.8)^3$

32. **(A)** The critical t-scores for 99% confidence with $df = 6$ are ± 3.707.

33. **(B)** The PTA survey has strong response bias in that students may not give truthful responses to a parent or teacher about their engaging in unprotected sex. The talk show survey results in a voluntary response sample, which typically gives too much emphasis to persons with strong opinions. The *Ladies Home Journal* survey has strong selection bias; that is, people who read the *Journal* are not representative of the general population.

34. **(D)** To divide the interval estimate by d, the sample size must be increased by a multiple of d^2.

35. **(B)** $P(0 \text{ pts}) = (.05)^2$, $P(\frac{1}{2} \text{ pt}) = 2(.05)(.15)$, $P(1 \text{ pt}) = (.15)^2 + 2(.05)(.8)$, $P(1\frac{1}{2} \text{ pts}) = 2(.15)(.8)$, and $P(2 \text{ pts}) = (.8)^2$.

36. **(A)** The *P*-value is the probability of obtaining a result as extreme as the one seen *given that* the null hypothesis is true; thus it is a conditional probability. The *P*-value depends on the sample chosen. The *P*-value in a two-sided test is calculated by doubling the indicated tail probability. *P*-values are not restricted to use with any particular distribution. With a small *P*-value, there is evidence to reject the null hypothesis, but we're not *proving* anything.

37. **(C)** A Type II error is a mistaken failure to reject a false null hypothesis or, in this case, a failure to realize that the machinery is turning out wrong size ball bearings.

38. **(E)** Using a measurement from a sample, we are never able to say *exactly* what a population mean is; rather we always say we have a certain *confidence* that the population mean lies in a particular *interval.* In this case we are 95% confident that Lipitor will outperform niacin by 18 to 41 points.

39. **(C)** The median number of hours is less than four for June, July, and August.

40. **(D)** A horizontal line drawn at the .9 probability level corresponds to roughly 19 rooms.

SECTION II

1. (a) In each block, two of the sites will be randomly assigned to receive Type A vegetation, while the remaining two sites in the block will receive Type B vegetation. Randomization of vegetation type to the sites within each block should reduce bias due to any confounding variables. In particular, the randomization in blocks in the first scheme should even out the effect of the distance vegetation in blocks is from the industrial park.

   (b) The first scheme creates homogeneous blocks with respect to distance from the river, while the second scheme creates homogeneous blocks with respect to the industrial park. Randomization of vegetation types to sites within blocks in the first scheme should even out effects of distance from the industrial park, while randomization of vegetation types to sites within blocks in the second scheme should even out effects of distance from the river.

Scoring

It is important to explain why randomization is important within the context of this problem, so use of terms like bias and confounding must be in context. One must explain the importance of homogeneous experimental units (sites, not vegetation types) within blocks.

4 Complete Answer	Correct explanation both of use of blocking and of use of randomization in context of this problem.
3 Substantial Answer	Correct explanation of use of either blocking or randomization in context and a weak explanation of the other.
2 Developing Answer	Correct explanation of use of either blocking or randomization in context OR weak explanations of both.
1 Minimal Answer	Weak explanation of either blocking or randomization in context.

2. (a)

Region A		Region B
5	3	5 7
5	4	1 3 3 8
8 6 6 2	5	0 4
9 6 5 3 0	6	5
7 4 2 1	7	1 8
2	8	2 3 7 9
2	9	1 2

   (b) While the percentages in both regions have roughly symmetric distributions, the percentages from Region A form a distinctly unimodal pattern, while those from Region B are distinctly bimodal. That is, in Region B the countries tended to show either a very low or a very high percentage, while in Region A most countries showed a percentage near the middle one.

(c) Either an argument can be made for Region A because Region B has so many countries with high percentages (for primary-school-age children enrolled in school), or an argument can be made for Region B because Region B has so many countries with low percentages.

(d) Data are already given on *all* the countries in the two regions. In doing inference, one uses sample statistics to estimate population parameters. If the data are actually the whole population, there is no point of a *t*-test.

Scoring

Part (a) is essentially correct for a correct stemplot (numbers in each row do not have to be in order) with labeling as to which side refers to which region. Part (a) is partially correct for a correct stemplot missing the labeling.

Part (b) is essentially correct for noting that the percentages from Region A form a distinctly unimodal pattern, while those from Region B are distinctly bimodal. Part (b) is partially correct for noting either the unimodal pattern from the Region A data *or* the bimodal pattern from the Region B data.

Part (c) is essentially correct for choosing either region and giving a reasonable argument for the choice. Part (c) is partially correct if the argument is weak. Part (c) is incorrect for no argument or an incorrect argument for the choice given.

Part (d) is essentially correct for explaining that inference is not proper when the data are the whole population. Part (d) is partially correct if the explanation is correct but weak.

Count partially correct answers as one-half an essentially correct answer.

4 Complete Answer	Four essentially correct answers.
3 Substantial Answer	Three essentially correct answers.
2 Developing Answer	Two essentially correct answers.
1 Minimal Answer	One essentially correct answer.

Use a holistic approach to decide a score totaling between two numbers.

3. (a) Note that there are a great number of values between 2.0 and 2.4, fewer values between 2.4 and 2.8, on down to the fewest values being between 3.6 and 4.0. Thus the distribution is skewed right (skewed toward the higher values).

   (b) Note that the greatest five values, out of the 100 values, are above a point roughly at 3.4 or 3.5, so the 95th percentile is roughly 3.4 or 3.5.

   (c) In a normal distribution, the 95th percentile has a *z*-score of 1.645. Reading up from the *x*-axis to the normal line and across to the *y*-axis gives a GPA of approximately 3.2.

Scoring

Part (a) is essentially correct if the shape is correctly identified as skewed right, and a correct explanation based on the given normal probability plot is given. Part (a) is partially correct for a correct identification with a weak explanation. Part (a) is incorrect if a correct identification is given with no explanation.

Part (b) is essentially correct for correctly noting 3.4 or 3.5 as the 95th percentile with a correct explanation. Part (b) is partially correct for giving 3.4 or 3.5 with a weak explanation or no explanation.

Part (c) is essentially correct for noting approximately 3.2 as the GPA and stating the method clearly. Part (c) is partially correct for giving 3.2 with a weak explanation or no explanation.

4 Complete Answer	All three parts essentially correct.
3 Substantial Answer	Two parts essentially correct and one part partially correct.
2 Developing Answer	Two parts essentially correct OR one part essentially correct and one or two parts partially correct OR all three parts partially correct.
1 Minimal Answer	One part essentially correct OR two parts partially correct.

4. (a) There are four elements to this solution.

State the hypotheses:

H_0: The distribution of blood types among patients at this hospital is in the ratios of 9:8:2:1 for types O, A, B, and AB, respectively.

H_a: The distribution of blood types among patients at this hospital is not in the ratios of 9:8:2:1 for types O, A, B, and AB, respectively.

Identify the test by name or formula and check the assumptions:

Chi-square goodness-of-fit test:

$$\chi^2 = \sum \frac{(\textbf{observed} - \textbf{expected})^2}{\textbf{expected}}$$

Check the assumptions:

1. The researcher claims that the data are from a random sample of patient records.
2. The ratios 9:8:2:1 give expected cell frequencies of $\frac{9}{20}(500) = 225$, $\frac{8}{20}(500) = 200$, $\frac{2}{20}(500) = 50$, $\frac{1}{20}(500) = 25$, each of which is at least 5.

Demonstrate correct mechanics:

$$\chi^2 = \frac{(253-225)^2}{225} + \frac{(194-200)^2}{200} + \frac{(38-50)^2}{50} + \frac{(15-25)^2}{25} = 10.544$$

With $df = 4 - 1 = 3$, the *P*-value is .0145.

State the conclusion in context with linkage to the *P*-value:

With this small a *P*-value (for example, less than $\alpha = .05$), there is evidence that the distribution of blood types among patients at this hospital is *not* in the ratios of 9:8:2:1 for types O, A, B, and AB, respectively.

(b) In this case the mechanics give:

$$\chi^2 = \frac{(230-225)^2}{225} + \frac{(196-200)^2}{200} + \frac{(48-50)^2}{50} + \frac{(26-25)^2}{25} = 0.311$$

With $df = 4 - 1 = 3$, the probability that χ^2 is *less than or equal to* 0.311 is .0421. With such a small probability, there is evidence that the data are too good to be true; that is, there is evidence that Researcher 2 made up the data to fit the 9:8:2:1 model.

Scoring

Part (a1) is essentially correct if the hypotheses are given, the test is identified, and the assumptions are checked. Part (a1) is partially correct for two of these three steps.

Part (a2) is essentially correct for correct mechanics and the conclusion given in context with linkage to the *P*-value. Part (a2) is partially correct if there is a minor error in mechanics OR the conclusion is not in context OR there is no linkage to the *P*-value.

Part (b) is essentially correct for noting that the probability of a χ^2 value of only 0.311 is very small, and thus the data are suspicious. Part (b) is partially correct for a correct idea but with a weak explanation.

4 Complete Answer	All three parts essentially correct.
3 Substantial Answer	Two parts essentially correct and one part partially correct.
2 Developing Answer	Two parts essentially correct OR one part essentially correct and one or two parts partially correct OR all three parts partially correct.
1 Minimal Answer	One part essentially correct OR two parts partially correct.

5. (a) The regression line is $\widehat{mpg} = 40.2462 - 4.61969\ liter$.
   Then $40.2462 - 4.61969(1.5) \approx 33.32$.
   Residual = actual – predicted = 35 – 33.32 = 1.68
   (b) The predicted value for mpg is $40.2462 - 4.61969(5.0) \approx 17.15$, which is very close to the actual value of 17. The added point is consistent with the linear pattern given by the computer output, so the new slope should be about the same as the old slope.
   (c) Since the new point fits the old linear pattern so well, and has an x-value (*liter*) much greater than $\bar{x} = 2.52$, the absolute value of the correlation will increase.

Scoring

Part (a) is essentially correct for a correct calculation of the residual. Part (a) is partially correct for noting that residual = actual – predicted, but obtaining an incorrect answer because of an incorrect derivation of the regression line.

Part (b) is essentially correct for showing that the actual value is very close to the predicted value and concluding that the new slope should be about the same. Part (b) is partially correct if the predicted value is miscalculated but a proper conclusion based on the miscalculation is reached.

Part (c) is essentially correct for concluding that the absolute value of the correlation will increase because the new point fits the old linear pattern so well. Part (c) is partially correct for concluding that the absolute value of the correlation will increase but giving a weak explanation.

4 Complete Answer	All three parts essentially correct.
3 Substantial Answer	Two parts essentially correct and one part partially correct.
2 Developing Answer	Two parts essentially correct OR one part essentially correct and one or two parts partially correct OR all three parts partially correct.
1 Minimal Answer	One part essentially correct OR two parts partially correct.

6. (a) State the hypotheses: H_0: $\mu_d = 0$ and H_a: $\mu_d > 0$ where μ_d is the mean difference (2000 – 1999) in the populations of the southern states.

   Identify the test: paired t-test.

   Calculate the test statistic t and the P-value: $t = \dfrac{1600 - 0}{\frac{1758}{\sqrt{5}}} = 2.035$ with $df = 5 - 1 = 4$, $P = .056$

   Conclusion in context: Since $P = .056 > .05$, at the $\alpha = .05$ significance level, there is no evidence to conclude that southern states grew between 1990 and 2000.

(b) Identify the confidence interval: two-sample t-interval for $\mu_N - \mu_S$, the difference in mean populations between northern and southern states in 1990.

Calculate the confidence interval: with $df = (5 - 1) + (5 - 1) = 8$,

$$(8600 - 7929) \pm 2.306\sqrt{\frac{6755^2}{5} + \frac{6665^2}{5}} = 680 \pm 9{,}786$$

Interpretation in context: We are 95% confident that the true difference in the average populations of northern and southern states in 1990 was between –9,106 and 10,466 thousand people.

(c) State the hypotheses: H_0: $\mu_d = 0$ and H_a: $\mu_d > 0$ where μ_d is the difference in the mean growths of populations of southern and northern states.

Identify the test: two-sample t-test.

Calculate the test statistic t and the P-value: $t = \dfrac{1600 - 608}{\sqrt{\frac{407^2}{5} + \frac{1758^2}{5}}} = 1.229$ and with $df = \min\{5 - 1, 5 - 1\} = 4$ we have $P = .14$.

State the conclusion in context: With this large a P value, there is no evidence that southern states grew faster than northern states between 1990 and 2000.

Scoring

Conditions for inference are assumed to be met and don't have to be mentioned.

Part (a) is essentially correct if all four steps are correct and partially correct if three steps are correct.

Part (b) is essentially correct if all three steps are correct and partially correct if two steps are correct. Full credit for $df = 4$ or 8.

Part (c) is essentially correct if all four steps are correct and partially correct if three steps are correct. Full credit for $df = 4$, 4.43, or 8.

4 Complete Answer	All three parts essentially correct.
3 Substantial Answer	Two parts essentially correct and one part partially correct.
2 Developing Answer	Two parts essentially correct OR one part essentially correct and one or two parts partially correct OR all three parts partially correct.
1 Minimal Answer	One part essentially correct OR two parts partially correct.

Relating Multiple-Choice Problems to Review Book Topics

The table below gives the above Topic corresponding to each MC question.

Question	Exam 1	Exam 2	Exam 3	Exam 4	Exam 5
1	1, 2	4	7, 8	4	12
2	4, 6	7, 8	2	7	3
3	11	11	8	14	8
4	13	4	13	11	13
5	6	2	8	1, 2	9, 11
6	9	9	15	14	9
7	14	13	4	8	13, 14
8	9, 12	14	10	7, 8	2
9	9	7	7	4	7, 8
10	9	4	11	7	12
11	14	3	11	6, 14	13
12	8	2	7	1, 2	2
13	1, 2	10	9	7	13
14	2	9	14	12	9
15	14	13	2	9	10
16	2	6	13	9, 11	10
17	11, 15	2	7	13	7
18	7	8	9	4	1, 2
19	2	14	11	15	11, 12
20	10	1, 2	3	9	9
21	4	9	12, 13	9	7
22	13	11	9	7, 8	11
23	14	8	9	11	4
24	2	2	9	3	4
25	7	4	7	4	13
26	13	4	4	9	7, 8
27	4	9	7	13	4
28	9	7	1, 2	12	14
29	14	7	8	14	12
30	3	1, 2	13	13	12
31	9, 11	14	14	9	9
32	14	11	2, 3	4	13
33	7	13	14	2	7
34	7	12	7, 8	7, 8	13
35	13	3	15	9	9
36	14	12	14	14	14
37	12	14	4	14	14
38	4	1, 2	9	2	13
39	13	14	3	8	3
40	13	15	13	3	1

Appendix

CHECKING ASSUMPTIONS FOR INFERENCE

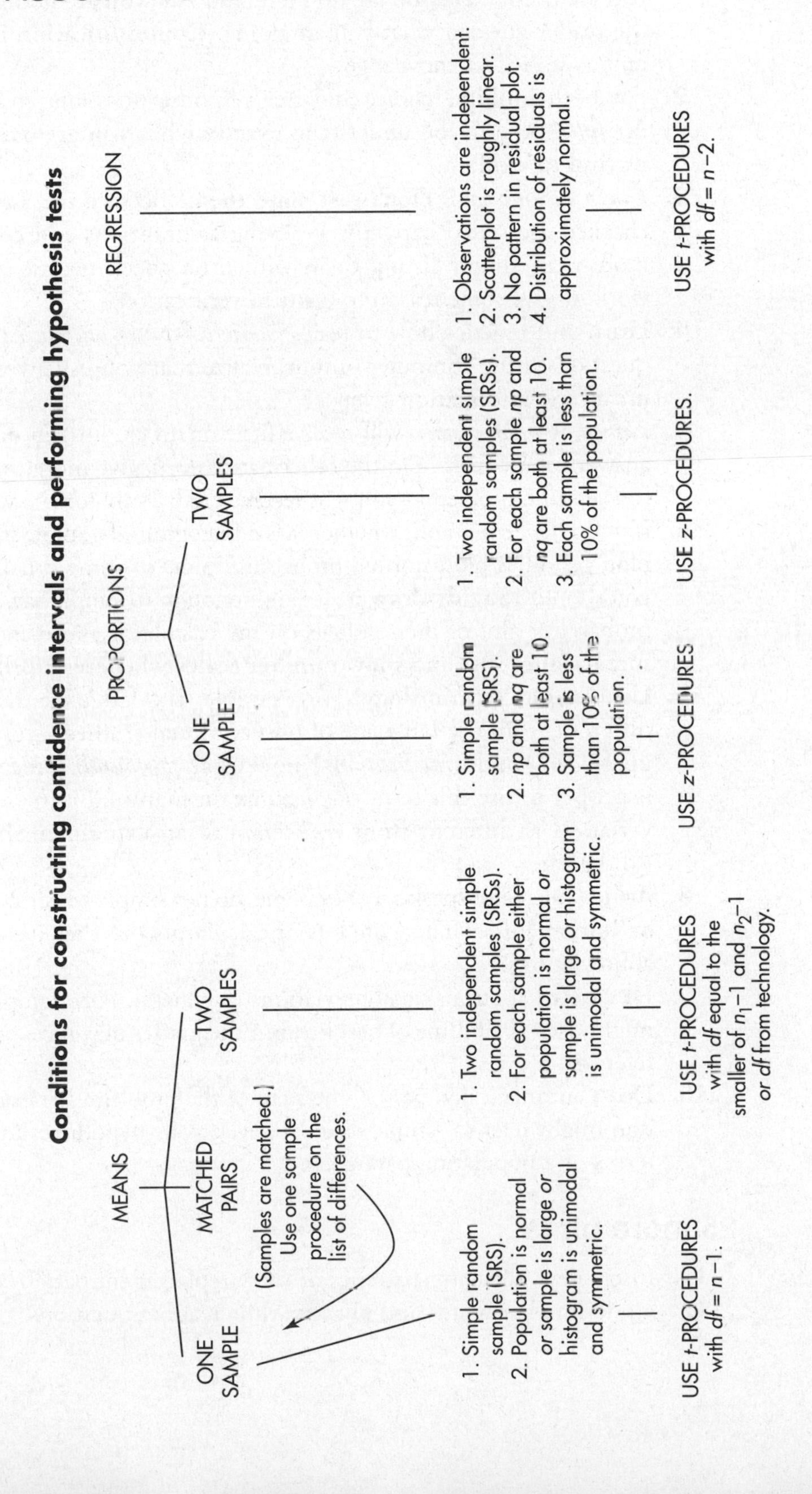

Conditions for constructing confidence intervals and performing hypothesis tests
MEANS
ONE SAMPLE
MATCHED PAIRS
TWO SAMPLES
(Samples are matched.) Use one sample procedure on the list of differences.
1. Simple random sample (SRS).
2. Population is normal or sample is large or histogram is unimodal and symmetric.
USE t-PROCEDURES with df = n−1.
1. Two independent simple random samples (SRSs).
2. For each sample either population is normal or sample is large or histogram is unimodal and symmetric.
USE t-PROCEDURES with df equal to the smaller of n1−1 and n2−1 or df from technology.
PROPORTIONS
ONE SAMPLE
TWO SAMPLES
1. Simple random sample (SRS).
2. np and nq are both at least 10.
3. Sample is less than 10% of the population.
USE z-PROCEDURES
1. Two independent simple random samples (SRSs).
2. For each sample np and nq are both at least 10.
3. Each sample is less than 10% of the population.
USE z-PROCEDURES.
REGRESSION
1. Observations are independent.
2. Scatterplot is roughly linear.
3. No pattern in residual plot.
4. Distribution of residuals is approximately normal.
USE t-PROCEDURES with df = n−2.

35 AP EXAM HINTS

The top ten:

1. Graders want to give you credit—help them! Make them understand *what* you are doing, *why* you are doing it, and *how* you are doing it. Don't make the reader guess at what you are doing. **Communication** is just as important as statistical knowledge!
2. For both multiple choice and free response questions, read the question *carefully!* Be sure you understand **exactly what you are being asked to do or find or explain**.
3. *Check assumptions.* **Don't just state them!** Be sure the assumptions to be checked are stated correctly. Verifying assumptions and conditions means more than simply listing them with little check marks—you must show work or give some reason to confirm verification.
4. Learn and practice how to read *generic computer output.* And in answering questions using computer output, realize that you usually will **not** need to use all the information given.
5. *Naked* or *bald answers* will receive little or **no** credit! You must show where answers come from. On the other hand, don't give more than one solution to the same problem—you will receive credit only for the weaker one.
6. If you refer to a graph, whether it is a histogram, boxplot, stemplot, scatterplot, residuals plot, normal probability plot, or some other kind of graph, you should **roughly draw it**. It is not enough to simply say, "I did a normal probability plot of the residuals on my calculator and it looked linear." Be sure to **label axes** and **show number scales** whenever possible.
7. **Use proper terminology!** For example, the language of experiments is different from the language of observational studies—you shouldn't mix up *blocking* and *stratification.* Know what *confounding* means and when it is proper to use this term. *Replication* on many subjects to reduce chance variation is different from *replication* of an experiment itself to achieve validation.
8. **Avoid** "calculator speak"! For example, do not simply write "2-SampTTest . . ." or "binomcdf. . ." There are lots of calculators out there, each with its own abbreviations.
9. **Be careful** about using abbreviations in general. For example, your teacher might use LOBF (line of best fit), but the grader may have no idea what this refers to.
10. **Don't** automatically "parrot" the stem of the problem! For example, the question might refer to sample data; however, your hypotheses must be stated in terms of a population parameter.

25 more hints:

11. Insert new calculator batteries or carry replacement batteries with you.
12. Underline key words and phrases while reading questions.

13. Some problems look scary on first reading but are not overly difficult and are surprisingly straightforward if you approach them systematically. And some questions might take you beyond the scope of the AP curriculum; however, remember that they will be phrased in such ways that you should be able to answer them based on what you have learned in your AP Stat class.
14. Read through all six free-response questions, unerlining key points. Go back and do those questions you think are easiest (this will usually, but not always, include Problem 1). Then tackle the heavily weighted Problem 6 for a while. Finally, try the remaining problems. If you have a few minutes at the end to try some more on Problem 6, that's great!
15. Answers do not have to be in paragraph form. Symbols and algebra are fine. Just be sure that your method, reasoning, and calculations will be clear to the reader, and that the explanations and conclusions are given *in context of the problem.*
16. If you show calculations carefully, a wrong answer due to a computational error might still result in full credit.
17. If you can't solve part of a problem, but that solution is necessary to proceed, make up a reasonable answer to the first part and use this to proceed to the remaining parts of the problem.
18. When using a formula, write down the formula and then substitute the values.
19. Conclusions should be stated in proper English. For example, don't use double negatives.
20. Read carefully and recognize that sometimes very different tests are required in different parts of the same problem.
21. Realize that there may be several reasonable approaches to a given problem. In such cases pick either the one you feel most comfortable with or the one you feel will require the least time.
22. Realize that there may not be one clear, correct answer. Some questions are designed to give you an opportunity to creatively synthesize a relationship among the problem's statistical components.
23. Punching a long list of data into the calculator doesn't show statistical knowledge. Be sure it's necessary!
24. Remember that uniform and symmetric are not the same, and that not all symmetric unimodal distributions are normal.
25. When making predictions and interpreting slopes and intercepts, don't become confused between your original data and your regression model.
26. If the slope is close to ± 1, it does *not* follow that the correlation is strong.
27. When describing residual graphs, "randomly scattered" does *not* mean the same as "half below and half above." You should comment on whether or not there are nonlinear patterns, and increasing or decreasing spread, and on whether the residuals are small or large compared to the associated y-values.
28. Simply using a calculator to find a regression line is not enough; you must understand it (for example, be able to interpret the slope and intercepts in context).
29. With regard to hypothesis test problems, (1) the hypotheses must be stated in terms of population parameters with all variables clearly defined; (2) the test must be specified by name or formula, and you must show how the test

assumptions are met; (3) the test statistic and associated *P*-value should be calculated; (4) the *P*-value must be *linked* to a decision; and (5) a conclusion must be stated *in the context* of the problem.

30. For any inference problem, be sure to ask yourself whether the variable is categorical (usually leading to proportions) or quantitative (usually leading to means), whether you are working with raw data or summary statistics, whether there is a single population of interest or two populations being compared, and in the case of comparison whether there are independent samples or a paired comparison.

31. A simple random sample (SRS) and a random assignment of treatments to subjects both have to do with randomness, but they are not the same. Understand the difference!

32. Simply saying to "randomly assign" subjects to treatment groups is usually an incomplete response. You need to explain how to make the assignments, for example, using a random number table or through generating random numbers on a calculator.

33. Blinding and placebos in experiments are important but are not always feasible. You can still have "experiments" without these.

34. The distribution of a particular sample (data set) is not the same as the sampling distribution of a statistic. Understand the difference!

35. Just because a sample is large does not imply that the distribution of the sample will be close to a normal distribution.

AP SCORING GUIDE

The Multiple-Choice and Free-Response sections are weighted equally.

There is no penalty for guessing in the Multiple-Choice section.

The Investigative Task counts for 25% of the Free-Response section. Each of the Free-Response questions has a possible four points.

To find your score use the following guide:

Multiple-Choice section (40 questions)

Number correct × 1.25 = __________

Free-Response section (5 Open-Ended questions plus an Investigative Task)

Question 1 __________ (out of 4) × 1.875 = __________

Question 2 __________ (out of 4) × 1.875 = __________

Question 3 __________ (out of 4) × 1.875 = __________

Question 4 __________ (out of 4) × 1.875 = __________

Question 5 __________ (out of 4) × 1.875 = __________

Question 6 __________ (out of 4) × 3.125 = __________

Total points from Multiple-Choice and Free-Response sections = __________

Conversion chart based on a recent AP exam

Total points	AP Score
70–100	5
57–69	4
44–56	3
33–43	2
0–32	1

In the past, roughly 10% of students scored 5, 20% scored 4, 25% scored 3, 20% scored 2, and 25% scored 1 on the AP Statistics exam. Colleges generally require a score of at least 3 for a student to receive college credit.

BASIC USES OF THE TI-83/TI-84

There are many more useful features than introduced below – see the guidebook that comes with the calculator. For reference, following is a listing of the basic uses with which all students should be familiar. However, always remember that the calculator is only a tool, that it will find minimal use in the multiple-choice section, and that "calculator talk" (calculator syntax) should NOT be used in the free-response section.

Plotting statistical data:

> **STAT PLOT** allows one to show scatterplots, histograms, modified boxplots, and regular boxplots of data stored in lists. Note the use of **TRACE** with the various plots.

Numerical statistical data:

> **1-Var Stats** gives the mean, standard deviation, and 5-number summary of a list of data.

Binomial probabilities:

> **binompdf** (n, p, x) gives the probability of exactly x successes in n trials where p is the probability of success on a single trial.
> **binomcdf** (n, p, x) gives the cumulative probability of x or fewer successes in n trials where p is the probability of success on a single trial.

Geometric probabilities:

> **geometpdf** (p, x) gives the probability that the first success occurs on the x-th trial, where p is the probability of success on a single trial.
> **geometcdf** (p, x) gives the cumulative probability that the first success occurs on or before the x-th trial, where p is the probability of success on a single trial.

The normal distribution:

> **normalcdf** (lowerbound, upperbound, μ, σ) gives the probability that a score is between the two bounds for the designated mean μ and standard deviation σ. The defaults are $\mu = 0$ and $\sigma = 1$.
> **InvNorm** (area, μ, σ) gives the score associated with an area (probability) to the left of the score for the designated mean μ and standard deviation σ. The defaults are $\mu = 0$ and $\sigma = 1$.

The t-distribution:

> **tcdf** (lowerbound, upperbound, df) gives the probability a score is between the two bounds for the specified df (degrees of freedom).
> **invT**(area, df) gives the t-score associated with an area (probability) to the left of the score under the student t-probability function for the specified df (degrees of freedom). [Note: this is available on the new operating system for the TI-84+.]

The chi-square distribution:

> **χ^2cdf** (lowerbound, upperbound, df) gives the probability a score is between the two bounds for the specified df (degrees of freedom).

χ^2**GOF-Test** is a chi-square goodness-of-fit test to confirm whether sample data conforms to a specified distribution. [Note: this is available on the new operating system for the TI-84+.]

Linear regression and correlation:

LinReg (ax + b) fits the equation $y = ax + b$ to the data in lists L1 and L2 using a least-squares fit. When **DiagnosticOn** is set, the values for r^2 and r are also displayed.

Confidence intervals:

For proportions—

1-PropZInt gives a confidence interval for a proportion of successes.
2-PropZInt gives a confidence interval for the difference between the proportion of successes in two populations.

For means—

TInterval gives a confidence interval for a population mean (use the t-distribution because population variances are never really known).
2-SampTInt gives a confidence interval for the difference between two population means.

Hypothesis tests:

For proportions—

1-PropZTest
2-PropZTest compares the proportion of successes from two populations (making use of the pooled sample proportion).

For means—

T-Test
2-SampTTest

For chi-square test for association—

χ^2**-Test** gives the χ^2-value and P-value for the null hypothesis H_0: no association between row and column variables, and the alternative hypothesis H_a: the variables are related. The observed counts must first be entered into a matrix.

For linear regression—

LinRegTTest calculates a linear regression and performs a t-test on the null hypothesis H_0: $\beta = 0$ (H_0: $\rho = 0$). The regression equation is stored in **RegEQ** (under **VARS Statistics EQ**) and the list of residuals is stored in **RESID** (under **LIST NAMES**).

Catalog help:

To activate Catalog Help, press APPS, choose CtlgHelp, and press ENTER. Then, for example, if you press 2nd, DISTR, arrow down to normalcdf, and press +, you are prompted to insert (lowerbound, upperbound, [μ,σ]), that is, to insert the bounds and, optionally, the mean and SD.

FORMULAS

The following are in the form and notation as recommended by the College Board.

Descriptive Statistics

$$\bar{x} = \frac{\Sigma x_i}{n}$$

$$s = \sqrt{\frac{1}{n-1}\Sigma\left(x_i - \bar{x}\right)^2}$$

$$s_p = \sqrt{\frac{\left(n_1 - 1\right)s_1^{\,2} + \left(n_2 - 1\right)s_2^{\,2}}{\left(n_1 - 1\right) + \left(n_2 - 1\right)}}$$

$$\hat{y} = b_0 + b_1 x$$

$$b_1 = \frac{\Sigma\left(x_i - \bar{x}\right)\left(y_i - \bar{y}\right)}{\Sigma\left(x_i - \bar{x}\right)^2}$$

$$b_0 = \bar{y} - b_1\bar{x}$$

$$r = \frac{1}{n-1}\sum\left(\frac{x_i - \bar{x}}{s_x}\right)\left(\frac{y_i - \bar{y}}{s_y}\right)$$

$$b_1 = r\frac{s_y}{s_x}$$

$$s_{b1} = \frac{\sqrt{\frac{\Sigma\left(y_i - \hat{y}_i\right)^2}{n-2}}}{\sqrt{\Sigma\left(x_i - \bar{x}\right)^2}}$$

Probability

$$P(A \cup B) = P(A) + P(B) - P(A \cap B)$$

$$P\left(A \mid B\right) = \frac{P\left(A \cap B\right)}{P\left(B\right)}$$

$$E\left(X\right) = \mu_x = \sum x_i p_i$$

$$\operatorname{var}\left(X\right) = \sigma_x^2 = \sum\left(x_i - \mu_x\right)^2 p_i$$

If X has a binomial distribution with parameters n and p, then:

$$P\left(X = k\right) = \binom{n}{k} p^k \left(1 - p\right)^{(n-k)}$$

$$\mu_x = np$$

$$\sigma_x = \sqrt{np\left(1 - p\right)}$$

$$\mu_{\hat{p}} = p$$

$$\sigma_{\hat{p}} = \sqrt{\frac{p\left(1 - p\right)}{n}}$$

If X has a normal distribution with mean μ and standard deviation σ, then:

$$\mu_{\bar{x}} = \mu$$

$$\sigma_{\bar{x}} = \frac{\sigma}{\sqrt{n}}$$

Inferential Statistics

Standardized statistic: $\dfrac{\text{Estimate} - \text{parameter}}{\text{Standard deviation of the estimate}}$

Confidence interval:
Estimate ± (critical value) · (standard deviation of the estimate)

SINGLE SAMPLE

Statistic	Standard Deviation
Mean	$\dfrac{\sigma}{\sqrt{n}}$
Proportion	$\sqrt{\dfrac{p(1-p)}{n}}$

TWO SAMPLE

Statistic	Standard Deviation
Difference of means (unequal variances)	$\sqrt{\dfrac{\sigma_1^2}{n_1} + \dfrac{\sigma_2^2}{n_2}}$
Difference of means (equal variances)	$\sigma\sqrt{\dfrac{1}{n_1} + \dfrac{1}{n_2}}$
Difference of proportions (unequal variances)	$\sqrt{\dfrac{p_1(1-p_1)}{n_1} + \dfrac{p_2(1-p_2)}{n_2}}$
Difference of proportions (equal variances)	$\sqrt{p(1-p)}\sqrt{\dfrac{1}{n_1} + \dfrac{1}{n_2}}$

$$\text{Chi-square statistic} = \sum \frac{(\text{observed} - \text{expected})^2}{\text{expected}}$$

GRAPHICAL DISPLAYS

Right-skewed | Symmetric | Left-skewed

Dotplots

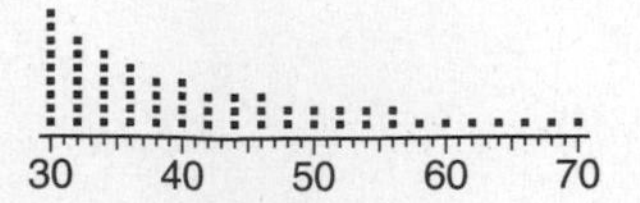

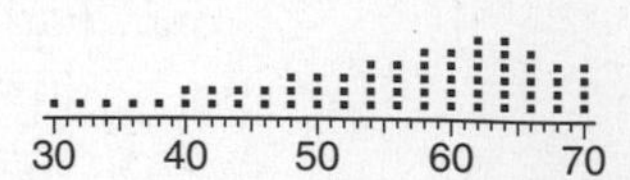

Stemplots

Right-skewed:

```
3 | 000000000222222244444466666668888
4 | 000022244466688
5 | 002244668
6 | 02468
7 | 0
```

Symmetric:

```
3 | 0246688
4 | 0002224444666666888888
5 | 000000222222444446666888
6 | 000224468
7 | 0
```

Left-skewed:

```
3 | 02468
4 | 00224466888
5 | 000222444666688888
6 | 00000222222444444666668888
7 | 0000
```

Histograms

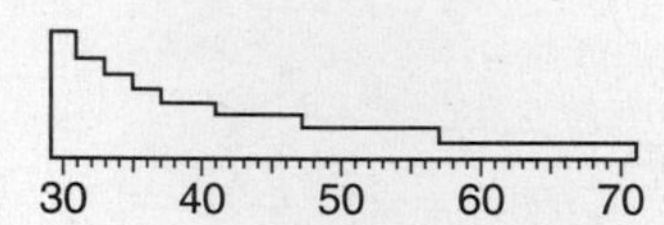

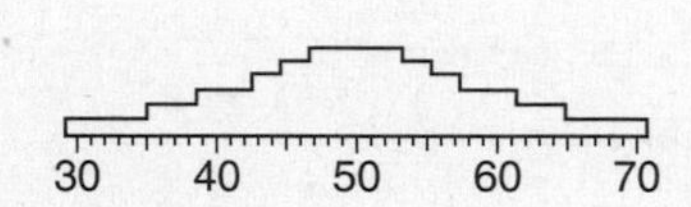

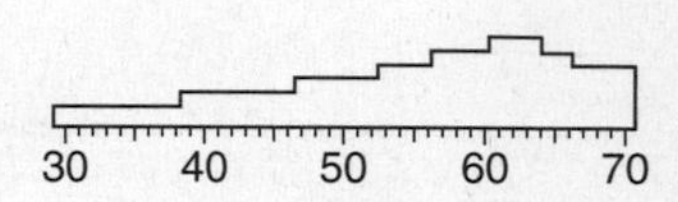

Cumulative Frequency Plots

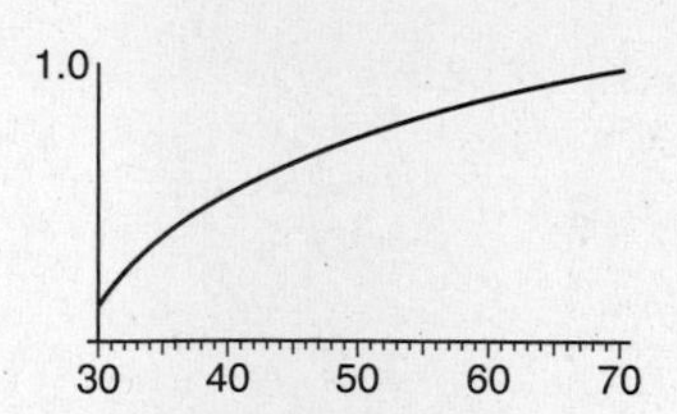

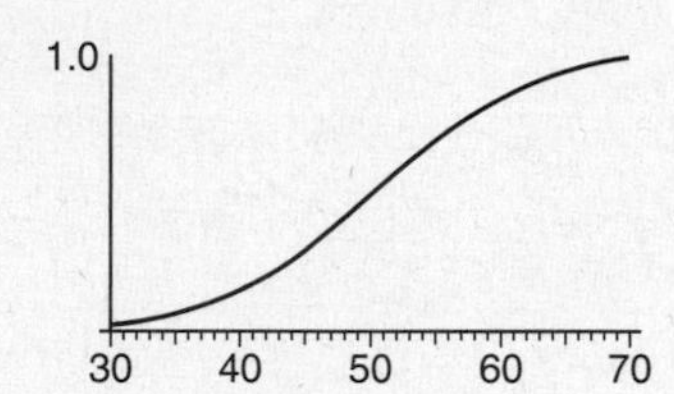

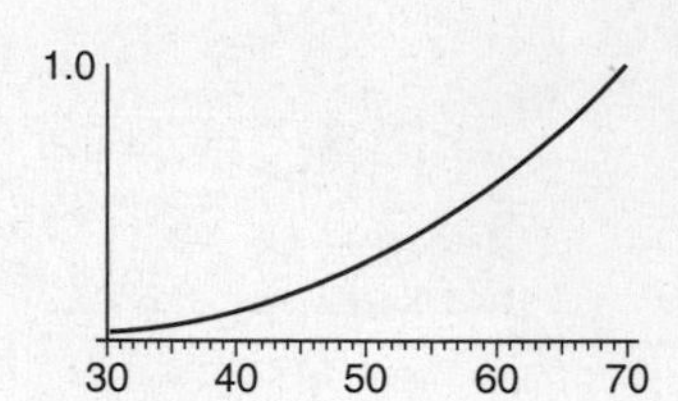

Boxplots

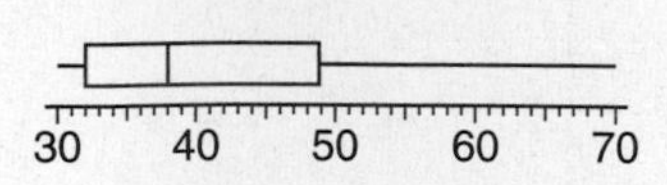

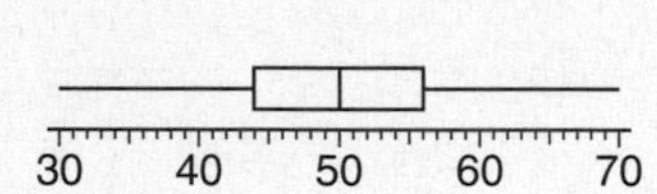

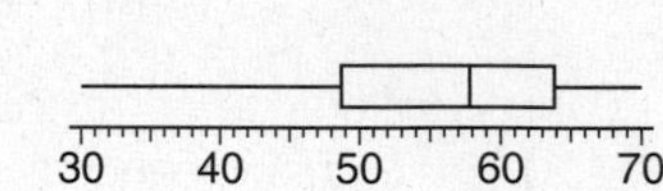

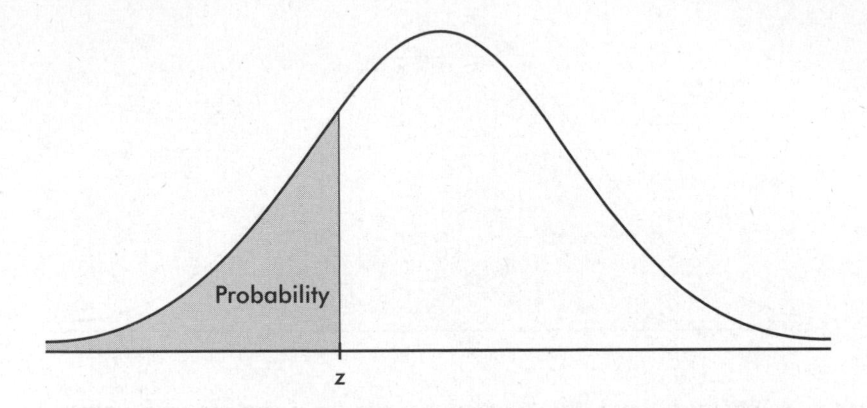

TABLE A: STANDARD NORMAL PROBABILITIES

z	.00	.01	.02	.03	.04	.05	.06	.07	.08	.09
–3.4	.0003	.0003	.0003	.0003	.0003	.0003	.0003	.0003	.0003	.0002
–3.3	.0005	.0005	.0005	.0004	.0004	.0004	.0004	.0004	.0004	.0003
–3.2	.0007	.0007	.0006	.0006	.0006	.0006	.0006	.0005	.0005	.0005
–3.1	.0010	.0009	.0009	.0009	.0008	.0008	.0008	.0008	.0007	.0007
–3.0	.0013	.0013	.0013	.0012	.0012	.0011	.0011	.0011	.0010	.0010
–2.9	.0019	.0018	.0018	.0017	.0016	.0016	.0015	.0015	.0014	.0014
–2.8	.0026	.0025	.0024	.0023	.0023	.0022	.0021	.0021	.0020	.0019
–2.7	.0035	.0034	.0033	.0032	.0031	.0030	.0029	.0028	.0027	.0026
–2.6	.0047	.0045	.0044	.0043	.0041	.0040	.0039	.0038	.0037	.0036
–2.5	.0062	.0060	.0059	.0057	.0055	.0054	.0052	.0051	.0049	.0048
–2.4	.0082	.0080	.0078	.0075	.0073	.0071	.0069	.0068	.0066	.0064
–2.3	.0107	.0104	.0102	.0099	.0096	.0094	.0091	.0089	.0087	.0084
–2.2	.0139	.0136	.0132	.0129	.0125	.0122	.0119	.0116	.0113	.0110
–2.1	.0179	.0174	.0170	.0166	.0162	.0158	.0154	.0150	.0146	.0143
–2.0	.0228	.0222	.0217	.0212	.0207	.0202	.0197	.0192	.0188	.0183
–1.9	.0287	.0281	.0274	.0268	.0262	.0256	.0250	.0244	.0239	.0233
–1.8	.0359	.0351	.0344	.0336	.0329	.0322	.0314	.0307	.0301	.0294
–1.7	.0446	.0436	.0427	.0418	.0409	.0401	.0392	.0384	.0375	.0367
–1.6	.0548	.0537	.0526	.0516	.0505	.0495	.0485	.0475	.0465	.0455
–1.5	.0668	.0655	.0643	.0630	.0618	.0606	.0594	.0582	.0571	.0559
–1.4	.0808	.0793	.0778	.0764	.0749	.0735	.0721	.0708	.0694	.0681
–1.3	.0968	.0951	.0934	.0918	.0901	.0885	.0869	.0853	.0838	.0823
–1.2	.1151	.1131	.1112	.1093	.1075	.1056	.1038	.1020	.1003	.0985
–1.1	.1357	.1335	.1314	.1292	.1271	.1251	.1230	.1210	.1190	.1170
–1.0	.1587	.1562	.1539	.1515	.1492	.1469	.1446	.1423	.1401	.1379
–0.9	.1841	.1814	.1788	.1762	.1736	.1711	.1685	.1660	.1635	.1611
–0.8	.2119	.2090	.2061	.2033	.2005	.1977	.1949	.1922	.1894	.1867
–0.7	.2420	.2389	.2358	.2327	.2296	.2266	.2236	.2206	.2177	.2148
–0.6	.2743	.2709	.2676	.2643	.2611	.2578	.2546	.2514	.2483	.2451
–0.5	.3085	.3050	.3015	.2981	.2946	.2912	.2877	.2843	.2810	.2776
–0.4	.3446	.3409	.3372	.3336	.3300	.3264	.3228	.3192	.3156	.3121
–0.3	.3821	.3783	.3745	.3707	.3669	.3632	.3594	.3557	.3520	.3483
–0.2	.4207	.4168	.4129	.4090	.4052	.4013	.3974	.3936	.3897	.3859
–0.1	.4602	.4562	.4522	.4483	.4443	.4404	.4364	.4325	.4286	.4247
–0.0	.5000	.4960	.4920	.4880	.4840	.4801	.4761	.4721	.4681	.4641

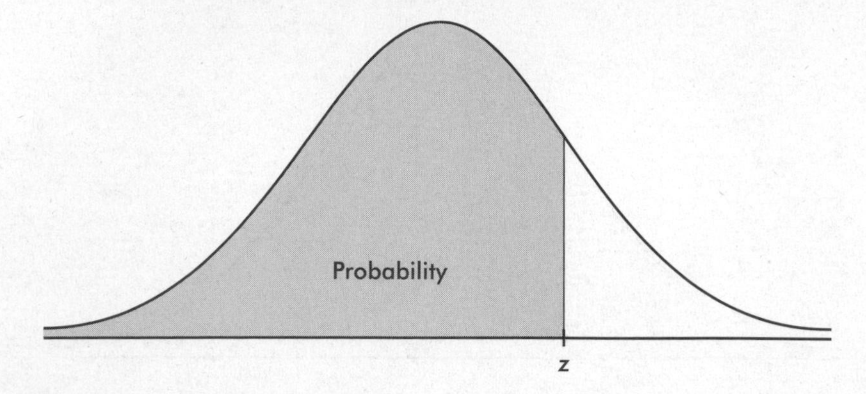

TABLE A: STANDARD NORMAL PROBABILITIES (CONTINUED)

z	.00	.01	.02	.03	.04	.05	.06	.07	.08	.09
0.0	.5000	.5040	.5080	.5120	.5160	.5199	.5239	.5279	.5319	.5359
0.1	.5398	.5438	.5478	.5517	.5557	.5596	.5636	.5675	.5714	.5753
0.2	.5793	.5832	.5871	.5910	.5948	.5987	.6026	.6064	.6103	.6141
0.3	.6179	.6217	.6255	.6293	.6331	.6368	.6406	.6643	.6480	.6517
0.4	.6554	.6591	.6628	.6664	.6700	.6736	.6772	.6808	.6844	.6879
0.5	.6915	.6950	.6985	.7019	.7054	.7088	.7123	.7157	.7190	.7224
0.6	.7257	.7291	.7324	.7357	.7389	.7422	.7454	.7486	.7517	.7549
0.7	.7580	.7611	.7642	.7673	.7704	.7734	.7764	.7794	.7823	.7852
0.8	.7881	.7910	.7939	.7967	.7995	.8023	.8051	.8078	.8106	.8133
0.9	.8159	.8186	.8212	.8238	.8264	.8289	.8315	.8340	.8365	.8389
1.0	.8413	.8438	.8461	.8485	.8508	.8531	.8554	.8577	.8599	.8621
1.1	.8643	.8665	.8686	.8708	.8729	.8749	.8770	.8790	.8810	.8830
1.2	.8849	.8869	.8888	.8907	.8925	.8944	.8962	.8980	.8997	.9015
1.3	.9032	.9049	.9066	.9082	.9099	.9115	.9131	.9147	.9162	.9177
1.4	.9192	.9207	.9222	.9236	.9251	.9265	.9279	.9292	.9306	.9319
1.5	.9332	.9345	.9357	.9370	.9382	.9394	.9406	.9418	.9429	.9441
1.6	.9452	.9463	.9474	.9484	.9495	.9505	.9515	.9525	.9535	.9545
1.7	.9554	.9564	.9573	.9582	.9591	.9599	.9608	.9616	.9625	.9633
1.8	.9641	.9649	.9656	.9664	.9671	.9678	.9686	.9693	.9699	.9706
1.9	.9713	.9719	.9726	.9732	.9738	.9744	.9750	.9756	.9761	.9767
2.0	.9772	.9778	.9783	.9788	.9793	.9798	.9803	.9808	.9812	.9817
2.1	.9821	.9826	.9830	.9834	.9838	.9842	.9846	.9850	.9854	.9857
2.2	.9861	.9864	.9868	.9871	.9875	.9878	.9881	.9884	.9887	.9890
2.3	.9893	.9896	.9898	.9901	.9904	.9906	.9909	.9911	.9913	.9916
2.4	.9918	.9920	.9922	.9925	.9927	.9929	.9931	.9932	.9934	.9936
2.5	.9938	.9940	.9941	.9943	.9945	.9946	.9948	.9949	.9951	.9952
2.6	.9953	.9955	.9956	.9957	.9959	.9960	.9961	.9962	.9963	.9964
2.7	.9965	.9966	.9967	.9968	.9969	.9970	.9971	.9972	.9973	.9974
2.8	.9974	.9975	.9976	.9977	.9977	.9978	.9979	.9979	.9980	.9981
2.9	.9981	.9982	.9982	.9983	.9984	.9984	.9985	.9985	.9986	.9986
3.0	.9987	.9987	.9987	.9988	.9988	.9989	.9989	.9989	.9990	.9990
3.1	.9990	.9991	.9991	.9991	.9992	.9992	.9992	.9992	.9993	.9993
3.2	.9993	.9993	.9994	.9994	.9994	.9994	.9994	.9995	.9995	.9995
3.3	.9995	.9995	.9995	.9996	.9996	.9996	.9996	.9996	.9996	.9997
3.4	.9997	.9997	.9997	.9997	.9997	.9997	.9997	.9997	.9997	.9998

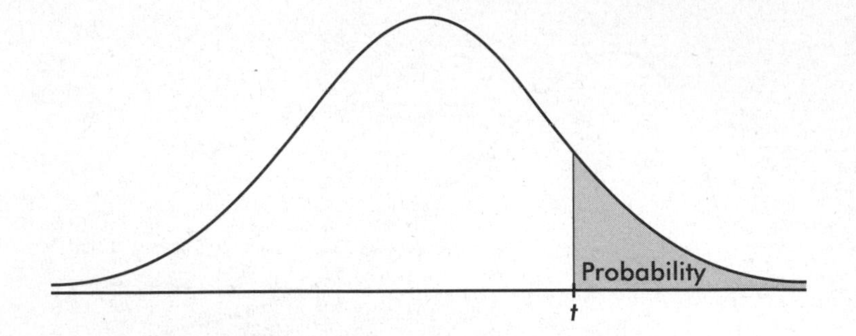

TABLE B: *t*-DISTRIBUTION CRITICAL VALUES

	Tail probability *p*											
df	.25	.20	.15	.10	.05	.025	.02	.01	.005	.0025	.001	.0005
1	1.000	1.376	1.963	3.078	6.314	12.71	15.89	31.82	63.66	127.3	318.3	636.6
2	.816	1.061	1.386	1.886	2.920	4.303	4.849	6.965	9.925	14.09	22.33	31.60
3	.765	.978	1.250	1.638	2.353	3.182	3.482	4.541	5.841	7.453	10.21	12.92
4	.741	.941	1.190	1.533	2.132	2.776	2.999	3.747	4.604	5.598	7.173	8.610
5	.727	.920	1.156	1.476	2.015	2.571	2.757	3.365	4.032	4.773	5.893	6.869
6	.718	.906	1.134	1.440	1.943	2.447	2.612	3.143	3.707	4.317	5.208	5.959
7	.711	.896	1.119	1.415	1.895	2.365	2.517	2.998	3.499	4.029	4.785	5.408
8	.706	.889	1.108	1.397	1.860	2.306	2.449	2.896	3.355	3.833	4.501	5.041
9	.703	.883	1.100	1.383	1.833	2.262	2.398	2.821	3.250	3.690	4.297	4.781
10	.700	.879	1.093	1.372	1.812	2.228	2.359	2.764	3.169	3.581	4.144	4.587
11	.697	.876	1.088	1.363	1.796	2.201	2.328	2.718	3.106	3.497	4.025	4.437
12	.695	.873	1.083	1.356	1.782	2.179	2.303	2.681	3.055	3.428	3.930	4.318
13	.694	.870	1.079	1.350	1.771	2.160	2.282	2.650	3.012	3.372	3.852	4.221
14	.692	.868	1.076	1.345	1.761	2.145	2.264	2.624	2.977	3.326	3.787	4.140
15	.691	.866	1.074	1.341	1.753	2.131	2.249	2.602	2.947	3.286	3.733	4.073
16	.690	.865	1.071	1.337	1.746	2.120	2.235	2.583	2.921	3.252	3.686	4.015
17	.689	.863	1.069	1.333	1.740	2.110	2.224	2.567	2.898	3.222	3.646	3.965
18	.688	.862	1.067	1.330	1.734	2.101	2.214	2.552	2.878	3.197	3.611	3.922
19	.688	.861	1.066	1.328	1.729	2.093	2.205	2.539	2.861	3.174	3.579	3.883
20	.687	.860	1.064	1.325	1.725	2.086	2.197	2.528	2.845	3.153	3.552	3.850
21	.686	.859	1.063	1.323	1.721	2.080	2.189	2.518	2.831	3.135	3.527	3.819
22	.686	.858	1.061	1.321	1.717	2.074	2.183	2.508	2.819	3.119	3.505	3.792
23	.685	.858	1.060	1.319	1.714	2.069	2.177	2.500	2.807	3.104	3.485	3.768
24	.685	.857	1.059	1.318	1.711	2.064	2.172	2.492	2.797	3.091	3.467	3.745
25	.684	.856	1.058	1.316	1.708	2.060	2.167	2.485	2.787	3.078	3.450	3.725
26	.684	.856	1.058	1.315	1.706	2.056	2.162	2.479	2.779	3.067	3.435	3.707
27	.684	.855	1.057	1.314	1.703	2.052	2.158	2.473	2.771	3.057	3.421	3.690
28	.683	.855	1.056	1.313	1.701	2.048	2.154	2.467	2.763	3.047	3.408	3.674
29	.683	.854	1.055	1.311	1.699	2.045	2.150	2.462	2.756	3.038	3.396	3.659
30	.683	.854	1.055	1.310	1.697	2.042	2.147	2.457	2.750	3.030	3.385	3.646
40	.681	.851	1.050	1.303	1.684	2.021	2.123	2.423	2.704	2.971	3.307	3.551
50	.679	.849	1.047	1.299	1.676	2.009	2.109	2.403	2.678	2.937	3.261	3.496
60	.679	.848	1.045	1.296	1.671	2.000	2.099	2.390	2.660	2.915	3.232	3.460
80	.678	.846	1.043	1.292	1.664	1.990	2.088	2.374	2.639	2.887	3.195	3.416
100	.677	.845	1.042	1.290	1.660	1.984	2.081	2.364	2.626	2.871	3.174	3.390
1000	.675	.842	1.037	1.282	1.646	1.962	2.056	2.330	2.581	2.813	3.098	3.300
∞	.674	.841	1.036	1.282	1.645	1.960	2.054	2.326	2.576	2.807	3.091	3.291
	50%	60%	70%	80%	90%	95%	96%	98%	99%	99.5%	99.8%	99.9%
	Confidence level *C*											

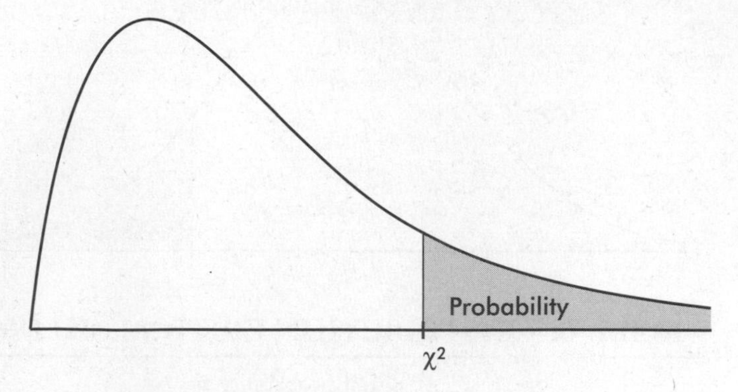

TABLE C: χ^2 CRITICAL VALUES

	Tail probability p										
df	.25	.20	.15	.10	.05	.025	.02	.01	.005	.0025	.001
1	1.32	1.64	2.07	2.71	3.84	5.02	5.41	6.63	7.88	9.14	10.83
2	2.77	3.22	3.79	4.61	5.99	7.38	7.82	9.21	10.60	11.98	13.82
3	4.11	4.64	5.32	6.25	7.81	9.35	9.84	11.34	12.84	14.32	16.27
4	5.39	5.99	6.74	7.78	9.49	11.14	11.67	13.28	14.86	16.42	18.47
5	6.63	7.29	8.12	9.24	11.07	12.83	13.39	15.09	16.75	18.39	20.51
6	7.84	8.56	9.45	10.64	12.59	14.45	15.03	16.81	18.55	20.25	22.46
7	9.04	9.80	10.75	12.02	14.07	16.01	16.62	18.48	20.28	22.04	24.32
8	10.22	11.03	12.03	13.36	15.51	17.53	18.17	20.09	21.95	23.77	26.12
9	11.39	12.24	13.29	14.68	16.92	19.02	19.68	21.67	23.59	25.46	27.88
10	12.55	13.44	14.53	15.99	18.31	20.48	21.16	23.21	25.19	27.11	29.59
11	13.70	14.63	15.77	17.28	19.68	21.92	22.62	24.72	26.76	28.73	31.26
12	14.85	15.81	16.99	18.55	21.03	23.34	24.05	26.22	28.30	30.32	32.91
13	15.98	16.98	18.20	19.81	22.36	24.74	25.47	27.69	29.82	31.88	34.53
14	17.12	18.15	19.41	21.06	23.68	26.12	26.87	29.14	31.32	33.43	36.12
15	18.25	19.31	20.60	22.31	25.00	27.49	28.26	30.58	32.80	34.95	37.70
16	19.37	20.47	21.79	23.54	26.30	28.85	29.63	32.00	34.27	36.46	39.25
17	20.49	21.61	22.98	24.77	27.59	30.19	31.00	33.41	35.72	37.95	40.79
18	21.60	22.76	24.16	25.99	28.87	31.53	32.35	34.81	37.16	39.42	42.31
19	22.72	23.90	25.33	27.20	30.14	32.85	33.69	36.19	38.58	40.88	43.82
20	23.83	25.04	26.50	28.41	31.41	34.17	35.02	37.57	40.00	42.34	45.31
21	24.93	26.17	27.66	29.62	32.67	35.48	36.34	38.93	41.40	43.78	46.80
22	26.04	27.30	28.82	30.81	33.92	36.78	37.66	40.29	42.80	45.20	48.27
23	27.14	28.43	29.98	32.01	35.17	38.08	38.97	41.64	44.18	46.62	49.73
24	28.24	29.55	31.13	33.20	36.42	39.36	40.27	42.98	45.56	48.03	51.18
25	29.34	30.68	32.28	34.38	37.65	40.65	41.57	44.31	46.93	49.44	52.62
26	30.43	31.79	33.43	35.56	38.89	41.92	42.86	45.64	48.29	50.83	54.05
27	31.53	32.91	34.57	36.74	40.11	43.19	44.14	46.96	49.64	52.22	55.48
28	32.62	34.03	35.71	37.92	41.34	44.46	45.42	48.28	50.99	53.59	56.89
29	33.71	35.14	36.85	39.09	42.56	45.72	46.69	49.59	52.34	54.97	58.30
30	34.80	36.25	37.99	40.26	43.77	46.98	47.96	50.89	53.67	56.33	59.70
40	45.62	47.27	49.24	51.81	55.76	59.34	60.44	63.69	66.77	69.70	73.40
50	56.33	58.16	60.35	63.17	67.50	71.42	72.61	76.15	79.49	82.66	86.66
60	66.98	68.97	71.34	74.40	79.08	83.30	84.58	88.38	91.95	95.34	99.61
80	88.13	90.41	93.11	96.58	101.9	106.6	108.1	112.3	116.3	120.1	124.8
100	109.1	111.7	114.7	118.5	124.3	129.6	131.1	135.8	140.2	144.3	149.4

TABLE D: RANDOM NUMBER TABLE

84177	06757	17613	15582	51506	81435	41050	92031	06449	05059
59884	31180	53115	84469	94868	57967	05811	84514	75011	13006
63395	55041	15866	06589	13119	71020	85940	91932	06488	74987
54355	52704	90359	02649	47496	71567	94268	08844	26294	64759
08989	57024	97284	00637	89283	03514	59195	07635	03309	72605
29357	23737	67881	03668	33876	35841	52869	23114	15864	38942

Index

Parentheses indicate problem numbers

THE FOLLOWING DOCUMENTATION APPLIES IF YOU PURCHASED

AP Statistics, 7th Edition with CD-ROM.

How to Use the CD-ROM

The software is not installed on your computer; it runs directly from the CD-ROM. Barron's CD-ROM includes an "autorun" feature that automatically launches the application when the CD is inserted into the CD-ROM drive. In the unlikely event that the autorun feature is disabled, follow the manual launching instructions below.

Windows®

1. Click on the Start button and choose "My Computer."
2. Double-click on the CD-ROM drive, which will be named **AP_Statistics.exe**.
3. Double-click **AP_Statistics.exe** to launch the program.

MAC®

1. Double-click the CD-ROM icon.
2. Double-click the **AP_Statistics** icon to start the program.

SYSTEM REQUIREMENTS

(Flash Player 10.2 is recommended)

Microsoft® Windows®
Processor: Intel Pentium 4 2.33GHz, Athlon 64 2800+ or faster processor (or equivalent).
Memory: 128MB of RAM.
Graphics Memory: 128MB.
Platforms:
Windows 7, Windows Vista®, Windows XP, Windows Server® 2008, Windows Server 2003.

MAC® OS X
Processor: Intel Core™ Duo 1.33GHz or faster processor.
Memory: 256MB of RAM.
Graphics Memory: 128MB.
Platforms:
Mac OS X 10.6, Mac OS X 10.5, Mac OS X 10.4 (Intel) and higher.

Linux® and Solaris™
Processor: Intel Pentium 4 2.33GHz, AMD Athlon 64 2800+ or faster processor (or equivalent).
Memory: 512MB of RAM.
Graphics Memory: 128MB.
Platforms:
Red Hat® Enterprise Linux (RHEL) 5 or later, openSUSE® 11 or later, Ubuntu 9.10 or later.
Solaris: Solaris™ 10.